AF344432

Constrained Optimization and Image Space Analysis

Constrained Optimization and Image Space Analysis

Volume 1: Separation of Sets and Optimality Conditions

Franco Giannessi

University of Pisa
Pisa, Italy

Springer

Library of Congress Control Number: 2005922927

ISBN-10: 0-387-24770-X (Hardbound)   Printed on acid-free paper.
ISBN-13: 978-0387-24770-0

Printed in the United States of America

9 8 7 6 5 4 3 2 1

springeronline.com

For since the fabric of the universe is the
most perfect and the work of a most wise
Creator, nothing at all takes place in the
universe in which some rule of maximum
or minimum does not appear.

Leonhard Euler

To Alessandro, Stefano and Luca, my sons,
and to Rita

PREFACE

The book is devoted to the basic theory of smooth and nonsmooth constrained extremum problems and variational inequalities. The main feature consists of a uniform and general treatment of these theories, which quickly leads to the common framework of them. This approach is based on the analysis in the image space, namely the space where the images of the involved functions run; its development began about three decades ago. Indeed, the study of the properties of the image of a real-valued function is an old one; however, in most cases the properties of the image have not been the purpose of the study and their investigation has occurred as an auxiliary step toward other achievements. The analysis in the image space is viewed as a preliminary and auxiliary step for studying extremum and equilibrium problems, and not as a competitor of the analysis in the given space.

The analysis in the image space is strongly based on separation theorems and on theorems of the alternative. These mathematical tools have been used, in the field of constrained extremum problems, for a long time in the given space mainly as a step of a proof. Here it is shown that their use in the image space leads to an acknowledgement of them as foundations of the theory of constrained extrema, at least of Lagrangian type, and not only as auxiliary tools; they appear as a natural language for expressing Lagrange ideas.

The first chapter is devoted to some mathematical models which are studied in the field of Optimization and to some real problems. Chapter 2 contains some basic concepts of Convex Analysis. Chapter 3 deals with an introduction to Image Space Analysis for constrained extremum problems; this is preceded by an extension of the classic differentiability and stationarity, and followed by several illustrative examples. Chapter 4 deals with theorems of the alternative and separation theorems. Chapter 5 is concerned with preliminary results on optimality conditions. Each chapter contains examples, comments and suggestions for further investigations.

I want to express my sincere gratitude to my Colleagues Drs. A.Antoni, F.Gori, K.Madani, G.Mastroeni, M.Passacantando, Qinghua Zhang and Profs. M.Pappalardo, L.Pellegrini, T.Rapcsák, Xiao Qi Yang, who have contributed in several ways to the existence of this book, any mistakes or omissions in which are due to me only. The contribution of the President of Consiglio Nazionale delle Ricerche is gratefully acknowledged. Thanks are due also to Kluwer Academic/Plenum Publishers for their unfailing cooperation and patience in waiting this book for a long time, particularly to Mrs. Ana Bozicevic, and to my son Luca for the typing.

Pisa, Italy   F.Giannessi
September, 2004

CONTENTS

CHAPTER 1. INTRODUCTION

1.1. Constrained Extremum Problems

Assume we are given the integers m, n and p with $m \geq 0$, $0 \leq p \leq m$, $n > 0$, the nonempty set $X \subseteq \mathbb{R}^n$ and the functions $f : X \to \mathbb{R}$, $g_i : X \to \mathbb{R}$, $i \in \mathfrak{I}:=\{1,.....,m\}$. We consider problems of the following kind:

$$f^{\downarrow} := \min \ f(x_1, ..., x_n), \tag{1.1.1a}$$

$$\text{s.t.} \quad g_i(x_1, ..., x_n) = 0, \quad i \in \mathfrak{I}^0 := \{1, ..., p\}, \tag{1.1.1b}$$

$$g_i(x_1, ..., x_n) \geq 0, \quad i \in \mathfrak{I}^+ := \{p+1, ..., m\}, \tag{1.1.1c}$$

$$x = (x_1, ..., x_n) \in X \subseteq \mathbb{R}^n, \tag{1.1.1d}$$

where $p{=}0 \Rightarrow \mathfrak{I}^0{=}\varnothing$, $p{=}m \Rightarrow \mathfrak{I}^+{=} \varnothing$, $m = 0 \Rightarrow \mathfrak{I} = \mathfrak{I}^0 \cup \mathfrak{I}^+{=} \varnothing$. Unless differently stated, we will assume that card $X > 1$. The *feasible region* of (1.1.1) is the set

$$R := \{x \in X : g(x) \in D\}, \tag{1.1.2}$$

where $g(x) := (g_1(x),.....,g_m(x))$, $D := O_p \times \mathbb{R}_+^{m-p}$ with $O_p :=(0,.....,0){\in} \mathbb{R}^p$; we stipulate that $D = \mathbb{R}_+^m$ when $p{=}0$ and $D = O_m := (0, ..., 0) \in \mathbb{R}^m$ when $p{=}m$; $m{=}0$ does not require to define D. The constraints (1.1.1b) and (1.1.1c) are called *bilateral* and *unilateral*, respectively. A concise form of (1.1.1) is the following one:

$$f^{\downarrow} := \min f(x), \quad \text{s.t.} \ \ x \in R; \tag{1.1.3}$$

it will be used also by considering R as any subset of $\mathbb{R}^n$.

Definition 1.1.1. An element $\bar{x} \in R$ is said to be a *global minimum point* of problem (1.1.1) iff $f(x) \geq f(\bar{x})$, $\forall x \in R$. Iff this inequality is strictly verified for $x \neq \bar{x}$, then a minimum point is said to be *strict*. Iff there exists a neighbourhood $N(\bar{x})$ of $\bar{x}$, such that the above inequality is (strictly) satisfied $\forall x \in R \cap N(\bar{x})$, then $\bar{x}$ is said to be a *local (strict) minimum point*. Iff $\exists N(\bar{x})$ such that $\bar{x}$ is the unique local minimum point of (1.1.1) in $R \cap N(\bar{x})$, then $\bar{x}$ is said to be *isolated*. Iff $\exists k \in \mathbb{R}_+ \backslash \{0\}$ such that $f(x) \geq f(\bar{x}) + k||x - \bar{x}||^2$, $\forall x \in R$ (or $\forall x \in R \cap N(\bar{x})$), then $\bar{x}$ is said to be *strong* global (or local) minimum point of (1.1.1). $\bar{x}$ is called *lower (upper) semistationary* iff

$$\liminf_{x \to \bar{x}} [f(x) - f(\bar{x})]/||x - \bar{x}|| \geq 0 \quad (\text{or} \limsup_{x \to \bar{x}} [f(x) - f(\bar{x})]/||x - \bar{x}|| \leq 0);$$

it is called *stationary* iff it is both lower and upper semistationary or

$$\lim_{x \to \overline{x}}[f(x) - f(\overline{x})]/\|x - \overline{x}\| = 0$$

(in the above limits, $x \in X \cap N(\overline{x})\backslash\{\overline{x}\}$).

Unless it is explicitly said, the operator "min" is meant in the sense of finding one minimum point (for short, m.p.) and not all of them. When $m=0$ and X is open, then (1.1.1) is called *unconstrained*; otherwise it is *constrained*. Due to a tradition established in the applications, an element of R is called often a *feasible solution* of (1.1.3) − even if such a terminology may appear to contain a redundancy − and R the *feasible region*. Local and global maximum points and the operator "max" are defined in a quite similar way; the obvious relation $\max f = -\min(-f)$ holds. f is often called *objective function*.

A m.p. of (1.1.1) may be strict and even strong, but not isolated, as shown by the following example with only one bilateral constraint: $p = m = 1$, $n = 1$, $X = \mathbb{R}$, $f(x) = x^2$, $g(x) = x^2 \sin^2(\frac{1}{x})$ if $x \neq 0$ and $g(0) = 0$; $\overline{x} = 0$ is evidently (local and global) strict and strong (for $k = 1$) m.p. of (1.1.1); as local m.p., $\overline{x}$ is not isolated, since in every $N(\overline{x})$ there are points different from $\overline{x}$ ($x = \pm\frac{1}{h\pi}$, $h \in \mathbb{N}$) which are local m.p. Another example is the following unconstrained problem: $m = 0$, $n = 1$, $X = \mathbb{R} = R$, $f(x)$ equals the previous $g(x)$; as local m.p., $\overline{x} = 0$ is not isolated for the same preceeding reason; now is neither strict nor strong. Obviously, a strict global m.p. is unique; an isolated global m.p. is not necessarily unique; a strong m.p. is also strict, but not vice versa (*e.g.*, $m = 0, n = 1, X = \mathbb{R} = R, f(x) = e^{-1/x}$ if $x \neq 0$ and $f(0) = 0$).

Besides (1.1.1), we consider the following isoperimetric-type problems. Let $\mathbb{V}$ be the subset of $C^0(T)^n$ with continuous derivatives $x'(t) = (x'_1(t), \ldots, x'_n(t))$, $t \in T$, except at most a finite number of points $\overline{t}$ at which exist and are finite $\lim_{t \downarrow \overline{t}} x'(t)$ and $\lim_{t \uparrow \overline{t}} x'(t)$; $x'(\overline{t}) := \lim_{t \downarrow \overline{t}} x'(t)$. $\mathbb{V}$ forms a vector space on the set of real numbers. X is defined as the subset of $\mathbb{V}$, whose elements satisfy a boundary condition, for instance fixed endpoints conditions $x(t_0) = x^0$ and $x(t_1) = x^1$, x^0 and x^1 being given vectors of $\mathbb{R}^n$, or merely an initial (or a final) condition only. X is a subset of the Banach space $B = C^0(T)^n$; $T:=[t_0, t_1] \subset \mathbb{R}$, $-\infty <t_0 < t_1 < +\infty$;

$$\|x\| = \|x\|_\infty := \max_{t \in T} \|x(t)\|_2.$$

$$f^\downarrow := \min \left[f(x) := \int_T \psi_0(t,\ x(t),\ x'(t))dt \right], \tag{1.1.4a}$$

$$\text{s.t.} \qquad g_i(x) := \int_T \psi_i(t,\ x(t),\ x'(t))\ dt = 0, \quad i \in \mathfrak{I}^0, \tag{1.1.4b}$$

$$g_i(x) := \int_T \psi_i(t,\ x(t),\ x'(t))\ dt \geq 0, \quad i \in \mathfrak{I}^+, \tag{1.1.4c}$$

$$x \in X \subseteq C^0(T)^n, \tag{1.1.4d}$$

where ψ_0, $\psi_i : \mathbb{R} \times \mathbb{R}^n \times \mathbb{R}^n \to \mathbb{R}$. We set $\psi = (\psi_1, ..., \psi_m)$. (1.1.4) will be expressed, in a concise form, as (1.1.3); when this will be done, it will be explicitly noticed.

Even if most of the attention will be addressed to (1.1.1) and (1.1.4), in the sequel, we will consider also the following geodesic-type minimization problem:

$$f^{\downarrow} := \min \left[f(x) := \int_T \psi_0(t,\ x(t),\ x'(t))dt \ \right], \tag{1.1.5a}$$

$$\text{s.t.} \qquad g_i(x) := \psi_i(t,\ x(t),\ x'(t)) = 0, \quad \forall t \in T,\ \ i \in \mathcal{J}^0, \tag{1.1.5b}$$

$$g_i(x) := \psi_i(t,\ x(t),\ x'(t)) \geq 0, \quad \forall t \in T,\ \ i \in \mathcal{J}^+, \tag{1.1.5c}$$

$$x \in X \subseteq C^0(T)^n, \tag{1.1.5d}$$

where the symbols are as in (1.1.4).

Problems (1.1.1) and, if the $\psi_i{'}s$ are continuous, (1.1.4) and (1.1.5) are special cases of the following formulation. Here and throughout the entire book, B and $\mathcal{B}$ denote Banach spaces. Let X be a subset of the Banach space B, $\mathcal{B}$ be another normed linear space, and $\mathcal{B}_+$ a nonempty, closed and convex cone of $\mathcal{B}$ with apex at the origin 0. Consider the problem:

$$f^{\downarrow} := \min f(x), \tag{1.1.6a}$$

$$\text{s.t.} \quad g_i(x) = 0, \quad i \in \mathcal{J}^0, \tag{1.1.6b}$$

$$g_i(x) \subset \mathcal{B}_+, \quad i \subset \mathcal{J}^+, \tag{1.1.6c}$$

$$x \in X \subseteq B. \tag{1.1.6d}$$

The feasible region of (1.1.6) can be expressed by (1.1.2), where now $D = O_p \times \mathcal{B}_+^{m-p}$ with $O_p = (0, ..., 0) \in \mathcal{B}^p$ and $D = \mathcal{B}_+^m$ if $p = 0$, $D = O_m = (0, ..., 0) \in \mathcal{B}^m$ if $p=m$. For $B = \mathbb{R}^n$ and $\mathcal{B}_+ = \mathbb{R}_+$ (or $D = O_p \times \mathbb{R}_+^{m-p} \subseteq \mathbb{R}^m$), (1.1.6) shrinks to (1.1.1). For $B = C^0(T)^n$ and (as above) $\mathcal{B}_+ = \mathbb{R}_+$ (or $D = O_p \times \mathbb{R}_+^{m-p} \subset \mathbb{R}^m$), (1.1.6) collapes to (1.1.4). For $B = C^0(T)^n$ and $\mathcal{B}_+ = \{\phi : T \to \mathbb{R} \text{ s.t. } \phi(t) \geq 0, \quad \forall t \in T\}$, (1.1.6) collapses to (1.1.5). Of course, a more general case would be that where $\mathcal{B}_+$ in (1.1.6c) is replaced by any convex cone.

Definition 1.1.1 is given, without any change, for (1.1.4), (1.1.5) and (1.1.6). In the sequel, when a definition is given (or a proposition is stated) for (1.1.1), its extension to (1.1.4), (1.1.5) will be discussed, if any and if necessary.

The feasible regions of (1.1.4), (1.1.5) and (1.1.6) will be denoted by R as for (1.1.1) and, when there is no fear of confusion, the same concise form (1.1.3) will be adopted.

If we look at the space B where the unknown runs, then (1.1.1) turns out to be finite dimensional, while both (1.1.4) and (1.1.5) are infinite dimensional; this is the usual classification of minimization problems. If we look at the space $\mathcal{B}$ where the images of f and g run, then the classification changes: (1.1.1) and (1.1.4) share the characteristic of having a finite dimensional image (see Fig. 1.1.1); hence, in the Image Space (for short, IS), (1.1.5) continues to be classified as infinite dimensional (we have $\mathcal{B} \neq \mathbb{R}$), while both (1.1.1) and (1.1.4) are finite dimensional and can be treated with the same kind of mathematical arguments.

In the above formats there are both *bilateral* and *unilateral* constraints. Trivial devices allow us to equivalently reduce the former constraints to the latter ones and vice versa. For instance, for (1.1.1) a bilateral constraint can be turned into a unilateral one, since (1.1.1b) is equivalent to the system $g_i(x) \geq 0$, $-g_i(x) \geq 0$, $i \in \mathcal{I}^0$, or to the system $g_i(x) \geq 0$, $i \in \mathcal{I}^0$, $-\sum_{i \in \mathcal{I}^0} g_i(x) \geq 0$; vice versa, a unilateral constraint can be turned to

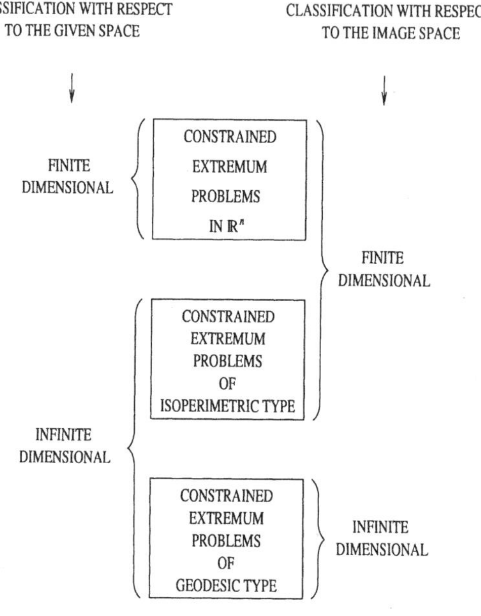

Fig. 1.1.1

a bilateral one, since (1.1.1c) is equivalent to the system $g_i(x) - y_i^2 = 0$, $i \in \mathcal{I}^+$, where the new unknown y_i is real.

From a formal point of view, it might be enough to develop the theory and methods of solution for only one type of constraint; in the past, there has been the belief that this was enough. Nowadays, it is clear that, if we want to deepen the analysis and understand the structure underlying the problem, we must consider explicitly both types of constraints.

When $X \subseteq \mathbb{Z}^n$, then (1.1.1) is called *discrete optimization problem* and embraces the *combinatorial optimization problems* (when card $X < +\infty$); the vice versa is not true, as simple examples show (for instance, (1.1.3), where R is replaced by $\mathbb{R}^n_+ \cap \mathbb{Z}^n$). A particularly interesting case is that where $X = \mathbb{B}^n$, which will be considered explicitly:

$$\min f(x), \quad \text{s.t.} \quad g(x) \in D, \quad x \in \mathbb{B}^n. \tag{1.1.7}$$

In several applications of optimization models, one is faced by more than one objective. In this case, f of (1.1.1), (1.1.4), (1.1.5) must be replaced by $f : X \to \mathbb{R}^\ell$ with ℓ being a positive integer. As a consequence, we lose the total ordering which has been used in Definition 1.1.1; hence, a partial ordering must be introduced. Assume that a cone

$C \subset \mathbb{R}^{\ell}$, with apex at the origin, be given; we assume that C is convex, closed, pointed (see Sect. 2.2), and with int $C \neq \varnothing$, even if some of the propositions which will be established do not require all these assumptions on C. We will consider two subcones, namely $C_0 := C \setminus \{O\}$ and $\overset{o}{C} := \text{int } C$, and two corresponding vector optimization problems (for short, VOP). The first VOP is:

$$\min_{C_0} \; f(x), \quad \text{s.t.} \;\; x \in R, \tag{1.1.8}$$

where $\min_{C_0}$ denotes the vector minimum with respect to the cone C_0:

Definition 1.1.2. An element $\overline{x} \in R$ is said to be a *global vector minimum point* (for short, v.m.p.) of (1.1.8) iff

$$f(\overline{x}) \not\geq_{C_0} f(x), \quad \forall x \in R, \tag{1.1.9}$$

where the inequality means $f(\overline{x}) - f(x) \notin C_o$; the ℓ-vector $f(\overline{x})$ is the *vector minimum* of (1.1.8).

The second VOP is

$$\min_{\overset{o}{C}} \; f(x), \quad \text{s.t.} \;\; x \in R, \tag{1.1.10}$$

where $\min_{\overset{o}{C}}$ denotes the vector minimum with respect to the cone $\overset{o}{C}$.

Definition 1.1.3. An element $\overline{x} \in R$ is said to be a *global* v.m.p. of (1.1.10) iff

$$f(\overline{x}) \not\geq_{\overset{o}{C}} f(x), \quad \forall x \in R, \tag{1.1.11}$$

where the inequality means $f(\overline{x}) - f(x) \notin \overset{o}{C}$; the ℓ-vector $f(\overline{x})$ is the *vector minimum* of (1.1.10).

Local v.m.p. are obviously defined by intersecting R with a neighbourhood $N(\overline{x})$ *of* $\overline{x}$, as in Definition 1.1.1. At $\ell=1$, we have $C_o = \overset{o}{C} = \mathbb{R}_+ \setminus \{0\}$, so that Definitions 1.1.2 and 1.1.3 collapse to Definition 1.1.1 and both problems (1.1.8) and (1.1.10) to (1.1.3). (1.1.8) and (1.1.10) are generalizations of classic problems: respectively, the so-called *vector Pareto* and *weak vector Pareto* problems, which are recovered at $C = \mathbb{R}_+^{\ell}$.

Problem (1.1.10) is obviously different from (1.1.8), since different cones identify different vector problems. The term "weak" comes from the following tradition. Notwithstanding the fact that (1.1.8) and (1.1.10) be distinct problems, since the v.m.p. of (1.1.8) are v.m.p. also of (1.1.10) (but not necessarily vice versa), then the v.m.p. of (1.1.10) are often called "weak solutions" of (1.1.8). It is better to say that a v.m.p. of (1.1.10) is *necessary* (but not sufficient) to be a v.m.p. of (1.1.8); "relaxed" is a more appropriate term than "weak", which is misleading here and in contrast with its traditional use in several other branches of mathematics. From the same cone C, several cones different from C_o and $\overset{o}{C}$ might be derived; however, C_o and $\overset{o}{C}$ cover most of applications.

In some applications, it happens that the feasible region of (1.1.1) depends on the unknown itself. Such a situation conceals a fixed point condition. To achieve the formulation of this case, with obvious notation, let us consider the following parametric problem:

$$\min f(x;\xi), \quad \text{s.t.} \quad x \in R(\xi), \tag{1.1.12}$$

where

$$R(\xi) := \{x \in X(\xi) \ : g(x;\xi) \in D\}, \tag{1.1.13}$$

and $\xi \in \Xi$ is a parameter; it is assumed that $\exists \hat{\xi} \in \Xi$ such that $X(\hat{\xi}) = X$, $f(x;\hat{\xi}) = f(x)$, $g(x;\hat{\xi}) = g(x)$, so that (1.1.1), (1.1.4) and (1.1.5) are special cases of (1.1.12); namely, (1.1.12) represents an embedding of them.

Denote by $x(\xi)$ the set of all m.p. of (1.1.12); in general, $x(\xi)$ is a point-to-set map; without any fear of confusion, $x(\xi)$ will denote both the set and one of its elements. We call (1.1.12) *Quasi-minimum Problem* when we require that a m.p. of it be a fixed point of the map $x(\xi)$, or when we define $\overline{x}$ to be a *quasi-minimum point* of (1.1.12) iff it is a m.p. of (1.1.12) for $\xi = \overline{x}$. The term "quasi" is borrowed from variational inequalities field (see Sect. 1.3).

Since the purpose of this book is to give an instance of the kind of analysis which can be done through the IS, and not to have a classic treatise on optimization, several usual topics are understood or merely recalled. Existence and uniqueness of m.p. are topics of this kind and here are only shortly recalled; we are referred to the excellent books quoted in the references for details and proofs. In Sect. 3.2, a general existence theorem will be proved.

The most classic and important existence condition is expressed by the extended Weierstrass Theorem (see Corollary 3.2.2)

Theorem 1.1.1. Let R be a compact subset of B (in particular of $\mathbb{R}^n$), and f be lower (upper) semicontinuous on X. Then, the minimum (or the maximum; in the previous problems) exists.   □

This theorem can be extended to the case where R is unbounded, by merely assuming the existence of a nonempty and compact level set of f on R. To this end, a coerciveness assumption is classically made on f.

Theorem 1.1.2. Let B be a reflexive Banach space, R be a convex and closed subset of B, and f be a weakly lower semicontinuous function, such that:

$$\lim_{\substack{||x||\to\infty \\ x\in R}} f(x) = +\infty. \tag{1.1.14}$$

Then, the minimum (in the previous problems) exists.   □

When $B = \mathbb{R}^n$, then the convexity assumption on R can be dropped in Theorem 1.1.2. In order to obtain the uniqueness, we can add the assumption of strict (quasi)convexity of f on the convex set R (see Chapter 2).

In Vol. 2, we will see that the replacement of a VOP with a family of scalar problems allows one to reduce the existence and uniqueness of solutions to (1.1.8) and (1.1.10) to those of (1.1.3).

In some applications, as well as in theoretical topics, we are faced with problems of the following kind, which are known, respectively, as *minimax* and *maxmin* problems:

$$\min_{x \in X} \max_{y \in Y} F(x, y), \tag{1.1.15a}$$

and

$$\max_{y \in Y} \min_{x \in X} F(x, y), \tag{1.1.15b}$$

where X, Y are subsets of given spaces and $F : X \times Y \to \mathbb{R}$. With suitable positions — setting f as max-function; see (2.3.20) — the minimax problem (1.1.15a) becomes a special case of (1.1.6); analogous transformation holds for (1.1.15b). However, such changes may hide the structure of (1.1.15).

Problems (1.1.15) are special cases of a wide cluster of problems termed partial maxima and minima. Let us recall a classic result.

Theorem 1.1.3. Let $R \subseteq B$ and $f : R \to \mathbb{R}$. **(i)** If $\{R(\xi)\}_{\xi \in \Xi}$ is a cover of R, then we have:

$$\inf_{x \in R} f(x) = \inf_{\xi \in \Xi} \inf_{x \in R(\xi)} f(x) \tag{1.1.16}$$

If, furthermore, there exist

$$f^{\downarrow} := \min_{x \in R} f(x), \quad f^{\downarrow}(\xi) := \min_{x \in R(\xi)} f(x), \quad \forall \xi \in \Xi, \tag{1.1.17}$$

or if R and $R(\xi)$, $\xi \in \Xi$, are compact, and f is l.s.c., then we have:

$$\min_{x \in R} f(x) = \min_{\xi \in \Xi} \min_{x \in R(\xi)} f(x). \tag{1.1.18}$$

(ii) If R is the Cartesian product of sets $R_1, ..., R_k$, then we have:

$$\inf_{x \in R} f(x) = \inf_{x^1 \in R_1} \inf_{x^2 \in R_2} \ .... \ \inf_{x^k \in R_k} f(x^1, x^2, ..., x^k), \tag{1.1.19}$$

where x^i runs in the subspace which, in the decomposition of B induced by the above product, corresponds to R_i, $i = 1, ..., k$.

Proof. **(i)** $\forall \xi \in \Xi$, $\ell(\xi) := \inf_{x \in R(\xi)} f(x) \geq \ell := \inf_{x \in R} f(x)$. Hence $\ell^0 := \inf_{\xi \in \Xi} \ell(\xi) \geq \ell$. On the other side, since $x \in R \Rightarrow \exists \xi_x \in \Xi$ s.t. $x \in R(\xi_x)$, we have $f(x) \geq \ell(\xi_x)$ and thus $\ell \geq \ell^0$, so that (1.1.16) follows. In order to achieve (1.1.18), it is enough to note that (1.1.16) and (1.1.17) imply the existence of $\xi \in \Xi$ s.t. $f^{\downarrow}(\xi) = f^{\downarrow}$. The assumptions of compactness and that lower semicontinuity imply the existence of minima (1.1.17). **(ii)** $\forall y \in R_1$, consider the set $S(y) := \{x \in B : x^1 = y, x^i \in R_i, i = 2, ..., k\}$. Since $\{S(y)\}_{y \in R_1}$ is a partition of R, because of (1.1.16), we have:

$$\inf_{x \in R} f(x) = \inf_{x^1 \in R_1} \inf_{x \in S(x^1)} f(x).$$

By applying the above reasoning to $\inf\limits_{x \in S(x^1)}$, and so on, we easily achieve (1.1.19).   □

In some fields of Engineering, especially in Electrotechnics and Aerodynamics, the unknown of the previous problems belongs to the Complex Space. Let ζ^* denote the conjugate of ζ. By restricting ourselves to (1.1.1), let us suppose now that $X \subseteq \mathbb{C}^n$, $D \subseteq \mathbb{C}^m$ be a convex cone with apex at the origin, $f : X \times X \to \mathbb{C}$, $g : X \times X \to C^m$ be analytic, and

$$R = \{\zeta \in X : \ g(\zeta, \zeta^*) \in D\}. \tag{1.1.20}$$

As for problems (1.1.8) and (1.1.10), we have not a total order for the values of f. Therefore, we must introduce either a function which goes from the image of f to $\mathbb{R}$, or a partial order.

A classic instance of the former case consists in considering the problem:

$$f^{\downarrow} := \min \ \mathrm{Re} \ f(\zeta, \zeta^*), \quad \text{s.t.} \ \ \zeta \in R, \tag{1.1.21}$$

where R is given by (1.1.20), and Re denotes the real part of ζ. In the particular case where f and g are functions of n real variables (with values, respectively, in $\mathbb{R}$ and $\mathbb{R}^m$), then (1.1.21) becomes (1.1.3) by setting $\zeta = x$. In the general case, the minimum, minimum points, semistationary points and stationary points follow the Definition 1.1.1, where, of course, f must be replaced with Re f.

In the latter case, we must introduce a partial order, namely a cone C, as for (1.1.8) and (1.1.10); since the image of f is in $\mathbb{C}$ (the extension to $\mathbb{C}^\ell$ is straightforward), then C is a cone or $\mathbb{R}^2_+$. The two main problems we are faced with correspond to (1.1.8) and (1.1.10); they are:

$$\min_{C_0} \ f(\zeta, \zeta^*), \quad \text{s.t.} \ \ \zeta \in R, \tag{1.1.22}$$

and

$$\min_{\overset{\circ}{C}} \ f(\zeta, \zeta^*), \quad \text{s.t.} \ \ \zeta \in R. \tag{1.1.23}$$

Consider a generic positive integer r. Between $\mathbb{C}^r$ and $\mathbb{R}^{2r}$ a natural one-to-one correspondence is that which associates every $\eta \in \mathbb{C}^r$ with the $2r$-tuple $y \in \mathbb{R}^{2r}$, whose first r elements are given by Re η, and whose remaining elements are given by Im η, so that:

$$\eta = (y_1, ..., y_r) + (y_{r+1}, ...y_{2r})\mathbf{i}, \tag{1.1.24}$$

where $\mathbf{i} := \sqrt{-1}$ is the imaginary unit. By means of such a correspondence, problems (1.1.22) and (1.1.23) can be equivalently brought to the formats (1.1.8) and (1.1.10), respectively. Therefore, the definitions of *complex minimum point* and *complex minimum* can be done through Definitions 1.1.2 and 1.1.3. Beginning with these definitions, the entire development of the theory for (1.1.1), which will be considered in the sequel, can be extended to problems (1.1.22) and (1.1.23).

1.2. Special Extremum Problems

Some classic and less classic instances of (1.1.1), (1.1.4) and (1.1.5) are now considered. Here, we briefly describe the topic, which asks for an optimization formulation, until it assumes the format (1.1.1) or (1.1.4); its analysis will be continued in subsequent chapters. The symbols of this section must be considered independent of those of the other sections, if they overlap.

Example 1.2.1 Let A and B be matrices with real entries, A symmetric, and of dimensions $n \times n$ and $r \times n$, respectively. We want to find a necessary and sufficient condition for the restriction of the quadratic form $\langle x, Ax \rangle$ to the linear variety $Bx = 0$ to be (strictly) convex. This happens iff the minimum of the problem

$$\min \ \langle x, Ax \rangle, \quad \text{s.t.} \quad Bx = 0, \quad \langle x, x \rangle - 1 = 0 \tag{1.2.1}$$

is non$-$negative (positive). The "only if" is trivial. The "if" follows immediately by noting that the solutions x of $Bx = 0$ can be expressed as $x = \alpha y$ with $\alpha \geq 0$ and y a feasible solution of (1.2.1), so that $\langle x, Ax \rangle = \alpha^2 \langle y, Ay \rangle$. (1.2.1) is a special case of (1.1.1), where we must set $X = \mathbb{R}^n$, $p = m = r + 1$, $f(x) = \langle x, Ax \rangle$, and (1.1.1b) is the pair $Bx = 0$, $\langle x, x \rangle - 1 = 0$. $\qquad\qquad\square$

Examples 1.2.2 (see also Examples 2.5.4, 5.4.1 and 5.8.1). Among all hyperrectangles having the same given volume $V > 0$, find the one such that the sum of the edgelengths is minimum. If $x_1, ..., x_n$ denote the lengths of the edges, we are led to the problem

$$\min \ f(x) = \sum_{j=1}^{n} x_j, \quad \text{s.t.} \quad g(x) = \prod_{j=1}^{n} x_j - V = 0, \ x_j > 0, \ j = 1, ..., n, \tag{1.2.2a}$$

which is identified easily with (1.1.1) for $X = \text{int } \mathbb{R}^n_+$, $p = m = 1$. A reciprocal version (see Sect. 5.8) of this problem consists in finding, among all hyperrectangles having the same given sum ℓ of edgelengths, the one whose volume is maximum. We are led to the problem:

$$\max \ \phi(x) = \prod_{j=1}^{n} x_j, \quad \text{s.t.} \quad \sum_{j=1}^{n} x_j - \ell = 0, \ x_j > 0, \ j = 1, ..., n, \tag{1.2.2b}$$

which is identified easily with (1.1.1) for $X = \text{int } \mathbb{R}^n_+$, $p = m = 1$, $f(x) = -\phi(x)$.
The infinite dimensional version in $\mathbb{R}^2$ of (1.2.2a) consists in finding a closed curve encircling a given area V and having minimum length. A precise formulation of this problem requires to represent the curve in parametric form; this leads to (1.1.4) at $n = 2$ (where the unknown is a vector of 2 functions). A simple and classic device allows us to circumvent the introduction of 2 unknown functions and achieve the same result. To this end, consider the following problem:

$$f^\downarrow(V) := \min \ f(x) = \int_T \sqrt{1 + x'^2(t)} \ dt, \quad \text{s.t.} \quad g(x) = \int_T x(t)dt - V = 0,$$

$$\tag{1.2.3a}$$

$$x(t_0) = x^0, x(t_1) = x^1,$$

which is identified easily with (1.1.4) for $p = m = 1$, $n = 1$, $\psi_0 = \sqrt{1 + x'^2}$, $\psi = \psi_1 = x - V/(t_1 - t_0)$. A reciprocal version of (1.2.3a) (see Sect. 5.8) is:

$$\phi^\uparrow(\ell) := \max \phi(x) = \int_T x(t)\, dt, \quad s.t. \quad g(x) = \int_T \sqrt{1 + x'^2(t)}\, dt - \ell = 0,$$

$$(1.2.3b)$$

$$x(t_0) = x^0, x(t_1) = x^1,$$

where ℓ is the given curvelength; (1.2.3b) is identified easily with (1.1.4) for $p = m = 1$, $n = 1$, $\psi_0 = -x$, $\psi = \psi_1 = \sqrt{1 + x'^2} - \ell/(t_1 - t_0)$. (1.2.2) and (1.2.3) are the most classic isoperimetrical problems; their solutions — the (regular) n-hypercube for (1.2.2), and an arc of circumference (of the circle) for (1.2.3) — were known to the ancients (Dido's problem; 880 B.C; see Sect. 1.9); general methods for this kind of problems are due to Leonhard Euler and Giuseppe Luigi Lagrange. Some topics, apparently far from the isoperimetric idea, can be reduced to the formats (1.2.2) or (1.2.3), as we will see later.   □

Example 1.2.3. Given the set $R \subseteq \mathbb{R}^n$ and the points $x^i \in \mathbb{R}^n$, $i = 1, ..., s$, find $x^0 \in R$ which minimizes the sum of its distances from the s points. This problem can be formulated as

$$\min f(x) = \sum_{i=1}^{s} \| x - x^i \|_2 , \quad s.t. \quad x \in R. \tag{1.2.4}$$

When $n = 2$, $s = 3$, $R = \mathbb{R}^2$ and x^1, x^2, x^3 are not collinear, then (1.2.4) is the famous Fermat$-$Torricelli problem (formulated by P. de Fermat in 1643 [22], brought to Italy by M.Mersenne, and solved in a very elegant geometrical way by E. Torricelli around 1645 [59]; the solution is called *Torricelli point*). (1.2.4) is immediately identified with (1.1.1) for $X = R$ and in absence of (1.1.1b,c). When $s = n + 1$ and $x^1, ..., x^{n+1}$ are affinely independent, then we can consider a generalization of (1.2.4):

$$\min f(x) = \frac{\sum\limits_{i=1}^{n+1} \|x - x^i\|_2}{\sum\limits_{i=1}^{n+1} \text{dist}(x,\ F_i)}, \quad s.t. \quad x \in R, \tag{1.2.5}$$

where F_i is a $((n - 1)-$dimensional) facet of the simplex $S := \text{conv}\{x^1, ..., x^{n+1}\}$ (see Sect. 2.1). (1.2.5) is identified immediately with (1.1.1). When $n = 2$ and $R = \mathbb{R}^2$, then (1.2.5) is the generalization of Fermat-Torricelli problem formulated by P.Erdös [21]. If $R = \mathbb{R}^n$ and S is regular (in the sense that all its $n + 1$ facets are equal), then the denominator in (1.2.5) is constant; in fact, whatever x may be, the volume V of S is the sum of the volumes, say $V_1, ..., V_{n+1}$, of the simplices $S_i := \text{conv}(\{x, x^1, ..., x^{n+1}\}\setminus\{x^i\}), i = 1, ..., n+1$. If V_0 is the volume of each facet F_i, then we have [17] $V = \sum\limits_{i=1}^{n+1} V_i = \sum\limits_{i=1}^{n+1} \frac{1}{n} V_0\, \text{dist}(x,\ F_i)$, so that $\sum\limits_{i=1}^{n+1} \text{dist}(x,\ F_i) = \dfrac{nV}{V_0}$. This quantity is the distance of each x^i from the opposite facet, i.e. $\text{conv}(\{x^1, ..., x^{n+1}\}\setminus\{x^i\})$, and hence is the *altitude* of S.   □

Examples 1.2.4. To find a point $x \in \mathbb{R}^n$ which minimizes the maximum of the distances between x and each of the given points $x^i \in \mathbb{R}^n$, $i = 1, ..., N$. This problem is of type (1.1.15a); however, it can be easily reduced to (1.1.1). If the distance is the Euclidean one, then this problem is equivalent to finding, among the n−dimensional spheres, say S_n, which contain the N given points, one having the minimum radius. If $\sqrt{r_n}$ with $r_n \geq 0$ denotes the radius, then the problem can be formulated as

$$\min\ (r_n), \quad \text{s.t.} \quad r_n - \sum_{j=1}^{n}(x_j - x_j^i)^2 \geq 0, \quad i = 1, ..., N, \tag{1.2.6a}$$

where $x^i := (x_1^i, ..., x_n^i)$. (1.2.6a) is identified easily with (1.1.1) at $p = 0$, $m = N$, $X = \mathbb{R}^{1+n}$, and x replaced by (r_n, x); the constraint $r_n \geq 0$ is obviously redundant. The existence and uniqueness of the solution to (1.2.6a) (which is a possible definition of centre of the set of given points; see $(2.1.8)^*$) are given as consequences of Theorem 2.1.3.

With the same data as above, we can define another problem : among all the spheres S_n having the centre in $x \in \text{conv}\ \{x^1, ..., x^N\}$ and the radius $\sqrt{r_n}$ not greater than the distances between the centre and each point x^i, find the one which has the maximum radius. This problem can be formulated as

$$\max\ (r_n), \quad \text{s.t.} \quad r_n - \sum_{j=1}^{n}(x_j - x_j^i)^2 < 0, \quad i = 1, ..., N, \quad x = \sum_{i=1}^{N}\alpha_i\ x^i,$$

$$\tag{1.2.6b}$$

$$\sum_{i=1}^{N}\alpha_i = 1, \quad \alpha_i \geq 0, \quad i = 1, ..., N,$$

where the unknowns are r_n, $x = (x_1, ..., x_n)$ and $\alpha = (\alpha_1, ..., \alpha_N)$. Taking into account the obvious and already mentioned relation $\max f = -\min\ (-f)$, (1.2.6b) is identified easily with (1.1.1) for $X = \mathbb{R}^{1+n+N}$, $p = 2$, $m = 2 + 2N$, and with obvious positions for f and g.

A further problem consists in finding an n−dimensional tetrahedron (simplex), say TH_n, contained in a given sphere S_n and having maximum (positive) volume. Denote by $x^1, ..., x^{n+1} \in \mathbb{R}^n$ the vertices of TH_n, and by $\sqrt{r_n}$ with $r_n \geq 0$ the radius of S_n. It is well known that the volume of TH_n is given by [35] :

$$V(TH_n) = \frac{1}{n!}\ |D(x^1, ..., x^{n+1})|,$$

where :

$$D(x^1, ..., x^{n+1}) := \det \begin{pmatrix} 1 & x_1^1 & ..... & x_n^1 \\ \vdots & \vdots & & \vdots \\ 1 & x_1^{n+1} & ..... & x_n^{n+1} \end{pmatrix}.$$

If the coordinates of the centre of S_n are denoted again by $x = (x_1, ..., x_n) \in \mathbb{R}^n$, then the problem can be formulated as :

$$\max \ \frac{1}{n!} \ |D(x^1,...,x^{n+1})|, \quad \text{s.t.} \ \sum_{j=1}^{n}(x_j^i - x_j)^2 - r_n \leq 0, \quad i = 1,...,n+1, \qquad (1.2.7)$$

where, unlike (1.2.6), the unknowns are now $x^1,...,x^{n+1}$, while r_n and x are given. (1.2.7) is identified easily with (1.1.1) at $X = \mathbb{R}^{n(n+1)}$, $p = 0$, $m = n + 1$, and with obvious positions for f and g. $\qquad\square$

Example 1.2.5. We are given a physical obstacle in the ordinary space and must draw a path of minimum length between two given points by either overcoming the obstacle or by penetrating (e.g., by excavating) it under a constraint on the curvature of the path. If the path lies in a vertical plane and is smooth, a simplified version of such a problem can be formulated as:

$$\min \int_T \sqrt{1 + x'(t)^2} \ dt, \quad \text{s.t.} \quad x(t) - a(t) - kc(t;x) \geq 0, \ \forall t \in T,$$

$$(1.2.8)$$

$$x(t_1) = x(t_2) = 0, \ x \in C^2(T),$$

where $a(t) = 0$ at $t_0 \leq t \leq \tau_0$, and at $\tau_1 \leq t \leq t_1$ with $t_0 \leq \tau_0 < \tau_1 \leq t_1$, and $a(t)$ is strictly concave and positive in $]\tau_0, \tau_1[$; $c(t; \mathrm{x}):=\frac{x''}{[1+x'^2]^{\frac{3}{2}}}$ is the curvature of x, $k \gtrless 0$ is a parameter, and x is the unknown curve. (1.2.8) is easily identified with (1.1.5) for $p = 0$, $m = 1$ and with obvious positions for X, ψ_0 and ψ. For $k = 0$, the path x must overcome the obstacle and (1.2.8) is a classic problem. At $k < 0$ (or $k > 0$), the path x can penetrate the obstacle in the subintervals (of $[\tau_0, \tau_1]$) where it is strictly convex (or strictly concave). $\qquad\square$

Example 1.2.6. The problem of allocating given resources (such as raw materials, energy, capitals, manpower) to competing activities, in such a way to maximize or minimize a suitable measure of the total output or return, is a classic one. In the finite case, if there are n activities and m resources, if x is the unknown vector of the levels of the activities and $g_i(x)$ is the difference between the quantity of the ith resource absorbed by the production strategy x and the available quantity, if f is the cost implied by x and X a set of subsidiary constraints, then (1.1.1) is the formulation of the above problem. A special, but interesting, case is that where the activities are independent of each other, in the sense that the cost and the consumption of the resources in an activity do not depend on those of the others. This means that the production problem can be expressed by the following particular case of (1.1.1):

$$\min \sum_{j=1}^{n} f_j(x_j), \ \text{s.t.} \ \sum_{j=1}^{n} g_{ij}(x_j) = 0, \ i \in \mathfrak{I}^0, \ \sum_{j=1}^{n} g_{ij}(x_j) \geq 0, \ i \in \mathfrak{I}^+, \ x \in X. \quad (1.2.9a)$$

In other words, f and g can be assumed to be separable. A further particular case is obtained from (1.2.9a) for $p = 1$, $m = 1 + n$, $X = \mathbb{R}^n$, $g_{ij}(x_j) = x_j - \frac{b}{n}$, $i \in \mathfrak{I}^0 = \{1\}$, $j = 1,...,n$, where b is a positive constant, and , $\forall i \in \mathfrak{I}^+ = \{2,...,1+n\}$ and $\forall j = 1,...,n$, $g_{ij}(x_j) = 0$ if $j \neq i - 1$, $g_{jj}(x_j) = x_j$ if $j = i - 1$; hence, the particular case becomes:

$$\min \sum_{j=1}^{n} f_j(x_j), \quad \text{s.t.} \quad \sum_{j=1}^{n} x_j - b = 0, \quad x_j \geq 0, \quad j = 1, ..., n . \tag{1.2.9b}$$

$\square$

Example 1.2.7. Consider the sets (the present notation is independent of that used elsewhere):

$$Y := \{y \in \mathbb{R}^r_+ : \sum_{i=1}^{r} y_i = 1\}, \quad Z := \{z \in \mathbb{R}^s_+ : \sum_{j=1}^{s} z_j = 1\},$$

and the function $F : Y \times Z \to \mathbb{R}$. We search for $(\bar{y}, \bar{z}) \in Y \times Z$, such that

$$\max_{z \in Z} \min_{y \in Y} F(y, z) = F(\bar{y}, \bar{z}) = \min_{y \in Y} \max_{z \in Z} F(y, z). \tag{1.2.10a}$$

This includes the following application: there are two players whose strategies are represented by y and z, respectively. F is the function of payments between the two players; $(\bar{y}, \bar{z})$ is called the saddle point (see Definition 5.2.1) of F and $F(\bar{y}, \bar{z})$ the saddle value. A special case is that where $F(y, z) = \langle y, Az \rangle$, $A = (a_{ij})$ being a matrix of dimension $r \times s$ with real entries. (1.2.10a) is verified if A has an element, say $a_{\bar{i}\bar{j}}$, s.t.

$$\max_{j=1,...,s} \min_{i=1,...,r} a_{ij} = a_{\bar{i}\,\bar{j}} = \min_{i=1,...,r} \max_{j=1,...,s} a_{ij}. \tag{1.2.10b}$$

In this case, $\bar{y}$ and $\bar{z}$ have, respectively, the $\bar{i}$th entry and the $\bar{j}$th entry $=1$ and all the others $=0$ (so−called pure strategies); $a_{\bar{i}\,\bar{j}}$ is called the saddle value of the matrix A. See also Example 5.2.2. $\square$

Example 1.2.8. The problem consists in drawing, within a given triangle T of the plane, whose sides are denoted by s_1, s_2, s_3, N circles with centres in $C^1, ..., C^N$ and corresponding radii $r_1, ..., r_N$, each two of them with disjoint interiors, and such that the sum of their areas be a maximum. A formulation of the problem as (1.1.1) is the following one ($X = \mathbb{R}^2$, $p = 0$, $m = \frac{1}{2}N(N + 7)$, $n = 3N$; $x = (C^1, ..., C^N, r_1, ..., r_N)$; f and g are defined in an obvious way):

$$\pi \cdot \max \sum_{i=1}^{N} r_i^2 \tag{1.2.11a}$$

$$\text{s.t.} \quad C^i \in T, \quad i = 1, ..., N, \tag{1.2.11b}$$

$$\text{dist}\,(C^i, C^j) \geq r_i + r_j, \quad i, j = 1, ..., N, \quad i \neq j, \tag{1.2.11c}$$

$$\text{dist}\,(C^i, s_j) \geq r_i, \quad i = 1, ..., N, \quad j = 1, 2, 3. \tag{1.2.11d}$$

The above problem is an instance of real problems which are met when a hard stone must be cut into circular columns (or other forms). In these cases, there might be additional constraints, as bounds on the radii.

Example 1.2.9. In a huge variety of real situations, there are unknown quantities which can take only two possible values; these can be assumed to be either zero or one.

This happens, for instance, in the scheduling of activities, in the timetable problems, design of computer memories, of electric/electronic circuits, crew scheduling, locating facilities, assignment problems, and so on. If a variable can take more than two integer values, then a suitable transformation turns it into binary variables. These problems can be reduced to the format (1.1.7).   □

Example 1.2.10. In many real situations, there is more than one objective or criterion; to reduce them to just one, as it is done often, is unsatisfactory. This can happen in the field of industrial production (see Example 1.2.6), where quantity of production, cost, gain, reliability of production may be in competition. The presence of several objectives occurs frequently in engineering design [58].   □

Example 1.2.11. Consider the following problem:

$$f^{\downarrow} := \min \left[f(x) := \int_T f^0(t, x(t), \xi(t))dt \right], \tag{1.2.12a}$$

$$\text{s.t.} \qquad \frac{dx_i}{dt} = f^i(x; \xi), \qquad i \in I := \{1, ..., n\}, \tag{1.2.12b}$$

$$(x, \xi) \in X \times \Xi, \tag{1.2.12c}$$

where $x = (x_1, ..., x_n)$ and $\xi = (\xi_1, ..., \xi_q)$ are vectors of functions. The interval $T := [t_1, t_2] \subset \mathbb{R}$, with $t_1 < t_2$, and the functions $f^0 : \mathbb{R}^{1+n+q} \to \mathbb{R}$, $f^i : \mathbb{R}^{n+q} \to \mathbb{R}$, $i \in I$, are given, as well as $x^0, x^1 \in \mathbb{R}^n$. f^0 is assumed to be integrable and $f^1, ..., f^n$ to be continuous with respect to $(x; \xi)$ and continuously differentiable with respect to x. ξ is a parameter, which acts as a "control", while x is a "state variable", called also "trajectory". It is assumed that, $\forall \xi \in \Xi$, the differential system (1.2.12b) admits one and only one solution, such that $x(t_0) = x^0$. The pair (x, ξ) is feasible, iff ξ is such that the (unique) solution x of the differential system (1.2.12b) (which satisfies the initial condition $x(t_0) = x^0$) meets the boundary condition $x(t_1) = x^1$; with a slight abuse of notation, X denotes the set of such vectors x (indeed, the present X differs from that of (1.1.4) and depends on Ξ, which, in its turn, must guarantee the above feasibility); under this proviso, we can say that (1.2.12b,c) give feasibility.

In the applications, the controls are, in general, at least piecewise continuous functions (with range in Ξ) having a finite number of discontinuities of 1st kind (like for x in (1.1.4)). If this number is positive, and $\tau_1, ..., \tau_r \in T$ are the points of discontinuity, then (1.2.12b) must be considered in each of the $\tau + 1$ subintervals of T: the solution of (1.2.12b) in $[t_0, \tau_1]$, with the initial condition $x(t_0) = x^0$, leads to $x(\tau_1)$; this vector becomes the initial condition for (1.2.12b) in $[\tau_1, \tau_2]$; and so on. Consequently, the trajectory will turn out to be continuous and piecewise differentiable; to be feasible, the $\tau + 1$ controls must be chosen in such way that x will achieve x^1 at $t = t_1$. When f^0 is independent of t, then (1.2.12) is called *autonomous* problem; it is called *nonautonomous* problem in the general case. The above problem is, more or less, the classic Mayer's problem [III24, 6]. (1.2.12b) represents the equations of motion (in physical or late sense) depending on a vector ξ of parameters (such as acceleration, direction, and so on). It seems that some terminology in terms of optimal control has been introduced in 1933 by L.M.Graves (see [6]). With a slight abuse of terminology, we can say that (1.2.12) is of type (1.1.5). Indeed, it would be a special case of (1.1.5), if in (1.1.5) the derivability of the components of x were required only for a part of them; in such a case, x of (1.1.5) should be identified with the pair (x, ξ) of (1.2.12); (1.1.5b) should be identified with (1.2.12b); $p = m$ (of course, the present n corresponds to m of Sect. 1.1), so that (1.1.5c) do not exist; (1.1.5d) would become (1.2.12c). $\qquad\square$

1.3. Variational Inequalities, Complementarity Problems and Generalized Systems

Since the pioneering Work of G. Stampacchia [5, 42, 57], there has been a great development of mathematical models for equilibrium problems. Indeed, in some cases where it is desirable to study the equilibrium of a system, it is difficult or arbitrary to postulate the existence of a functional f whose constrained minimum gives equilibrium; in other words, (1.1.6) or (1.1.8) or (1.1.10) cannot be always adopted to represent the equilibrium of a system. The symbols of this section must be considered independent of those of the other sections, if they overlap. Since the present scope is to give an instance on how IS analysis can be extended also to variational inequalities and to generalized systems, and not to have a treatise on the subject, then, for the sake of simplicity, the following formats are given in a finite dimensional setting. However, some of the following formats will be considered also in a Hilbert space; to this end, it is enough to replace $\mathbb{R}^n$ with B and assume that it is equipped with a scalar product. Let $\langle\,\cdot\,,\,\cdot\,\rangle$ denote the scalar product.

Assume that we are given the multifunctions $\mathcal{F} : \mathbb{R}^n \rightrightarrows \mathbb{R}^n$, $\mathbb{K} : \mathbb{R}^n \rightrightarrows \mathbb{R}^n$, and the function $\phi : \mathbb{R}^n \to \mathbb{R}$. Consider the problem of finding $\bar{x} \in \mathbb{K}(\bar{x})$, and $F \in \mathcal{F}(\bar{x})$ such that

$$\langle F, x - \bar{x}\rangle + \phi(x) - \phi(\bar{x}) \geq 0, \quad \forall x \in \mathbb{K}(\bar{x}). \tag{1.3.1}$$

If $\mathbb{K}$ does not depend on $\bar{x}$, then we have a particular case of (1.3.1) which consists of finding $\bar{x} \in \mathbb{K}$ and $F \in \mathcal{F}(\bar{x})$ such that:

$$\langle F, x - \bar{x}\rangle + \phi(x) - \phi(\bar{x}) \geq 0, \quad \forall x \in \mathbb{K}. \tag{1.3.2}$$

If $\mathcal{F}(\bar{x})$ is a singleton ($\mathcal{F}$ is single-valued), then (1.3.1) becomes: find $\bar{x} \in \mathbb{K}(\bar{x})$ such that

$$\langle F(\bar{x}), x - \bar{x}\rangle + \phi(x) - \phi(\bar{x}) \geq 0, \quad \forall x \in \mathbb{K}(\bar{x}). \tag{1.3.3}$$

If $\mathbb{K}$ does not depend on $\bar{x}$ and $\mathcal{F}(\bar{x})$ is a singleton, then (1.3.1) becomes: find $\bar{x} \in \mathbb{K}$ such that:

$$\langle F(\bar{x}), x - \bar{x} \rangle + \phi(x) - \phi(\bar{x}) \geq 0, \quad \forall x \in \mathbb{K}. \tag{1.3.4}$$

For ϕ constant, (1.3.4) becomes: find $\bar{x} \in \mathbb{K}$ such that

$$\langle F(\bar{x}), x - \bar{x} \rangle \geq 0, \quad \forall x \in \mathbb{K}. \tag{1.3.5}$$

If $F(x) \equiv 0$, then (1.3.4) expresses $\bar{x}$ as global m.p. of ϕ on $\mathbb{K}$ (see Definition 1.1.1). If $\mathbb{K} = \mathbb{R}^n$ and ϕ is the indicator function of $\mathbb{K}$ (see (2.3.16)), then (1.3.4) is equivalent to (1.3.5), as it is easy to show.

The inequality (1.3.4) − and hence (1.3.5) − is called a *Variational Inequality* (for short, VI); the other formats are generalizations of a VI; when the domain $\mathbb{K}$ depends on the unknown $\bar{x}$, then the inequality is called *Quasi-Variational*.

A VI, whose initial development is due to G. Stampacchia [57], can be associated with another type of inequality: find $\bar{x} \in \mathbb{K}$ such that

$$\langle F(x),\ \bar{x} - x \rangle + \phi(\bar{x}) - \phi(x) \leq 0, \quad \forall x \in \mathbb{K}. \tag{1.3.6}$$

When (1.3.4) and (1.3.6) are considered together, then they are called the *Stampacchia* VI (for short, SVI) and *Minty* VI (for short, MVI), respectively. Indeed, G. J. Minty proved an important proposition, the so-called Minty Lemma [5, 42, 51], which has been instrumental in the early development of VI; however, it has turned out to be more than a lemma (see Vol. 2).

Almost in the same period when a VI has been conceived, another type of mathematical model for equilibrium problems has been proposed: *complementarity systems* (for short, CS). Assuming that $\mathbb{K}$ be a closed and convex cone in $\mathbb{R}^n$ with apex at the origin, and denoting by $\mathbb{K}^*$ its (positive) polar, a CS consists in seeking $\bar{x}$ such that:

$$\bar{x} \in \mathbb{K}, \quad F(\bar{x}) \in \mathbb{K}^*, \ \langle F(\bar{x}), \bar{x} \rangle = 0. \tag{1.3.7}$$

A CS was introduced initially in a finite dimensional setting to study an orthogonality relation appearing in the stationarity conditions for optimality (see Sect. 5.3), and later was applied to equilibrium problems.

As in the field of optimization problems, also in that of equilibrium ones we may be faced with real situations having a vector nature. For this reason, both VI and CS have been considered in the more general context of vector equilibrium. For the sake of simplicity, we will consider only finite dimensional settings; for a more general treatment we refer the reader to [27] . Here, the operator is assumed to be matrix−valued: $F : \mathbb{R}^n \to \mathbb{R}^{\ell \times n}$. A *Vector Variational Inequality* (for short, VVI) consists in finding $\bar{x} \in \mathbb{K}$ such that:

$$F(\bar{x})(x - \bar{x}) \nleq_{C_0} O_\ell, \quad \forall x \in \mathbb{K}, \tag{1.3.8}$$

where C_0 and $\nleq$ are as in Sect. 1.1. For $\ell=1$ and $C = \mathbb{R}_+$, (1.3.8) collapses to (1.3.5).

(1.3.8) is the *Stampacchia* VVI (for short, SVVI). Analogously to what happens in the scalar case, (1.3.8) can be associated with another inequality, which we call the *Minty Vector Variational Inequality* (for short, MVVI); it consists in finding $\bar{x} \in \mathbb{K}$ such that:

$$F(x)(\bar{x} - x) \not\geq_{C_0} O_\ell, \quad \forall x \in \mathbb{K}. \tag{1.3.9}$$

At $\ell{=}1$ and $C = \mathbb{R}_+$, (1.3.9) collapses to (1.3.6) with φ constant. The above VVI have been defined by means of the cone C_0 (which generalizes that of Pareto); the cone $\overset{o}{C}$ can be adapted — instead of C_0 — to obtain different VVI, which will turn out to be *relaxations* of (1.3.8) and (1.3.9). Hence, the former — which is a relaxation of SVVI — consists in finding $\bar{x} \in \mathbb{K}$ such that:

$$F(\bar{x})(x - \bar{x}) \not\leq_{\overset{o}{C}} O_\ell, \quad \forall x \in \mathbb{K}. \tag{1.3.10}$$

The latter consists in finding $\bar{x} \in \mathbb{K}$, such that:

$$F(x)(\bar{x} - x) \not\geq_{\overset{o}{C}} O_\ell, \quad \forall x \in \mathbb{K}, \tag{1.3.11}$$

Just as a VI has been associated with a CS, a VVI can be associated with a *vector complementarity system* (for short, VCS). Assuming that $\mathbb{K}$ be a closed and convex cone in B, a VCS consists in finding $\bar{x}$ such that

$$\bar{x} \in \mathbb{K}, \quad F(\bar{x}) \in \mathbb{K}^*_{C_o}, \quad F(\bar{x})\bar{x} = O_\ell, \tag{1.3.12}$$

where F is as in (1.3.8), and

$$\mathbb{K}^*_{C_o} := \{M \in \mathbb{R}^{\ell \times n} : Mx \not\leq_{C_o} O_\ell, \quad \forall x \in \mathbb{K}\} \tag{1.3.13}$$

is the *vector polar* of $\mathbb{K}$ with respect to C_o (see Sect. 2.2).

Proposition 1.3.1. If $\bar{x}$ is a solution of VCS (1.3.12), then it is a solution of VVI (1.3.8).

Proof. It is enough to show that

$$F(\bar{x})x \not\leq_{C_o} O_\ell, \quad \forall x \in \mathbb{K}.$$

This holds, since $F(\bar{x}) \in \mathbb{K}^*_{C_o}$. $\qquad\square$

A result quite analogous to Proposition 1.3.1 holds if in (1.3.13) the inequality is replaced by $Mx \geq_C O_\ell$.

In order to invert the above proposition we need to assume that the condition $F(\bar{x})\bar{x} = O_\ell$ be fulfilled by a solution $\bar{x}$ of VVI.

Proposition 1.3.2. Suppose that $\bar{x}$ be a solution of VVI (1.3.8) such that $F(\bar{x})\bar{x} = O_\ell$. Then $\bar{x}$ is a solution of VCS (1.3.12).

Proof. We only need to prove that

$$F(\bar{x}) \in \mathbb{K}^*_{C_o}. \tag{1.3.14}$$

From (1.3.8) we have $F(\bar{x})x \not\leq_{C_o} O_\ell, \quad \forall x \in \mathbb{K}$, which gives (1.3.14) $\qquad\square$

When a VI or a CS is the formulation of some classic problems of Mechanics, often the solution is unique. However, in several problems — for instance, in the fields of

Structural Mechanics, or in the design of networks, or in the analysis of economic equilibrium $-$ the lack of uniqueness is a normal situation, in the sense that reality admits several equilibrium shapes. When these equilibria are not indifferent, we may be given a function to be minimized (or maximized) on the set of equilibrium points. Therefore, we may be faced with problems of the following kind:

$$\min f(x), \quad \text{s.t.} \quad x \in R \cap \mathbb{K}^0, \tag{1.3.15}$$

where f and R are as in Sect. 1.1, and $\mathbb{K}^0$ is the set of solutions to (1.3.4) or (1.3.7) (with obvious change of notations); in the former case, (1.3.15) is called *minimization problem with equilibrium constraints* (for short, MPEC); in the latter case, it is called *complementarity problem* (for short, CP).

In the present section and in the previous one, we have introduced some formats of extremum problems and of equilibrium ones. The fact that they might appear different has a foundation; however, they have also something in common. To show this, we introduce the following notation.

Definition 1.3.1. Let ν be a positive integer, $\mathcal{H} \subset \mathbb{R}^\nu$ a convex cone with apex at the origin, Ξ a set of parameters, $X \subseteq \mathbb{R}^n$ and $A : X \times \Xi \to \mathbb{R}^\nu$ a mapping. The relation

$$A(x;\xi) \in \mathcal{H}, \quad x \in X, \qquad (\xi \in \Xi), \tag{1.3.16}$$

is a *parametric generalized system* in the variable x.

In the sequel, instead of $A(x;\xi)$, we will use also the notation $A_\xi(x)$ $-$ especially when ξ belongs to a discrete set $-$ and even $A(x)$ when the parameter ξ will not play any substantial role.

Now, it is shown that most of the mathematical models, which are used in optimization and related fields can be reduced to find $\xi \in \Xi$ s.t. (1.3.16) be impossible.

Example 1.3.1. Consider (1.1.6) in the case $\mathcal{B} = \mathbb{R}$, so that it embraces (1.1.1) and (1.1.4) but not (1.1.5). Let X be that of (1.1.6), and set:

$$\nu = 1 + m, \quad \mathcal{H} = (\mathbb{R}_+ \backslash \{0\}) \times D, \quad \Xi = X, \tag{1.3.17a}$$

$$\xi = \overline{x}, \quad A(x;\overline{x}) = (f(\overline{x}) - f(x), g(x)). \tag{1.3.17b}$$

Then, according to Definition 1.1.1, $\overline{x}$ is a global m.p. of (1.1.6) iff (1.3.16) is impossible. This case includes (1.1.7) too. $\square$

Example 1.3.2. Consider (1.1.8). Let X be as in the above example, and set:

$$\nu = \ell + m, \quad \mathcal{H} = C_o \times D, \quad \Xi = X, \tag{1.3.18a}$$

$$\xi = \overline{x}, \quad A(x;\overline{x}) = (f(\overline{x}) - f(x), g(x)). \tag{1.3.18b}$$

Then, according to Definition 1.1.2, $\overline{x}$ is a v.m.p. of (1.1.8) iff (1.3.16) is impossible. A quite analogous claim holds for (1.1.10).

Example 1.3.3. Consider (1.3.4), and set:

$$\nu = 1, \quad \mathcal{H} = \mathbb{R}_+ \backslash \{0\}, \quad \Xi = X = \mathbb{K}, \tag{1.3.19a}$$

$$\xi = \overline{x}, \quad A(x;\overline{x}) = \langle F(\overline{x}), \overline{x} - x \rangle + \phi(\overline{x}) - \phi(x). \tag{1.3.19b}$$

Then $\overline{x}$ is a solution of (1.3.4) iff (1.3.16) is impossible.

Even if VI is a recent branch of mathematics, the related literature is wide; in special, existence and uniqueness theorems have been established for the several formats mainly in a Hilbert space. Here, we merely recall a few classic results and generalizations concerning (1.3.5) in the finite dimensional case; for details, see for instance [5,13,27,42]. The following statement is the finite dimensional version of the well known Hartman-Stampacchia Theorem.

Theorem 1.3.1. If $\mathbb{K}$ is convex and compact, and F is continuous on $\mathbb{K}$, then (1.3.5) has solutions. $\square$

This theorem can be extended in several ways to the case where $\mathbb{K}$ is not compact. The most classic way consists in adding coerciveness and monotonicity (isotonicity) assumptions on the operator F. See [5,42], Sect. 2.3, and Vol. 2.

Theorem 1.3.2. Let $\mathbb{K}$ be convex. If **(i)** F is continuous and monotone (isotone) on $\mathbb{K}$; and **(ii)** $\exists y \in \mathbb{K}$ such that:

$$\lim_{\substack{||x|| \to +\infty \\ x \in \mathbb{R}^n}} \frac{\langle F(x), x - y \rangle}{||x||} = +\infty, \tag{1.3.20}$$

then (1.3.5) has solutions. $\square$

Theorem 1.3.1 can be obtained also as a corollary of a very general existence theorem stated, in a Banach space, for the generalized VI (1.3.1) for ϕ constant [5,42]. See also (3.2.7) for a more general setting in the IS.

Theorem 1.3.3. Let $X \subseteq \mathbb{R}^n$ be compact, and $\mathcal{F} : X \rightrightarrows X$, $\mathbb{K} : X \rightrightarrows X$. Suppose that $\mathbb{K}$ be continuous with convex and compact values, and $\mathcal{F}$ be upper semicontinuous with strongly compact and convex values. Then (1.3.1) for ϕ constant has solutions. $\square$

With regard to the uniqueness of solutions to (1.3.5), one of the most classic conditions consists in assuming the strict monotonicity (isotonicity) of the operator F. Therefore, if we replace, in (i) of Theorem 1.3.2, the assumption of monotonicity (isotonicity) with that of strict monotonicity (isotonicity), then we achieve uniqueness besides the existence of solutions to (1.3.5). Some results about VVI and VCS are in Vol. 2.

1.4. Optimal Design of an Underwater Pipeline

Around 1970 G. Maier has initiated an original approach to both classic and new problems in the field of Structural Mechanics by exploiting optimization models and complementarity systems. See [10, 23, 44] for an overview. Here and in Sect. 1.5, we consider some simplified instances. The symbols of this section are independent of those of the others, if they overlap.

An underwater pipeline laid to rest freely on a rough seabottom can be considered as a slender beam subjected to vertical loads and supported by a distribution of rigid unilateral constraints. The equilibrium configuration of the pipeline can be determined by solving a particular contact (free boundary) problem. Such an analysis of the pipeline may reveal excessive bending moments due to the irregularity of the seafloor. Sometimes these excesses can be avoided by suitably modifying the laying route. In relatively shallow waters, it may be preferable to regularize the supporting profile along the route. Modern offshore technology allows one to excavate trenches and build artificial supports (trestles) at considerable waterdepths. Such provisions interrupt suspended spans and reduce the curvatures of the pipe below preassigned limits (variations of the flexural strength along the pipe are not feasible). Since underwater operations are very costly, it is necessary to minimize the total cost of the seabottom modifications, subject to a constraint on the bending moment in the pipe.

In order to achieve a mathematical formulation of the above optimal design problem, let us adopt the following assumptions, which are regarded as acceptable for practical engineering purposes:

(a) the pipeline is a linear elastic beam deflected by vertical loads within a vertical plane; (b) the deformations are small, in the sense that the equilibrium configuration of the pipe can be defined by vertical displacements (with respect to a horizontal straight line, say to sealevel) on which the curvatures depend linearly; (c) the seabottom is a rigid and frictionless profile, which can provide at contact upward vertical reactions; (d) the cost of trenching per unit length depends quadratically on the excavation depth; (e) the deformed pipe configuration is assumed to be piecewise linear. In other words, the flexural deformability is thought of as concentrated in m elastic hinges or nodes, connecting $m + 1$ rigid elements of length ℓ_i. At both ends, the pipe is considered to be hinged to the ground in fixed points at the reference level. The rotational stiffness attributed to the ith elastic hinge is

$$K_i = (EI)_i \bar{\ell}_i^{-1} := 2(EI)_i(\ell_i + \ell_{i+1})^{-1}, \qquad (1.4.1)$$

$\bar{\ell}_i$ being the tributary length of section i and $(EI)_i$ being the bending stiffness of the pipe supposed constant along ℓ_i. EI will be assumed constant along the whole span under consideration. Contact between pipeline and seabottom is concentrated at the elastic hinges, i.e., the reactions provided by the seabottom are concentrated forces w_i, $i = 1, ..., m$, acting on the nodes. Consistent with (c), these reactive forces will be assumed to be vertical and positive, if directed upward. Similarly, the external loads will be considered to be a set of vertical, positive downward forces applied to the nodes $F_i = g\bar{\ell}_i$, where g (assumed constant) represents the weight of the pipe and its contents minus the Archimedes lifting thrust per unit length.

The trench section is assumed to be trapezoidal and piecewise constant along the sealine, namely constant over the tributary length $\bar{\ell}_i$ of each nodal section. Hence, the total cost Φ^+ of the trenching operations, assumed proportional to the soil volume removed, turns out to depend quadratically on the m sign-constrained design variables x_i^+ which denote the excavation depth at the nodes,

$$\Phi^+ := \sum_{i=1}^{m} c_i^+ x_i^+ + \sum_{i=1}^{m} \hat{c}_i^+ (x_i^+)^2, \quad x_i^+ \geq 0, \quad i = 1, ..., m, \tag{1.4.2}$$

where the positive coefficients c_i^+ and $\hat{c}_i^+$ depend on the tributary length $\bar{\ell}_i$, the slope of the trench walls α_i, and the nature of the soil in the neighbourhood of node i. Analogously, the total cost of artificial supports is expressed as a quadratic function of the heights x_i^- of trestles (or embankments) at the nodes,

$$\Phi^- := \sum_{i=1}^{m} c_i^- x_i^- + \sum_{i=1}^{m} \hat{c}_i^- (x_i^-)^2, \quad x_i^- \geq 0, \quad i = 1, ..., m, \tag{1.4.3}$$

where the positive coefficients c_i^- and $\hat{c}_i^-$ are analogous to the previous ones.

The original stressless configuration of the pipe is conceived as straight at the reference level. By virtue of the small deformation hypothesis (b), the geometric compatibility can be expressed as

$$e = Cu, \tag{1.4.4a}$$

where e is the $m-$vector of the (relative) rotations at the elastic hinges, u is the $m-$vector of the depths of the final equilibrium configuration of the pipeline, and the compatibility matrix C depends only on the location of the nodes along the span and is nonsingular and tridiagonal. Let Q denote the vector of bending moments in the nodes, F the vector of loads, and w the vector of the support reactions supplied by the seabottom; the equilibrium equation reads :

$$C^T Q = F - w. \tag{1.4.4b}$$

The generalized stress—strain relation for all deformable elements is

$$Q = ke, \tag{1.4.4c}$$

where $k := \text{diag } |k_i|$ is the matrix which contains all the stiffnesses (1.4.1) as diagonal entries. Let z denote the $m-$vector of the vertical distances between the final seabottom and the pipe in its equilibrium configuration; for all nodes the relation between the reactions w and distances z can be represented in the form:

$$w \geq 0, \quad z \geq 0, \quad \langle w, z \rangle = 0 \ . \tag{1.4.4d}$$

The last (nonlinear, orthogonality) relation, since it concerns non-negative vectors, holds componentwise, i.e. implies the complementarity relations $w_i z_i = 0$, $i = 1, ..., m$, which are required by the mechanical meaning of the variable involved. Let r denote the $m-$vector of the depths of the original, undisturbed seabed, with respect to the reference level ; the vector u of displacements can be expressed as the sum

$$u = r + x^+ - x^- - z, \tag{1.4.4e}$$

where $x^+ := (x_1^+, ..., x_m^+)$, $x^- := (x_1^-, ..., x_m^-)$. For given x^+ and x^-, the relation set (1.4.4) governs the configuration of the pipeline model on the seabottom.

The ratio $e_i/\bar{\ell}_i$ between elastic rotation and tributary length at each node represents the approximation, consistent with the present model, of the curvature in the actual pipe. Hence, if $\beta > 0$ is the maximum admissible absolute value of the curvature for the pipe, the behaviour constraints to be complied with by its configuration on the modified seabed , taking into account (1.4.4a), are

$$-\ell\beta \; < \; Cu \; < \; \ell\beta. \tag{1.4.5}$$

Through (1.4.4 a,c), (1.4.4b) becomes

$$Su = F - w, \tag{1.4.6}$$

with $S := C^T kC$, S being the positive definite, symmetric stiffness matrix of the assembled structural model of the pipeline ; S proves to be pentadiagonal. Substituting (1.4.4a) into (1.4.5) and (1.4.6), and setting $d := F - Sr$, $b^+ := \ell\beta - Cr$, $b^- := \ell\beta + Cr$, one obtains the following formulation of the optimal design problem:

$$\min \; (\langle c^+, x^+ \rangle + \langle x^+, C^+ x^+ \rangle + \langle c^-, x^- \rangle + \langle x^-, C^- x^- \rangle) \tag{1.4.7a}$$

$$\text{s.t.} \quad w + S(x^+ - x^-) - Sz = d, \tag{1.4.7b}$$

$$-b^- \leq C(x^+ - x^-) - Cz \leq b^+, \tag{1.4.7c}$$

$$x^+, x^-, w, z \geq 0, \tag{1.4.7d}$$

$$\langle w, z \rangle = 0. \tag{1.4.7e}$$

The above problem is identified easily with (1.1.1) for $X = \mathbb{R}^n$, $x = (x^+, x^-, w, z)$, and further obvious positions. However, the presence of the orthogonality constraint (1.4.7e) requires considerations of (1.4.7b-e) also as CS like (1.3.7). In other words, it is suitable to consider (1.4.7) as MPEC (Sect. 1.3).

The assumptions made on excavation and embankment correspond to the reality of engineering operations; hence, the objective function in (1.4.7a) represent the real situation and is not an approximation of the real function, as happens often when a quadratic function is adopted. The system (1.4.7 b,d,e) is a complementarity system (see Sects. 1.3 and 5.8; when x^+ and x^- are fixed, it is of type (1.3.7)); a complementarity system has been considered as an important property of constrained extrema (see Sect. 4.3) ; however, it is to G.Maier's credit that he conceived the possibility of formulating important real life, practical problems as the minimization of a convex quadratic function under constraints represented by a complementarity system, such as the present one [10].

The above optimal design problem can be formulated also in an infinite dimensional space. Let $T = [t_0, t_1]$ be the sea-level corresponding to the pipe (again supposed to lie in a vertical plane), and let $r(t)$, $x(t)$, $z(t)$, $u(t)$, $w(t)$ be, respectively, the depth of the original seabed, excavation depth, vertical distance between the final seabottom and the pipe, support reaction supplied by the seabottom at point $t \in T$. Assuming again the excavation to be trapezoidal, $a(t)x(t)$ and $b(t)x^2(t)$ denote the intensities of cost of the rectangular and triangular parties of the excavation, respectively. E and I are the constant elastic module of the steel and the constant inertia moment, respectively; $c(t)$

is the weight of the pipe at t. Neglecting the embankment, the optimal design problem can be formulated as [25, 29] :

$$min \int_T [a(t)x(t) + b(t)x^2(t)] \, dt, \qquad (1.4.8a)$$

$$\text{s.t.} \qquad E \cdot I \cdot \frac{d^4 u}{du^4} = c(t) - w(t), \quad u(t) = r(t) + x(t) - z(t), \qquad (1.4.8b)$$

$$-\beta \leq \frac{\ddot{u}}{(1 + \dot{u}^2)^{3/2}} \leq \beta, \quad \forall t \in T, \qquad (1.4.8c)$$

$$x(t), \; w(t), \; z(t), \; u(t) \geq 0, \quad \forall t \in T, \qquad (1.4.8d)$$

$$w(t) \cdot z(t) = 0, \quad \forall t \in T. \qquad (1.4.8e)$$

Equations (1.4.8b) represent the mechanical equilibrium, (1.4.8e) the unilateral frictionless contact. (1.4.8) is easily identified with (1.1.6) for $n = 4, p = 3, m = 5$ and with $X = \{(x, w, z, u) \in [C^4(T)]^4 : x(t), w(t), z(t), u(t) \geq 0, \; \forall t \in T\}$; (1.1.6b) is identified with (1.4.8b,e) and (1.1.6c) with (1.4.8c). However, the orthogonality constraint (1.4.8e) suggests consideration of (1.4.8b-e) also as a CS like (1.3.7).

Note that Example 1.2.5 can be seen as a preliminary and simplified version of (1.4.8); they both can be considered as elements of a family of obstacle problems where the obstacles can be penetrated.

1.5. Further Problems in Applied Mechanics

We will now outline a few further problems, which have a great importance in the applications to real life: **(i)** elastoplastic and poroplastic behaviour of materials; **(ii)** quasi-brittle fracture process; **(3i)** a problem in Flight Mechanics; **(4i)** a problem in Astrodynamics.

In the field of Structural Mechanics and related applications to Engineering, there are several situations which lead to mathematical formulations. As it happens in most fields, there is not always a unique way of formulating a mathematical model of a real problem. Consequently, the success of the application of the mathematical theories depends strongly on the kind of model which is formulated and proposed for the given engineering problem.

The equilibrium models of complementarity type and the constrained optimization models conceived by G.Maier to treat problems arising in Structural Mechanics, like those outlined in the previous section, turned out to be appropriate in the above sense; they have had the merit to give both an insight into the real problems and to allow numerical solutions and often effective applications. As an instance, we briefly describe below a few more problems.

(i) In computational mechanics, under suitable idealizations, the elastoplastic and poroplastic behaviour of materials can be described by means of a linear complementarity system of type (1.3.7) with F affine manifold.

In a first stage, the problem is formulated by a system of nonlinear equations and inequalities, adopting a piecewise linear approximation for the yield conditions and for the hardening laws, and considering a space discretization in the framework of mixed finite element multi-field modelling, using the so-called Prager's generalized variables [11]. Then, the relationships that characterize the model can be cast into the form of the following linear CS. The symbols of this section are independent of those of the others, if they overlap.

$$\overline{\varphi} := \overline{N}_\sigma^T(\mu\overline{\sigma}^E + \overline{\sigma}_o^E) - \overline{A}\,\overline{\lambda} - \overline{Y} \le 0, \quad \overline{\lambda} \ge 0, \quad \langle\overline{\varphi},\overline{\lambda}\rangle = 0, \tag{1.5.1}$$

where $\overline{N}_\sigma$ collects all the normal unit vectors to the piecewise-linearized yield surfaces; μ is the load multiplier; $\overline{\sigma}^E$ and $\overline{\sigma}_o^E$ represent the elastic stresses in response to live and dead loads, respectively; $\overline{\lambda}, \overline{\varphi}$ and $\overline{Y}$ denote the vectors of all plastic multipliers, yield functions and yield limits, respectively, in the space-discrete model; $\overline{A}$ is a matrix which contains, for the whole finite element aggregate, the influence matrix relating (generalized) plastic strains to consequent self-stresses and the hardening matrix.

By identifying $\overline{\lambda}, -\overline{\varphi}, \overline{\lambda} \ge 0$ and $\overline{\varphi} \le 0$, respectively, with $\overline{x}, F(x), \overline{x} \in \mathbb{K}$ and $F(\overline{x}) \in \mathbb{K}^*$, (1.5.1) turns out to be a special case of (1.3.7).

It is worth noting that the multiplier μ plays the role of a parameter in (1.5.1). Mechanical arguments allow us to claim that (1.5.1) admits a solution iff $\mu \le s$, where s is the limit load multiplier or "safety factor": in fact, amplifying the live load over the collapse threshold makes it impossible to simultaneously satisfy plastic admissibility and equilibrium enforced in the above formulation through the preliminary computation of the elastic stress response to the basic loads and of the influence matrix relating selfstresses to plastic strains.

(ii)Another instance is offered by the analysis of quasi-brittle fracture process. The idealization of crack propagation process rests on the following assumptions: let Ω denote the open domain occupied by the quasi-brittle structure; the location of the potential fracture surface Γ (called interface) is a priori known; cohesive tractions are transmitted across the two faces of crack surface (process zone) up to some critical value of the displacement jumps; opening mode only is considered (i.e., relative displacements and tractions can be reduced to their normal components with respect to Γ), but the generalization to "mixed mode"interface model as linear CS is already available; in the domain Ω outside Γ a linear elastic behaviour is assumed. For each point $x \in \Gamma$, we denote by $w(x)$ the opening displacement across Γ and by $t(x)$ the tractions at x. The interface Γ can be conceived as the union of true crack Γ_c (where $t = 0$), undamaged material Γ_e (where $w = 0$) and process zone Γ_z (where $w \ne 0$ and $t \ne 0$). The behaviour on Γ_z may be described by a constitutive law (called interface law) relating tractions t to displacements discontinuities w, which implies traction softening, i.e. tractions decrease with the increase of opening displacements.

The linear elastic behaviour of the material outside Γ allows us to write:

$$t(x) = \int_{\Gamma} G(x,\xi)w(\xi)d\Gamma + \mu t^E(x), \quad x,\xi \in \Gamma, \tag{1.5.2}$$

where G denotes the Green function which relates tractions t to displacement discontinuites w in the otherwise unloaded body, t^E denotes tractions generated on Γ by external actions in the absence of displacements jumps and μ is a load factor.

The interface law gives us a further relationship between $t(x)$ and $w(x)$.

For progressive fracture phenomena the interface law can be expressed by the following (holonomic or reversible) relations into the form of a CS:

$$\begin{cases} \phi_w(x) = f(w(x)) - f(\lambda(x) + w_c) - t(x) \geq 0, \\ \phi_\lambda(x) = w_c - w(x) + \lambda(x) \geq 0, \\ w(x) \geq 0, \quad \lambda(x) \geq 0, \\ \phi_w(x)w(x) = 0, \quad \phi_\lambda(x)\lambda(x) = 0, \end{cases} \tag{1.5.3}$$

where f is a given function such that $t = f(w)$ in the process zone Γ_z, $f(0) = t_c$ and $f(w_c) = 0$; t_c and w_c represent, respectively, the material tensile strength and the critical opening displacement, λ is an auxiliary variable.

Note that $\lambda(x) = \max\{0, w(x) - w_c\}$; namely, λ expresses the gap between w and critical value w_c, when w exceeds w_c.

The above relationships allow us to represent the material behaviour in all the points of interface Γ: those belonging to the undamaged material ($w = \lambda = 0$ and $t \leq t_c$), or to the process zone ($0 < w \leq w_c, \lambda = 0, t = f(w)$), or to the true fracture ($w > w_c, \lambda(x) = w(x) - w_c, t = 0$).

By combining equations (1.5.2) and (1.5.3) and by identifying (w, λ), (ϕ_w, ϕ_λ), $(w, \lambda) \geq 0$ and $(\phi_w, \phi_\lambda) \geq 0$ with $\overline{x}, F(\overline{x}), \overline{x} \in \mathbb{K}$ and $F(\overline{x}) \in \mathbb{K}^*$, respectively, then (1.5.3) turns out to be a special case of (1.3.7).

When the material behaviour is nonholonomic (irreversible), the interface law on the process zone Γ_z can be described as follows:

$$\begin{cases} \dot{\phi}(x) = h_w(x)\dot{w}(x) - \dot{t}(x) \geq 0, \\ \dot{w}(x) \geq 0, \\ \dot{\phi}(x)\dot{w}(x) = 0, \end{cases} \tag{1.5.4}$$

where $h_w(x) = f'(w(x))$, that is the slope of the $f(x)$ at the considered point and dots mark derivatives with respect to the time. Along Γ, but outside Γ_z, the material behaviour is assumed to be reversible: in the undamaged zone, one has $\dot{w} = 0$ and $t < t_c$, while in the true fracture $\dot{t} = 0$. (1.5.4) take into account irreversibility.

As above remarked, the nonholonomic interface relation (1.5.4) coupled with the corresponding rate form of the equation (1.5.2) turns out to be into the form of a CS.

(3i) Another field of Applied Mechanics, where the mathematical models of optimization and those of equilibrium have shown to be useful, is that of flight control. In the 1960s, A.Miele has given a fundamental contribution to the introduction of constrained extremum problems in the field of Flight Mechanics [49]. As an instance, here we shortly mention one of the many problems which have been reduced to an optimization model (see [50] and the references therein); namely, the climb problem for a constant mass aircraft flying in a vertical plane. More precisely, we consider the motion of an aircraft under the following hypotheses: the flight is in a vertical plane over a flat Earth, the acceleration of gravity g is constant, the change of mass m due to final consumption is neglected and the thrust is assumed tangent to the flight path. The state variables are:

$$
\begin{aligned}
x(t) &= \text{horizontal distance} \\
h(t) &= \text{altitude} \\
V(t) &= \text{aircraft velocity (airspeed)} \\
\gamma(t) &= \text{inclination of the flight path w.r.t. the horizon}
\end{aligned}
$$

The equations of motion (see [49]) are the following:

$$\dot{x} = V \cos \gamma, \tag{1.5.5}$$

$$\dot{h} = V \sin \gamma, \tag{1.5.6}$$

$$\dot{V} = \frac{T - D}{m} - g \sin \gamma, \tag{1.5.7}$$

$$\dot{\gamma} = \frac{L}{mV} - g \cos \gamma. \tag{1.5.8}$$

(1.5.5) and (1.5.6) are the kinematic equations in the horizontal and vertical direction, while (1.5.7) and (1.5.8) are the dynamic equations on the tangent and the normal to the flight paths. We assume that the thrust $T = T(h, V, \pi)$, the drag $D = D(h, V, \alpha)$ and the lift $L = L(h, V, \alpha)$ are known functions of the controls $\alpha(t)$ (angle of attack, i.e. the angle which the velocity vector forms w.r.t. the aircraft reference line) and $\pi(t)$ (thrust setting of the engines). Usually, there are some constraints on the controls, for instance:

$$\alpha_{\min} \leq \alpha \leq \alpha_{\max}, \tag{1.5.9}$$

$$\pi_{\min} \leq \pi \leq \pi_{\max}. \tag{1.5.10}$$

The independent variable is the time t, which varies in the interval $[0, \tau]$, where 0 is the initial time and τ is the final time. The final time τ is generally an unknown parameter to be optimized. Moreover, we impose suitable boundary conditions: if the problem is to transfer an aircraft from a given combination of altitude and velocity in level flight, then the boundary conditions have the following form:

$$h(0) = \text{given}, \quad V(0) = \text{given}, \quad \gamma(0) = 0; \tag{1.5.11}$$

$$h(\tau) = \text{given}, \quad V(\tau) = \text{given}, \quad \gamma(\tau) = 0. \tag{1.5.12}$$

The climb problem consists in determining the state variables $x(t)$, $h(t)$, $V(t)$, $\gamma(t)$, the control variables $\alpha(t)$, $\pi(t)$ and the parameter τ, such that some performance index is minimized. In a minimum time problem, the performance index is τ.

We now suppose that the boundary conditions $\gamma(0) = 0$ and $\gamma(\tau) = 0$ be omitted in (1.5.11)-(1.5.12), and let they be replaced by $\gamma(0) = \text{free}$ and $\gamma(\tau) = \text{free}$. Moreover, let the centripetal acceleration $V\dot\gamma$ be neglected in the equation of motion (1.5.8) on the tangent to the flight paths and let the hypothesis $\cos\gamma \cong 1$ be used. Then (1.5.8) is replaced by

$$L - mg \cong 0.$$

Also, let inequality (1.5.9) be disregarded, and let (1.5.10) be replaced by $\pi = \text{const}$ (climb with constant thrust setting). With these hypotheses, one obtains the simplified climb problem. While the original climb problem is nonsingular, the simplified climb problem is singular.

(**4i**) In the field of Astrodynamics, another interesting problem deals with the transfer of a spacecraft from a low orbit to a high orbit under the following assumptions: (**i**) the low orbit and the high orbit are circular and coplanar; (**ii**) there is only one source of gravitational attraction along the entire trajectory; (**3i**) circularization of the motion is assumed at both the departure and arrival; (**4i**) velocity impulses are applied at only the terminal points of the trajectory and tangentially to the trajectory; (**5i**) the motion of the spacecraft is in a central gravitational field (two-bodies model).

Let μ denote the gravitational constant of the actracting body. Let r denote the radial distance of the spacecraft from the centre of actraction and θ the phase angle (angle of the radius vector with respect to a reference direction), so that the pair (r, θ) gives the spacecraft position. Let V be the velocity modulus of the spacecraft and let γ be the angle between the velocity vector and the local horizon (perpendicular to the radius vector), so that the pair (V, γ) gives the spacecraft velocity.

In 1925, W.Hohmann proved that, energetically speaking, the most efficient trajectory for transferring a spacecraft from a circular orbit to another circular orbit (in a central gravitational field) is the elliptic trajectory bitangent to the terminal orbits.

For an ascending Hohmann transfer, the path inclination γ vanishes at the endpoints and is positive elsewhere. Therefore, there is a point on the Hohmann transfer trajectory, where the path inclination γ attains the maximum. One way to find such a maximum is to make use of the energy and angular momentum integrals. This lead to the following problem in the unknown (r, V, γ) :

$$\min \cos\gamma \tag{1.5.13}$$

$$\text{s.t.} \quad \frac{1}{2}V^2 - \frac{\mu}{r} - E = 0, \tag{1.5.14}$$

$$rV\cos\gamma - M = 0, \tag{1.5.15}$$

where E denotes the total energy (kinetic plus potential energies) per unit mass of the spacecraft, and M the angular momentum (moment of the velocity vector with respect to the centre of attraction). Of course, (1.5.13)-(1.5.15) is a special case of (1.1.1) for $p = m = 2$, $n = 3$, $X = \mathbb{R}^3$ and with obvious position for f and g.

To establish that the Hohmann transfer trajectory is not only feasible but optimal, it is necessary to embed it into a larger class of trajectories by assuming that the departure and arrival velocity impulses are not necessarily tangential. This is the same as assuming the possible presence of discontinuities in the path inclination at the departure and arrival. Consequently, we meet the following problem in the unknowns $V_1, V_2, \gamma_1, \gamma_2$:

$$\min \left(\sqrt{(V_1 - V_0)^2 + 2V_0V_1(1 - \cos\gamma_1)} + \sqrt{(V_3 - V_2)^2 + 2V_2V_3(1 - \cos\gamma_2)} \right), \quad (1.5.16)$$

$$\text{s.t.} \quad V_1^2 - V_2^2 + 2(V_3^2 - V_0^2) = 0, \quad (1.5.17)$$

$$V_3^2 V_1 \cos\gamma_1 - V_0^2 V_2 \cos\gamma_2 = 0, \quad (1.5.18)$$

where the meanings of the symbols are as before with the proviso that the subscripts 0 and 1 denote the spacecraft conditions before and after the application of the accelerating velocity impulse at departure; the subscripts 2 and 3 denote the spacecraft conditions before and after the application of the accelerating velocity impulse at arrival. (1.5.17) is an alternative form of the energy relation, and (1.5.18) is an alternative form of the angular momentum relation; (1.5.16) requires to minimize the total characteristic velocity. The circular velocities V_0 and V_3 are known quantities; the velocities V_1, V_2 and the path inclinations γ_1, γ_2 are unknowns quantities.

Details on these kind of problems can be found in the paper: A.Miele, M.Ciarcia and J.Mathwig, "Reflections on the Hohmann Transfer", Jou. Optimiz. Theory Appls., Vol.123, No.2, 2004, pp.223-253.

1.6. Equilibrium Flows in a Network

The study of a network is an old one, especially in some fields of engineering, as electrical, hydraulic and transportation. Since the 1950s, the great development of communications and related complex activities has induced a fast increase of studies in the field. Indeed, a large number of real situations have joined the classic ones in asking for a network formulation. Here, we will consider only the aspect of equilibrium flows; of course, in a real situation there are several other aspects, which cannot be taken into account in a simplified treatise like the present one. The symbols of this section are independent of those of the others, if they overlap.

Assume we are given a digraph, say $G = (N, A)$, where $N := \{N_1, ..., N_m\}$ is the set of nodes and $A := \{A_1, ..., A_n\}$ is the set of arcs; an arc is identified, of course, by an ordered pair of nodes. For each arc (N_i, N_j) we are given a capacity, say c_{ij}, with $0 \le c_{ij} \le +\infty$. We suppose that in each arc a substance (road traffic, oil, water, information strings, economic investments, goods, etc.) might flow; denote by x_{ij} the flow of an arc (N_i, N_j). Each node can generate or/and absorb flow; we will not distinguish between the two opposite flows, and consider only the difference, say q_j, at node N_j; hence N_j will be a "source" or a "sink" or a "transit" node, according to $q_j < 0$ or $q_j > 0$ or $q_j = 0$, respectively; $q_j = 0$ expresses the classic conservation of flow. The sets:

$$I_j := \{i : (N_i, N_j) \in A\}, \qquad O_j := \{i : (N_j, N_i) \in A\}$$

identify, for node N_j, the incoming and outgoing arcs, respectively. The network is now identified with the digraph G.

Among the several aspects of a network, the analysis of the equilibrium of flows is one of the most important. Of course, this presupposes to have a definition of equilibrium of flows which mirrors the real situation. Since the early 1950s, the equilibrium flows have been identified as those which maximize or minimize a suitable function, as a utility of the set of the users of the network, the global energy, the global cost. Apart from the fact that for some of them — as the global utility — it is arbitrary to claim they exist, the real situations where the flows imply the maximum or the mimimum of a functional are — with the exception of some classic networks — very few. Recently, it has been recognized that a variational inequality may overcome the above drawback [15, 32, 47]; this is here shortly described.

Let $x \in \mathbb{R}^n$ be the vector of flows on the arcs of G. A flow x is said *feasible*, iff

$$\sum_{r \in I_j} x_{rj} - \sum_{s \in O_j} x_{js} = q_j, \quad j = 1, ..., m, \tag{1.6.1}$$

$$0 \le x_{ij} \le c_{ij}, \quad \forall (N_i, N_j) \in A. \tag{1.6.2}$$

Equations (1.6.1) represent the balance (or conservation) of flow at each node, and (1.6.2) are obvious capacity constraints. Let $\mathbb{K}$ denote the set of $x = (x_{ij} : (N_i, N_j) \in A)$ which fulfil (1.6.1)-(1.6.2). By introducing the node-arc incidence-matrix $\Gamma = (\gamma_{ij})$ with

$$\gamma_{ij} := \begin{cases} -1, & \text{if } \exists \, k \ \text{s.t.} \ A_j = (N_i, N_k), \\ +1, & \text{if } \exists \, k \ \text{s.t.} \ A_j = (N_k, N_i), \\ 0, & \text{otherwise,} \end{cases}$$

the set of feasible flows can be expressed as:

$$\mathbb{K} = \{x \in \mathbb{R}^n : \ \Gamma x = q, \ 0 \le x \le c\}.$$

where $q = (q_1, ..., q_m)$ and $c = (c_{ij} \ \text{s.t.} \ (N_i, N_j) \in A)$.

Now, we assume that each arc be associated with a weight which depends on the flows in the network. Hence, we assume that, $\forall (N_i, N_j) \in A$, a function $w_{ij}(x)$ be given which expresses such a weight. Then, we define the weight of the network as:

$$w(x) := \sum w_{ij}(x) \cdot x_{ij}$$

where the summation is over all the arcs of the network; we assume that $w(x)$ can mirror the weight of the real network. A typical expression of the weight is the cost per unit of flow.

We are now in the position to give a definition of equilibrium flow of variational kind.

Definition 1.6.1. A vector $\bar{x} \in \mathbb{K}$ is an *equilibrium flow*, if and only if

$$\sum w_{ij}(\bar{x}) \cdot \bar{x}_{ij} \ \le \ \sum w_{ij}(\bar{x}) \cdot x_{ij}, \quad \forall x \in \mathbb{K}, \tag{1.6.3}$$

where the summations are extended to all pairs (i, j), such that $(N_i, N_j) \in A$.

If for each pair (i, j) the function $w_{ij}(x)$ does not depend on the flow x (constant weights), then $\overline{x}$ is a m.p. of the classic *minimum weight flow problem*:

$$\min \sum w_{ij}x_{ij}, \quad \text{s.t.} \quad x \in \mathbb{K}. \tag{1.6.4}$$

According to Definition 1.6.1, the problem of finding the equilibrium flows consists in searching for the vector $\overline{x}$ such that

$$\sum w_{ij}(\overline{x})(x_{ij} - \overline{x}_{ij}) \geq 0, \quad \forall x \in \mathbb{K}, \tag{1.6.3)'}$$

which is identified easily as a special case of VI (1.3.5); the operator $F(x)$ of (1.3.5) is now the vector

$$w(x) := (w_{ij}(x), \quad \forall(i, j) \quad \text{s.t.} \quad (N_i, N_j) \in A). \tag{1.6.5}$$

Let $w(x)$ be continuously differentiable on an open and convex set containing $\mathbb{K}$. Then, there exists a functional, say $W(x)$, whose gradient is the very $w(x)$, if and only if the Jacobian matrix $w'(x)$ is symmetric for all x of the domain (symmetry principle, [53]). The assumptions of this property are trivially satisfied when $w_{ij}(x)$ is constant and the above functional is the objective function of (1.6.4). These facts let us understand better the remarks made to achieve (1.6.1)-(1.6.2). Even if the objective function were any nonlinear functional, to define an equilibrium flow as a m.p. of (1.6.4) — like as been done classically — should be much restrictive. Indeed, only rarely a real network enjoys a symmetry. On the other side, the VI formulation $(1.6.3)'$ does not require postulation of the existence of any functional of type $W(x)$, without excluding the possibility it exists.

1.7. Testing Statistical Hypotheses

The problem of testing simple statistical hypotheses is a classic one [53]. Here, it will be considered from a slightly different point of view [17, 46] and formulated as constrained minimization. Notation and symbols adopted in this section are independent of those used elsewhere, if they overlap.

Let $\mathcal{T}$ be a continuous real random variable, defined on a suitable measurable space, with values in the interval $T \subset \mathbb{R}$ endowed with the Borel σ-algebra. Suppose that the distribution of $\mathcal{T}$ be absolutely continuous with respect to the Lebesgue measure, so that it admits a density function that is denoted by $p(\mathcal{T})$. Consider the following simple hypotheses on the density function:

$$H_i: \quad p(\mathcal{T}) = p_i(\mathcal{T}), \quad i = 1, ..., n + 1.$$

Given k independent observations, say $t_1, ..., t_k$, of the random variable $\mathcal{T}$, we want to decide which hypothesis to accept, with zero the cost of a right decision and one that of a wrong decision. For $i = 1, ..., n + 1$, let

- $v_i(t) := \prod_{j=1}^{k} p_i(t_j)$ be the likelihood function of $t = (t_1, ..., t_k)$ under (H_i);

- $V_i(t) := \pi_i v_i(t)$, where $\pi_i \in [0,1]$ s.t. $\sum_{i=1}^{n+1} \pi_i = 1$ are the *a priori* probabilities on the possible choices;

- $x_i(t)$ be the probability of accepting hypothesis H_i in the presence of the observation t.

The probability of rejecting a generical hypothesis H_i, when it is true, is given by:

$$\pi_i[1 - \int_{T^k} v_i(t)x_i(t)dt],$$

where $T^k := T \times \cdots \times T$ (k times) is a Cartesian product. Therefore, the probability of taking a wrong decision is given by

$$\sum_{i=1}^{n+1} \pi_i[1 - \int_{T^k} v_i(t)x_i(t)dt].$$

The present approach consists in minimizing the above functional, under the condition that all the terms in the sum be equal, in order not to privilege any hypothesis with respect to another. Hence, the problem that we will be concerned with consists in finding the functions $x_1(t), ..., x_{n+1}(t)$, such that

$$\pi_1[1 - \int_{T^k} v_1(t)x_1(t)dt] = \pi_2[1 - \int_{T^k} v_2(t)x_2(t)dt] = ...$$

$$... = \pi_{n+1}[1 - \int_{T^k} v_{n+1}(t)x_{n+1}(t)dt = \min$$

The obvious equality

$$x_{n+1}(t) = 1 - \sum_{i=1}^{n} x_i(t)$$

allows us to avoid considering $x_{n+1}(t)$. Let us set $x(t) = (x_1(t), ..., x_n(t))$ and assume that x belongs to the set:

$$Y := \{y : T \to \mathbb{R}^n \text{ s.t. } y \text{ is bounded and continuous a.e. on } T^k\}.$$

The statistical decision problem can be formulated as

$$\max f(x) = \int_{T^k} V_1(t)x_1(t)dt, \tag{1.7.1a}$$

$$\text{s.t.} \qquad \int_{T^k} \{[V_1(t) + V_{n+1}(t)]x_1(t) + V_{n+1}(t)\sum_{j=2}^{n} x_j(t)\}dt = \pi_1, \tag{1.7.1b}$$

$$\int_{T^k} [V_1(t)x_1(t) - V_i(t)x_i(t)]dt = \pi_1 - \pi_i, \ i = 2, ..., n \ (n > 1), \tag{1.7.1c}$$

$$0 \le x_j(t) \le 1, \ \forall t \in T^k, \ j = 1, ..., n, \tag{1.7.1d}$$

$$\sum_{j=1}^{n} x_j(t) \le 1, \ \forall t \in T^k. \tag{1.7.1e}$$

The problem of testing simple hypotheses can be treated by means of the Neyman-Pearson Lemma [53] in the case $n = 1$. However, problem (1.7.1) is formulated from a point of view that is different from the one followed by Neyman and Pearson. The constraint (1.7.1b) states that the probabilities of error under the two different hypotheses must be equal, while, in the classic theory, we fix one of them (size of test) and we minimize the other (probability of error of the second kind).

Problem (1.7.1) is identified easily with (1.1.4), where X is defined by (1.7.1d, e), $p = m = n$, and with further obvious positions. However, it might be reduced also to a format having both constraints (1.1.4b) and (1.1.6c).

1.8. Vector Problems from Industry

In the last decade, in the field of industrial design, there has been a remarkable development of techniques for handling several objectives at the same time (without combining them into one objective only). This has happened almost independently of the existing mathematical theory; perhaps, because of its dishomogeneity, farraginous status, and lack of efficient methods of solution. Such a gap between theory and applications is negative for both. The following two problems, which come from industry (see, for instance, [58]) aim to stress the importance, for the mathematical research, to try to decrease such a gap.

One of the main parts of an aircraft is the gas turbine engine (for short, GTE). There are many types of aircraft with different purposes and performance characteristics and each of them requires its own optimal engine. In this section we will describe only universal performance criteria to evaluate the GTE quality, without considering specific features of the aircraft.

The essential characteristic for any gas turbine engine is the capability of creating thrust. We want a GTE with the highest possible thrust in order to increase the aircraft manoeuverability and, at the same time, the engine must be able to develop the lowest possible thrust in the idle mode, in order to start the engine without moving the aircraft and to perform low speed manoeuvres in the air. Therefore the main requirement of an aircraft gas turbine engine is to provide the maximum thrust range.

The production of any feasible thrust with a minimum fuel consumption is another important requirement for a GTE. We consider as a second performance criterion the specific fuel consumption, defined as the ratio between fuel consumption to thrust, which must be minimized for all operating modes.

Other important efficiency criteria for a GTE are the specific mass of the engine, that is the ratio of the engine's mass to its maximum thrust, and the specific thrust, defined as the ratio of the maximum thrust to the maximum air flow. Other things being equal, the air flow is obtained by the cross-sectional area of the GTE air intake and, hence, by the engine's geometric dimensions. We require the maximum specific thrust and the minimum specific mass for a GTE.

The engine's reliability and safety depend on some criteria. The first one is the gas dynamic stability margin of compressor stages, which must be as large as possible in order to provide stable working of the engine. We require also to increase as much as possible the strength margin of the engine's structural elements, in order to prevent destruction of the engine when it is working. Moreover, the sensitivity of the GTE units to high temperatures and to repeated temperature changes must be as low as possible. Safety and reliability depend on the engine stability features in long-term operating conditions.

The gas dynamic characteristics in a GTE include: the gas temperature at the turbine intake characterizing the air heat ratio, the total pressure ratio in the compressors, the bypass ratio, that is the ratio of the flow rate of the air passing by the turbine to the flow rate through the main combustion chamber and the turbines.

The gas dynamic features and performance criteria depend on all the design variables, which are parameters characterizing the engine type and layout (number of shafts, the presence or absence of a bypass, a mixing chamber and an afterburner), types of engine functional units (compressors, combustion chambers, turbines, intake and output devices), geometric characteristics of the GTE passages (passage cross-section areas, characteristic size of the turbine and compressor blade rows), and others. The designer has to find the design variables values providing the highest possible efficiency over the entire thrust range of the engine.

Designers increase the pressure ratio and the gas temperature at the turbine intake in order to decrease the specific fuel consumption and to increase the engine's specific thrust. To ensure a high pressure ratio they increase either the number of compressor stages, which leads to an increase in the mass of compressor and to a reduction in the gas dynamic stability margin for lower operating modes, or the rotors rotation speed, which leads to a reduction in the strength margin. On the other hand, an increase in the gas temperature at the turbine intake reduces the GTE's reliability. Hence they have to use new heat proof materials or cool the turbine passages, which also increase the cost of developing the engine. This shows that a multicriteria (or vector) analysis is necessary in order to design an aircraft gas turbine engine.

Two of the most important characteristics which influence the safety of a transportation vehicle are its handling and stability. We consider a model of a small front wheel drive car in which there are six degrees of freedom for the sprung mass and to simplify we suppose that: the car moves over a horizontal plane; there is a reference frame rigidly associated to the car and its origin is located in the car body center of gravity; the projection of the car velocity on the road surface is constant.

The mathematical model of the car is based on Euler-Lagrange equations of the second kind in generalized coordinates, which include position of the origin of the movable reference frame with respect to a fixed one and the orientation angles of the car body. We use the following notations:

- $R_{x_1} - R_{x_4}$ are the longitudinal reactions of the car's wheels;

- $R_{y_1} - R_{y_4}$ are the lateral reactions of the car's wheels;

- $R_{z_1} - R_{z_4}$ are the vertical reactions of the car's wheels;

- $M_{\delta_1} - M_{\delta_4}$ are the stabilizing moments of the car's wheels;

- V_α is the linear velocity of the car;

- X, Y, Z are the fixed reference frame;

- X_1, Y_1, Z_1 are the movable reference frame connected to the car's center of gravity;

- U, V, W are longitudinal, lateral and vertical components of the car's linear velocity in the reference frame X_1, Y_1, Z_1;

- p, q, r are the yaw, roll and pitch velocities of the car;

- β is the sideslip angle;

- δ_1, δ_2 are the steering angles of the left and right wheels;

- a, b are the distances from the centres of the front and rear axles to the center of gravity of the car;

- h_g is the distance from the road surface to the vehicle's centre of gravity.

We consider for tyres the so called brush models, in which the tyre carcass is considered as a row of elastic bars of constant length attached to a circular band; the bar can undergo deformation under vertical and lateral loads, we assume linear dependence between the strain and the load.

In this model we consider the following design variables: the mass of the car, the distance from the road surface to the car's centre of gravity (h_g), the distance from the front axle to the vehicle's centre of gravity (a), the roll stiffness of the rear suspension, the pitch stiffness of the suspension system, the transfer functions relating the relative displacement of the car body to the displacements of the wheels in linear and angular coordinates.

As functional constraints we impose no sideslip of the wheels and no separation of the wheels from the road surface. Finally, we expose the following performance criteria: ϕ_1, ϕ_2, ϕ_3 are control moment margins, respectively, for lateral accelerations of 2,4 and 6 m/s^2; ϕ_4, ϕ_5, ϕ_6 are stabilizing moment margins, respectively, for lateral accelerations of 2,4 and 6 m/s^2; ϕ_7, ϕ_8, ϕ_9 are the values of the vehicle static turnability, respectively, for lateral accelerations of 2,4 and 6 m/s^2; $\phi_{10}, \phi_{11}, \phi_{12}$ are the values of the car sensitivity to the control input, respectively, for lateral accelerations of 2,4, and 6 m/s^2 ($\phi_7 - \phi_{12}$ are measured in road tests under a constant control input). Criterion ϕ_{13} is the maximum yaw velocity of the car when tested under step control input; ϕ_{14} is the time during which the yaw velocity reaches its maximum value under step control input; ϕ_{15} is the time by which the yaw velocity reaches 90% of its maximum value; ϕ_{16} is the ratio of the amplitude of the vehicle sideslip angle to the amplitude of the control input at a fixed frequency of 0.75 Hz; ϕ_{17} is the phase shift between the control input and car sideslip angle at the frequency of 0.75 Hz (ϕ_{16} and ϕ_{17} are measured in road tests under a sinusoidal control input). The car's stability is characterized by ϕ_{18}, defined as the value of the sideslip angle for a fixed lateral acceleration of 6 m/s^2. Criteria $\phi_{13}, \phi_{14}, \phi_{15}, \phi_{17}$ and ϕ_{18} are to be minimized whereas the other ones are to be maximized.

1.9. Comments

1. The formats considered in Sect. 1.1 do not cover all the problems, which are studied and applied within optimization and related fields; recently introduced formulations, such as extremum problems in a complex space or bottleneck extremum problems, are interesting and promising. Problems (1.1.4) and (1.1.5) have been chosen as typical extremum problems of the Calculus of Variations. Of course, they are not the most general formats. This is due to the fact that this volume does not aim to be a treatise on Calculus of Variations, but wants to show the particularities of IS analysis. Its extension to other formats, as well as its exploitation to deepen the several topics of the Calculus of Variations and more generally of optimization, should appear clear; they go beyond the scope of this book and require further investigation.

2. The topic of Example 1.2.1 might be extended, by considering the restriction of a function f (not necessarily quadratic), such that $f(0) = 0$, to a convex cone (not necessarily a linear variety) with apex at the origin, in particular a polyhedral cone defined by $Bx \geq 0$. Within such a more general scheme, one might consider, beside convexity, also quasiconvexity, pseudoconvexity, and so on. Furthermore, the same reasoning might be applied to the monotonicity of operators: to find a necessary and sufficient condition for the restriction of an operator to a convex cone to be strictly or strongly isotone. If the operator is linear and the cone is a linear variety, then the topic is equivalent to that of Example 1.2.1. Besides, some generalizations of isotonicity, symmetry and coerciveness would be of great importance. The above topics have been considered for $B = \mathbb{R}^n$; of course, they can be extended to a Banach space ($B \neq \mathbb{R}^n$).

3. The oldest trace of the isoperimetrical problems of Examples 1.2.2 (and of constrained extrema) is, perhaps, the problem met by Dido about 880 B.C.: according to the Virgil tale, Dido, escaping from Phoenicia, reached North Africa, where she was offered whatever land she could enclose with an oxhide; then, she cut a hide into very narrow strips, joined them to make a continuous thread more than two-and-a-half miles long, and enclosed the land with an arc of circumference of this length; this originated Carthage and Dido was its first queen (we can say that Dido had also the perception of the unboundedness of the functional of the length). However, the perception of isoperimetrical properties is very old and coming from the observation of physical facts. For instance, the use of leather bags should have led to understand that the maximum of water corresponded to circular cross-section of the bag or even the spherical shape of it. This can be given in modern terms as follows, and represents an useful interpretation

of (1.2.3). Let a closed curve γ of a given length ℓ be drawn on a horizontal table, and let a cylindrical vessel be constructed by letting a vertical segment (generator) go through the points of γ. The (truncated) cylinder produced by such a construction — and hence γ — are assumed to be perfectly flexible, the length of γ being constant and the generators remaining vertical. Let us pour a certain quantity of water in the vessel, which consequently will take an equilibrium form; this will be s.t. the potential energy of the water be a minimum, so that the centre of gravity of the water, and hence of the water surface, will be as low as possible. Then, the volume of the water being constant, the area of a horizontal cross-section of the vessel must be as great as possible. At each point of the boundary of such a section, the pressure of the water, and hence the curvature (which is determined by the pressure), must be the same. It follows that such a boundary must be a circumference. Zenodorus (who lived in Alexandria 2 centuries B.C.) posed such an isoperimetrical problem, perhaps for the first time, in mathematical terms and found explicitly the circumference as solution. Therefore, problems of type (1.1.4) are very old, while there is no similar evidence for those of type (1.1.5). Of course, to meet the birth of the theory of the Calculus of Variations, we must wait for Jakob and Johann Bernoulli, Newton, Euler, and Lagrange. However, the gestation of this field is due to Galilei: in 1638, he posed the *brachistochrone problem* and several related questions; see "Opere di Galileo Galilei" (Galileo Galilei's works; in Italian) by Franz Brunetti, Vol. II, published by UTET, Torino, 1999, pp.766-767, where Theorem 22 says "If in a circle, upright with respect to the horizon, we raise from its lowest point an inclined plane which subtends an arc less than or equal to a quadrant, and if from the extrema of such a plane we draw two other inclined planes ending in any point of the arc, then the descent along the system of these last two inclined planes will occur in a shorter time than along the only first inclined plane or along only one of the last inclined planes (more precisely, the lower one)". From this and several other propositions and comments (see also Galileo Galilei, "Dialog über beide hauptsächlichsten Weltsysteme, 1630 (trans. Strauss), pp. 471-472, and "Discorso intorno a due nuove scienze (Dialogues concerning two new sciences)", in Italian (trans. Crew and De Salvio), 1938, in the above Vol. II published by UTET, pp. 553-811) we can argue that the brachistochrone problem was clear to Galilei; besides this, he had discovered that the straight line is not optimal and found a method (a sequence of convex piecewise linear paths) for going down from a feasible solution (the line) toward the optimal one. In this sense, he seems to have been also the founder of *descent methods* for the search of a minimum. Galilei should have realized the importance of the circle for finding the solution; however, to achieve it (cycloid) we must wait for Jakob and Johann Bernoulli (around 1696). Apart from rare exceptions (see Encyclopedia of Mathematics, Vol. 1, pag. 465, Reidel, Kluwer, Dordrecht, 1988), they are considered to have posed first the problem, even if it is unbelievable to think that they did not know of Galilei's works.

4.Problems (1.2.4) and (1.2.5) can be treated by means of general algebraic methods, as will be seen in Chapter 5. However, when $n = 2$, the challenge to solve them by using only straightedge and compasses or by elementary geometry is still alive. Indeed, it is just in this sense that Fermat posed the problem and Torricelli solved it, and that many questions are still open. Among them, we can mention the constrained (weighted) Fermat problem: given a triangle of the plane and a segment joining any two points of the boundary of the triangle, find a point x^0 of the segment which minimizes the sum of the (weighted) distances between x^0 and the vertices of the triangle (the weights being given); see [4, 6]. The constrained (weighted) Erdős problem is defined in quite a similar way. An interesting generalization of the above problem is obtained when the triangle is replaced by a polygon: some results have been obtained for some special cases, such as the quadrilateral and pentagon. Further generalizations consist in searching for x^0 which minimizes the sum of (weighted) distances between x^0 and s given polygons; when these are singletons we recover the preceding generalization, and we shrink to the Fermat problem at $s = 3$. The "inverse" problems are also of interest; for instance, with regard to the Fermat problem, one can assume that x^0 be given and look for 3 points belonging to a given set of the plane and such that the sum of distances between them and x^0 be a minimum (see also the subsequent comment about the centre of a convex set). During the 18th and 19th centuries there was a continuous interest in the Fermat-Torricelli problem and contributions have been given by several scholars, as B.F. Cavalieri, V. Viviani, T. Simpson, J. Steiner, E. Fasbender, and A. Weber. The first two problems of Examples 1.2.4 can be seen also from a combinatorial point of view, by considering the $\binom{N}{n+1}$ n-dimensional spheres identified by $n + 1$ of the N points (if, of course, $N > n + 1$). With regard to the third problem, note that, without any loss of generality, we can assume that $x^1, ..., x^n$ lie in an $(n-1)$-dimensional plane, in particular in the coordinate subspace $x_n = 0$, and that x^{n+1} belongs to the halfspace $x_n > 0$; therefore, we can assume that $x_n^1 = ... = x_n^n = 0$ and $x_n^{n+1} > 0$. As a consequence, the following recurrence equality:

$$V(TH_n) = \frac{1}{n}\left[\frac{1}{(n-1)!}|D(x^1, ..., x^n)|\right]x_n^{n+1} = \frac{1}{n}V(TH_{n-1})x_n^{n+1} \qquad (1.9.1)$$

holds; for $n = 2$, it shrinks to the area of a triangle; $V(TH_{n-1})$ and x_n^{n+1} play the role of the "base" and "altitude" of the simplex, respectively. (1.9.1) can be useful for the methods of solutions as well as for applications. If $N = n + 1$ and $x^1, ..., x^N$ define a simplex, then the solution of (1.2.6a) generalizes the circumcircle ($N = 3$); if in addition the simplex is regular, then (1.2.6) is trivially connected with (1.2.7). Problems (1.2.6) can be considered in more general forms, by adopting non-Euclidean distances or weighted distances. The first of the cases of Example 1.2.4 has been formulated as a minimization problem; indeed, we could have expressed it as "minmax" [20], like Example 1.2.7; this shows once more that the mathematical formulation of a problem is not necessarily unique.

5.Isoperimetrical problems can be considered as part of a family, which can be named *isoproblems*. In this family, it is interesting to consider and study the *isodiametrical problems*; the distance function is not necessarily the Euclidean one. Let $K \subset \mathbb{R}^n$ denote a convex body (see Definition 2.1.1). The problem consists in finding, among all K having diameter equal to 1, one with maximum (hyper)volume. A special case consists in finding, among all polytopes (see Definition 2.1.3), in particular polygons, which have diameter equal to 1, one having maximum (hyper)volume (or area). This problem can be generalized in several ways. A natural one leads to consider K as a subset of a Banach space. Another interesting way consists in defining K as a geodesic convex [56] subset of a connected Riemannian manifold; $\forall x^1, x^2 \in K$, $\eta(x^1, x^2)$ denotes the geodesic joining x^1 and x^2, meas denotes measure and

$$d_K := \sup_{x^1, x^2 \in K} \text{meas } \eta(x^1, x^2)$$

is the diameter of K. The problem consists in finding, among all K having $d_K = 1$, one for which meas K is maximum. The interesting development carried out in [56] should be the natural context for such a problem. Connections with the notion of centre of a set (see $(2.1.8)^*$) might be interesting. The study of the reciprocal problems of the diametrical ones is also of interest.

6.From a pure mathematical point of view, it is interesting to investigate the problems of Examples 1.2.4 − and their extensions − with different, not necessarily Euclidean, norms; namely, in the case where the norm depends on the point x^i.

7.Example 1.2.5 is a simplified version of a family of problems, which have come out within the IS analysis of (1.1.5), as will be seen in Sect. 3.2. Only subsequently, it has been recognized to be a very special case of an excavation problem, like (1.4.8). The complete solution of (1.2.8) − and, a fortiori, of (1.4.8) − is still an open question.

8.The problem of Example 1.2.6 is one of the most popular in mathematical programming, whose applications occupy a huge literature. Here, we deal mainly with the implications that (1.2.9) have had on the development of the Lagrange theory.

9.The problem of Example 1.2.7 belongs to the field of game theory, whose initial development is mainly due to J. von Neumann [26]. Such a theory has stimulated theoretical research, like that on saddle points of a manifold (see Sect. 4.2), and has influenced some applied fields, like statistics (the problem of Sect. 1.7 follows a minimax approach).

10.At $N = 3$, (1.2.11) becomes the famous problem posed by Malfatti in 1803. He wrote: "Given a right prism of any raw material like marble, to extract from it 3 (rightcircular) cylinders with the same altitude of the prism and having as much total bigness as possible, and hence with as small residue of raw material as possible corresponding to the achieved bigness" (this will be quoted as Malfatti original problem; for short, MOP); see: Gianfrancesco Malfatti, "Memoria sopra un problema stereotomico (Memoir on a stereotomic problem)" (in Italian), Journal "Memorie di Matematica e Fisica della Societá Italiana", Vol.10, Parte 1^a, 1803, pp.235-244; anastatic reprint by Unione Matematica Italiana as "Opere by G.Malfatti", Vol.II, Cremonese Publisher, Bologna, 1981, pp. 703-713. In the same paper, Malfatti claims that this problem can be reduced to inscribe in the given triangle, 3 circles in such a way that each circle be tangent to the two other circles and to two sides of the triangle (this will be quoted as Malfatti alternative problem; for short, MAP). Malfatti solves MAP in a very elegant way. In the literature, as "Malfatti Problem", we find MAP and not MOP (see, for instance, "100 Great Problems of Elementary Mathematics" by H.Dörrie, Dover Publ.Inc., New York, 1965, page 147). This is due, perhaps, to Malfatti's claim of equivalence between MOP and MAP. Unfortunately, in general, MOP and MAP are not equivalent. Let T be an equilateral triangle ABC, $\ell > 0$ the length of its sides, and O its incentre. For this special case, the solution of MAP is obtained this way (see page 710 of the anastatic reprint; Luigi Gozzi, pupil of Malfatti, is indicated as author of this procedure): let H be the midpoint of side AB, C' be the point of CH (the bisector of $A\hat{C}B$ and altitude of T) such that $\overline{C'H} = \overline{AH} = \frac{1}{2}\ell$. Let A' and B' be the intersections of the circonference, with centre O and radius $\overline{OC'} = \frac{1}{6}(3 - \sqrt{3})\ell$, and, respectively, the segments OA and OB (parts of the other bisectors and altitudes). The circles, with centres in A', B', C' and common radius equal to

$$\frac{1}{2}\overline{AA'} = \frac{1}{2}\overline{BB'} = \frac{1}{2}\overline{CC'} = \frac{1}{4}(\sqrt{3} - 1)\ell,$$

are the solution of MAP for the equilateral triangle, and their total area is $\frac{3}{8}(2 - \sqrt{3})\pi\ell^2$. Now, consider the incircle of T (with centre in O and radius $\frac{1}{6}\sqrt{3}\ell$), whose area is $\frac{1}{12}\pi\ell^2$. In the segment CO, consider the point C'' such that $\overline{CC''} = \frac{1}{6}\sqrt{3}\ell$, and the circle with centre in C'' and radius $\frac{1}{18}\sqrt{3}\ell$ (which is tangent to the incircle and to the sides CA and CB), whose area is $\frac{1}{108}\pi\ell^2$. In the segment AO, consider the point A'' such that $\overline{AA''} = \frac{1}{6}\sqrt{3}\ell$, and the circle with centre in A'' and radius $\frac{1}{18}\sqrt{3}\ell$, whose area is $\frac{1}{108}\pi\ell^2$. The sum of the areas of the incircle and of the last two circles is $\frac{11}{108}\pi\ell^2$, which is greater than $\frac{3}{8}(2 - \sqrt{3})\pi\ell^2$. This shows that the solution of MAP does not solve MOP. Is $\frac{11}{108}\pi\ell^2$ the maximum of MOP? Which is a solution of MOP in the general case? Are MOP and MAP never equivalent? Extension to regular or general polygons, with $N \geq 3$, would be interesting. Note that the MOP and, more generally, (1.2.11) are special cases of the so-called packing problems; the ratio of the total measure of the packed objects to that of the space in which the objects are packed is called *density of the packing* (the function of (1.2.11a) is proportional to it). In the above case, the density of Malfatti-Gozzi triplet and of the last triplet are, respectively,

$$\frac{1}{2}(2\sqrt{3} - 3)\pi \simeq 0.72 \quad \text{and} \quad \frac{11}{81}\sqrt{3}\pi \simeq 0.73 \,.$$

11. The real problems, which lead to the format (1.1.7) (a few are mentioned in Example 1.2.9), and the related mathematical treatments, form — like those of Example 1.2.6 — a huge field, known as Combinatorial Optimization, and the reader is referred to the existing excellent books. Here we are concerned only with the connections between (1.1.7) and (1.1.1). Indeed, till now, the connections between Combinatorial Optimization and the so-called Continuous Optimization (with reference to (1.1.1)) are not many to great disadvantage for both fields.

12. The format (1.3.16) is not the most general that one can adopt. Indeed, it does

not embrace (1.3.1)-(1.3.3), since these types of VI require a formulation where the parameter set is not constant. A more general format of (1.3.16) is conceivable and its development would be interesting.

13. To stress the importance of a VI as a mathematical model of equilibrium problems in the sense that the equilibrium of a system may not be expressed necessarily by the extremum point of any functional, let us take an example from the world of birds. Consider a flock of birds which fly in a "V formation". It is generally believed that the bird, which flies in the advanced position, be the leader. This does not correspond to reality. Such a formation is the outcome of a set of individuals which connect each other according to a law of harmony without any need for a leader. What happens in a network (Sect. 1.6) is related to the above example.

14. The formulation (1.4.7) is quite close to reality, and it has been used as a first, good approximation for concrete design. However, it would be useful to introduce some improvements. The most important concerns a fixed-charge cost. At each node i, suitable equipment is transferred from the sealevel to the seabottom, independently of the size of excavation or embankment; at the end of this, it is brought again to the sealevel and, remaining at sealevel, it is transferred from node i to node $i + 1$. Therefore, there is a cost which does not depend on the unknowns x_i^+ and x_i^-. It would be interesting also to include this improvement in the smooth formulation (1.4.8).

15. Apart from the fact that, in general, (1.6.3)$'$ is much more representative of reality than (1.6.4), both models can be improved, as is now shortly outlined. First of all, we should associate each node with a capacity, as done for the arcs. Both (1.6.3)$'$ and (1.6.4) assume that a digraph G be given. This makes sense when one wants to evaluate the status of an existing network; while, for designing a network or rationalize an existing one such an assumption is unfounded. Therefore, one should assume one is given merely an undirected graph $G_0 = (X, E)$, where X is the set of vertices and E is that of edges (pairs of vertices). To the same graph G_0 we can associate a finite number of digraphs which are "projected" into the same G_0; such a number is, in general, huge. In graph theory, a condition is known, which gives the existence of strongly connected digraphs G, under the assumption that G_0 be connected. Not much more exists. Formulations of the orientation problem to be joined to (1.6.3)$'$ or (1.6.4) would be very interesting.

16. When $c_{ij} = +\infty$ for each pair (i, j), then Definition 1.6.1 is equivalent to one which was given before VI had developed [15, 48, 60]; the latter, called the Wardrop Principle, says: a vector $\bar{x}$ is an equilibrium pattern flow iff for every pair (N_i, N_j) and for every two paths, says P_r and P_s, which connect N_i to N_j, we have:

$$W_r(\bar{x}) > W_s(\bar{x}) \;\Rightarrow\; \bar{x}_{P_r} = 0, \tag{1.9.2}$$

where $W_r(\bar{x})$ and $W_s(\bar{x})$ are, respectively, the sums of the weights associated with the arcs of P_r and of P_s; $\bar{x}_{P_r}$ is the subvector of flows of the arcs of P_r. (1.6.3) is, of course, more general than (1.9.2), so that it might seem more adequate than (1.9.2) for the applications. This can be accepted if we adopt a static view of the network (we are satified by only a photogram of the film "living network"). However, without introducing explicitly a dynamic model, (1.6.3)$'$ can surrogate the dynamic behaviour. Let $\overline{x}$ be a solution of (1.6.3)$'$ without the capacity constraints ($c_{ij} = +\infty$). If $\overline{x} \leq c$, then $\overline{x}$ fulfils also (1.9.2) and then represents real equilibrium also in terms of some stability and not only in static terms. If $\overline{x} \not\leq c$, then a certain instability of the network is pointed out, and then a change of the network (either the orientation or the capacities, or the weights) must be performed until a solution fulfils the capacity constraints. This dynamic approach to a static model does not exclude the formulation of a dynamic model and strengthens the need for it. The availability of a dynamic model of a variational type for the equilibrium flow in a network would be important. Research in this direction should be of great interest from a mathematical point of view independently of any application. Indeed, a dynamic model for equilibrium flow must have a time-lag; therefore, there is the need to begin the study of variational inequalities with a time-lag [26]. Besides network, other fields (perhaps, the mechanics of vibrations) might benefit from this new branch of mathematics. As an instance and preliminary stimulation to this, consider the following Quasi-Variational Inequality. To this end, assume we are given $T = [t_0, t_1]$, $\tau_0, \tau_1 \in \mathbb{R}$ with $\tau_0 < \tau_1$, the point-to-set mapping $X : B \rightrightarrows B$, the functions $F : B \to B$, $\psi_i : \mathbb{R}^n \times \mathbb{R} \times \mathbb{R}^{2n} \to \mathbb{R}, i \in \mathbb{J}$, and $\Delta \in \mathbb{R}$. Consider the domain for $\xi \in B$:

$$\mathbb{K}(\xi) := \left\{ x \in X(\xi) \cap C^2(T)^n : \int_{\tau_0}^{\tau_1} \psi_i(\xi(\tau); t, x(t), x'(t))dt \begin{bmatrix} = 0, & \text{if} & i \in \mathbb{J}^0 \\ \geq 0, & \text{if} & i \in \mathbb{J}^+ \end{bmatrix} \right\}.$$

The Quasi-Variational Inequality consists in finding, $\overline{x} \in \mathbb{K}(\overline{x})$, such that:

$$\int_T \langle F(\overline{x}(t - \Delta)), x(t) - \overline{x}(t - \Delta)\rangle dt, \quad \forall x \in \mathbb{K}(\overline{x}). \tag{1.9.3}$$

Δ, which may or may not depend on the time t, represents a delay with which the users of the network react to what happens. The extrema of integration in (1.9.3) may also contain a time-lag not necessarily equal to Δ. Furthermore, the operator F may depend also on the gradient of the vector $\overline{x}$ of flows.

17.It has been noted in Sect. 1.3 that the inequality contained in Minty Lemma plays an important role in the study of equilibria (see also Sect. 5.5). Indeed, $MVI \Rightarrow SVI$ but not vice versa, as simple examples show (in (1.3.2) and (1.3.6) take, e.g., $\mathbb{K} = \mathbb{R}$, $\phi \equiv 0, \overline{x} = 0$ and **(i)** $F(x) = 3x^2$, **(ii)** $F(x) = -2x$, and **(iii)** $F(x) = 2x$; (1.3.2) is satisfied in all 3 cases, while the 3rd only fulfils (1.3.6)). Therefore, among the equilibria characterized by SVI, those which verify MVI are "qualified". This shows the importance of introducing MVI (besides, of course, SVI) in the field of networks (as well as in all other fields where equilibrium is of interest). Hence, it is of great importance to develop MVI side by side with SVI, to deepen the connections between them, with extremum problems, and with the theory of dynamical systems. With regard to the field of networks, it would be interesting to have a definition of "qualified equilibrium", which were equivalent to (1.3.6) (specialized to networks), like Wardrop's definition is equivalent to $(1.6.3)'$ (for $c = +\infty$).

18.In Sect. 1.6, each arc has been associated with a weight, which has been introduced as an ordinary function of the vector of flows. In some real networks this is not enough, since the arc is associated with several quantities (such as sizes, time, energy, cost, and so on) whose composition into a unique quantity may be unsatisfactory. In such cases, the vector models (1.3.8)-(1.3.11) are more adequate; their investigation has begun recently [27].

19.Complementarity problems are mathematical formulations of equilibrium problems as well as VI; indeed, under very general conditions, they are equivalent. However, their history is somewhat special. These two branches of mathematical science were born independently of each other. VI began early in the 1960s as an evolution of calculus of variations and on the stimulation of an important problem of mathematical physics, known as the Signorini Problem. Despite some notable exceptions, VI have been carried on by people educated and working in classic mathematics, as differential equations and topological vector spaces, or in mathematical physics. CS have had a completely different origin: the orthogonality relation, which is their main feature, was born as orthogonality between the constraining function and the associated Lagrange multiplier in a constrained extremum problem; this was certainly known in the 1920s within the calculus of variations, but did not produce the CS field. Perhaps independently of this, such an orthogonality relation (whose precise history would be of great interest) has been studied by people who — again with some exceptions — lie closer to other branches of mathematical science, such as linear algebra, mathematical programming and combinatorics, and work mainly in finite dimensional spaces. The fact that the motivations to their investigations were coming often (though not always) from operations research, management or economic problems, has led to study a CS in a Euclidean space. The interactions between the two fields have been almost non-existent, so that in 1977 G. Stampacchia, observing that the same results had been achieved in the two fields independently of each other, conceived the first meeting on both VI and CS, which was held in 1978 [13].

20.Consider again the network of Sect. 1.6, and the problem of finding equilibrium flows, defined as a solution of the VI (1.6.3). The operator of the weights, namely $w(x)$, has been defined as a point-to-point map; this means that the weights are uniquely determined by the flows. Reality may require a more general model. Indeed, in the real situations, the weight operator depends on several quantities, like, for instance, cost of fuel, travelling time, green-red ratios in traffic-lights, tolls; most of them are control parameters, which are fixed by the management of the network. Therefore, in real applications, the management can only limit the variability of such parameters, so that $w(x)$ is a point-to-set map. Alternatively, $w(x)$ can be considered a point-to-point map depending on some parameters. The study of these two VI and the comparison of their solutions — when interpreted as equilibrium flows — are interesting.

References

[1] Abadie J.(Ed.), "Nonlinear Programming". North-Holland, Amsterdam, 1967.

[2] Abadie J.(Ed.), "Integer and Nonlinear Programming". North-Holland, Amsterdam, 1970.

[3] Aigner M. and Ziegler G.M., "Proofs from THE BOOK". Springer-Verlag, Berlin, 1999.

[4] Antoni C. and Dalena A., "The Fermat-Weber problem and the Lagrangian duality theory". In "New Trends in Mathematical Programming", Applied Optimization Series, Vol.13, Kluwer Academic Publ., Dordrecht, Boston, 1998, pp.5-12.

[5] Baiocchi C. and Capelo A., "Variational and Quasivariational Inequalities. Applications to free-boundary problems". J.Wiley, Chichester, 1984.

[6] Boltyanski V., Martini H. and Soltan V., "Geometric Methods and Optimization Problems". Series "Combinatorial Problems", D.-Z.Du and P.M.Pardalos (Eds.), Kluwer, Dordrecht, Boston, 1999.

[7] Céa J., "Optimisation. Théorie et Algorithmes". Dunod, Paris, 1971.

[8] Charnes A. and Greenberg II.J., "Plastic collapse and linear programming. Preliminary report". Bull.Amer.Math.Soc., Vol.57, No.6, 1951, p.480.

[9] Clarke F.H., Dem'yanov V.F. and Giannessi F.(Eds.), "Nonsmooth Optimization and Related Topics". Plenum Press, New York, 1989.

[10] Cohn M.Z. and Maier G.(Eds.), "Engineering Plasticity by Mathematical Programming". Pergamon Press, New York, 1979.

[11] Comi C., Maier G. and Perego U., "Generalized variable finite element modeling and extremum theorems in stepwise holonomic elastoplasticity with internal variable". Comput.Methods Appl.Mech.Eng., Vol.96, 1992, pp.213-237.

[12] Conti R., De Giorgi E. and Giannessi F. (Eds.), "Optimization and Related Fields". Springer-Verlag, Lecture Notes in Mathematics, No.1190, Berlin, 1986.

[13] Cottle R.W., Giannessi F. and J.-L. Lions (Eds.), "Variational Inequalities and Complementarity Problems". J.Wiley, Chichester,1980.

[14] Cottle R.W., Pang J.-S. and Stone R.E., "The Linear Complementarity Problem". Academic Press, Boston, 1992.

[15] De Luca M. and Maugeri A., "Quasi-variational Inequalities and applications to equilibrium problems with elastic demand". In [9], pp.61-77.

[16] Dem'yanov V.F. and Rubinov A., "Quasidifferential Calculus". Optimization Soft.Inc.,New York, 1986.

[17] Di Bacco M., Giannessi F. and Naddeo A., "On testing simple statistical hypotheses". Annals of Univ. of Venice, Vol.6, Canova Publishing House, Treviso, Italy, 1972, pp.1-61.

[18] Di Pillo G. and Giannessi F.(Eds.), "Nonlinear Optimization and Applications". Plenum Press, New York, 1996.

[19] Di Pillo G. and Giannessi F.(Eds.), "Nonlinear Optimization and related topics". Kluwer Acad.Publishers, Dordrecht, Boston, 2000.

[20] Du D.-Z. and Pardalos P.M.(Eds.), "Minimax and applications". Kluwer Acad. Publishers, Dordrecht, Boston, 1995.

[21] Erdös P., "Open problems". Jou. of Mathematics and Physics for High School Students, Bolyai Math. Soc., vol.43, n.10, Dec.1993, pag.444, (in Hungarian).

[22] de Fermat P., "Oevres", Vol.I, H. Tannery, Paris, 1891, supplement: Paris, 1922. Deutsche Ubersetzug: Abhandlungen über Maxima und Minima, Hrsg. M.Miller, Ostwalds Klassiker der exakten Wissenschaften, No.238, Leipzig, 1934.

[23] Ferrero O., "On the convexity of the restriction of a function to an affine manifold and applications". Rendiconti seminario matematico, Università di Torino, Vol.45, No.2, 1987, pp.25-39, (in Italian).

[24] Giannessi F., "Metodi Matematici della Programmazione. Problemi Lineari e non Lineari (Mathematical methods of programming. Linear and nonlinear problems)". Monograph No.23 of Italian Mathem. Soc.(UMI), Pitagora Publisher, Bologna, 1982, (in Italian).

[25] Giannessi F.(Ed.), "Nonsmooth Optimization. Methods and applications". Gordon and Breach, U.K., 1992.

[26] Giannessi F., "A remark on infinite-dimensional variational inequalities". Le matematiche, Vol.XLIX, fasc.II, Univ.of Catania, 1994, pp.243-247.

[27] Giannessi F.(Ed.), "Vector Variational Inequalities and Vector Equilibria. Mathematical Theories". Series "Nonconvex Optimization and its Applications", Vol.38, Kluwer Acad.Publishers, Dordrecht, Boston, 2000.

[28] Giannessi F., Jurina L. and Maier G., "Optimal excavation profile for a pipeline freely resting on the sea floor". Engineering Structures, Vol.1, 1979, pp.81-91.

[29] Giannessi F. and Maier G., "Complementarity Systems and Optimization Problems in Structural Engineering". Engineering Optimization, Vol.18, 1991, pp.43-66.

[30] Giannessi F., Maier G. and Jurina L., "A quadratic complementarity problem related to the optimal design of a pipeline freely resting on a rough sea bottom". Engineering Structures, Vol.4, 1982, pp.186-196.

[31] Giannessi F., Maier G. and Nappi A., "Indirect identification of yeld limits by Mathematical Programming". Engineering Structures, Vol.4, 1982, pp.186-196.

[32] Giannessi F. and Maugeri A.(Eds.), "Equilibrium problems with side constraints. Lagrangian theory and Duality". Le Matematiche, Dept. of Mathematics, Univ. of Catania, Vol.XLIX, Fasc.II, 1994.

[33] Giannessi F., Rapcsàk T. and Komlosi S.(Eds.), "New Trends in Mathematical Programming". Kluwer Acad. Publishers, Dordrecht, Boston, 1998.

[34] Girsanov I.V., "Lectures on mathematical theory of extremum problems". Lecture Notes in Ec. and Math. Syst., No.67, Springer-Verlag, Berlin and New York, 1972.

[35] Gruber P.M. and Wills J.M.(Eds.), "Handbook of Convex Geometry". Vol.A and Vol.B, North-Holland, Amsterdam, 1993.

[36] Hammer P.L. and Rudeanu S., "Méthodes Boolèennes en recherche opèrationelle". Dunod, Paris, 1970.

[37] Hancock H., "Theory of maxima and minima". Dover Publ., New York, 1960.

[38] Horst R. and Tuy H., "Global Optimization". Springer-Verlag, Berlin, 1990.

[39] Ioffe A.D. Tikhomirov V.M., "Theory of Extremal Problems". North-Holland, Amsterdam, 1979.

[40] Isac G., "Complementarity Problems". Springer-Verlag, Berlin, 1991.

[41] Jung H.W.E., "Über die kleinste Kugel, die eine räumliche Figur einschliesst". Jou.Reine Angew.Math., Vol.123, 1901, pp.241-257.

[42] Kinderlehrer D. and Stampacchia G., "An introduction to Varational Inequalities and their Application". Academic Press, New York, 1980.

[43] Kuhn H.W. and Tucker A.W.(Eds), "Linear inequalities and related systems", Annals of Mathematics Studies, No.38, Princeton Univ. Press, Princeton, N.I., 1956.

[44] Maier G., Giannessi F., Andreuzzi F., Jurina L. and Taddei F., "Unilateral Contact, elastoplasticity and complementarity with reference to offshore pipeline design". Series "Computer Methods in Applied Mechanics and Engineering", Vol.17-18, North-Holland, Amsterdam, pp.469-495.

[45] Mangasarian O.L., "Nonlinear Programming". SIAM Monograph in Applied Mathematics, Philadelphia, 1994.

[46] Mastroeni G., "Application of an extremization method to a linear integral of a statistical decision problem". Jou. of Optimiz. Theory Appls., Vol.109, No.3 , 2001, pp.539-556.

[47] Maugeri A., "Variational and Quasi-Variational Inequalities in network models. Recent developments on theory and algorithms". In [V1], pp.195-211.

[48] Maugeri A., "Monotone and Nonmonotone Variational Inequalities".In [V 4], pp.179-184.

[49] Miele A., "Flight Mechanics, Vol.I, Theory of Flights Paths". Addison-Wesley Publ.Co., Reading, Massachussetts, 1962.

[50] Miele A. and Salvetti A.(Eds.), "Applied Mathematics in Aerospace Science and Engineering". Plenum Press, New York, 1994.

[51] Minty G.J., "Monotone (nonlinear) operator in Hilbert space". Duke Math. Jou., Vol.29,1962, pp.341-346.

[52] Neyman J. and Pearson E.S., "On the problem of the most efficient tests of statistical hypotheses". Philosophical Transactions of the Royal Society of London, Vol.231, pp.289-337, 1933.

[53] Ortega J.M. and Rheinboldt W.C., "Iterative Solution of Nonlinear Equation in Several Variables". Academic Press, New York, 1970.

[54] Pappalardo M., "On the connections between optimality conditions, varational inequalities and equilibrium problems". In [32], pp.333-339.

[55] Ponstein J., "Approaches to the Theory of Optimization". Cambridge Univ. Press, 1980.

[56] Rapcsák T., "Smooth Nonlinear Optimization in $\mathbb{R}^n$". Kluwer Acad. Publishers, Dordrecht, Boston, 1997.

[57] Stampacchia G. "Formes bilinéares coercitives sur les ensembles convexes". Comptes Rendus Acad.Sci.Paris, Vol.258, 1964, pp.4413-4416,.

[58] Statnikov R.B., "Multicriteria Design. Optimization and Identification". Kluwer Acad.Publishers, Dordrecht, Boston, 1999.

[59] Torricelli E., "Opere, Vol.I, Part 2", Faenza, 1919, pp.90-97; "Opere, Vol.III", Faenza, 1919, pp.426-431. Also "De maximis et minimis" (in Latin), published in the collection "Pupils of Galileo", Vol.26.

[60] Wardrop J.C., "Some theoretical aspects of road traffic research". Proc. Just. Civil Eng., Part II, Vol. 1, 1952, pp. 325-378.

CHAPTER 2. ELEMENTS OF CONVEX ANALYSIS AND SEPARATION

2.1. Convex Sets and Cones

The concept of convexity is one of the most important and popular in Mathematics. The reason is that, in the presence of convexity, a problem enjoys nice properties. This is generally believed, but not always true; indeed, we will show in the next section that, in some cases, — even if very few — convexity may behave very badly. The beginning of modern Convex Analysis starts with the work of Minkowski in the years 1897-1904, when he proved important theorems; see [36]; since then the development continues with important contributions by many scholars, like Radon [42], Helly [27], Bonnesen-Fenchel [6, 23], Fan [20], and Valentine [51].

Several contributions to Convex Analysis have come from other fields of Mathematics, like Calculus of Variations [III5,III24,V61]. There exist several very excellent books on Convex Analysis; see, for instance, [6,19,20,22,23,26,27,29,38,43,48,53,I35]; here we will consider only the basic concepts and some extensions which are necessary for the other chapters. The properties will be established in $\mathbb{R}^n$ even if most of them hold in a Hilbert or Banach space, since they will be exploited prevalently in a Euclidean space; the applications in B will be considered case by case. Indeed, this chapter does not aim to be a treatise, even if introductory, on Convex Analysis, but merely to give the minimum body of notions for reading the remaining part of this volume, and the subsequent one.

The symbols of each section of this chapter are independent of those of the other sections or chapters, if they overlap.

Definition 2.1.1. A nonempty set $K \subseteq \mathbb{R}^n$ is *convex*, iff

$$(1 - \alpha)x^1 + \alpha x^2 \in K, \quad \forall \alpha \in [0, 1], \quad \forall x^1, x^2 \in K.$$

We stipulate to consider convex both a singleton and $\varnothing$. K is *concave*, iff $\sim K$ is convex. K is a *convex body*, iff it is convex, $cl\, K$ is compact and $int\, K \neq \varnothing$. If $int\, K \neq \varnothing$, K is *strictly convex*, iff

$$(1 - \alpha)x^1 + \alpha x^2 \in int\, K, \quad \forall \alpha \in\,]0, 1[, \quad \forall x^1, x^2 \in cl\, K, \quad x^1 \neq x^2.$$

K is *strictly concave* iff $\sim K$ is strictly convex.

Examples of convex sets are $\mathbb{R}^n$; every subspace; an affine manifold (i.e., one which contains the line through any two distinct points of the set); a halfspace; an ellipsoid (i.e., $\{x : \langle x, Ax \rangle \leq 1\}$ with A positive definite matrix) of $\mathbb{R}^n$ and in particular a sphere; the interior of a convex set ; the set of solutions to a system of linear algebraic equalities in $\mathbb{R}^n$; the image or the inverse image of a convex set through an affine transformation ; the set $\alpha K_1 + \beta K_2$ whatever the convex set K_1, K_2 and $\alpha, \beta \in \mathbb{R}$ may be. A halfspace and the entire space are both convex and concave. Such claims are easily proved by direct use of Definition 2.1.1. If $n = 1$, obviously, every convex set, but a singleton, is strictly convex. Among the several properties enjoyed by the convex bodies, there are the isoperimetric inequalities: for instance, if K is a convex and closed body of $\mathbb{R}^2$ (or of $\mathbb{R}^3$), and ℓ (or s) denotes the measure of the boundary of K, and V that of the area (or of the volume), then the *isoperimetrical inequality* $\ell^2 \geq 4\pi V$ (or $s^3 \geq 36\pi V^2$) holds; see Example 5.4.1.

The strict convexity of a set K having empty interior is of some interest. It can be defined by excluding those K, which are contained in a line of $\mathbb{R}^n$ and fulfil the above relation with int replaced by ri.

In the present section we will consider only a very few properties of convex sets which are necessary for the theory of extrema.

Proposition 2.1.1. The intersection of any family of convex sets is convex.

Proof. Let $\{K(\xi),\ \xi \in \Xi\}$ denote any finite or infinite family of convex sets. If the family is empty or a singleton, then the claim is trivial. Otherwise, because of the convexity of every $K(\xi)$, we have $x', x'' \in \underset{\xi \in \Xi}{\cap} K(\xi) \Rightarrow x', x'' \in K(\xi),\ \forall \xi \in \Xi \Rightarrow$ $(1 - \alpha)x' + \alpha x'' \in K(\xi),\ \forall \alpha \in [0, 1],\ \forall \xi \in \Xi \Leftrightarrow (1 - \alpha)x' + \alpha x'' \in \underset{\xi \in \Xi}{\cap} K(\xi).$ $\qquad\square$

Definition 2.1.2. Let $x, x^1, ..., x^r \in \mathbb{R}^n$. x is called *convex combination* of $x^1, ..., x^r$ iff $\exists \alpha_1, ..., \alpha_r \in [0, 1]$ with $\sum\limits_{i=1}^{r} \alpha_i = 1$, such that:

$$x = \sum_{i=1}^{r} \alpha_i\, x^i \ . \tag{2.1.1}$$

The convex combination is called *proper*, iff $\alpha_i > 0$, $i = 1, ..., r$, $r \geq 2$. Given any $K \subseteq \mathbb{R}^n$, the intersection of all convex sets of $\mathbb{R}^n$ containing K is called *convex hull* of K and denoted by conv K. An element of K is called *extreme point*, iff it cannot be expressed as proper convex combination of elements of K.

Sometimes, an extreme point is defined − with regard to convex set − as a point which cannot be expressed as a proper convex combination of two elements of K (see, for instance, [2,38]). Indeed, this definition is equivalent to that in Definition 2.1.2, if K is convex. While, if K is not convex, such an equivalence may fail, as simple examples show (for instance, let K be the set of vertices of a triangle of $\mathbb{R}^2$ and of a point of its interior).

As the next proposition shows, the two concepts given by Definition 2.1.2 express the same thing and differ from each other in language only; however, as it happens not infrequently, both are useful. The coming and going between them is fundamental to deepen the nature of sets, especially the convex ones. Fig. 2.1.1 shows two instances of convex hull. The convex combination has several interesting interpretations, the most classic of which is as the centre of gravity, where x^i is the location of a mass and α_i the mass relative to the sum of r masses.

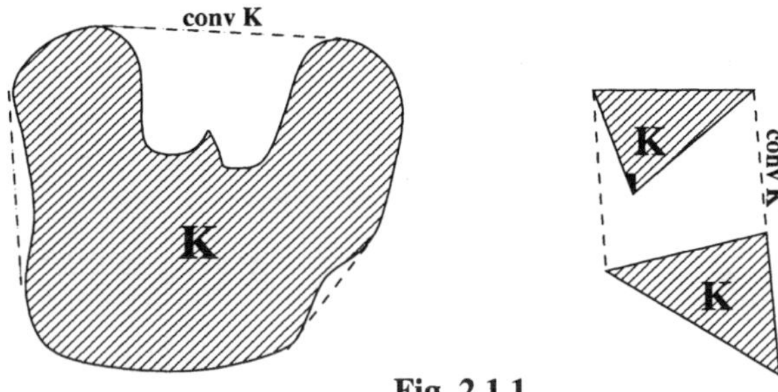

Fig. 2.1.1

Proposition 2.1.2. The convex hull of $K \subseteq \mathbb{R}^n$ equals the set of all convex combinations of elements of K.

Proof. Without any loss of generality, only proper convex combinations can be considered. Denote by X_r the set of all proper convex combinations of r elements of K. Obviously, $X_r \subseteq \operatorname{conv} K$ at $r \leq 2$. It will be shown that $X_r \subseteq \operatorname{conv} K \Rightarrow X_{r+1} \subseteq \operatorname{conv} K$. By setting $\hat{x} := \sum\limits_{i=1}^{r} \frac{\alpha_i}{1-\alpha_{r+1}} x^i$, noting that (in order to have $\hat{x} \in K$) $\alpha_i > 0$, $i = 1, ..., r$ imply $\alpha_{r+1} < 1$, and that $\hat{x}, x^{r+1} \in \operatorname{conv} K$ (because of the inductive assumption and of the 2nd part of Definition 2.1.2), we have $\sum\limits_{i=1}^{r+1} \alpha_i x^i = (1 - \alpha_{r+1})\hat{x} + \alpha_{r+1}\, x^{r+1} \in \operatorname{conv} K$, so that $X_{r+1} \subseteq \operatorname{conv} K$. It follows that $X_r \subseteq \operatorname{conv} K$, whatever r may be. The reverse inclusion is achieved by observing that the set of all convex combinations of elements of K, being convex (as is easily proved by a direct use of Definition 2.1.1), contains conv K. $\square$

Proposition 2.1.3. $K \subseteq \mathbb{R}^n$ is convex, if and only if $K = \operatorname{conv} K$, or if and only if every convex combination of any number of elements of K belongs to K.

Proof. If. Obvious consequence of Proposition 2.1.1. **Only if.** Since we have: conv $K = \bigcap\limits_{\xi \in \chi} K(\xi)$, $\{K(\xi), \xi \in \chi\}$ being the set of all convex sets containing K, then $K \subseteq \operatorname{conv} K$. K convex $\Rightarrow \exists \hat{\xi} \in \chi$, $s.t.$ $K = K(\hat{\xi}) \Rightarrow K \supseteq \operatorname{conv} K$. The last part of the claim is a trivial consequence of Proposition 2.1.2 and of the 1st part. $\square$

Because of Propositions 2.1.2 and 2.1.3, $K \subseteq B$ is convex, iff it coincides with the set of all convex combinations of its elements. The following 3 theorems are cornerstones of Convex Analysis.

Lemma 2.1.1. Let $x, x^1, ..., x^r \in \mathbb{R}^n$ and x be a proper convex combination of $x^1, ..., x^r$ with coefficients $\alpha_1, ..., \alpha_r$. If $r > n + 1$, then we have:

$$x \in \operatorname{conv}(\{x^1, ..., x^r\} \setminus \{x^p\}), \qquad (2.1.2)$$

where p is such that:

$$\frac{\alpha_p}{\beta_p} = \min_{\beta_i > 0}\{\frac{\alpha_i}{\beta_i}, \quad i = 1, ..., r; \quad i \neq q\}, \qquad (2.1.3)$$

and where q is any integer from $\{1, ..., r\}\backslash\{p\}$, and β_i $i = 1, ..., r$, $i \neq q$, are reals, not all zero — at least one of the which can be assumed to be positive —, such that

$$\sum_{\substack{i=1 \\ i \neq q}}^{r} \beta_i(x^i - x^q) = 0.$$

Then, the coefficients which give (2.1.2) are:

$$\hat{\alpha}_i := \alpha_i - \frac{\alpha_p}{\beta_p}\beta_i, \ i = 1, ..., r, \ i \neq q; \quad \hat{\alpha}_q := \alpha_q + \frac{\alpha_p}{\beta_p}\sum_{\substack{i=1 \\ i \neq q}}^{r}\beta_i,$$

where $\hat{\alpha}_i = 0$ for at least $i = p$.

Proof. The existence of the β_i's is implied, of course, by $r > n + 1$. It is not restrictive to assume $\beta := \sum_{\substack{i=1 \\ i \neq q}}^{r} \beta_i > 0$. From (2.1.1), $\forall \gamma \in \mathbb{R}$, we have:

$$x = \sum_{i=1}^{r}\alpha_i x^i = \sum_{i=1}^{r}\alpha_i x^i - \gamma\sum_{\substack{i=1 \\ i \neq q}}^{r}\beta_i(x^i - x^q) = \sum_{i=1}^{r}\hat{\alpha}_i x^i,$$

where $\hat{\alpha}_i := \alpha_i - \gamma\beta_i$, $i = 1, ..., r$, $i \neq q$; $\hat{\alpha}_q := \alpha_q + \gamma\beta$. Of course, $\sum_{i=1}^{r}\hat{\alpha}_i = 1$, $\forall \gamma \in \mathbb{R}$. Being $\alpha_i > 0$, $i = 1, ..., r$, we have $\frac{\alpha_p}{\beta_p} > 0$. If we set $\gamma = \frac{\alpha_p}{\beta_p}$, then because of (2.1.3), we have $\hat{\alpha}_i \geq 0$, $i = 1, ..., r$ and $\hat{\alpha}_i = 0$ for, at least, $i = p$, so that (2.1.2) follows. $\quad\square$

The above Lemma, which of course cannot be inverted, allows one to eliminate the redundant elements in a convex combination, by exploiting it iteratively. Indeed, such a reasoning is the basis of the next fundamental theorem. An useful mechanical interpretation of Lemma 2.1.1 is obtained by looking at x^i as a force applied to the origin: the composition of a system of forces, namely $\{\alpha_1 x^1, ..., \alpha_r x^r\}$, does not change, if we add (or substract) to the system another system of forces which be in equilibrium; if the latter system is suitable, then the former system is simplified.

Theorem 2.1.1. (C.Carathéodory, 1907). Let $K \subseteq \mathbb{R}^n$. Every element of conv K can be expressed as convex combination of at most $n + 1$ elements of K.

Proof. Let $x \in$ conv K. Because of Proposition 2.1.2, (2.1.1) holds, where we can assume $\alpha_i > 0$, $i = 1, ..., r$. If $r > n + 1$, then we can apply Lemma 2.1.1 and eliminate at least one point in (2.1.1). The thesis is achieved by applying iteratively such a Lemma until $r \leq n + 1$. $\quad\square$

The above theorem can be read as a necessary (and — the inverse claim being trivial — sufficient) condition for a point of $\mathbb{R}^n$ to belong to conv K. Its extension to a locally convex space (a nonempty, compact and convex set is the convex hull of its extreme points) is the well known Krein-Milman Theorem [30,48]. Lemma 2.1.1 and Theorem

2.1.1 can be easily refined by replacing $\mathbb{R}^n$ with aff K, so that we obtain $1+\dim$ aff K as an upper bound for the number of necessary points.

Lemma 2.1.2. Let $x^1, ..., x^r \in \mathbb{R}^n$. If $r \geq n + 2$, then there exist both a partition of $I := \{1, ..., r\}$ into two nonempty subsets, say I_1 and I_2, and non-negative reals $\alpha_1, ..., \alpha_r$ with $\sum_{i \in I_1} \alpha_i = \sum_{i \in I_2} \alpha_i = 1$, such that:

$$\sum_{i \in I_1} \alpha_i x^i = \sum_{i \in I_2} \alpha_i x^i . \tag{2.1.4}$$

Proof. Set $x^i = (x_1^i, ..., x_n^i)$, $i \in I$. The linear homogeneous algebraic system (in the unknowns $\beta_1, ..., \beta_r$):

$$\sum_{i=1}^{r} \beta_i x_j^i = 0, \quad j = 1, ..., n; \quad \sum_{i=1}^{r} \beta_i = 0 \tag{2.1.5}$$

has, obviously, nontrivial solutions; call $\hat{\beta} := (\hat{\beta}_1, ..., \hat{\beta}_r)$ one of them. Set $I_1 := \{i \in I : \hat{\beta}_i \geq 0\}$, $I_2 := \{i \in I : \hat{\beta}_i < 0\}$. Since $\hat{\beta}$ satisfies the last of (2.1.5), I_1 and I_2 are nonempty and it is not restrictive to assume that $c := \sum_{i \in I_1} \hat{\beta}_i > 0$. Hence $\sum_{i \in I_2} \hat{\beta}_i = -c$, so that $\sum_{i \in I_1} (\hat{\beta}_i/c) x^i = \sum_{i \in I_2} (-\hat{\beta}_i/c) x^i$. Therefore, (2.1.4) follows at $\alpha_i := \hat{\beta}_i/c$, $i \in I_1$, $\alpha_i := -\hat{\beta}_i/c$, $i \in I_2$. $\qquad \square$

Lemma 2.1.2, which is instrumental for next theorem, has an useful interpretation in terms of forces.

Theorem 2.1.2 (J.Radon, 1921). If a set $K \subseteq \mathbb{R}^n$ has at least $n + 2$ elements, then there exists a partition of K into two nonempty subsets, say K_1 and K_2, such that:

$$(\text{conv } K_1) \cap \text{conv } K_2 \neq \varnothing \tag{2.1.6}$$

Proof. Because of Lemma 2.1.2, every set of $r \geq n+2$ points $x^i \in K$, $i \in I := \{1, ..., r\}$, can be partitioned into two subsets $X_1 := \{x^i, i \in I_1\}$ and $X_2 := \{x^i, i \in I_2\}$, such that (2.1.4) holds. Hence, any partition of K into two subsets, say K_1 and K_2, such that $X_1 \subseteq K_1$ and $X_2 \subseteq K_2$, fulfils (2.1.6). $\qquad \square$

Theorem 2.1.2, which of course cannot be inverted, holds for a partition of K into any number of subsets [50].

Theorem 2.1.3 (E.Helly, 1913). Let $\mathcal{F}$ be a finite family of at least $n + 1$ convex sets of $\mathbb{R}^n$. If the intersection of every $n + 1$ elements of $\mathcal{F}$ is nonempty, then also the intersection of all the elements of $\mathcal{F}$ is nonempty.

Proof. If card $\mathcal{F} = n+1$, then the thesis is trivial. By induction, suppose that it holds for every family of $r - 1$ convex sets of $\mathbb{R}^n$ with $r \geq n + 2$, and consider any family, say $\mathcal{F}$, of r convex sets, for which (by assumption) the intersection of the elements of every subfamily, having the cardinality equal to $n + 1$, be nonempty. Because of the inductive assumption, $\forall K \in \mathcal{F}$, the intersection of the elements of $\mathcal{F} \setminus \{K\}$ is nonempty; let p_K denote a generic element of such an intersection, and consider the set $P := \{p_K : K \in \mathcal{F}\}$. If card $P < r$, then $\exists K_1, K_2 \in \mathcal{F}$ with $K_1 \neq K_2$, such that $p_{K_1} = p_{K_2}$; set $p := p_{K_1} = p_{K_2}$. Since $p_{K_2} \in K_1$ and $p_{K_1} \in \bigcap_{K \in \mathcal{F} \setminus \{K_1\}} K$, then $p \in \bigcap_{K \in \mathcal{F}} K$ and the thesis is achieved. If card $P = r \geq n + 2$, then, because of Theorem 2.1.2, there exists a partition of P into two nonempty subsets, say P_1 and P_2, such that $S := (\operatorname{conv} P_1) \cap \operatorname{conv} P_2 \neq \varnothing$. Set $\mathcal{F}_i := \{K \in \mathcal{F} : p_K \in P_i\}$, $i = 1, 2$. Let $\hat{p} \in S$. We have:

$$P_1 \subset \bigcap_{K \in \mathcal{F}_2} K, \quad P_2 \subset \bigcap_{K \in \mathcal{F}_1} K \ ,$$

so that, due to Proposition 2.1.1,

$$\operatorname{conv} P_1 \subset \bigcap_{K \in \mathcal{F}_2} K, \quad \operatorname{conv} P_2 \subset \bigcap_{K \in \mathcal{F}_1} K \ .$$

It follows that $\hat{p} \in \bigcap_{K \in \mathcal{F}} K$. $\qquad\qquad\qquad\qquad\qquad\qquad\qquad\qquad\qquad\qquad\qquad\qquad\square$

Theorem 2.1.3 was conceived by Helly around 1913, but published later [27]; see also [29,43]. The present proof is due to Radon and was published together with Theorem 2.1.2 [42]. The assumption of Theorem 2.1.3 cannot be weakened: in fact, the family $\mathcal{F} = \{K_1, K_2, K_3\} \subset 2^{\mathbb{R}^2}$ of Fig. 2.1.2 shows that the thesis of Theorem 2.1.3 may be false, if we merely assume that the intersection of every set of $s < n + 1$ elements of $\mathcal{F}$ is nonempty; $\mathcal{F} = \{K_i, \ i = 1, 2, 3, 4\} \subset 2^{\mathbb{R}^2}$ of Fig. 2.1.3 shows that we cannot remove the assumption of convexity, without introducing further assumptions. Theorems 2.1.1, 2.1.2 and 2.1.3 express three properties which are strongly related each other (see [I35], Vol. A, Ch. 2.1, for details).

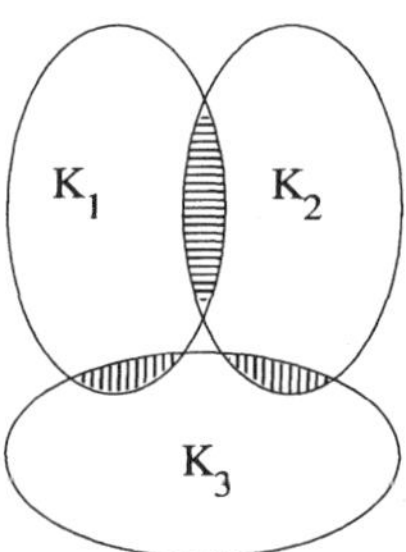

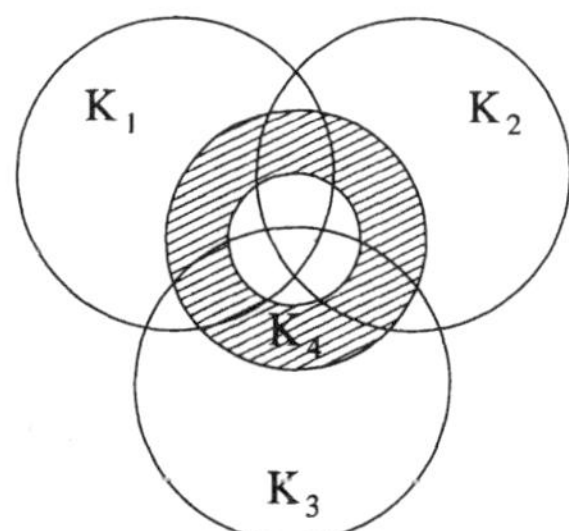

Fig. 2.1.2   Fig. 2.1.3

Definition 2.1.3. A set $K \subseteq \mathbb{R}^n$ is called *affine*, iff it is the intersection of a finite number (possibly, zero; in this case $K = \mathbb{R}^n$) of hyperplanes. The *affine hull* of K, denoted by aff K, is the intersection of all the affine sets, which contain K. The vectors $k_1, ..., k_{m+1} \in K$, with $m \leq n$, are *affinely independent*, iff dim aff $\{k_1, ..., k_{m+1}\} = m$. K is called *polytope*, iff it is the convex hull of a finite number of elements of $\mathbb{R}^n$. If $m + 1$, with $m \leq n$, of such elements are enough to identify the polytope and they are affinely independent, then the polytope is called *simplex m-dimensional* or *m-simplex*;

if $m = n$, it is called *simplex*. A *subsimplex* of a m-simplex is a r-simplex defined by $r + 1$ of the $m + 1$ given points with $r \leq m$; it will be called r-dimensional *face* of the m-simplex; it will be called *improper face* iff $r = m$, *facet* iff $r = m - 1$, *vertex* iff $r = 0$. A m-simplex is called *regular* iff all its facets are equal.

Among the several properties enjoyed by the polytopes, there are the isoperimetric inequalities; for istance, if ℓ and V denote, respectively, the (length of the) perimeter and the (measure of the) area of a n-gon (of $\mathbb{R}^2$) with $n \geq 3$, then the *isoperimetric inequality* (tn denotes trigonometric tangent) $\ell^2 \geq 4n V \mathrm{tn}(\pi/n)$ holds; as $n \to +\infty$, we find that mentioned after Definition 2.1.1; see Example 5.4.1. For $n = 3$ and $m = 0, 1, 2, 3$, a m-simplex is a point, a segment, a triangle, a tetrahedron, respectively. Every element of a m-simplex can be expressed, in a unique way, as convex combination of its vertices.

Theorem 2.1.1 claims that every element of conv K belongs to an m-simplex whose vertices lie in K and with $m \leq n$.

Theorem 2.1.3 has had many consequences and many are yet to come. Indeed, theorems, which apparently have nothing to share with its subject, can be formulated as Helly type theorems; see, for instance, Theorem 3.4.1. An excellent overview of the applications of the Helly Theorem is offered by [14]. The Helly Theorem has shown to be useful to simplify the proofs of already stated propositions. Let us consider an important instance. By means of Helly Theorem, it is possible to prove [14, p.132] *Jung Inequality* (stated by H.W.Jung in 1901, Jou. Reine Angew. Math., Vol.130, pp. 310-313): if $K \subset \mathbb{R}^n$ has diameter d, then there exists a (unique) n-dimensional sphere containing K of radius $r \leq d\,[n/2(n+1)]^{1/2}$; the equality holds iff K is a regular simplex (see Example 1.2.4).

Theorem 2.1.1 can be formulated this way: *if $x \in$ conv K, with $K \subseteq \mathbb{R}^n$, then x belongs to a simplex whose vertices belong to K.* From Theorem 2.1.1 we can draw many consequences. For instances, conv K *is the union of all the simplices whose vertices belong to K; every polytope is the union of a finite number of simplices; if K is compact, then also* conv K *is compact.* The first two claims are obvious. To show the third, consider the $n \times (n + 1)$-dimensional vector $y := (x^1, ..., x^{n+1}) \in K^{n+1}$, where $x^i \in K$, $i = 1, ..., n + 1$, the set $S := \{\alpha = (\alpha_1, ..., \alpha_{n+1}) \in \mathbb{R}^{n+1}_+ : \sum_{i=1}^{n+1} \alpha_i = 1\}$, and the function $f(\alpha, y) := \sum_{i=1}^{n+1} \alpha_i x^i$. Since $S \times K^{n+1}$ is compact and f is continuous, then $f(S, K^{n+1})$ is compact. Because of Proposition 2.1.2, we have conv $K = f(S, K^{n+1})$, and hence the conclusion.

A geometric interpretation of Theorem 2.1.2 is quite obvious: for instance, at $n = 2$, if card $K = 4$, then it claims that either one of the points of K belongs to the $-$ possibly degenerate $-$ triangle defined by the 3 other points, or the segment defined by 2 points intersects that of the 2 other points.

Theorem 2.1.3 has many important applications to Combinatorics; recently, it has been recognized that the Helly Theorem might play an interesting role also in the field of constrained extremum problems, especially those of type (1.1.7). Indeed, as it will be seen in Sect. 3.2 one of the most challenging problems of the entire theory of constrained extrema consists in proving the impossibility of a system or the disjunction of two or more sets; therefore, every theorem, which aims to show empty intersection of sets, may be extremely useful.

Consider the particular case $n = 1$ and let $\mathcal{F} = \{[a_i, b_i], i = 1, ..., r\}$ with $a_i, b_i \in \mathbb{R}$ and $a_i \leq b_i$ be a family of segments of the real line. Theorem 2.1.3 says that, if every pair of segments has nonempty intersection, then the same happens to the entire family; the intersection of all segments being, obviously, the segment $[\max_i(a_i), \min_i(b_i)]$. As an application of this case, we have that, given a set of parallel segments of the Euclidean plane, there exists a line, orthogonal to the segments, which intersects all of them, if for each pair of them there exists a line, orthogonal to the segments, which intersects both segments.

In Theorem 2.1.3, the assumption that $\mathcal{F}$ be finite can be removed, if its elements are supposed to be either compact or closed and at least one is bounded; if none of these assumptions is taken up, then the thesis may not hold, as shown by the following families of segments of the real line: $\mathcal{F} = \{[n, +\infty[, n \in \mathbb{N}\}$, $\mathcal{F} = \{]0, \frac{1}{n}], n \in \mathbb{N}\}$.

Definition 2.1.4. Let $x', x'' \in K \subseteq \mathbb{R}^n$. We say that x' *sees* x'', iff $[x', x''] \subseteq K$. The *star* of K is the set of elements of K, each of which sees all the elements of K. K is called *star-shaped*, iff its star is nonempty.

Definition 2.1.4 contains a concept of local convexity: " is convex at $\overline{x}$, iff $\overline{x}$ belongs to the star of K". This differs from the traditional way: "is convex at $\overline{x}$, iff there exists a neighbourhood $N(\overline{x})$ of $\overline{x}$, such that $K \cap N(\overline{x})$ is convex". For instance, the union of closed halflines having a common origin fullfils the former, but not the latter. To adopt one or the other depends on the purpose one pursuits.

A consequence of Theorem 2.1.3, is that the star − which is obviously convex − of a compact and infinite set $K \subseteq \mathbb{R}^n$ is nonempty, if for every set of $n + 1$ elements of K there exists an element of K which sees all of them. Indeed, $\overline{x}$ belongs to the star of K, iff the elements of the family $\mathcal{F} = \{[\overline{x}, x], x \in K\}$ (which are obviously convex) have nonempty intersection; the application of Helly Theorem leads to the conclusion.

Let $K_i \subset \mathbb{R}^n, i \in I := \{1, ..., r\}$, with $r \geq n + 1$, be convex (so that each $\sim K_i$ is concave), and suppose that the intersection of every set of $n + 1$ of the $K_1, ..., K_r$ be empty. Then $S := \bigcup_{i \in I}(\sim K_i) = \mathbb{R}^n$. In fact, ab absurdo, suppose that $\exists \hat{k} \in \mathbb{R}^n \setminus S$, so that $\hat{k} \notin \sim K_i, \forall i \in I$. This implies $\hat{k} \in \bigcap_{i \in I} K_i$. But this is impossible, because of Theorem 2.1.3. This property can be, of course, achieved without exploiting Theorem 2.1.3.

Several other properties can be deduced from the above theorems. For instance, let $X \subseteq \mathbb{R}^n$; if there exists a convex body K, such that every subset of X with cardinality $\leq n + 1$ is contained in K (or in a translation of it), then $X \subseteq K$ (or in its translation). In fact, $\forall x \in \text{conv} X$, because of Theorem 2.1.1, $\exists x^1, ..., x^r \in X$, with $r \leq n + 1$, s.t. $x \in \text{conv}\{x^1, ..., x^r\} \subseteq K \Rightarrow X \subseteq K$. Another instance is the following: let $\mathcal{F}$ be a family of closed halfspaces of $\mathbb{R}^n$, and let their intersection, say $\mathcal{F}^0$, be bounded; if there exists a convex body K, such that the intersection of any set S of $n+1$ elements of $\mathcal{F}$ contains K (or a traslation of it), then K (or its traslation) is contained in $\mathcal{F}^0$. The claim holds since K, being contained in every element of S, and hence of $\mathcal{F}$, is contained in $\mathcal{F}^0$; the emptiness of $\mathcal{F}^0$ is given by Theorem 2.1.3, whose assumption is guaranteed by the inclusion of K in the intersection of the elements of S.

Theorem 2.1.3 and its consequences have stimulated the formulation of a general statement: "if a family of (mathematical) objects is such that each subfamily of cardinality ν enjoys a certain property P, then P is enjoyed by the entire family". If this statement is proved to be true for a specific case, then we say that a *Helly-type theorem* has been established. This has opened the question of finding, for a given family, the *Helly number of the family*, namely the smallest number ν for which the statement holds. Theorem 2.1.3 claims that $n+1$ is the Helly number of the family of all convex sets of $\mathbb{R}^n$. Theorem 2.1.3 can be interpreted in terms of Graph Theory (a graph is a pair of sets, say (V, E), where V is a set of elements called vertices and E is a subset of $V \times V$, whose elements are called edges): the family $\mathcal{F}$ can be associated with a graph G, whose vertices are in one-to-one correspondence with the elements of $\mathcal{F}$, and an edge exists iff the intersection of the elements of $\mathcal{F}$ corresponding to its vertices is nonempty; such a graph is called *intersection graph*.

Definition 2.1.5. The *convex closure* of $K \subseteq \mathbb{R}^n$ is the intersection of all the closed and convex sets which contain K.

Proposition 2.1.4. The convex closure of K equals cl conv K.

Proof. cl conv K, being closed and convex, obviously contains K. Ab adsurdo, suppose that there exists a closed and convex set $\hat{K}$, s.t. $K \subseteq \hat{K} \subset$ cl conv$K$.This implies that $\exists \hat{x} \in (\text{conv}K) \backslash \hat{K}$. The contradiction comes from the fact that $\hat{x} \in \text{conv}K$ implies that $\hat{x}$ belongs to every convex set containing K and hence also to $\hat{K}$. $\square$

Easy instances show that the operators cl and conv are not commutative. If K is convex and open, then $K = \text{conv}K \subset$ cl conv$K$. Another example is offered by Fig. 2.1.4, where the set K is the union of a point and of a halfline without its origin (dashed means open).

Fig. 2.1.4

Another interesting aspect is the frontier. For instance, it is trivial to show that, if K is convex, then frt $K =$ frt conv K. As Fig. 2.1.5 shows, the vice versa is not true even if cl K is convex.

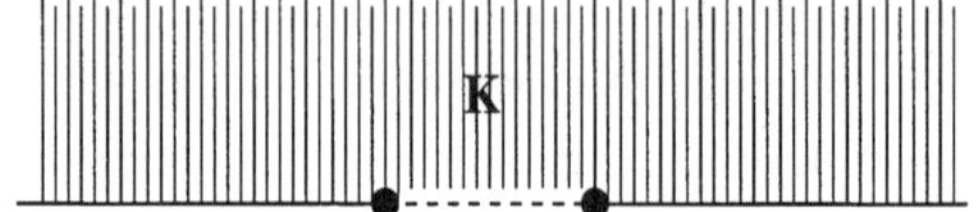

Fig. 2.1.5

The above concept, together with other ones, like interior, has led to state several properties, the most intuitive of which is the following one.

Proposition 2.1.5. Let $K, K_1, K_2 \subseteq \mathbb{R}^n$ be nonempty and convex. **(i)** ri K, int K and cl K are convex. **(ii)** ri $(K_1 + K_2)$ = ri K_1+ri K_2. **(iii)** $\hat{x} \in$ ri K and $x^\circ \in$ cl K with $x^\circ \neq \hat{x}$ imply $]x^\circ, \hat{x}] \subset$ ri K. **(4i)** cl ri K =cl K. **(5i)** ri cl K =ri K.

Proof. Call $S_\varepsilon(x)$ the intersection between aff K and a neighbourhood of x with radius $\varepsilon > 0$. **(i)** It is obvious, if card $K = 1$. Let card $K > 1$, and $x^1, x^2 \in$ ri K with $x^1 \neq x^2$. There exists a sufficiently small radius $\varepsilon > 0$ s.t. $S_\varepsilon(x^1)$, $S_\varepsilon(x^2) \subset$ ri K. Therefore, we have:

$$[x^1, x^2] \subset \text{conv}\,\{S_\varepsilon(x^1), S_\varepsilon(x^2)\} \subset \text{ri}\,K,$$

which (account taken that conv $\{S_\varepsilon(x^1),\ S_\varepsilon(x^2)\}$ is an open set of aff K) shows the convexity of ri K. Either int $K = \varnothing$ or the above reasoning can be repeated with the entire space in place of aff K. Ab adsurdo, suppose that cl K be not convex, so that (card $K > 1$ and) $\exists x^1, x^2 \in$ cl K s.t. $[x^1, x^2] \not\subseteq$ cl K. Then, $\exists x \in]x^1, x^2[$ s.t. $x \notin$ cl K. If, $\forall \varepsilon > 0$, $S_\varepsilon(x) \cap K \neq \varnothing$, then x is an accumulation point of K, so that $x \in$ cl K, which contradicts $x \notin$ cl K. Otherwise, $\exists \bar{\varepsilon} > 0$ s.t. $S_{\bar{\varepsilon}}(x) \cap K = \varnothing$, and then $S_{\bar{\varepsilon}}(x) \cap$ cl $K = \varnothing$. Let $S_\varepsilon(x^1)$ and $S_\varepsilon(x^2)$ be as above with $x^1 \neq x^2$. Whatever the radius ε may be, $\exists y^1 \in S_\varepsilon(x^1)$ and $\exists y^2 \in S_\varepsilon(x^2)$, so that $[y^1, y^2] \subseteq K$. By chosing ε small enough, we achieve an absurdo: $[y^1, y^2] \cap S_{\bar{\varepsilon}}(x) \neq \varnothing$. **(ii)**. $\forall S_1, S_2 \subseteq \mathbb{R}^n$, cl S_1+cl $S_2 \subseteq$ cl $(S_1 + S_2)$. Then, because of (4i),

$$\text{cl}\,(\text{ri}\,K_1 + \text{ri}\,K_2) \supseteq \text{cl ri}\,K_1 + \text{cl ri}\,K_2 = \text{cl}\,K_1 + \text{cl}\,K_2 \supseteq K_1 + K_2 \supseteq \text{ri}\,K_1 + \text{ri}\,K_2,$$

so that

$$\text{aff cl}\,(\text{ri}\,K_1 + \text{ri}\,K_2) \supseteq \text{aff}\,(K_1 + K_2) \supseteq \text{aff}\,(\text{ri}\,K_1 + \text{ri}\,K_2).$$

From the other side, we evidently have:

$$\text{aff cl}\,(\text{ri}\,K_1 + \text{ri}\,K_2) = \text{aff}\,(\text{ri}\,K_1 + \text{ri}\,K_2) = \text{aff}\,(K_1 + K_2);$$

$$\text{ri cl}\,(\text{ri}\,K_1 + \text{ri}\,K_2) \supseteq \text{ri}\,(K_1 + K_2); \quad \text{ri cl}\,(\text{ri}\,K_1 + \text{ri}\,K_2) = \text{ri}\,(\text{ri}\,K_1 + \text{ri}\,K_2).$$

Therefore,

$$\text{ri}\,(\text{ri}\,K_1 + \text{ri}\,K_2) \supseteq \text{ri}\,(K_1 + K_2),$$

so that

$$\text{ri}\,K_1 + \text{ri}\,K_2 \supseteq \text{ri}\,(K_1 + K_2).$$

The reverse inclusion is achieved this way. $k \in$ ri K_1 + ri K_2 implies $k = k^1 + k^2$ with $k^1 \in$ ri K_1, $k^2 \in$ ri K_2. For $i = 1, 2$, as above, call S_i an open set (of aff K_i) containing

k^i and contained into ri K_i. We find $k \in S_1 + S_2 \subset \text{ri}\,(K_1 + K_2)$. The last inclusion being consequence of the obvious equality $(k^1 + \varepsilon^1) + (k^2 + \varepsilon^2) = (k^1 + k^2) + \varepsilon$, where $\varepsilon := \varepsilon^1 + \varepsilon^2$, $\varepsilon^i \in \text{aff}\,K_i$, $||\varepsilon^i||$ is small enough, $i = 1, 2$. **(iii)** We have to show that:

$$x := (1 - \alpha)x^0 + \alpha\hat{x} \in \text{ri}\,K, \quad \forall \alpha \in]0, 1[,$$

or that $\exists \tau > 0$ s.t. $N_\tau(x) \subset K$; call y a generic element of $S_\tau(x)$ (the dependences on α are understood). $x^0 \in \text{cl}\,K$ implies that, $\forall \varepsilon > 0$, $S_\varepsilon(x^0)$ contains elements of K; call y^0 one of them. $\hat{x} \in \text{ri}\,K$ implies that $\exists \rho > 0$ s.t. $S_\rho(\hat{x}) \subset K$; call $\hat{y}$ any element of $S_\rho(\hat{x})$. By disposing of ε and τ, we will show that, $\forall y \in S_\tau(x), \exists y^0 \in S_\varepsilon(x^0)$, s.t.

$$y = (1 - \alpha)y^0 + \alpha\hat{y}.$$

This being done, since $y^0, \hat{y} \in K$, the thesis will follow. ρ being given, we choose ε and τ in such way that:

$$\frac{1 - \alpha}{\alpha}\varepsilon + \frac{1}{\alpha}\tau < \rho.$$

Consequently, $\exists y^0 \in S_\varepsilon(x^0)$ s.t. $||\,y^0 - x^0\,|| < \varepsilon$, and then, $\forall y \in S_\tau(x)$, we have:

$$\tfrac{1}{\alpha}||y - x|| + \tfrac{1-\alpha}{\alpha}||y^0 - x^0|| < \rho.$$

From this inequality, we draw the inequality:

$$||\hat{y} - \hat{x}|| = ||\tfrac{1}{\alpha}y - \tfrac{1-\alpha}{\alpha}y^0 - \hat{x}|| = \tfrac{1}{\alpha}||y - (1 - \alpha)y^0 - \alpha\hat{x}||$$

$$= \tfrac{1}{\alpha}||y - (1 - \alpha)y^0 - [x - (1 - \alpha)x^0]||$$

$$\leq \tfrac{1}{\alpha}||y - x|| + \tfrac{1-\alpha}{\alpha}||x^0 - y^0|| < \rho,$$

which shows that $\hat{y} \in S_\rho(\hat{x})$ and hence $\hat{y} \in K$. **(4i)**. Since ri $K \subseteq K$, cl ri $K \subseteq$ cl K and the inclusion $\subseteq$ holds. The reverse inclusion holds, since (i) implies that, $\forall x \in$ cl K, $\exists \hat{x} \in K \setminus \{x\}$ s.t. $[\hat{x}, x[\subseteq$ ri K, so that $x \in \text{cl}([\hat{x}, x[) \subseteq$ cl ri K. **(5i)** Since cl $K \supseteq K$, so that ri cl $K \supseteq$ ri K, then the inclusion $\supseteq$ holds. With regard to the reverse inclusion, note that, $\forall x \in$ ri cl K, $\exists \varepsilon > 0$ s.t. $S_\varepsilon(x) \subseteq$ cl K. Because of (4i), we have $x \in$ cl ri K. By definition of closure, $S_\varepsilon(x) \cap$ ri $K \neq \varnothing$. Let $y^1 := x + \varepsilon z \in S_\varepsilon(x) \cap$ ri K, where $z \in$ aff K, $||z|| \leq 1$. If $z = O$, then $y^1 = x$ and the thesis is achieved. Otherwise, set $y^2 := x - \varepsilon z \in S_\varepsilon(x) \subseteq$ cl K. Since $x = \tfrac{1}{2}y^1 + \tfrac{1}{2}y^2$, because of (iii), we have $x \in [y^1, y^2[\subseteq$ ri K. $\qquad\qquad\square$

Several special classes of convex sets are of great interest. Among them, polyhedra are instrumental for both theory and applications.

Definition 2.1.6. A set $K \subseteq \mathbb{R}^n$ is called *polyhedral* or simply a *polyhedron*, iff it is the intersection of a finite number (possibly zero; in this case $K = \mathbb{R}^n$) of closed halfspaces. An extreme point of K is called also *vertex* or *corner point*; the set of vertices of K is denoted by vert K.

It is easy to prove that a polyhedron is a *polytope*, iff it is bounded. A simplex is obviously a special case of polytope. More generally, a polyhedron is the convex hull of a finite number of points and halflines. From an algebraic point of view, a polyhedron is — with the coordinates of its points — the set of solutions of a linear algebraic system of equalities or inequalities. The obvious convexity of a halfspace and Proposition 2.1.1 give, of course, the convexity of a polyhedron. An n-dimensional cube (of $\mathbb{R}^n$), namely $\{x \in \mathbb{R}^n : a \leq x \leq b\}$, with $a, b \in \mathbb{R}^n$ and $a \leq b$, is an instance of a polytope. Iff $O \in \mathrm{aff}\, K$, K is a subspace; aff K is a translation of a subspace.

A seemingly obvious statement — which indeed is all but trivial —, due to Weyl [53], is: *if $x^1, ..., x^r \in \mathbb{Z}^n$, then there exists a matrix A and a vector b, with entries in $\mathbb{Q}$, such that :*

$$\mathrm{conv}\,\{x^1, ..., x^r\} = \{x \in \mathbb{R}^n : Ax \geq b\}.$$

A special type of convex set is a convex cone, which plays an important role in the entire field of constrained extrema. Indeed, also the nonconvex cones are of great utility. Therefore, the general concept of cone is considered first.

Definition 2.1.7. A set $K \subseteq \mathbb{R}^n$ is a *cone with apex at* $\overline{x} \in \mathrm{cl}\, K$ iff

$$x \in K, \quad \alpha \in \,]0, +\infty[\quad \Rightarrow \quad \overline{x} + \alpha(x - \overline{x}) \in K. \tag{2.1.7.a}$$

For $\overline{x} = O$, (2.1.7a) becomes:

$$x \in K, \quad \alpha \in \,]0, +\infty[\quad \Rightarrow \quad \alpha x \in K, \tag{2.1.7.b}$$

and K is called a cone with apex at the origin or simply a *cone*. Iff furthermore K is convex, it is called a *convex cone*. Iff

$$(\mathrm{cl}\, K) \cap (-\mathrm{cl}\, K) = \{O\}, \tag{2.1.8}$$

a cone (with apex at O) is called *pointed* (or *acute*, if it is also convex); when the apex is at $\overline{x}$, then (2.1.8) must be verified by $K - \overline{x}$. A cone K is *properly pointed*, iff K and conv K are pointed. K is said to be *solid* (or a body; see Definition 2.1.1), iff int $K \neq \varnothing$. A cone K is associated with another cone, called *positive polar* of K (which, in Sect.2.2, will be obtained as a special case of a more general concept; see Definition 2.2.3) and given by

$$K^* := \{y \in \mathbb{R}^n : \langle y, x \rangle \geq 0\,, \ \forall x \subset K\}.$$

The entire space $\mathbb{R}^n$, a halfspace, a hyperplane, a line, a dihedron, their relative interiors are cones which possess infinite apices: every element of the space, of the boundary of the halfspace, of the hyperplane, of the line can play the role of the apex. Obviously, up to a translation, (2.1.7a) and (2.1.7b) identify the same thing; therefore, when suitable, the analysis will be carried on for a cone (with apex at the origin).

The above definition does not require any change for a set $K \subseteq B$. An instance is $K = C^\circ([a, b])$.

Examples 2.1.1. The sets K and $K \backslash \{O\}$, where $K = \mathbb{R}^2_+ \cup \{(x_1, x_2) \in \mathbb{R}^2 : x_1 < 0, x_2 > 0\}$, are convex cones which are not pointed. See Fig. 2.1.6.(a),(b). $\qquad \square$

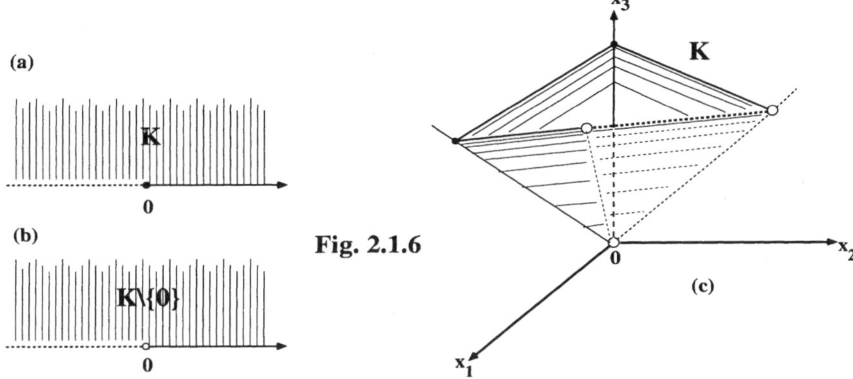

Fig. 2.1.6

Example 2.1.2. The set:

$$K = \mathbb{R}^3_+ \cap \{x \in \mathbb{R}^3 : -x_1 - x_2 + x_3 \geq 0\} \setminus \{x \in \mathbb{R}^3 : -x_1 - x_2 + x_3 = 0, \ x_2 \geq x_1\}$$

is a convex cone which is properly pointed. See Fig. 2.1.6(c). $\square$

Examples 2.1.3. The set:

$$K = \{(x_1, x_2, x_3) \in \mathbb{R}^3 : x_1 \geq 0, x_3 \geq 0\}$$

is convex but not pointed. The set

$$K = \{\alpha(1,1), \alpha(-1,1), \alpha(0,-1), \alpha \geq 0\} \subset \mathbb{R}^2,$$

(which can be expressed as cone $\{(1,1), (-1,1), (0,-1)\}$; see (2.1.13b)) is pointed, but not properly; $(0,0)$ is the only apex and is extreme point. This example shows that *a (not convex) pointed cone is not contained necessarily in a halfspace.* $\square$

If K is a cone, then

$$x \in \mathrm{cl}\, K, \ \alpha \in]0, +\infty[\ \Rightarrow \ \alpha x \in \mathrm{cl}\, K. \tag{2.1.9}$$

In fact, $\forall \varepsilon > 0, \exists x_\varepsilon \in K$ s.t. $||x - x_\varepsilon|| < \varepsilon$; this inequality together with $\alpha x_\varepsilon \in K$ (for each fixed $\alpha > 0$) imply $||\alpha x - \alpha x_\varepsilon|| = \alpha ||x - x_\varepsilon|| < \alpha \varepsilon$, which shows that $\alpha x \in \mathrm{cl}\, K$. It follows also that *the apex of a cone belongs to its boundary.* From (2.1.8) it is immediate to see that a convex cone is pointed, iff it does not contain any subspace other than the origin.

Further examples of cones are an orthant, its interior, in particular $\overset{\circ}{\mathbb{R}}^n_+$, $\mathbb{R}^n_+ \setminus \{O\}$, an halfline, an octant. All these cones possess only one apex: $\bar{x} = O$. Indeed, they are pointed. It is easy to show that *a pointed cone has one and only one apex.* In fact, ab adsurdo, suppose that the pointed cone K, besides $\bar{x} = O$ have also the apex $\hat{x} \neq 0$. Because of the definition of apex (not necessarily the origin), taking into account that, if (but not only if) (2.1.7) hold, also (1.2.7) with K replaced by $\mathrm{cl}\, K$ hold, then from

(2.1.7a) with $\operatorname{cl} K$ in place of K for $\alpha = 1$ and $x = \hat{x}$ we draw $\hat{x} - \overline{x} \in (\operatorname{cl} K) - \overline{x}$, while for $\overline{x}$ replaced by $x, \alpha = 2$ and $x = \overline{x}$ we draw $2\overline{x} - \hat{x} \in \operatorname{cl} K$ or $\hat{x} - \overline{x} \in -[(\operatorname{cl} K) - \overline{x}]$. Thus, (2.1.8) is contradicted.

An apex of a cone is not necessarily an extreme point, even if it is unique, as simple examples show. For instance, let K be union of the axes of $\mathbb{R}^2$ so that K is not pointed; O is the only apex, but not extreme point.

Further examples of cones are:

$$\{x \in \mathbb{R}^n : Ax \geq O\}, \tag{2.1.10}$$

where A is an m×n matrix with real entries, and

$$\{\overline{x} + \alpha(x - \overline{x}) : \alpha \in \,]0, +\infty[\}, \quad \overline{x}, x \in C^0([a, b]) \tag{2.1.11}$$

with $a, b \in \mathbb{R}$, $a < b$.

When a cone enjoys the convexity, we have the very important case of *convex cone*, which is characterized by the following:

Proposition 2.1.6. $K \subseteq \mathbb{R}^n$ is a convex cone, if and only if

$$x^1, x^2 \in K, \ (\alpha, \beta) \in \mathbb{R}_+^2 \backslash \{O\} \ \Rightarrow \ \alpha x^1 + \beta x^2 \in K. \tag{2.1.12}$$

Proof. **If.** By setting $\alpha = 1 - \beta$ with $\beta \in [0, 1]$, (2.1.12) shows the convexity of K; by setting $x^2 = x^1$, (2.1.12) becomes (2.1.7b). **Only if.** Set $\delta := \alpha + \beta$ and $\gamma := \beta/\delta$. $\forall x^1, x^2 \in K$, the convexity of K implies $\widetilde{x} := (1 - \gamma)x^1 + \gamma x^2 \in K$; K being a cone, the last relation implies $\delta \widetilde{x} \in K$. (2.1.2) follows. $\qquad\square$

Proposition 2.1.6 characterizes the *convex cones as the sets which are closed under vector addition and non-negative scalar multiplication.*

Obviously, *a convex cone K is contained in a closed halfspace, say H^+; if, furthermore, K is pointed, then $K \backslash \{O\} \subset \operatorname{int} H^+$.* In fact, if ab absurdo we deny this inclusion, so that $\exists \hat{x} \in (K \cap \operatorname{frt} K) \backslash \{O\}$, then $\hat{x} \in (\operatorname{cl} K) \cap (-\operatorname{cl} K)$ and (2.1.8) is contradicted. This is the reason why a convex and pointed cone is called acute (Definition 2.1.7), in agreement with the classic Geometry.

A convex cone K remains convex (besides, of course, a cone) if it loses the (unique; since $\hat{x} \in \operatorname{vert} K \Rightarrow \hat{x}$ is apex, but not necessarily vice versa) extreme point (if any) or $\hat{x} \in \operatorname{vert} K$ and K convex cone imply $K \backslash \{\hat{x}\}$ convex cone. In fact, let x be s.t. $\hat{x} \neq x \in K \backslash \{O\}$. If, ab adsurdo, $\exists \alpha > 0$ s.t. $\alpha x = \hat{x}$, then $x^1 := \frac{\alpha}{2}x \in K$ and $x^2 := 2\alpha x \in K$ would imply $\frac{2}{3}x^1 + \frac{1}{3}x^2 = \alpha x = \hat{x}$ and contradict the extremality of $\hat{x}$.

If the cone K is convex and pointed, then its (unique) apex is extreme point. In fact, ab adsurdo, suppose that $\exists x^1, x^2 \in (\operatorname{cl} K) \backslash \{O\}$ and $\exists \alpha \in \,]0, 1[$ s.t. $0 = (1 - \alpha)x^1 + \alpha x^2$. It follows that:

$$x^1 \in \operatorname{cl} K, \ \frac{1 - \alpha}{\alpha} > 0 \ \Rightarrow \ \frac{1 - \alpha}{\alpha}x^1 \in \operatorname{cl} K \ \Rightarrow \ -x^2 = \frac{1 - \alpha}{\alpha}x^1 \in \operatorname{cl} K,$$

and hence $x^2 \in (\operatorname{cl} K) \cap (-\operatorname{cl} K)$ which contradicts (2.1.8).

A convex cone can have a unique apex without being contained in an open halfspace, as shown by the following example:

$$K = \{(x_1, x_2) \in \mathbb{R}^2 : x_1, x_2 \geq 0 \ \ \text{or} \ \ x_1 < 0, x_2 > 0\}.$$

If a cone K is convex and pointed, then $K \backslash \{O\}$ is contained in an open halfspace. In fact, given $\hat{x}$ s.t. $O \neq \hat{x} \in K \cap \text{frt } K$, because of (2.1.8), $-\hat{x}$ cannot enjoy the same property. Because of this property, a pointed and convex cone turns out to be also properly pointed.

A cone of $\mathbb{R}^n$ is called *polyhedral cone* iff it is the intersection of a finite number of halfspaces through the origin. Because of Proposition 2.1.1, a *polyhedral cone is convex.* The set of solutions to a system of linear homogeneous algebraic inequalities of equalities (each of which can be replaced equivalently by 2 inequalities) is a polyhedral cone; (2.1.10) is an instance. Of course, a polyhedral cone may be identified also by a nonhomogeneous algebraic system; for instance, $\{x \in \mathbb{R} : x + 1 \geq 0, \ x \geq 0\}$.

Sometimes, it is useful to consider a cone as generated by a set. To this end, consider the following:

Definition 2.1.8. Let $\bar{x} \in \mathbb{R}^n$ and $X \subset \mathbb{R}^n$ with $X \neq \{\bar{x}\}$. The *cone generated by X from $\bar{x}$* is the set:

$$\text{cone } (\bar{x}; X) := \{x \in \mathbb{R}^n : \ x - \bar{x} + \alpha(y - \bar{x}), \ y \in X, \ \alpha \in]0, +\infty[\ \}, \qquad (2.1.13\text{a})$$

which, for $\bar{x} = O$, becomes:

$$\text{cone } X := \{x \in \mathbb{R}^n : \ x = \alpha y, \ y \in X, \ \alpha \in]0, +\infty[\ \}. \qquad (2.1.13\text{b})$$

Of course, (2.1.13) are cones; their convex hulls are called *cone spanned by X from $\bar{x}$* and *cone spanned by X*, respectively.

By setting $X = \{x \in K : \|x - \bar{x}\| = 1\}$, (2.1.7) turn out to be equivalent to (2.1.13) (x of (2.1.7) corresponds to y of (2.1.13)). This shows how to consider any cone as generated by a set, and how to consider it as the union of halflines having a common origin at $\bar{x}$ and another element in X; any of these halflines is called *generatrix* and X the *generating set*. Cone (2.1.11) by itself is a halfline of the Banach space $C^0([a, b])$ generated by $X = \{x\}$. If X is a disk, whose centre is the orthogonal projection of $\bar{x}$ on it, then K is called *right circular cone*; it is called merely *circular cone* when the centre of X does not enjoy the above property. If X is a polyhedron, then K is called *pyramid* (in its turn, a special case of polyhedron); a pyramid achieves a more familiar aspect, if X is a polytope.

The above definition induces, in a natural way, that of *quasirelative interior* of a set X, denoted by qri X. More precisely,

$$\text{qri } X := \{x \in \mathbb{R}^n : \text{cl cone } (X - x) \text{ is a linear subspace}\}.$$

Of course,

$$\text{ri } X \neq \varnothing \ \Rightarrow \ \text{qri } X = \text{ri } X.$$

However, it may happen that $\operatorname{ri} X = \varnothing$, while $\operatorname{qri} X \neq \varnothing$, so that the qri X may be a good "surrogate" of $\operatorname{ri} X$. For instance, let us consider the set

$$X = \{x \in \mathbb{R}^3 : x_1^2 + x_2^2 \leq 1,\ x_3 = 0,\ \text{and}\ x_i \in \mathbb{R}\backslash\mathbb{Q}\ \text{for either}\ i = 1\ \text{or}\ i = 2\}.$$

Obviously $\operatorname{int} X = \varnothing$; we have also $\operatorname{ri} X = \varnothing$, since, whatever $\overline{x} \in X$ may be, every its neighbourhood intersects $\sim X$. It is easy to see that $\operatorname{qri} X \neq \varnothing$ and

$$\operatorname{qri}\ X = \{x \in \mathbb{R}^3 : x_1^2 + x_2^2 < 1,\ x_3 = 0\} := X^0.$$

In fact, $\forall \overline{x} \in X^0$ and $\forall x \in X^0\backslash\{\overline{x}\}$, it is easy to show that

$$]\overline{x}, x] \cap (\mathbb{R}^3\backslash\mathbb{Q}^3) \neq \varnothing,$$

so that every halfline of the plane $x_3 = 0$ with apex at $\overline{x}$ intersects an element of X different from $\overline{x}$; this means that cone $(X - \overline{x})$ is a linear subspace. To prove the above relation, set $y := (x_1, x_2)$ and consider the possible cases. If $\overline{y}, y \in \mathbb{Q}^2$, then, by choosing any $\alpha \in \mathbb{R}\backslash\mathbb{Q}$, we have $\overline{y} + \alpha(y - \overline{y}) \in \mathbb{R}^2\backslash\mathbb{Q}^2$, as it is easy to show ab absurdo. If at least one component of y is irrational, then the thesis is trivial. Let $\overline{y} \notin \mathbb{Q}^2$ and $y \in \mathbb{Q}^2$; we can suppose that $\overline{y}_1 \in \mathbb{R}\backslash\mathbb{Q}$ and $y_1 \in \mathbb{Q}$. Then the thesis is shown by $\frac{1}{2}(\overline{y}_1 + y_1)$ as it is easy to prove ab absurdo.

Corollary 2.1.1. Let $K \subset \mathbb{R}^n$ be a cone with apex at the origin. Every ray of $\operatorname{conv} K$ can be expressed as convex combination of at most n rays of K.

Proof. Without any loss of generality, we can assume $n > 1$ and $K \neq \{O\}$. Let ρ be any ray of $\operatorname{conv} K$, and $\hat{x} \in \rho\backslash\{O\}$. Because of Theorem 2.1.1, $\exists x^1, ..., x^r \in K$ with $r \leq n + 1$, and $\exists \alpha_i \in [0, 1]$ with $\sum_{i=1}^{r} \alpha_i = 1$, s.t.

$$\hat{x} = \sum_{i=1}^{r} \alpha_i x^i.$$

Call ρ_i the ray generated by x^i (if $x^i = O$, then the thesis is immediate). Consider any $a \in \operatorname{int}(\operatorname{cone}\{x^1, ..., x^r\})^*$, where $*$ denotes positive polar which is considered with respect to $\operatorname{aff}\{x^1, ..., x^r\}$ as well as the interior. Since $x^1, ..., x^r$ cannot belong to a same hyperplane through the origin, we have $\langle a, x^i \rangle > 0, i = 1, ..., r$. Let H° be the hyperplane whose equation is $\langle a, x - \hat{x} \rangle = 0$, and set $\beta_i := \langle a, \hat{x} \rangle / \langle a, x^i \rangle, i = 1, ..., r$; we have $y^i := \beta_i x^i \in H^\circ$, $i = 1, ..., r$. Let $Y := \operatorname{aff}\{y^1, ..., y^r\}$. Since $\dim Y \leq n - 1$, by applying Theorem 2.1.1 to such an affine set, we draw that at most $r \leq n$ points $x^1, ..., x^r$ are enough to express $\hat{x}$, and therefore at most $r \leq n$ rays $\rho_1, ..., \rho_r$ are enough to express ρ as their proper convex combination. $\qquad\square$

Note that, in the above proof, cone $\{x^1, ..., x^r\}$ turns out to be properly pointed. Just as Theorem 2.1.1 has been specialized to cones, Lemma 2.1.1 could be specialized to cones giving an algorithm for expressing a ray as a convex combination of at most n rays of the given cone. The extension of Corollary 2.1.1 to a cone with apex at $\overline{x}$

is trivial. If we look at the cone K in the form (2.1.13b), then the thesis of Corollary 2.1.1 can be easily refined by obtaining dim X as an upper bound of the number of necessary rays.

An important class of convex cones (with apex at the origin) is defined by the following condition:

$$K + \mathrm{cl}\, K = K. \tag{2.1.14}$$

The above condition is satisfied if K is convex and either closed (in which case (2.1.14) is trivial) or open, as it is easy to show. *A convex cone, which is not closed and fulfils* (2.1.14), *cannot contain the origin*; in fact, assuming the contrary case, for $a = O \in K$, and for b s.t. $O \neq b \in (\mathrm{cl}\, K)\backslash K$ we would have $a + b \notin K$, which contradicts (2.1.14). The class (2.1.14), which will be used in Theorem 2.2.7 and exploited in Sect. 3.2 (Proposition 3.2.9) and in Sect.4.8, does not contain that of convex and pointed cones, as Example 2.1.2 shows. The class of convex and pointed cones does not contain the class (2.1.14), as the 1st of Examples 2.1.3 shows. However, the two above classes have nonempty intersection. More precisely, *if a convex cone K fulfils* (2.1.14) *and $K = (\mathrm{cl}\, K)\backslash\{O\}$, then K is pointed.* In fact, if ab absurdo we deny (2.1.8), so that $\exists k \in (\mathrm{cl}\, K) \cap (-\mathrm{cl}\, K)\backslash\{O\}$, then we have $k, -k \in K$ and, because of the convexity of K, $O = k + (-k) \in K$, which contradicts the assumption. Of course, the above intersection contains convex cones $K \neq (\mathrm{cl}\, K)\backslash\{O\}$; for instance, those that are closed and pointed.

One of the most important roles played by a cone occurs when it is used to replace locally a set. Unfortunately, in general, there is not a unique way to do it. As a consequence, several cones have been introduced for approximating a given set; the choice depends on the purpose which is pursuited. Here, only the main types are briefly analysed.

Definition 2.1.9. Let the nonempty set $X \subseteq \mathbb{R}^n$ and $\overline{x} \in \mathrm{cl}\, K$ be given. The set of $\overline{x} + x \in \mathbb{R}^n$ for which $\exists \{x^i\}_1^\infty \subseteq \mathrm{cl}\, X$ with $\lim\limits_{i \to +\infty} x^i = \overline{x}$, and $\exists \{\alpha_i\}_1^\infty \subset \mathbb{R}_+\backslash\{0\}$, such that:

$$\lim_{i \to +\infty} \alpha_i (x^i - \overline{x}) = x, \tag{2.1.15}$$

is called *tangent cone* to X at $\overline{x}$ and denoted by $TC(\overline{x}; X)$. We stipulate that $TC(\overline{x}; \varnothing) = \varnothing$. If $\overline{x} = O$, then the notation $TC(X)$ is used.

It is immediate to see that $TC(\overline{x}; X)$ is a cone with apex at $\overline{x}$, that $\overline{x} \in \mathrm{int}\, X \Rightarrow TC(\overline{x}; X) = \mathbb{R}^n$, and that $TC(\overline{x}; \mathbb{R}^n) = \mathbb{R}^n$, $TC(\overline{x}; \mathbb{Q}^n) = \mathbb{R}^n$, $TC(\overline{x}; \mathbb{Z}^n) = \{\overline{x}\}$ (of course, with $\overline{x} \in \mathbb{Z}^n$, since $\mathrm{cl}\, \mathbb{Z}^n = \mathbb{Z}^n$). Additional examples are offered by Figs. 2.1.7-2.1.14. Note that in Fig. 2.1.10 and Fig. 2.1.11 the tangent cone is the same, notwithstanding the fact that X be a continuous set in the former and a denumerable one in the latter. In Fig. 2.1.12, X be the complement of a circle (open or closed, no matter); in Fig. 2.1.13, X is an infinite sequence of boundaries of squares; in Fig. 2.1.14, $TC(\overline{x}; X) = \mathbb{R}^2$. It is easy to see that $X \neq \varnothing \Rightarrow TC(\overline{x}; X) \neq \varnothing$, and that $TC(\overline{x}; X) = TC(\overline{x}; \mathrm{cl}\, X) = TC(\overline{x}; \mathrm{ri}\, \mathrm{cl}\, X)$.

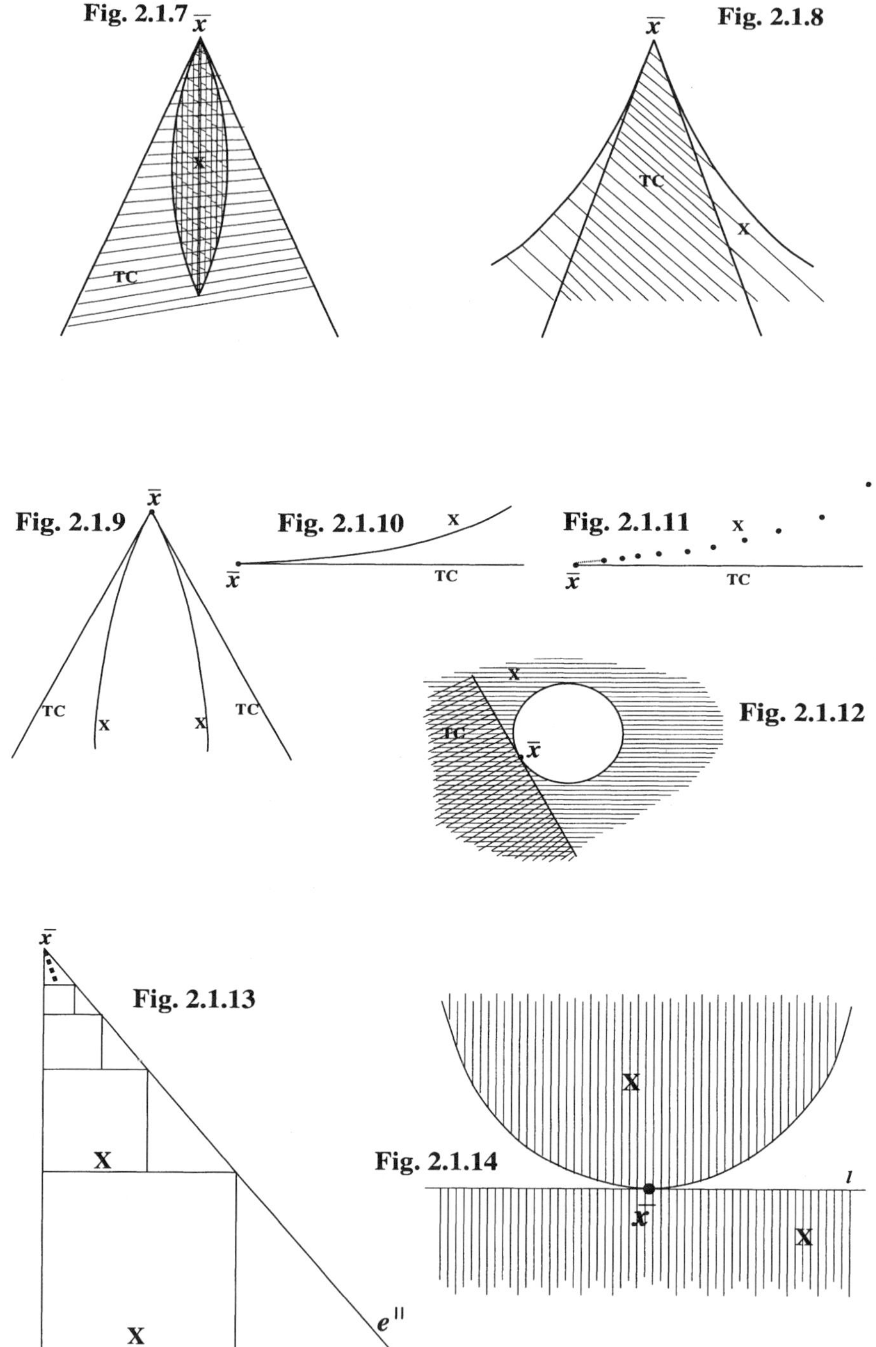

Fig. 2.1.7
Fig. 2.1.8
Fig. 2.1.9
Fig. 2.1.10
Fig. 2.1.11
Fig. 2.1.12
Fig. 2.1.13
Fig. 2.1.14
TC
X
x̄
l
e'
e''

It is immediate to see that, unless $\overline{x}$ be an isolated point of cl K, or $x = O$, (2.1.15) implies $\lim\limits_{i \to +\infty} \alpha_i = +\infty$.

Therefore, setting $\beta_i := 1/\alpha_i$, (2.1.15) can be equivalently written in the following way:

$$\lim_{\beta_i \downarrow 0} \tfrac{1}{\beta_i}(x^i - \overline{x}) = x, \tag{2.1.15}'$$

which shows x as a sort of generalized derivative, $(1/\beta_i)(x^i - \overline{x})$ being a sort of quotient ratio. Such an interpretation becomes closer to the classic concept of derivative, if we deal with a cone which enjoys more properties; this happens to the cone of the following Definition 2.1.10, even if (2.1.18) does not show explicitly a form like (2.1.15)$'$.

From the above examples, we see that the tangent cone may be a "bad representative" of the set. For instance, in Fig. 2.1.13, the tangent cone is the convex angle with edges e' and e''; and does not allow us to guess how X may be, even if we are in a neighbourhood of $\overline{x}$; indeed, the tangent cone does not change, if the interiors of the squares join the boundaries. However, for some classes of sets, like the convex ones, the tangent cone is an excellent local approximation; in general, it enjoys the nice property shown by the next theorem.

Theorem 2.1.4. Let $X \subset \mathbb{R}^n$ and $\overline{x} \in \text{cl}\, X$. **(i)** $TC(\overline{x}; X)$ is closed. **(ii)** If X is convex, then $TC(\overline{x}; X)$ contains cl X and is convex. **(iii)** $TC(\overline{x}; X)$ is isotone (see Definition 2.3.2), or $X_1 \subseteq X_2 \;\Rightarrow\; T(\overline{x}; X_1) \subseteq T(\overline{x}; X_2)$. **(iiii)** Consider the 2 opposite halfspaces $H^{\pm} := \{(x,y) \in \mathbb{R}^n \times \mathbb{R} : \pm y \geq 0\}$, and let $f : \mathbb{R}^n \to \mathbb{R}$ be continuous at O and such that $f(O) = 0$. If $TC(\text{gr}\, f) \cap \text{int}\, H^{+} = \varnothing$, then

$$\limsup_{x \to 0} \frac{f(x)}{||x||} \leq 0;$$

if $TC(\text{gr f}) \cap \text{int}\, H^{-} = \varnothing$, then

$$\liminf_{x \to 0} \frac{f(x)}{||x||} \geq 0.$$

Proof. Without any loss of generality, we can assume that $\overline{x} = O$. **(i)** $y \in \text{cl}\, TC(X)$ implies that, $\forall \varepsilon > 0$, $\exists x(\varepsilon) \in TC(X)$ s.t.

$$||x(\varepsilon) - y|| < \frac{\varepsilon}{2}. \tag{2.1.16}$$

Because of the Definition 2.1.9, $x(\varepsilon) \in TC(X)$ means that $\exists \{x^i(\varepsilon)\}_1^{\infty} \subseteq \text{cl}\, X$, with $\lim\limits_{i \to +\infty} x^i(\varepsilon) = O$ and $\exists \{\alpha_i(\varepsilon) > 0\}_1^{\infty}$, s.t. (the notation of those sequences is slightly improper: the dependence on ε should be referred, first of all, to the entire sequence):

$$\lim_{i \to +\infty} \alpha_i(\varepsilon)x^i(\varepsilon) = x(\varepsilon),$$

so that $\exists i(\varepsilon) \in \mathbb{N}$ s.t., $\forall i > i(\varepsilon)$, we have:

$$||\alpha_i(\varepsilon)\, x^i(\varepsilon) - x(\varepsilon)|| < \frac{\varepsilon}{2}. \tag{2.1.17}$$

From (2.1.16) and (2.1.17), $\forall i > i(\varepsilon)$, we draw:

$$||\alpha_i(\varepsilon)x^i(\varepsilon) - y|| = ||\alpha_i(\varepsilon)x^i(\varepsilon) - x(\varepsilon) + x(\varepsilon) - y|| \le$$

$$\le ||\alpha_i(\varepsilon)x^i(\varepsilon) - x(\varepsilon)|| + ||x(\varepsilon) - y|| < \frac{\varepsilon}{2} + \frac{\varepsilon}{2} = \varepsilon. \qquad (2.1.17)'$$

Let $n \in \mathbb{N}$. From the sequence $\{x^i(\frac{1}{n})\}_{i=1}^{\infty}$, let us extract the element which corresponds to $i = i_n := i(\frac{1}{n}) + 1$, namely $y^n := x^{i_n}(\frac{1}{n})$; analogously, from $\{\alpha_i(\frac{1}{n})\}_{i=1}^{\infty}$ let us extract $\beta_n := \alpha_{i_n}(\frac{1}{n})$. The pair of sequences $\{y^n\}_1^{\infty}$ and $\{\beta_n\}_1^{\infty}$ fulfils Definition 2.1.9. In fact, from $(2.2.17)'$, $\forall \varepsilon \in \mathbb{R}_+ \setminus \{0\}$ (as above), we have:

$$|| \beta_n y^n - y || \le || \beta_n y^n - x(\tfrac{1}{n}) || + || x(\tfrac{1}{n}) - y || < \varepsilon.$$

Hence $y \in TC(X)$. **(ii)** Let $x \in \mathrm{cl}X$. The convexity of X and Proposition 2.1.5 imply that, $\forall i \in \mathbb{N}$, $\frac{1}{i}x \in \mathrm{cl}\,X$. Then, $\forall \varepsilon > 0$, $\exists x^i(\varepsilon) \in X$ s.t. $||x^i(\varepsilon) - \frac{1}{i}x|| < \frac{\varepsilon}{i}$ or $||i \cdot x^i(\varepsilon) - x|| < \varepsilon$, which implies that $x \in TC(X)$, because of (2.1.15) for $\overline{x} = O$, $x^i = x^i(\varepsilon)$ and $\alpha_i = i$. **(iii)** Immediate consequence of (2.1.15). **(iiii)** With regard to the 1st part, ab absurdo, suppose that

$$\limsup_{x \to O} \frac{f(x)}{||x||} > 0.$$

Then, $\exists \alpha > 0$ and $\exists \{x^i\}_1^{\infty} \subset \mathbb{R}^n \setminus \{O\}$ with $\lim_{i \to +\infty} x^i = O$, s.t. $f(x^i) \ge \alpha ||x^i||$, $\forall i$. Let us set $\alpha_i := 1/||x^i||$ and $z^i := (1/||x^i||)x^i$; obviously $\alpha_i > 0$ and $\lim_{i \to +\infty} \alpha_i = +\infty$. We stipulate to replace, here and in the sequel, $\{x^i\}_1^{\infty}$ with a suitable subsequence, if necessary. Thus, we can suppose that $\exists \hat{z} := \lim_{i \to +\infty} z^i$ with $z \ne O$. The continuity of f implies that $\lim_{i \to +\infty} f(x^i) = 0$. If the sequence $\{\alpha_i f(x^i)\}_1^{\infty}$ contains a bounded subsequence, then it contains also a convergent subsequence, which we call $\{\alpha_{i_r} f(x^{i_r})\}_{r=1}^{\infty}$; of course, we have:

$$y := \lim_{r \to +\infty} \alpha_{i_r} f(x^{i_r}) \ge \alpha.$$

Then, we find :

$$\lim_{r \to +\infty} \alpha_{i_r}(x^{i_r}, f(x^{i_r})) = (\hat{z}, y) \in TC(\mathrm{gr}\, f) \cap H^+,$$

which contradicts the assumption. If $\{\alpha_i f(x^i)\}_1^{\infty}$ does not contain any bounded sequence, then $\lim_{i \to +\infty} \alpha_i f(x^i) = +\infty$. Set $\beta_i := 1/[\alpha_i f(x^i) || x^i ||]$. Without any loss of generality, we can assume that $\exists \tilde{z} := \lim_{i \to +\infty} \frac{1}{||x^i||}x^i$; of course, $||\tilde{z}|| < +\infty$. Then, we see that

$$\lim_{i \to +\infty} \beta_i x^i = \lim_{i \to +\infty} \frac{1}{\alpha_i f(x^i)} \cdot \frac{1}{||x^i||} \cdot x^i = 0 \cdot \tilde{z} = O.$$

Hence, we find:

$$\lim_{i \to +\infty} \beta_i(x^i, f(x^i)) = (0, 1) \in TC(\mathrm{gr}\, f) \cap \mathrm{int}\, H^+,$$

and again the assumption is contradicted. With regard to the 2nd point, it is enough to apply the 1st part to $-f$. $\qquad\qquad\qquad\qquad\square$

If in Definition 2.1.9 we require that, $\forall \{\alpha_i\}_1^\infty \subset \mathbb{R}_+\backslash\{0\}$ with $\lim\limits_{i\to+\infty} \alpha_i = +\infty$, $\exists\{x^i\}_1^\infty \subseteq \operatorname{cl} X$, such that (2.1.15) holds, then we have a strengthening of the tangent cone. For instance, in the example of Fig. 2.1.13, only the edge e' is admitted.

The fact that Definition 2.1.9 requires the existence of only a sequence of elements of cl X is responsible of the possible bad approximation, in the nonconvex case, of the given set. An improvement is obtained by requiring the existence of a "curve" instead of a sequence:

Definition 2.1.10. Let the nonempty set $X \subseteq \mathbb{R}^n$ and $\overline{x} \in \operatorname{cl} X$ be given. The set of $\overline{x} + x \in \mathbb{R}^n$ for which there exists $\overline{\alpha} \in \mathbb{R}_+\backslash\{0\}$ and a differentiable function $\tau : \mathbb{R}_+ \to \mathbb{R}^n$, such that

$$\tau(0) = \overline{x}, \quad \tau'(0) = x, \quad \tau(\alpha) \in X, \quad \forall \alpha \in\,]0,\overline{\alpha}], \tag{2.1.18}$$

is called *reachable cone* to X at $\overline{x}$ and denoted by $RC(\overline{x}; X)$. We stipulate that $RC(\overline{x}; \varnothing) = \varnothing$. If $\overline{x} = O$, then the notation $RC(X)$ is used. x is called *curvilinear tangent*.

It is immediate to see that $RC(\overline{x}; X)$ is a cone with apex at $\overline{x}$, that is convex and cl $RC(\overline{x}; X) = TC(\overline{x}; X)$ if X is convex, and that, in general, $RC(\overline{x}; X) \subseteq TC(\overline{x}; X)$. In the examples of Figs. 2.1.7-2.1.10, 2.1.12, 2.1.14 we have $RC(\overline{x}; X) = TC(\overline{x}; X)$; in that of Fig. 2.1.11, $RC(\overline{x}; X) = \varnothing$; in that of Fig. 2.1.13, $RC(\overline{x}; X) = e'$. Furthermore, $RC(\overline{x}; \mathbb{Q}^n) = RC(\overline{x}; \mathbb{Z}^n) = \varnothing$; $RC(\overline{x}; X) = \mathbb{R}^n$ if $\overline{x} \in \operatorname{int} X$.

It may happen, to both the tangent and reachable cones, that a ray intersects X at most in $\overline{x}$. This drawback is overcome by requiring, in Definition 2.1.10, that τ be affine. This leads to the following:

Definition 2.1.11. Let the nonempty set $X \subseteq \mathbb{R}^n$ and $\overline{x} \in \operatorname{cl} X$ be given. The set of $\{\overline{x} + x \in \mathbb{R}^n$ for which there exists $\overline{\alpha} \in \mathbb{R}_+\backslash\{0\}$, such that $\tau(\alpha) := \overline{x} + \alpha x \in X$, $\forall \alpha \in\,]0,\overline{\alpha}]$, is called *admissible cone* to X at $\overline{x}$ and denoted by $AC(\overline{x}; X)$. We stipulate that $AC(\overline{x}; \varnothing) = \varnothing$. If $\overline{x} = O$, the notation $AC(X)$ is used.

Obviously, $AC(\overline{x}; X)$ is a cone with apex at $\overline{x}$; it is convex and cl $AC(\overline{x}; X) = TC(\overline{x}; X)$ if X is convex; in the general case, $AC(\overline{x}; X) \subseteq RC(\overline{x}; X)$. In the examples of Figs. 2.1.8, 2.1.12 and 2.1.14, $AC(\overline{x}; X) = RC(\overline{x}; X)$; in that of Fig. 2.1.7, $AC(\overline{x}; X) = \operatorname{int} RC(\overline{x}; X)$; in those of Figs. 2.1.9-2.1.11, $AC(\overline{x}, X) = \varnothing$; in that of Fig. 2.1.13, $AC(\overline{x}; X) = e'$. Furthermore, $AC(\overline{x}; \mathbb{Q}^n) = AC(\overline{x}; \mathbb{Z}^n) = \varnothing$; $AC(\overline{x}; X) = \mathbb{R}^n$ if $\overline{x} \in \operatorname{int} X$.

The following definition is a further strengthening of the concept of the tangent cone, and extends to cones the notion of interior of a set.

Definition 2.1.12. Let the nonempty set $X \subseteq \mathbb{R}^n$ and $\overline{x} \in \operatorname{cl} X$ be given. The set of $\overline{x} + x \in \mathbb{R}^n$ for which there exist $\overline{\alpha} \in \mathbb{R}_+\backslash\{0\}$ and a neighbourhood $N_\varepsilon(x)$ of x with radius $\varepsilon > 0$, such that:

$$\overline{x} + \alpha y \in X, \qquad \forall \alpha \in\,]0,\overline{\alpha}[, \qquad \forall y \in N_\varepsilon(x), \tag{2.1.19}$$

is called the *interior cone* to the set X at $\overline{x}$ and denoted by $IC(\overline{x}; X)$. We stipulate that $IC(\overline{x}; \varnothing) = \varnothing$. If $\overline{x} = O$, the notation $IC(X)$ is used.

Of course, $IC(\overline{x}; X)$ is a cone with apex at $\overline{x}$; it is convex and cl $IC(\overline{x}; X) = AC(\overline{x}; X)$ if X is convex; in the general case, we have $IC(\overline{x}; X) \subseteq \text{int } AC(\overline{x}; X)$ where the equality may not occur as shown by the example of Fig. 2.1.14, where $IC(\overline{x}; X)$ is $\mathbb{R}^2 \backslash \{l\}$. In the examples of Figs. 2.1.7, 2.1.8 and 2.1.12, $IC(\overline{x}; X) = \text{int } AC(\overline{x}; X)$; in those of Figs. 2.1.9-2.1.11, 2.1.13, $IC(\overline{x}; X) = \varnothing$. Furthermore, $IC(\overline{x}; \mathbb{Q}^n) = IC(\overline{x}; \mathbb{Z}^n) = \varnothing$; $IC(\overline{x}; X) = \mathbb{R}^n$ if $\overline{x} \in \text{int } X$.

The previous remarks, even if few, show that, in general, when we go from the tangent cone up to the interior one, we gain reliability but lose properties. For instance, it is easy to prove that $TC(\overline{x}; X) \cup TC(\overline{x}; \sim X) = \mathbb{R}^n$, while this does not hold necessarily for the others: $RC(\overline{x}; \mathbb{Q}^n) = RC(\overline{x}; \sim \mathbb{Q}^n) = \varnothing$. In the convex case, the loss of properties is not very important. In fact, if X is convex, it is easy to show that $TC(\overline{x}; X) = \text{cl } RC(\overline{x}; X)$.

A comparison among the cones introduced with the Definitions 2.1.8-2.1.12 is useful. For instance, even if the cone (2.1.13) cannot be ordered with (2.1.15), as shown by the following two examples, it is easy to prove that $TC(\overline{x}; X) \subseteq \text{cl cone}(\overline{x}; X)$. In fact, setting $\overline{x} = 0$, $x \in TC(X)$ implies the existence of the sequences with the properties listed in Definition 2.1.9. $x^i \in X \Rightarrow \alpha_i x^i \in X$; (2.1.15) or $\lim\limits_{i \to +\infty} \alpha_i x^i = x$ means that $x \in \text{cl cone } X$. In the above inclusion, the equality may not hold, as Example 2.1.5 shows.

Example 2.1.4. Set $X = \{(x_1, x_2) \in \mathbb{R}^2 : x_2 = x_1^2\}$, $\overline{x} = (0, 0)$, We find $TC(X) = \{(x_1, x_2) \in \mathbb{R}^2 : x_2 = 0\}$ and cone $X = \{(x_1, x_2) \in \mathbb{R}^2 : x_2 > 0\} \cup \{O\}$. Therefore, $TC(X) \nsubseteq \text{cone } X$, but $TC(X) = \text{cl cone } X$. $\qquad\square$

Example 2.1.5. Set $X = \{(0, 0), (0, 1)\} \subset \mathbb{R}^2$, $\overline{x} = (0, 0)$. We find: $TC(X) = \{(0, 0)\}$ and cone $X = \{(x_1, x_2) \in \mathbb{R}^2 : x_1 = 0, x_2 = 0\}$. Therefore, $TC(X) \nsupseteq \text{cone } X$ and $TC(X) \neq \text{cl cone } X = \text{cone } X$. $\qquad\square$

Definition 2.1.13. Let the nonempty set $X \subset \mathbb{R}^n$ and $\overline{x} \in \text{cl } X$ be given. The set of $\overline{x} + x$, such that

$$\langle y - \overline{x}, x \rangle \leq 0, \qquad \forall y \in X, \tag{2.1.20}$$

is called *normal cone* to X at $\overline{x}$ and denoted by $NC(\overline{x}; X)$ (see also (2.2.12a)). We stipulate that $NC(\overline{x}; \varnothing) = \varnothing$. If $\overline{x} = O$, the notation $NC(X)$ is used. Each element of the normal cone is called *normal* to X at $\overline{x}$.

Definition 2.1.14. Let the nonempty set $X \subseteq \mathbb{R}^n$ and $\overline{x} \in \text{cl } X$ be given. The set of $\overline{x} + x \in \mathbb{R}^n$ for which there exist $\overline{\alpha} \in \mathbb{R}_+ \backslash \{0\}$ and a neighbourhood $N(\overline{x})$, both depending on x, such that, $\forall x' \in X \cap N(\overline{x})$ and $\forall \alpha \in]0, \overline{\alpha}[$, we have $x' + \alpha x \in X$, is called *hypertangent cone* to X at $\overline{x}$ and denoted by $HC(\overline{x}; X)$. We stipulate that $HC(\overline{x}; \varnothing) = \varnothing$. If $\overline{x} = O$, the notation $HC(X)$ is used.

Let us now consider a property of cones which is useful for establishing equivalence among different types of problems [V2].

Theorem 2.1.5. Let $K, \tilde{K} \subset \mathbb{R}^n$ be cones with apices at O, such that $\tilde{K}$ be closed, and

$$\varnothing \neq (\tilde{K} \backslash \{O\}) \subseteq \operatorname{int} K$$

Let $S(\rho) := \{x \in \mathbb{R}^n : ||x|| \leq \rho\}$ with $\rho \geq 0$, and $S_1 := \{x \in \mathbb{R}^n : ||x|| = 1\}$. Then, there exists $\eta_0 \in \mathbb{R}$, such that:

$$V_1 + \eta V_2 \in \operatorname{int} K, \quad \forall \eta > \eta_0, \quad \forall V_1 \in S(\rho), \quad \forall V_2 \in \tilde{K} \cap S_1. \tag{2.1.21}$$

Proof. Since $||V_2|| = 1$ $\forall V_2 \in \tilde{K} \cap S_1$, and since the scalar product of vectors of unitary norm is ≤ 1, then $\forall \eta > \rho$ we have:

$$1 \geq \frac{\langle V_1 + \eta V_2, V_2 \rangle}{||V_1 + \eta V_2|| \cdot ||V_2||} \geq \frac{\langle V_1, V_2 \rangle + \eta ||V_2||^2}{||V_1|| \cdot ||V_2|| + \eta ||V_2||^2} \geq \frac{-\rho + \eta}{\rho + \eta}, \quad \forall V_2 \in \tilde{K} \cap S_1, \tag{2.1.22}$$

where the 3rd inequality is a consequence of the inequalities:

$$||V_1|| \leq \rho \quad \text{and} \quad \langle V_1, V_2 \rangle \geq -||V_1|| \geq -\rho, \quad \forall V_2 \in \tilde{K} \cap S_1.$$

Since the scalar product of vectors of unitary norm is 1 iff they coincide, passing in (2.1.22) to the limit as $\eta \to +\infty$, we deduce that:

$$\lim_{\eta \to +\infty} \frac{1}{||V_1 + \eta V_2||} (V_1 + \eta V_2) = V_2, \quad \forall V_2 \in \tilde{K} \cap S_1.$$

Since $\tilde{K} \cap S_1$ is a compact set included in $\operatorname{int} K$ (so that $\exists \varepsilon > 0$ s.t. $\tilde{K} \cap S_1 + N_\varepsilon(O) \subset K$), then $(\sim K) \cap S_1$ and $\tilde{K} \cap S_1$ have distance (induced by the norm considered) greater than $\varepsilon > 0$. Hence, because of the last inequality, (2.1.21) follows. $\square$

Note that, if $n = 1$, the assumptions of Theorem 2.1.5 are fulfilled only by $K = \mathbb{R}_+$ and by $K = \mathbb{R}_+ \backslash \{0\}$ (and, of course, by their opposites); in both cases $\tilde{K} = \mathbb{R}_+$ (or $\tilde{K} = \mathbb{R}_-$) necessarily.

The many properties enjoyed by convex sets have led to define a huge number of generalizations of the notion of convexity. Several of them look, now-a-days, more like a mathematical formalism than strongly motivated mathematics. Some of them, whose excellent treatment can be found in [I 45], have shown to be fundamental concepts. Recently, a new generalization, called *geodesic convexity*, has turned out to be highly fruitful. A crucial problem for every class of functions consists, obviously, in having at our disposal numerically applicable conditions for stating whether or not a given function belong to the class. Unlike what happens often in the generalized convexity, for geodesic convexity it has been possible to prove properties analogous to those of differentiable convex functions. An excellent treatment of this concept and its applications to extremum problems is contained in [I56].

2.2. Linear Support and Separation

The concepts, which are briefly recalled in the present section, are fundamental for the theory of constrained extrema and related fields. Even if most of the propositions of this section hold in an infinite dimensional space, they are proved in $\mathbb{R}^n$. We aim to show that their finite dimensional versions may be useful also for infinite dimensional problems. Let $a \in \mathbb{R}^n \backslash \{O\}$ and $b \in \mathbb{R}$; in the sequel we will consider the hyperplane

$$H^0 := \{x \in \mathbb{R}^n : \langle a, x \rangle = b\},$$

and the related halfspaces

$$H^- := \{x \in \mathbb{R}^n : \langle a, x \rangle \leq b\}, \quad H^+ := \{x \in \mathbb{R}^n : \langle a, x \rangle \geq b\},$$

where the dependence on a and b will be taken for granted; the notation $H^0(a)$, $H^0(b)$, $H^-(a,b)$, $H^-(a)$, $H^-(b)$, $H^+(a)$, $H^+(b)$ and $H^+(a,b)$ will be used only when there will be any fear of confusion.

Definition 2.1.1. A hyperplane $H^0 \subset \mathbb{R}^n$ is called *supporting hyperplane* or merely *support* of $K \subset \mathbb{R}^n$, iff

$$K \subseteq H^+, (\text{or } K \subseteq H^-) \quad \text{and} \quad H^0 \cap \operatorname{cl} K \neq \varnothing. \tag{2.2.1}$$

H^+ (or H^-) is called *supporting halfspace of K*. A support is called either *proper* or *strict* according to

$$K \nsubseteq H^0 \quad \text{or} \quad \operatorname{card}(H^0 \cap \operatorname{cl} K) = 1, \tag{2.2.2}$$

respectively. Every element of $H^0 \cap \operatorname{cl} K$ is called *supporting point*.

It is immediate to see that (2.2.1) is equivalent to:

$$\langle a, x \rangle \geq b, \quad \forall x \in K, \qquad \inf_{x \in K} \langle a, x \rangle = b ; \tag{2.2.1$'$}$$

that the support is proper iff, beside (2.2.1)$'$, we have:

$$\inf_{x \in K} \langle a, x \rangle \; < \; \sup_{x \in K} \langle a, x \rangle; \tag{2.2.2$'$}$$

and that the support is strict iff, beside (2.2.1)$'$, we have:

$$\operatorname{card} \arg \inf_{x \in K} \langle a, x \rangle = 1. \tag{2.2.2$''$}$$

In fact, because of (2.2.1), the equality in (2.2.1)$'$ is trivial, if K is a closed set; otherwise, $\forall \varepsilon > 0$, $\exists x_\varepsilon \in K$ s.t. $b \leq \langle a, x_\varepsilon \rangle < b + \varepsilon$, which shows the equality.

The next theorem, which is one of the fundamental cornerstones of Functional Analysis, is here considered as the basis for the development of the Lagrangian theory of constrained extrema. Even if it holds in complex linear spaces, we consider its finite dimensional version; indeed — as previously said — we will reduce ourselves to exploit its finite dimensional version even for infinite dimensional problems like (1.1.5) and also (1.1.4), besides, of course, (1.1.1). The proof, which here need not use either Zorn's

Lemma (which instead is useful and fundamental for some separation theorems; see just after Theorem 2.2.4) or equivalent axioms, is performed, since its reasoning may be instructive.

Theorem 2.2.1. Let $K, S \subset \mathbb{R}^n$; let K be relatively open and convex, and S be affine. If they are nonempty and disjoint, then there exists a hyperplane $H^0 \subset \mathbb{R}^n$ such that

$$S \subseteq H^0, \qquad K \cap H^0 = \varnothing. \tag{2.2.3}$$

Proof. The cases $n = 1$ or $\dim S = n - 1$ are trivial. Let $n > 1$ and $\dim S < n - 1$. Without any loss of generality, we can assume that $O \in S$, so that $S^\perp$ is a coordinate subspace and

$$\dim S^\perp = n - \dim S \geq 2.$$

Let S^2 be a subspace of $S^\perp$ with dimension 2, and consider the sets $K + S$ and $K^2 := (K + S) \cap S^2$ which do not contain the origin of $\mathbb{R}^n$ and are relatively open because of Proposition 2.1.5 (ii). Therefore, being $K = \operatorname{ri} K$, we have that:

$$S = \operatorname{ri} S \Rightarrow \operatorname{ri}(K + S) = \operatorname{ri} K + \operatorname{ri} S = K + S.$$

Now we show that there exists a line ρ of $\mathbb{R}^n$ through the origin which does not intersect K^2. This is trivial, if $\dim K^2 = 0$. If $\dim K^2 = 1$, or aff K^2 is a line, then ρ can be chosen as the line (of S^2) which crosses the origin 0 and is either parallel or orthogonal to aff K^2, according to respectively $O \notin$ aff K^2 or $O \in$ aff K^2. If $\dim K^2 = 2$, we consider cone $(K^2 \setminus \{O\})$, which is the smallest open angle of S^2 containing K^2; every line, containing any of the edges of such an angle, can be chosen as ρ. It follows that

$$\rho \cap (K + S) = \rho \cap (K + S) \cap S^2 = \rho \cap K^2,$$

so that the subspace $S + \rho$, whose dimension is $1 + \dim S$, is disjoint from K (otherwise, $\rho \cap (K + S) \neq \varnothing$). By repeating the above construction at most $n - 1 - \dim S$ times, $S + \rho$ turns out to be a hyperplane. $\qquad\square$

The previous theorem, which was established in a linear normed space by Hahn [IV20] and Banach [IV2] independently of each other, is known as linear extension theorem. The above proof justifies this terminology. We must say that, from a formal point of view, Theorem 2.2.1 does not deal either with support or with separation. The substance is different. Indeed, the hyperplane H^0, claimed in the statement, immediately implies the existence of a hyperplane, say $\hat{H}^0$, parallel to H^0, disjoint from K and having zero distance from K. $\hat{H}^0$ fulfils Definition 2.2.1. Hence, if we replace K with its relative interior, Theorem 2.2.1 gives the existence of supporting hyperplanes for convex sets. Indeed, it gives much more. Among the many consequences, it is the basis for characterizing a convex set by means of its supporting hyperplanes or halfspaces, as the following Theorem 2.2.2 shows. As a consequence of Theorem 2.2.1, we have:

Corollary 2.2.1. Let $K, S \subset \mathbb{R}^n$ be a nonempty, closed and convex set, and an affine set, respectively. If they are disjoint, then there exists a hyperplane $H^0 \subset \mathbb{R}^n$ such that:

$$S \subseteq H^0, \quad K \cap H^0 = \varnothing. \tag{2.2.3}'$$

Proof. Without any loss of generality, we can assume that S be linear and card $K > 1$. Consider the sets $M(\alpha) := (1-\alpha)K + \alpha S$, $\alpha \in [-\frac{1}{2}, \frac{1}{2}]$. First of all, we prove that:

$$S \cap M(\alpha) = \varnothing \quad , \quad \forall \alpha \in \left[-\tfrac{1}{2}, \tfrac{1}{2}\right] . \tag{2.2.3}''$$

Ab absurdo, suppose that $\exists s \in S, \exists k' \in K, \exists s' \in S$, s.t.

$$s = (1-\alpha)k' + \alpha s' ,$$

so that, $\forall \alpha \in [-\frac{1}{2}, \frac{1}{2}]$, due to the linearity of S, we draw:

$$k' = \frac{1}{1-\alpha}s - \frac{\alpha}{1-\alpha}s' \in S,$$

which contradicts the disjunction between K and S. $(2.2.3)''$ and hence

$$S \cap \left[\bigcup_{\alpha \in [-\frac{1}{2}, \frac{1}{2}]} M(\alpha) \right] = \varnothing$$

follow. Now we prove also that:

$$S \cap \operatorname{conv} \bigcup_{\alpha \in [-\frac{1}{2}, \frac{1}{2}]} M(\alpha) = \varnothing . \tag{2.2.3}'''$$

To this end, consider the homothety given by:

$$M(\alpha\,;s) := (1-\alpha)K + \alpha s \quad , \quad \alpha \in]-\infty, 1] , \; s \in S .$$

Because of (2.1.13a) and of the convexity of K, we have:

$$\bigcup_{\alpha \in]-\infty, 1]} M(\alpha\,;s) = \operatorname{cone}(s\,;K),$$

where (due to $K \cap S = \varnothing$) s is the unique apex (obtained for $\alpha = 1$). Moreover, due to the linearity of S, $\bigcup_{s \in S} \operatorname{cone}(s;K)$ is convex, disjoint from S, and contains the convex hull of $(2.2.3)'''$, which therefore follows. From the above consideration, we have that:

$$K \subset M := \bigcup_{\alpha \in]-\frac{1}{2}, \frac{1}{2}[} M(\alpha),$$

and that M is a relatively open set. Then, the Theorem 2.2.1 can be applied, with K replaced by M, to achieve the thesis. $\qquad \square$

Of course, the above proof shows also the statement of Corollary 2.2.1, where K *is still nonempty, convex, but not necessarily closed, and S disjoint from its closure,* *while* $(2.2.3)'$ *remains unchanged.*

At first glance, it may seem that the general situation for a nonconvex set be that where there are both boundary points at which the set has no supporting hyperplane and boundary points at which at least one supporting hyperplane exists. Indeed, this situation occurs frequently but not always, as next examples show.

Examples 2.2.1. It is easy to see that, at every point of its boundary, the strictly concave set

$$K = \{(x_1, x_2) \in \mathbb{R}^2 : x_1^2 + x_2^2 \geq 1\}$$

has not any support line. The same happens to the epigraph of the function $f : \mathbb{R} \to \mathbb{R}$, given by $f(x) = \sqrt{x} - x, \quad x \in \mathbb{R}$. $\square$

Theorem 2.2.2. A nonempty and closed set $K \subset \mathbb{R}^n$ is convex, if and only if it is the insersection of all the (closed) halfspaces which contain it.

Proof. If. Obvious consequence of Proposition 2.1.1, since a halfspace is convex. **Only if.** Let us adopt the notation of the beginning of this section. Because of Theorem 2.2.1, $\forall x \notin \mathrm{ri}\, K$, there exists a (closed) halfspace, say H^+, s.t. $x \in H^0$, $\mathrm{ri}\, K \subseteq \mathrm{int}\, H^+$, and therefore $K \subseteq H^+$. (If, besides $x \notin \mathrm{ri}\, K$, we have $x \notin K$, then, by replacing x with $\hat{x}$ s.t. $\mathrm{dist}\,(\hat{x}, K) = \frac{1}{2}\mathrm{dist}\,(x, K)$, we achieve again an inclusion of type $K \subseteq H^+$, while $x \in H^-$). $\square$

The above theorem (where $K = \mathbb{R}^n$ and $K = \varnothing$ have been omitted, since in these case the convexity is trivial) was established by Weyl [53], even if a trace of it can be found in Minkowski [36]. Also Theorem 2.2.2 is not in terms of supporting halfspaces, even if it can be easily formulated in this way, since those halfspaces, which do not support K, are useless.

Definition 2.2.1 allows us to extend to any set the ancient concept of face of a polygon and of a polyhedron.

Definition 2.2.2. Let $K \subset \mathbb{R}^n$. $F \subseteq \mathrm{cl}\, K$ is a *face* of K iff it is the intersection of $\mathrm{cl}\, K$ with a supporting hyperplane H^0 of K, or

$$F := H^0 \cap \mathrm{cl}\, K. \tag{2.2.4}$$

F is called *proper* iff $F \neq \varnothing$ and $F \neq \mathrm{cl}\, K$, and *improper* otherwise. The *dimension* of F is that of $\mathrm{aff}\, F$. F is called *minimal* iff it does not contain any other nonempty face different from F. A face F is said to be *exposed by a function f*, iff F is the set of minimum (or infimum) points of f on K. A given polytope (or, more generally, a polyhedron) K can be associated with another polytope (or polyhedron), called *dual* of K and denoted by $K^\triangle$, iff there exists a bijective map $\psi : \mathcal{F} \to \mathcal{F}^\triangle$ where $\mathcal{F}$ and $\mathcal{F}^\triangle$ are the sets of all faces, respectively, of K and $K^\triangle$, such that:

$$F' \subset F'' \iff \psi(F') \supset \psi(F''), \quad \forall F', F'' \in \mathcal{F}.$$

It follows $\psi(\varnothing) = K^\triangle$, $\psi(K) = \varnothing$, and $\dim F + \dim \psi(F) = n - 1$, $\forall F \in \mathcal{F}$; therefore the $(n - k)$- dimensional faces of polytope K are in one-to-one correspondence with the

$(k-1)$-dimensional faces of $K^{\triangle}$, $\forall k = 1, ..., n$. In fact, if F_r denotes a faces of K of dimension r, then $\psi(F_{r+1}) > \psi(F_r), r = n - 2, ..., 0$, implies $\psi(F_r) = n - 1 - r$.

In Definition 2.2.2, unlike the usual style, a face is a subset of cl K, but not necessarily of K. To consider, in such a definition, either cl K or K has, of course, advantages and disadvantages. Here the former seems to be more than the latter. Note that, according to the latter, a convex and pointed cone should possess its apex necessarily, or it would not be a face.

From Theorem 2.2.2, we draw that each point of the boundary of a convex set belongs to a face. The existence of sets which have no face (apart from the empty one) is proved by Examples 2.2.1. In agreement with Definition 2.1.6, when dim $F = n - 1$, then sometimes F is called *facet*. When dim $F = 0, 1$, then F is called *vertex* (or *corner point*), *edge*, respectively. By elementary arguments, it can be shown that *the boundary of an n-dimensional polyhedron of* $\mathbb{R}^n$ *is the union of its facets, whose number is finite* (see a remark before Definition 2.2.3) *and no two of which lie in the same hyperplane.*

A special case is that where K is defined by a system of linear algebraic equations or inequalities. For instance,

$$K = \{x \in \mathbb{R}^n : \ Ax \geq b\}, \tag{2.2.5}$$

where A is a matrix of dimension $m \times n$ and b a column m-vector, both with real entries. Because of Definition 2.1.6, K identifies a polyhedron. Consider any partition $(\mathfrak{J}', \mathfrak{J}\backslash\mathfrak{J}')$ of $\mathfrak{J}$ (of Sect. 1.1; (2.2.5) is a particular case of (1.1.1c,d)); it induces a partition of A and of B which, with obvious notation, can be written as

$$A = \begin{pmatrix} A_{\mathfrak{J}'} \\ A_{\mathfrak{J}\backslash\mathfrak{J}'} \end{pmatrix}, \quad b = \begin{pmatrix} b_{\mathfrak{J}'} \\ b_{\mathfrak{J}\backslash\mathfrak{J}'} \end{pmatrix},$$

where the (same) partitions of A and b are improper if $\mathfrak{J}' = \varnothing$ or $\mathfrak{J}\backslash\mathfrak{J}' = \varnothing$. In this case, (2.2.4) shrinks to

$$F_{\mathfrak{J}'} = \{x \in \mathbb{R}^n : \ A_{\mathfrak{J}'}x = b_{\mathfrak{J}'}, \ A_{\mathfrak{J}\backslash\mathfrak{J}'}x \geq b_{\mathfrak{J}\backslash\mathfrak{J}'}\}, \tag{2.2.6}$$

and, for $n = 2, 3$, the concepts of elementary Geometry are recovered (indeed, there a polygon or a polyhedron are not necessarily convex). (2.2.6) enjoys several properties, which are easily proved. For instance, assuming $K \neq \varnothing$, if $F_{\mathfrak{J}'}, F_{\mathfrak{J}''}$ and $F_{\mathfrak{J}'''}$ are faces of (2.2.5), then:

$$\mathfrak{J}' \subseteq \mathfrak{J}'' \ \Rightarrow \ F_{\mathfrak{J}'} \supseteq F_{\mathfrak{J}''}, \tag{2.2.7a}$$

$$\mathfrak{J}' \cup \mathfrak{J}'' = \mathfrak{J}''' \ \Rightarrow \ F_{\mathfrak{J}'} \cap F_{\mathfrak{J}''} = F_{\mathfrak{J}'''}, \tag{2.2.7b}$$

$$\mathfrak{J}' \cap \mathfrak{J}'' = \mathfrak{J}''' \ \Rightarrow \ F_{\mathfrak{J}'} \cup F_{\mathfrak{J}''} \subseteq F_{\mathfrak{J}'''}, \tag{2.2.7c}$$

$$\text{rank } A = n \ \Rightarrow \ \text{vert } K \neq \varnothing. \tag{2.2.7d}$$

(2.2.7a,b) are obvious. With regard to (2.2.7c), it is trivial if $\mathfrak{J}''' = \varnothing$, since $F_{\varnothing} = K$; otherwise, because of (2.2.7a) we have:

$$\mathfrak{J}''' \subseteq \mathfrak{J}' \ \Rightarrow \ F_{\mathfrak{J}'''} \supseteq F_{\mathfrak{J}'}; \quad \mathfrak{J}''' \subseteq \mathfrak{J}'' \ \Rightarrow \ F_{\mathfrak{J}'''} \supseteq F_{\mathfrak{J}''};$$

and (2.2.7c) follows. Easy examples show that the inclusion in (2.2.7c) may be strict: for instance, let K be a cube ($n = 3$, $m = 6$), $F_{\jmath'}$ and $F_{\jmath''}$ be adjacent edges; then $F_{\jmath'''}$ is the facet containing the 2 edges, which form $F_{\jmath'} \cup F_{\jmath''}$. Being nonempty, K turns out to be the intersection between a cone (identified by n inequalities corresponding to a non-null minor of order n of matrix A) and a polyhedron; this proves (2.2.7d).

Since every polyhedron can be represented in the form (2.2.5), from (2.2.6) we easily deduce that *the number of faces of a polyhedron is finite.* This is not true, in general, for nonpolyhedral sets. Exceptionally, it may be true; in the 2nd of Examples 2.2.1, the epigraph of the restriction of f to [-1,1] is supported by the x-axis and by the lines $x = \pm 1$, and has only 5 faces (apart from $\varnothing$), namely the points (-1,0), (0,0), (1,0) and the halflines ($x = -1, y \geq 0$), ($x = 1, y \geq 0$). A trivial example of a set having an infinite number of faces is offered by a (closed or open) circle (of $\mathbb{R}^2$): every point of its boundary (circumference) is a face; however, the set of faces remains unchanged, if the set is no longer the circle (convex), but its circumference (nonconvex). This fact, which, because of Theorem 2.2.2, happens to all convex sets, does not occur necessarily to nonconvex sets, as the 1st of Examples 2.2.1 shows.

The concepts of pole, polarity, polarization are very old; dual, duality, dualization are alternative terms. The following definition introduces one of these concepts.

Definition 2.2.3. Given $K \subseteq \mathbb{R}^n$, the set

$$K^* := \{y \in \mathbb{R}^n : \langle y, x \rangle \leq 1, \ \forall x \in K\} \tag{2.2.8}$$

is called *polar* of K. We stipulate that $(\varnothing)^* = \mathbb{R}^n$.

The definition of polar of a subset of a Hilbert space is given as above; for normed spaces and complex vector spaces suitable changes are necessary.

Being the intersection of closed halfspaces, because of Proposition 2.1.1, K^* is *convex* and *closed*; furthermore it contains the origin. A straightforward interpretation of $K^* \backslash \{O\}$ is as set of gradients y of halfspaces of type $\langle y, x \rangle \leq 1$ which support K.

The polar of a subspace S is its orthogonal complement, or $S^* = S^\perp$; in particular, $(\mathbb{R}^n)^* = O$ and $O^* = \mathbb{R}^n$. Of course, $K^* \neq \varnothing$, since $O \in K^*$. Figs. 2.2.1-2.2.6 show some examples of sets K and corresponding polars K^*. In figs. 2.2.1 and 2.2.2, K can

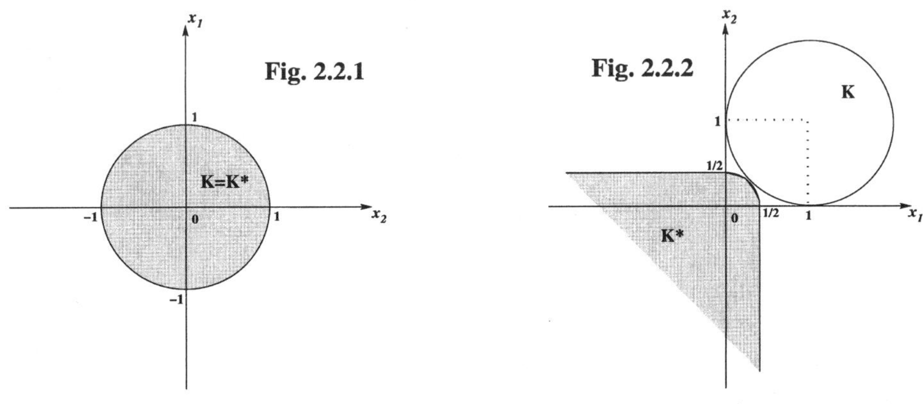

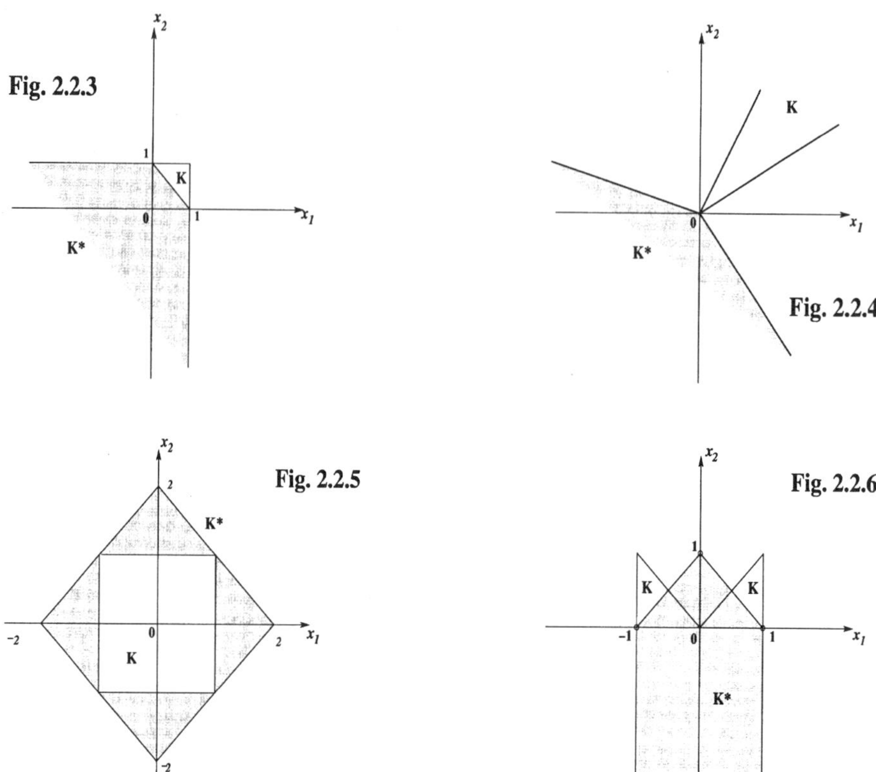

be considered, either as a circle or as a circumference of radius 1; K^* is the same for both cases. In Figs. 2.2.3 and 2.2.5, K can be considered again, either as a square of side 1/2 or as its boundary. In Fig. 2.2.4, K is an angle; in Fig. 2.2.6, K is the union of 2 right triangles; also in these cases, K or frt K lead to the same K^*. In fact, by using Definition 2.2.3, it is immediate to prove that:

$$K^* = (conv\ K)^* = (vert\ K)^* = (ri\ K)^*, \qquad (2.2.9)$$

since the inequality in (2.2.8) is satisfied by any convex combination of $x^1, x^2 \in K$, if it is satisfied by x^1 and x^2.

An useful interpretation of the polar is in terms of level set. Consider the function

$$\delta^*(x; K) := \sup_{y \in K} \langle x, y \rangle, \qquad (2.2.10)$$

which is called *support function* of K with respect to x. From (2.2.8) we have:

$$K^* = \text{lev}_{\leq 1}\delta^*(x; K). \qquad (2.2.11)$$

To think of polar in terms of level sets of a function suggests useful generalizations. For instance, instead of a linear function, in (2.2.8) we might consider any nonlinear function. In Chapter 4, these ideas will be carried out.

As we will see (Proposition 2.2.3), we might carry on the analysis in terms of supports, without introducing the concept of separation, leaving this as implicit form contained in Theorem 2.2.1. Being convinced that different languages, even if perfectly equivalent, may allow us to achieve more results than only one language, we prefer to postpone some properties (see Theorem 2.2.6) to the introduction of separation.

Fig. 2.2.4 shows the polar of a special set: a cone with apex at the origin. Because of its importance, let us consider Definition 2.2.3 in this case.

If K is a (not necessarily convex) *cone with apex at the origin, then its polar is a closed and convex cone, again with apex at the origin, given by*

$$K^* = \{y \in \mathbb{R}^n : \langle y, x \rangle \leq 0, \ \ \forall x \in K\}. \tag{2.2.12a}$$

In fact, because of (2.1.7b), the inequality in (2.2.8) becomes:

$$\langle y, \alpha x \rangle \leq 1, \ \ \forall \alpha > 0, \ \ \forall x \in K,$$

and, dividing both sides by α and letting $\alpha \to +\infty$, implies (2.2.12a).

The form (2.2.12a) is called *negative polar* of K (and, up to a translation, is the normal cone; see Definition 2.1.13). Indeed, the interpretation of the polar in terms of level sets lets us say that we would obtain an equivalent theory if in (2.2.8) the level 1 were replaced by any positive constant or the inequality ≤ 1 were replaced by ≥ -1. In the latter case, instead of (2.2.12a), we find:

$$K^* = \{y \in \mathbb{R}^n : \langle y, x \rangle \geq 0, \ \ \forall x \in K\}. \tag{2.2.12b}$$

This is called *positive polar of K*. Of course, these two polars are one the opposite of the other. $K^* \neq \{O\}$ iff K is contained into a halfspace. In the sequel, unless explicitly said, polar will mean positive polar.

When K is a cone of $\mathbb{C}^n$ with apex at the origin, its *complex positive polar* is defined by:

$$K^* := \{\eta \in \mathbb{C}^n : \mathrm{Re}\,\langle \eta, \zeta \rangle \geq 0 \ , \ \forall \zeta \in K\}, \tag{2.2.12c}$$

where $\langle \bullet, \bullet \rangle$ denotes scalar product in $\mathbb{C}^n$.

When K is a polyhedral cone, then, because of (2.2.9), in (2.2.12a,b) only a finite number of inequalities can be considered. This was exploited by J.Farkas in 1902 (see Corollary 4.5.7).

The previous concept of polarity is of fundamental importance for developing the theory of scalar extrema. This poses the natural question of whether or not the concept of polar can be extended to cover also vector problems, like (1.1.8) and (1.1.10). This is indeed possible.

For the vector case, the use of the concept of polar is, now-a-days, limited to cones. Therefore, in the following definition we do not consider any set K, but merely a cone; to simplify its use (which will be made in Vol. 2), taking into account that the notation for the vector case is necessarily heavy, the cone will be denoted with the same symbol D, which will be adopted in Vol. 2 .

Definition 2.2.4. Let ℓ, K be positive integers, and $C \subseteq \mathbb{R}^\ell$ and $D \subseteq \mathbb{R}^k$ be cones with apices at their origins and with $O_\ell \in C$. The *vector polar of D with respect to C* is given by

$$D_C^* := \{M \in \mathbb{R}^{\ell \times k} : \quad Md \geq_C 0, \quad \forall d \in D\}, \tag{2.2.13}$$

where $\mathbb{R}^{\ell \times k}$ denotes the set of matrices M of dimension $\ell \times k$ and with real entries, and where the inequality means $Md \in C$.

C being a cone, $M \in D_C^* \Rightarrow \alpha M \in D_C^*, \ \forall \alpha \in R_+\backslash\{0\}$; this allows us the use of the term *cone* for the vector polar; indeed, each row M_j of D_C^* describes a cone in the sense of Definition 2.1.7. For $\ell = 1$, D_C^* becomes either (2.2.12a) or (2.2.12b), according to $C = \mathbb{R}_-$ or $C = \mathbb{R}_+$, respectively. The case $C = \mathbb{R}_+^\ell$ is of interest for problems (1.1.8) and (1.1.10); in this case (2.2.13) becomes:

$$D_{\mathbb{R}_+^\ell}^* = \left\{ M = \begin{pmatrix} M_1 \\ \vdots \\ M_\ell \end{pmatrix} \in \mathbb{R}^{\ell \times k} : \ M_j \in D^*, \ \forall j \in J \right\}, \tag{2.2.14}$$

where $J := \{1, ..., \ell\}$ and M_j is the jth row of matrix M; (2.2.14) is the *positive vector polar* of D; the *negative vector polar* of D is obviously defined for $C = \mathbb{R}_-^\ell$.

If D is a subspace, then (2.2.13) collapses to the *vector orthogonal complement of D with respect to* the cone C:

$$D_C^\perp := \left\{ M = \begin{pmatrix} M_1 \\ \vdots \\ M_2 \end{pmatrix} \in \mathbb{R}^{\ell \times k} : \ M_j \in D^\perp, \quad \forall j \in J \right\}.$$

For $\ell = 1$ and $C = \mathbb{R}_+$, $\ D_C^\perp = D^\perp$.

Let C be defined by the following system of linear algebraic inequalities (so that C is convex):

$$\langle c_i, u \rangle \geq 0, \quad i \in I, \quad (u \in \mathbb{R}^\ell), \tag{2.2.15}$$

where $c_i = (c_{ij}, j \in J)$ and I is a finite or infinite set of indices. Then (2.2.13) becomes:

$$D_C^* = \{M \in \mathbb{R}^{\ell \times k} : \ c_i M \in D^*, \ \forall i \in I\}. \tag{2.2.16}$$

If card $I = \ell$ and $\begin{pmatrix} c_1 \\ \vdots \\ c_\ell \end{pmatrix} = I_\ell$, then (2.2.16) is equal to (2.2.14). Let card $I < +\infty$, and $d_1^*, ..., d_\ell^* \in D^*$. Then, an element M of D_C^* is found by solving the system:

$$\sum_{j \in J} c_{ij} M_j = d_i^*, \quad i \in I. \tag{2.2.17}$$

Example 2.2.2. Set $\ell = k = 2$, $I = \{1, 2\}$, and

$$C = \{(u_1, u_2) \in \mathbb{R}^2 : -u_1 + 2u_2 \geq 0, \ u_1 - u_2 \geq 0\},$$

$$D = \{(d_1, d_2) \in \mathbb{R}^2 : -d_1 + d_2 \geq 0, \ 2d_1 - d_2 \geq 0\}.$$

From (2.1.12b) we find:

$$D^* = \{(d_1^*, d_2^*) \in \mathbb{R}^2 : \tfrac{1}{2}d_1^* + d_2^* \geq 0, \ d_1^* + d_2^* \geq 0\}.$$

(2.2.17) becomes:

$$\begin{cases} -M_1 + 2M_2 = d_1^*, \\ M_1 - M_2 = d_2^*, \end{cases} \quad \text{or} \quad \begin{cases} M_1 = d_1^* + 2d_2^*, \\ M_2 = d_1^* + d_2^*. \end{cases}$$

Therefore, $\forall (d_1^*, d_2^*) \in D^* \times D^*$, the last system gives a pairs of vectors (M_1, M_2) and hence an element $M = \begin{pmatrix} M_1 \\ M_2 \end{pmatrix}$ of D_C^*. $\square$

If C is convex, then D_C^* is convex. In fact, $\forall M', M'' \in D_C^*$, $\forall (\alpha, \beta) \in \mathbb{R}_+^2 \setminus \{O\}$, from (2.2.13), $\forall d \in D$, we have $(\alpha M')d \in C$, $(\beta M'')d \in C$, and, beacuse of the convexity of C,

$$\alpha(M'd) + \beta(M''d) \in C,$$

or $(\alpha M' + \beta M'')d \in C$, which means $(\alpha M' + \beta M'') \in D_C^*$. Then, because of Proposition 2.1.6, we achieve the convexity of D_C^*.

If C is closed, then D_C^* is closed. In fact, let $M^0 = \lim\limits_{i \to +\infty} M^i$ with $\{M^i\}_1^\infty \subset D_C^*$. $M^i d \in C$ implies $\lim\limits_{i \to +\infty} M^i d = M^0 d$ which, because of the closure of C, gives $M^0 d \in C$ and hence $M^0 \in D_C^*$.

Since (2.2.13) implies

$$D \subseteq \operatorname{lev}_C Md, \tag{2.2.18a}$$

then D_C^* can be interpreted as set of Jacobians M of linear vector functions (namely, Md) whose vector level sets support D or M maps D onto C. This interpretation shrinks to that of (2.2.8) in case (2.2.12b) or K is a cone. For $\ell = 1$, M is a vector, Md a scalar product, and (2.2.18a) collapses to:

$$D \subseteq \operatorname{lev}_{\geq 0} \langle M, d \rangle, \tag{2.2.18b}$$

which is the supporting property met for (2.2.12b). In other words, D_C^* identifies (more precisely, is isomorphic to) the family of linear operators $M : \mathbb{R}^k \to \mathbb{R}^\ell$, such that $Md \in C$ $\forall d \in D$; when $\ell = 1$ and $C = \mathbb{R}_+$ or $C = \mathbb{R}_-$, then D^* identifies the family of linear functional $\langle M, d \rangle : \mathbb{R}^k \to \mathbb{R}$ (denoted by $\langle y, x \rangle$ in (2.2.8) and (2.2.12)), such that $\langle M, d \rangle \geq 0$, $\forall d \in D$. Further remarks about (2.2.13) are:

$$M \in D_C^*, \ u^* \in C^* \ \Rightarrow \ u^* M \in D^*; \tag{2.2.19a}$$

$$u \in C, \ d^* \in D^* \ \Rightarrow \ u d^* \in D_C^*, \tag{2.2.19b}$$

where ud^* is the $\ell \times k$ matrix obtained by multiplying the column u by the row d^*; if $\operatorname{int} C \neq \varnothing$, then

$$u^* \in C^* \setminus \{O\}, \ d^* \in D^* \ \Rightarrow \ \exists M \in D_C^* \ \text{s.t.} \ u^* M = d^*. \tag{2.2.19c}$$

In fact, from (2.2.13) and (2.2.12b), $\forall d \in D$, we draw that $Md \in C$ and $u^* \in C^*$ imply $0 \leq \langle Md, u^* \rangle = \langle u^*M, d \rangle$ and (2.2.19a) follows. $\forall d \in D$, since $\langle d^*, d \rangle \geq 0$, $u \in C$, and $O_\ell \in C$, we have:

$$(ud^*)d = \langle d^*, d \rangle u \in C,$$

so that $M = ud^* \in D_C^*$, and (2.2.19b) follows. Let $u \in \mathrm{int}\, C^*$. Then $\exists \hat{\varepsilon} \in \mathbb{R}^\ell$ such that, $\forall \|\varepsilon\| < \|\hat{\varepsilon}\|$, $\langle u^*, u + \varepsilon \rangle \geq 0$. By choosing $\varepsilon = -\alpha u^*$ with α positive and small enough so that $\|-\alpha u^*\| < \|\hat{\varepsilon}\|$, we have $\langle u^*, u \rangle \geq \alpha \langle u^*, u^* \rangle > 0$. Then $\hat{u} := (1/\langle u^*, u \rangle)u$ is s.t. $\langle u^*, \hat{u} \rangle = 1$. Set $M = \hat{u}d^*$ (where $\hat{u}$ and d^* are considered, respectively, as a row and a column). Since $d \in D \Rightarrow \langle d^*, d \rangle \geq 0$, $\forall d \in D$ we have $Md = \langle d^*, d \rangle \hat{u} \in C$, so that $M \in D_C^*$. Since $u^*M = \langle u^*, \hat{u} \rangle d^* = d^*$, (2.2.19c) follows.

With the notation of the beginning of this section, let us now consider the concept of separation.

Definition 2.2.5. The nonempty sets $K_1, K_2 \subset \mathbb{R}^n$ are *separable*, iff there exists a hyperplane $H^0 \subset \mathbb{R}^n$, such that:

$$K_1 \subseteq H^-, \quad K_2 \subseteq H^+, \tag{2.2.20}$$

where H^- and H^+ are the opposite, closed halfspaces defined by H^0, which is called *separating hyperplane*. The separation is: *strict,* iff

$$K_1 \subseteq \mathrm{int}\, H^-, \quad K_2 \subseteq \mathrm{int}\, H^+;$$

proper, iff besides (2.2.20) we have:

$$K_1 \cup K_2 \nsubseteq H^0; \tag{2.2.21}$$

disjunctive, iff besides (2.2.20) we have:

$$\text{either} \quad K_1 \cap H^0 = \varnothing \quad \text{or} \quad K_2 \cap H^0 = \varnothing; \tag{2.2.22}$$

strong (or *stable*), iff $\exists$ a sphere $N_\varepsilon \subset \mathbb{R}^n$ with centre in O and radius $\varepsilon > 0$, such that:

$$K_1 + N_\varepsilon \quad \text{and} \quad K_2 + N_\varepsilon \quad \text{are separable.} \tag{2.2.23}$$

It is immediate to see that K_1 and K_2 are *separable,* iff $\exists a \in \mathbb{R}^n \backslash \{O\}$ s.t.

$$\sup_{x \in K_1} \langle a, x \rangle \leq \inf_{x \in K_2} \langle a, x \rangle; \tag{2.2.20$'$}$$

and are *properly separable,* iff besides, (2.2.20)$'$ we have:

$$\inf_{x \in K_1} \langle a, x \rangle < \sup_{x \in K_2} \langle a, x \rangle. \tag{2.2.21$'$}$$

Disjunctive separation is equivalent to the existence of $a \in \mathbb{R}^n \backslash \{O\}$ and $b \in \mathbb{R}$ s.t.

$$\langle a, x \rangle \leq b, \quad \forall x \in K_1, \qquad \langle a, x \rangle > b, \quad \forall x \in K_2. \tag{2.2.22$'$}$$

Strong separation holds, iff

$$\sup_{x \in K_1} \langle a, x \rangle < \inf_{x \in K_2} \langle a, x \rangle. \tag{2.2.23$'$}$$

Other kinds of separation have been introduced. For instance, *strict separation*, which requires that K_1 and K_2 be included into opposite open halfspaces. Of course, strict separation implies disjunctive separation and is implied by the strong one.

Instead of a hyperplane, a nonlinear manifold can be introduced to define a more general concept of separation. This will be discussed in Chapter 4, where the concept of nonlinear weak and strong separation functions, will be introduced and analysed. When nothing is said, separability means linear separability. An obvious remark is that strong (or disjunctive) separation implies disjunctive (or proper) separation. The vice versa statement is not true, as shown by Figs. 2.2.7 and 2.2.8 where the dotted curves

Fig. 2.2.7

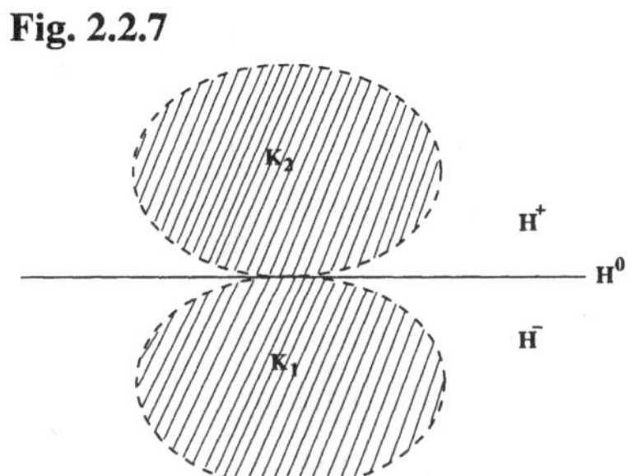

Fig. 2.2.8

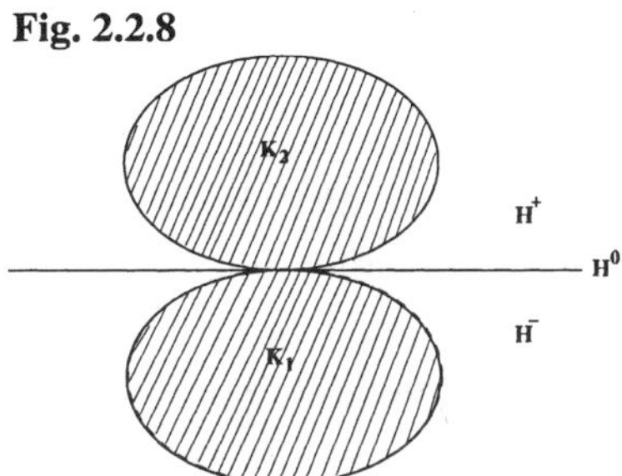

mean open). Moreover, the disjunction of K_1 and K_2 is neither sufficient, nor necessary for the proper separability, as Figs. 2.2.8 and 2.2.9 show, while it is necessary (but not sufficient) for disjunctive separation and, a fortiori, for a strong one. Fig. 2.2.10 shows

Fig. 2.2.9

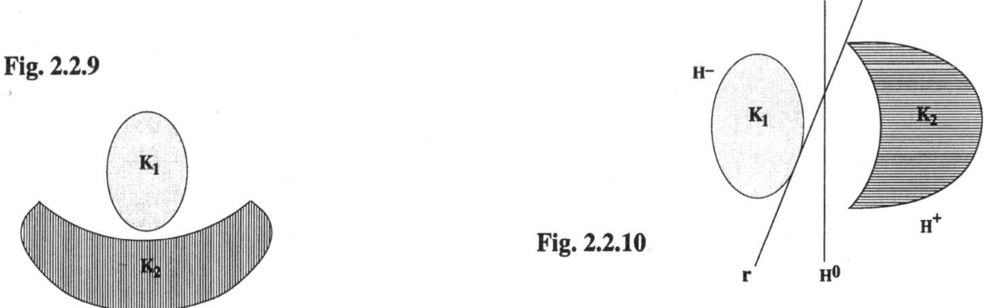

Fig. 2.2.10

an example of strong separation; the separation line r supports both sets. In general, this is not true, as shown by Fig. 2.2.11 where K_1 and K_2 are halfplanes. This example might lead one to think that two nonempty, closed and convex sets might be

Fig. 2.2.11

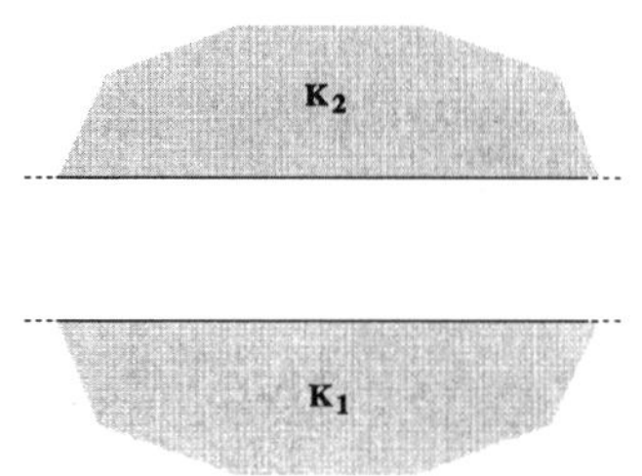

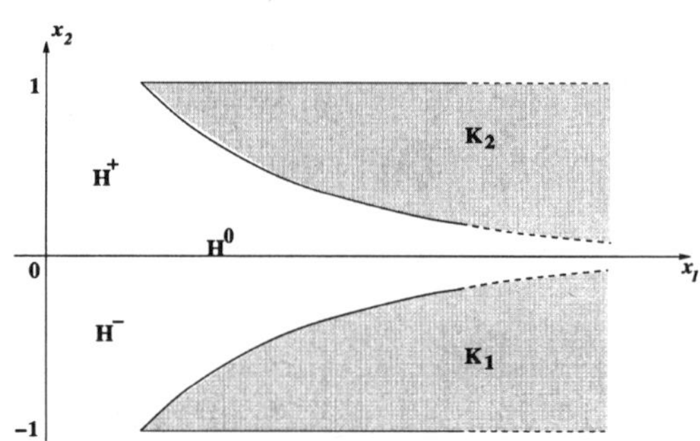

Fig. 2.2.12

strongly separable; this is not true, as shown by Fig. 2.2.12 where the dotted parts mean "continuation to infinity".

Before establishing a few fundamental separation theorems, let us show the previously mentioned equivalence between the concepts of support and separation.

Theorem 2.2.3. The sets $K_1, K_2 \subset \mathbb{R}^n$ are separable, if and only if $\exists a \in \mathbb{R}^n \backslash \{O\}$ such that:

$$\langle a, k_2 - k_1 \rangle \geq 0, \quad \forall k_1 \in K_1, \quad \forall k_2 \in K_2. \tag{2.2.24}$$

Proof. Only if. $\exists a \in \mathbb{R}^n \backslash \{O\}$ and $b \in \mathbb{R}$ such that:

$$\langle a, k_1 \rangle \leq b, \quad \langle a, k_2 \rangle \geq b, \quad \forall k_1 \in K_1, \ \forall k_2 \in K_2,$$

so that $\langle a, k_2 - k_1 \rangle \geq 0$, and (2.2.24) follows. **If.** It is enough to show that the sets

$$Y_1 := \{ y \in \mathbb{R} : y = \langle a, k_1 \rangle, \ k_1 \in K_1 \}, \ Y_2 := \{ y \in \mathbb{R} : y = \langle a, k_2 \rangle, \ k_2 \in K_2 \}$$

are separable. Ab absurdo, suppose that $\exists \hat{y}_1 \in Y_1$, $\exists \hat{y}_2 \in Y_2$, s.t. $\hat{y}_1 > \hat{y}_2$, so that $\exists \hat{k}_1 \in K_1$, $\exists \hat{k}_2 \in K_2$ s.t. $\langle a, \hat{k}_1 \rangle > \langle a, \hat{k}_2 \rangle$ or $\langle a, \hat{k}_2 - \hat{k}_1 \rangle < 0$ which contradicts the inequality (2.2.24). $\qquad \square$

Condition (2.2.24) is equivalent to claim that $"K_2 - K_1$ and O are separable $"$or $"O$ can be extended up to a hyperplane, which is separable from $K_2 - K_1"$. If $K_2 - K_1$ is convex and open, then the last sentence — and hence Theorem 2.2.3 — is precisely Hahn-Banach Theorem 2.2.1. $K_2 - K_1$ is open, iff dim $(K_1 \cup K_2) = n$ and, $\forall (k_1, k_2) \in K_1 \times K_2, \exists \varepsilon > 0$ s.t., $\forall d \in \mathbb{R}^n$, we have that, either $[k_1, k_1 + \varepsilon d[\subset K_1$, or $[k_2, k_2 - \varepsilon d[\subset K_2$. Such a condition is satisfied, if K_1 is open or if K_2 is open.

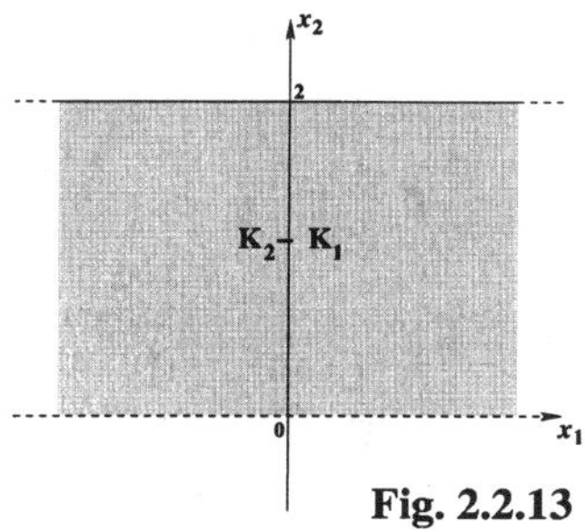

Fig. 2.2.13

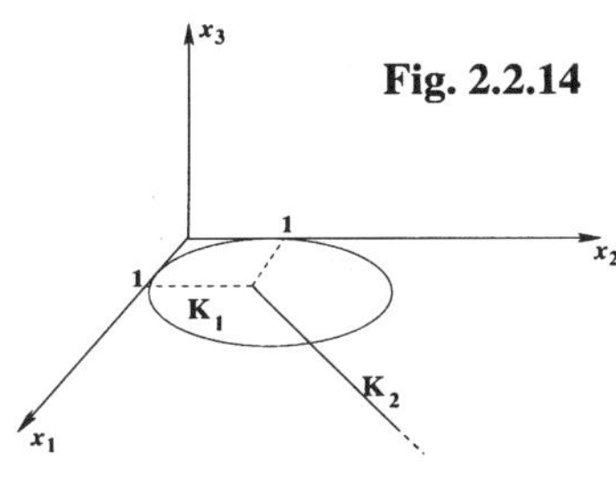

Fig. 2.2.14

Fig. 2.2.13 shows the set $K_2 - K_1$, K_1 and K_2 being those of Fig. 2.2.12. (2.2.24) is now $x_2 \geq 0$. Note that $K_2 - K_1$ is neither closed, nor open, notwithstanding the fact that both K_1 and K_2 be closed. If K_2 is deprived of its boundary, then $K_2 - K_1$ is that of Fig. 2.2.13 deprived of the line $x_2 = 2$, so that is open and is an example for Theorem 2.2.1.

Theorem 2.2.4. Let $K_1, K_2 \subset \mathbb{R}^n$ be nonempty and convex. **(i)** K_1 and K_2 are separable, if they are disjoint. **(ii)** K_1 and K_2 are properly separable, if and only if

$$\text{ri}\, K_1 \cap \text{ri}\, K_2 = \varnothing. \tag{2.2.25}$$

Proof. **(i)** $K_1 - K_2$ is convex and $O \notin K_1 - K_2$. By Theorem 2.2.1, there exists a hyperplane through the origin, say H^0, which does not intersect $\text{ri}\,(K_1 - K_2)$, so that $K_1 - K_2$ is contained in one of the closed halfspaces identified by H^0. Therefore, Theorem 2.2.3 leads to the conclusion. **(ii)** Set $K := K_1 - K_2$. Because of Proposition 2.1.5 (ii), we have $\text{ri}\, K = \text{ri}\, K_1 - \text{ri}\, K_2$, so that (2.2.25) holds iff $O \notin \text{ri}\, K$. This condition turns out to be equivalent to the proper separability of K_1 and K_2, since Theorem 2.2.1   gives   the existence of $a \in \mathbb{R}^n \backslash \{O\}$ s.t. $\langle a, k \rangle > 0$, $\forall k \in \text{ri}\, K$, or $\langle a, k_1 \rangle > \langle a, k_2 \rangle$, $\forall k_1 \in \text{ri}\, K_1$, $\forall k_2 \in \text{ri}\, K_2$, which implies (2.2.21). $\qquad\square$

Under the assumption (i) of Theorem 2.2.4, it is easy to prove that *there exists a partition of* $\mathbb{R}^n$, *namely* $\exists\, S_1, S_2 \subset \mathbb{R}^n$ *with* $S_1 \cup S_2 = \mathbb{R}^n$ *and* $S_1 \cap S_2 = \varnothing$, *such that* $K_1 \subseteq S_1$ *and* $K_2 \subseteq S_2$. In fact, if $k \notin K_1 \cup K_2$, then $K_1 \cup \{k\}$ and K_2 (or K_1 and $K_2 \cup \{k\}$) are again disjoint; this obvious fact and Zorn's Lemma (if X is a partially ordered set with the property that every totally ordered subset has an upperbound, then X contains a maximal element) allow one to prove the claim in full generality (and not only in $\mathbb{R}^n$).

Example 2.2.3. Consider the sets (see Fig. 2.2.14):

$$K_1 = \{x \in \mathbb{R}^3 : (x_1 - 1)^2 + (x_2 - 1)^2 \leq 1,\ x_3 = 0\},$$

$$K_2 = \{x \in \mathbb{R}^3 : x_1 \geq 1,\ x_1 = x_2,\ x_3 = 0\},$$

which are evidently separable, but not properly. Hence, condition (2.2.25) is not necessary for (not necessarily proper) separability. Of course, this remark must be dropped, if we restrict ourselves to the affine manifold (here $\mathbb{R}^2$). $\qquad\square$

From Theorem 2.2.4, several statements can be drawn. For instance, *if $K \subset \mathbb{R}^n$ is nonempty, closed and convex, then $\overline{x} \in \mathbb{R}^n$ and K admit disjunctive separation iff $\overline{x} \notin K$.* Several separation theorems can be established by exploiting the distance between two sets. To this end, consider the following fundamental:

Theorem 2.2.5. Let $K \subseteq \mathbb{R}^n$ be nonempty, closed and convex, and $\overline{x} \in \mathbb{R}^n$. **(i)** There exists a unique $x^0 \in K$, such that:

$$\|\overline{x} - x^0\| = \min_{x \in K} \|\overline{x} - x\|. \tag{2.2.26}$$

(ii) x^0 is a global minimum point of (2.2.26) if and only if

$$\langle x^0 - \overline{x}, \, x - x^0 \rangle \geq 0, \quad \forall x \in K. \tag{2.2.27}$$

Proof. **(i)** The thesis is trivial if $\operatorname{card} K = 1$ or if $\overline{x} \in K$. Let $\overline{x} \notin K$, and set $\overline{d} := \operatorname{dist}(\overline{x}, K) = \inf_{x \in K} \|\overline{x} - x\|$. From the very definition of $\overline{d}$, we deduce the existence of $\{x^i\}_1^\infty \subset K$ s.t. $\lim_{i \to +\infty} \|\overline{x} - x^i\| = \overline{d}$. By exploiting the parallelogram law (which is a consequence of the existence of the scalar product), we obtain:

$$\|x + y\|^2 + \|x - y\|^2 = 2(\|x\|^2 + \|y\|^2), \ \forall x, y \in \mathbb{R}^n,$$

$$\|x^i - x^j\|^2 = 2\|x^i - \overline{x}\|^2 + 2\|x^j - \overline{x}\|^2 - 4\|\overline{x} - \tfrac{1}{2}(x^i + x^j)\|^2.$$

The convexity of K implies $\tfrac{1}{2}(x^i + x^j) \in K$, so that $\|\overline{x} - \tfrac{1}{2}(x^i + x^j)\| \geq \overline{d}$. It follows that:

$$\|x^i - x^j\|^2 \leq 2\|x^i - \overline{x}\|^2 + 2\|x^j - \overline{x}\|^2 - 4\overline{d}^2,$$

so that $\lim_{i,j \to +\infty} \|x^i - x^j\| = 0$ or $\{x^i\}_1^\infty$ is a Cauchy sequence and hence admits a limit $x^0 \in K$; (2.2.6) is proved. Ab absurdo, suppose now that x^0 be not unique, so that $\exists \widetilde{x} \in K$ with $\widetilde{x} \neq x^0$ s.t. $\|\overline{x} - \widetilde{x}\| = \overline{d}$. By exploiting again the parallelogram law, we have:

$$\|x^0 - \widetilde{x}\|^2 + 4\|\overline{x} - \tfrac{1}{2}(x^0 + \widetilde{x})\|^2 = 2\|x^0 - \overline{x}\|^2 + 2\|\widetilde{x} - \overline{x}\|^2;$$

then, by setting $\hat{x} = \tfrac{1}{2}(x^0 + \widetilde{x})$, we obtain:

$$\|\overline{x} - \hat{x}\|^2 = \tfrac{1}{2}\overline{d}^2 + \tfrac{1}{2}\overline{d}^2 - \tfrac{1}{4}\|x^0 - \widetilde{x}\|^2 < \overline{d}^2,$$

so that, since the convexity of K implies $\hat{x} \in K$, we deny that $\overline{d}$ be minimum of (2.2.26). **(ii)** Let (2.2.27) hold. Since obviously $\|x^0 - x\| > 0$, $\forall x \neq x^0$, we have:

$$\|x - \overline{x}\|^2 = \|x - x^0 + x^0 - \overline{x}\|^2 = \|x - x^0\|^2 + 2\langle x - x^0, \, x^0 - \overline{x} \rangle + \|x^0 - \overline{x}\|^2 \geq$$

$$\geq \|x - x^0\|^2 + \|x^0 - \overline{x}\|^2 > \|x^0 - \overline{x}\|^2, \ \forall x \in K \backslash \{x^0\}.$$

This shows that x^0 is the unique minimum point of (2.2.26). Vice versa, let x^0 be the unique minimum point of (2.2.26) and, ab absurdo, suppose that (2.2.27) be false. Then, $\exists \hat{x} \in K$ s.t. $\langle x^0 - \overline{x}, \hat{x} - x^0 \rangle < 0$. From this inequality, account taken that

$x(\alpha) := (1 - \alpha)\hat{x} + \alpha x^0 \in K, \ \forall \alpha \in [0, 1]$, we deduce that:

$$||x(\alpha) - \overline{x}||^2 = ||x(\alpha) - x^0 + x^0 - \overline{x}||^2 = ||x(\alpha) - x^0||^2 + 2\langle x(\alpha) - x^0, \ x^0 - \overline{x}\rangle + ||x^0 - \overline{x}||^2 =$$

$$= (1 - \alpha)^2||\hat{x} - x^0||^2 + 2(1 - \alpha)\langle \hat{x} - x^0, \ x^0 - \overline{x}\rangle + ||x^0 - \overline{x}||^2 < ||x^0 - \overline{x}||^2,$$

if $1 - \alpha$ is small enough. Then, we contradict (2.2.26). $\qquad\square$

It is obvious to note that Theorem 2.2.5 and the related proof hold also if K is a subset of a Hilbert space, since this is a complete space. The set of $\overline{x} \notin \mathrm{ri}\, K$ and s.t. (2.2.26) holds with the same x^0, generates the *normal cone* to K at x^0.

Corollary 2.2.2. A nonempty set $K \subseteq \mathbb{R}^n$ is closed and convex, if and only if, $\forall \overline{x} \in \mathbb{R}^n$, there exists a unique $x^0 \in K$, which miminizes $f(x) := ||\overline{x} - x||$.

Proof. Only if. It is (i) of Theorem 2.2.5. **If.** The thesis is trivial, if $K = \mathbb{R}^n$ or if card $K = 1$. The assumption that K be not closed contradicts the existence of the minimum of $||\overline{x} - x||$, as is easily seen by considering $\overline{x} \in (\mathrm{cl}\, K)\backslash K$. Suppose that K be closed, but not convex, so that $\exists x^1, x^2 \in K$ s.t. $]x^1, x^2[\subset \sim K$. Set $\overline{x} := \frac{1}{2}(x^1 + x^2)$. Let $S_\rho(\overline{x})$ denote the closed hypersphere of $\mathbb{R}^n$ with centre at x and radius $\rho > 0$. Since $\overline{x} \in \mathrm{int}\,(\sim K), \ \exists \overline{\rho} > 0$ s.t.

$$K^0 := S_{\overline{\rho}}(\overline{x}) \cap \mathrm{frt}\, K \neq \varnothing, \qquad S_{\overline{\rho}}(\overline{x}) \cap \mathrm{ri}\, K = \varnothing.$$

If card $K^0 > 1$, then the assumption of the uniqueness is contradicted. Let card $K^0 = 1$ and set $K^0 = \{\widetilde{x}\}$. Consider the halfline, say r, described by $x(\alpha) := \widetilde{x} + \alpha(\overline{x} - \widetilde{x})$ and $\rho(\alpha) := \alpha||\overline{x} - \widetilde{x}||, \ \alpha \in [1, +\infty[$. Denote by $\mathcal{A}$ the set of $\alpha \in [1, +\infty[$ such that $S_{\rho(\alpha)}(x(\alpha)) \cap K = \{\widetilde{x}\}$. Of course, we have that $\mathcal{A}$ contains $1, S_{\rho(1)}(x(1)) = S_{\overline{\rho}}(\overline{x})$, and $\hat{\alpha} := ||x^1 - x^2||/2||\overline{x} - \widetilde{x}||$ is an upper bound for $\mathcal{A}$ (in fact, for $\alpha = \hat{\alpha}$, at least one of x^1, x^2 belongs to $S_{\rho(\hat{\alpha})}(x(\hat{\alpha}))$). Hence, $\exists \alpha^0 := \max \mathrm{cl}\,\mathcal{A}$, which corresponds to find the largest closed hypersphere with centre at $x(\alpha)$ and contained in $\mathrm{cl}\,(\sim K)$. It follows $\alpha^0 > 1$, so that $x(\alpha^0) \neq \widetilde{x}$, and

$$S_{\rho(\alpha^0)}(x(\alpha^0)) \cap \mathrm{frt}\, K \neq \varnothing, \qquad S_{\rho(\alpha^0)}(x(\alpha^0)) \cap \mathrm{ri}\, K = \varnothing,$$

which contradicts the assumption of uniqueness. $\qquad\square$

The property expressed by Corollary 2.2.2 is due to Motzkin [IV34].

Corollary 2.2.3. Let $K \subset \mathbb{R}^n$ be nonempty, closed and convex, and let $\overline{x} \notin K$. Then, **(i)** dist $(\overline{x}, K) > 0$, and **(ii)** $\overline{x}$ and K are strongly separable.

Proof. (i) $\overline{x} \notin K$ and the closure of K imply that $\overline{x} \in \mathrm{int}\,(\sim K)$, so that there exists a sphere, say $N_r(\overline{x})$, with centre at $\overline{x}$ and small enough radius $r > 0$, s.t. $N_r(\overline{x}) \subset \sim K$. Of course, dist $(N_r(\overline{x}), K) \geq 0$, so that dist $(\overline{x}, K) = r + \mathrm{dist}\,(N_r(\overline{x}), K) > 0$. **(ii)** Set $x^0 := \mathrm{proj}_K\overline{x}$. Because of (i), dist $(\overline{x}, x^0) = \mathrm{dist}\,(\overline{x}, K) > 0$. Let $\overline{H}$ denote the hyperplane through $\overline{x}$ and orthogonal to the segment $[\overline{x}, x^0]$, whose equation is $\langle x^0 - \overline{x}, \ x - \overline{x}\rangle = 0$, and H^0 the hyperplane through x^0 and parallel to $\overline{H}$, whose equation is

$$\langle x^0 - \overline{x}, \, x - x^0 \rangle = 0.$$

Because of Theorem 2.2.5(ii), K is contained in the halfspace (2.2.27). Therefore, the hyperplane $\frac{1}{2}\overline{H} + \frac{1}{2}H^0$, whose equation is $\langle x^0 - \overline{x}, \, x - \frac{1}{2}(\overline{x} + x^0) \rangle = 0$, fulfils (2.2.23).$\square$

Note that, in Corollary 2.2.3, (i) does not require the convexity of K; (ii) is a "closed version" of Theorem 2.2.1. Of course, *the theses of both (i) and (ii) of Corollary 2.2.3 remain true (with the same proof), if K is still nonempty and convex, but not necessarily closed, and $\overline{x} \notin clK$*. A quite analogous reasoning to that of the proof of (ii) of Corollary 2.2.3 enables one to easily show the following:

Corollary 2.2.4. The nonempty and convex sets $K_1, K_2 \subset \mathbb{R}^n$ are strongly separable, if and only if their distance is positive.

The thesis of Corollary 2.2.4 holds, if K_1 and K_2 are nonempty, disjoint and convex, with K_1 compact and K_2 closed, as is it easy to prove on the basis of Corollary 2.2.2. This statement is due to Minkowski [36].

Theorem 2.2.4 and its consequences allow one to establish several properties of the polar, besides other separation theorems. As previously noted, support and separation are different languages which express the same substance. Therefore, the statements are in terms of support or separation, according to convenience. In next theorem, for the sake of simplicity, we write K^{**}, instead of $(K^*)^*$, to denote the polar of the polar; here, polar of a cone means positive polar:

Theorem 2.2.6. Let $K, K_1, K_2 \subseteq \mathbb{R}^n$ be nonempty and convex. We have:

(i) $K^{**} = cl \ conv\,(K \cup \{O\})$;

      If $O \in K$, then $K^{**} = K$ if and only if K is closed.

(ii) $K_1 \subseteq K_2 \Rightarrow K_1^* \supseteq K_2^*$.

(iii) $(K_1 \cup K_2)^* = K_1^* \cap K_2^*$.

(4i)   If $O \in K_1 \cap K_2$, then $(K_1 \cap K_2)^* = cl \ conv\,(K_1^* \cup K_2^*)$.

(5i)   If $O \in K$, then K^* is bounded if and only if $O \in int\,K$.

(6i) $(\alpha K)^* = \frac{1}{\alpha}K^*, \ \ \forall \alpha \in \mathbb{R}_+ \backslash \{0\}$.

(7i)   If K is bounded, then $K^* = \{y \in \mathbb{R}^n : \langle x, y \rangle \le 1, \ \forall x \in vert\,K\}$.

(8i)   If K is a polytope and $O \in int\,K$, then K^* is the dual of K, namely $K^* = K^\triangle$.

(9i)   If K is a closed cone with apex at O, then $dim\,K^* = n$ (or K^* is solid), if and only if K is pointed.

(10i)   If K is a pointed cone with apex at O and $a \in int\,K^*$, then $K \cap lev_{\le \alpha}\langle a, x \rangle$ is bounded, $\forall \alpha \in \mathbb{R}_+ \backslash \{0\}$.

(11i)   If K_1 and K_2 are cones with apices at O, then $(K_1 \times K_2)^* = K_1^* \times K_2^*$.

(12i) Let K be a cone with apex at O. We have:

$$K^+ := \{y \in \mathbb{R}^n : \langle y, x \rangle > 0, \ \forall x \in K\} \subseteq \text{int } K^*,$$

where the inclusion becomes equality, if and only if $O \notin K$.

(13i) Let K be a nonempty and convex cone (with apex at the origin), with $K \neq \{O\}$ and $\text{int } K^* \neq \varnothing$. Then, for each $a \in \text{int } K^*$, we have $\langle a, x \rangle > 0$, $\forall x \in K \backslash \{O\}$. If, moreover, $K \cup \{O\}$ is closed, then the inequalities $\langle a, x \rangle > 0$, $\forall x \in K \backslash \{O\}$, imply $a \in \text{int } K^*$.

Proof. **(i)** Because of (2.2.8), $x \in K$ implies $\langle x, y \rangle \leq 1$, $\forall y \in K^*$, which implies $x \in K^{**}$. Since the polar of a set is closed, convex and contains O, it follows that $K^{**} \supseteq \hat{K} := \text{cl conv}\,(K \cup \{O\})$. Consider any $\overline{x} \notin \hat{K}$. Because of Corollary 2.2.3(ii), there is strong separation between $\overline{x}$ and $\hat{K}$; hence, being $O \in \hat{K}$, the equation of the separation hyperplane can be assumed to be $\langle a, x \rangle = 1$ and s.t. $\langle a, \overline{x} \rangle > 1$, $\langle a, x \rangle < 1$ for each $x \in \hat{K}$ and, a fortiori, for each $x \in K$. This shows that $a \in K^*$ and $\overline{x} \notin K^{**}$. If $O \in K$, then obviously $\hat{K} = K$ iff K is closed. **(ii)** From (2.2.8) we draw that $K_1 \subseteq K_2 \Rightarrow \langle x, y \rangle \leq 1$, $\forall x \in K_1$, $\forall y \in K_2^*$, and hence $K_2^* \subseteq K_1^*$. **(iii)** We have:

$$(K_1 \cup K_2)^* = \{y \in \mathbb{R}^n : \langle x^1, y \rangle \leq 1, \ \forall x^1 \in K_1; \ \langle x_1^2, y \rangle \leq 1, \ \forall x^2 \in K_2\} =$$

$$= \{y \in \mathbb{R}^n : \langle x^1, y \rangle \leq 1, \ \forall x^1 \in K_1\} \cap \{y \in \mathbb{R}^n : \langle x^2, y \rangle \leq 1, \ \forall x^2 \in K_2\} =$$

$$= K_1^* \cap K_2^*.$$

(4i) Because of (3i) and (i), we have:

$$(K_1 \cap K_2)^* = (K_1^{**} \cap K_2^{**})^* = [(K_1^* \cup K_2^*)^*]^* = \text{cl conv}\,(K_1^* \cup K_2^*).$$

(5i) If K^* is bounded, $\exists \rho > 0$ s.t. $||y|| \leq \rho$, $\forall y \in K^*$. Then, $\forall x$ s.t. $||x|| \leq \frac{1}{\rho}$, we have:

$$\langle x, y \rangle \leq ||x|| \cdot ||y|| \leq 1, \quad \forall y \in K^*,$$

so that $N_{1/\rho}(O) \subset K^{**}$. It follows that $O \in \text{int } K^{**}$ and hence $O \in \text{int } K$, since, being K^{**} closed, because of (i), $K^{**} = K$. Let K^* be unbounded and let $\hat{y} \in K^* \backslash \{O\}$ be s.t. $\alpha \hat{y} \in K^*, \forall \alpha \geq 0$. Ab absurdo, suppose that $O \in \text{int } K$, so that $\exists \hat{\alpha} > 0$ s.t. $\hat{\alpha} \hat{y} \in K$. Since $\forall x \in K$ and $\forall y \in K^* \ \langle x, y \rangle \leq 1$ holds, then we have:

$$1 \geq \langle \alpha \hat{y}, \hat{\alpha} \hat{y} \rangle = \alpha \hat{\alpha} ||\hat{y}||_2^2, \quad \forall \alpha \geq 0,$$

which, being $\hat{y} \neq O$, is false. **(6i)** $\forall \alpha \in \mathbb{R}_+ \backslash \{O\}$, we have:

$$(\alpha K)^* = \{y \in \mathbb{R}^n : \langle y, z \rangle \leq 1, \ \forall z \in \alpha K\} = \{y \in \mathbb{R}^n : \langle y, \alpha x \rangle \leq 1, \ \forall x \in K\} =$$

$$= \{y \in \mathbb{R}^n : \langle \alpha y, x \rangle \leq 1, \ \forall x \in K\} = \{\tfrac{1}{\alpha} t \in \mathbb{R}^n : \langle t, x \rangle \leq 1, \ \forall x \in K\} = \tfrac{1}{\alpha} K^*.$$

(7i) Since $\text{vert } K \subseteq K$, (ii) $\Rightarrow (\text{vert } K)^* \supseteq K^*$. Because of Theorem 2.1.1, $\forall x \in K$, $\exists x^1, ..., x^r \in K$ with $r \leq n+1$, and $\exists \alpha_1, ..., \alpha_r \in \mathbb{R}_+ \backslash \{0\}$ with $\alpha_1 + ... + \alpha_r = 1$, s.t. $x = \alpha_1 x^1 + ... + \alpha_r x^r$. Since K is bounded, it is not restrictive to assume that $x^1, ..., x^r \in \text{vert } K$ (otherwise, every $x^i \notin \text{vert } K$ can be eliminated by expressing it as convex combination of elements of $\text{vert } K$). Since we have:

$$\langle x^i, y \rangle \leq 1, \quad \forall y \in (\text{vert } K)^*,$$

then:

$$\langle x, y \rangle \leq 1, \quad \forall y \in (\text{vert } K)^*.$$

We have achieved that, $\forall y \in (\text{vert } K)^*$, we have:

$$\langle x, y \rangle \leq 1, \quad \forall x \in K,$$

or $y \in K^*$, and hence $(\text{vert } K)^* \subseteq K^*$. **(8i)** Since the number of (faces, in particular of) vertices of a polytope is finite, from (7i) we have that K^* is a polyhedron. Let $\mathcal{F}$ be the set of faces of K, consider any $F \in \mathcal{F}$, and the map ψ defined by:

$$\psi(F) = \{y \in K^* : \langle x, y \rangle = 1, \ \forall x \in F \neq \varnothing\}; \quad \psi(\varnothing) = K^*. \tag{2.2.28}$$

From Definition 2.2.2 in case (2.2.5), account taken of the fact that (7i) allows us to replace F with $\text{vert } F$ in (2.2.28), because of the very definition of face in the case (2.2.5), we have that $\psi(F)$ is a face of K^*. Because of (i), we have $K^{**} = K$; therefore, from (2.2.28) we deduce the relations:

$$\psi(\psi(F)) = \{z \in K^{**} : \langle z, y \rangle = 1, \ \forall y \in \psi(F)\}$$

$$= \{z \in K : \langle z, y \rangle = 1, \ \forall y \in \psi(F)\} \supseteq F.$$

Now, observe that $\psi(\psi(F)) - F \neq \varnothing \Rightarrow \exists \bar{z} \in K \backslash F$ and $\exists a \in \psi(F)$ s.t. $F = K \cap H^0$, where $H^0 := \{x \in \mathbb{R}^n : \langle a, x \rangle = 1\}$. This implies $\langle a, \bar{z} \rangle \neq 1$ and then $\bar{z} \notin \psi(\psi(F))$. The equality $\psi(\psi(F)) = F$ follows. To achieve the thesis, it is enough to observe that here the inclusion of Definition 2.2.2 (which defines the dual of a polytope) holds as equality. **(9i) If.** Ab absurdo, let $\dim K^* < n$. Then, there exists a linear manifold H (whose dimension belongs to $[1, n\text{-}1]$), s.t. $K^* \subseteq H$. By (ii) we have $K^{**} \supseteq H^* = H^\perp$ (where the dimension of $H^\perp$ belongs to $[1, n\text{-}1]$) or, by (i), $K \supseteq H^\perp$, which contradicts that K be pointed. **Only if.** Ab absurdo, let K be not pointed, so that $\exists \hat{x} \in K \cap (-K)$ with $\hat{x} \neq 0$; this implies $-\hat{x} \in (-K) \cap K$. Hence the line through $\hat{x}$ and $-\hat{x}$, call it ρ, belongs to K. Because of (ii), $\rho \subseteq K \Rightarrow K^* \subseteq \rho^* = \rho^\perp$. Since $\dim \rho = 1$, then $\dim \rho^\perp = n - 1$, which implies $\dim K^* \leq n - 1$, and contradicts $\dim K^* = n$. **(10i)** Ab absurdo, suppose that $\exists \alpha \in \mathbb{R}_+ \backslash \{0\}$ s.t. $H^-(\alpha) := K \cap \text{lev}_{\leq \alpha} \langle a, x \rangle$ be unbounded. Then, $\exists \{x^i\}_1^\infty \subset H^-(\alpha)$ s.t. $\lim_{i \to +\infty} ||x^i|| = +\infty$. Because of (2.1.7b), $y^i := (\frac{1}{||x^i||}) x^i \in K$, $\forall i = 1, 2, ...$; moreover, the sequence $\{y^i\}_1^\infty$ is bounded, so that it admits a convergent subsequence, call it $\{y^{i_r}\}_1^\infty$ and call $\bar{y}$ its limit. Of course, $||\bar{y}|| = 1$ and $\langle a, \bar{y} \rangle > 0$. From another side, we have:

$$\langle a, \bar{y} \rangle = \lim_{r \to +\infty} \langle a, y^{i_r} \rangle = \lim_{r \to +\infty} \frac{\langle a, x^{i_r} \rangle}{||x^{i_r}||} \leq \lim_{r \to +\infty} \frac{\alpha}{||x^{i_r}||} = 0,$$

which contradicts the above inequality. **(11i)** Because of (2.2.12b), we have:

$$(K_1 \times K_2)^* = \{(k_1^*, k_2^*) \in \mathbb{R}^n \times \mathbb{R}^n : \langle (k_1^*, k_2^*), (k_1, k_2) \rangle \geq 0, \ \forall (k_1, k_2) \in K_1 \times K_2\}$$

$$= \{(k_1^*, k_2^*) \in \mathbb{R}^n \times \mathbb{R}^n : \langle k_1^*, k_1 \rangle + \langle k_2^*, k_2 \rangle \geq 0, \ \forall k_1 \in K_1, \ \forall k_2 \in K_2\}$$

$$\supseteq \{(k_1^*, k_2^*) \in K_1^* \times K_2^* : \langle k_1^*, k_1 \rangle + \langle k_2^*, k_2 \rangle \geq 0, \ \forall k_1 \in K_1, \ \forall k_2 \in K_2\}$$

$$= K_1^* \times K_2^*,$$

where the last equality is consequence of the fact that, $\forall k_i \in K_i$, $\forall k_i^* \in K_i^*$, we have $\langle k_i^*, k_i \rangle \geq 0$, $i = 1, 2$. Now, ab absurdo, suppose that $\exists (\hat{k}_1, \hat{k}_2) \in (K_1 \times K_2)^* \backslash (K_1^* \times K_2^*)$, so that we have:

$$\langle \hat{k}_1, k_1 \rangle + \langle \hat{k}_2, k_2 \rangle \geq 0, \quad \forall k_1 \in K_1, \quad \forall k_2 \in K_2, \tag{2.2.29}$$

and either $\hat{k}_1 \notin K_1^*$ or $\hat{k}_2 \notin K_2^*$. In the former case, $\exists \tilde{k}_1 \in K_1$ s.t. $\langle \hat{k}_1, \tilde{k}_1 \rangle < 0$, and (2.2.29) implies:

$$\langle \hat{k}_2, k_2 \rangle \geq -\langle \hat{k}_1, \tilde{k}_1 \rangle > 0, \quad \forall k_2 \in K_2, \tag{2.2.30}$$

which leads to the contradiction $0 > 0$, if $O \in K_2$. If $O \notin K_2$, because of (2.1.7b), for $k_2 = \varepsilon \overline{k}_2$ with $\overline{k}_2 \in K_2$ and $0 < \varepsilon < -\langle \hat{k}_1, \tilde{k}_1 \rangle / \langle \hat{k}_2, \overline{k}_2 \rangle$, (2.2.30) is contradicted. In the latter case, we proceed in a quite analogous way by exchanging the role of K_1 and K_2. **(12i)** If $O \in K$, then the inclusion is trivial, and may be strict; for instance, if K is convex, closed and pointed. Let $O \notin K$, and consider 2 cases, according to $O \in \text{conv}\, K$ or not. In the former, $\exists k^1, ..., k^r \in K$ and $\exists \alpha_1, ..., \alpha_r \in]0, 1[$ with $\sum_{i=1}^{r} \alpha_i = 1$ and $r \geq 2$, s.t.

$$\sum_{i=1}^{r} \alpha_i k^i = 0.$$

By splitting the summand into 2 parts, we achieve the existence of $k', k'' \in \text{conv}\, K$, s.t. $k'' = -k'$. Therefore, because of (2.2.9), we have that K^* is contained in the intersection of 2 opposite halfspaces, and then $\text{int}\, K^* = \varnothing$. In this case, also $K^+ = \varnothing$, since K^+ is convex and O should satisfy its inequality. In the latter case, $O \notin \text{conv}\, K$ implies $K^+ \neq \varnothing$; furthermore, we have:

$$y \in \text{int}\, K^* \quad \Leftrightarrow \quad \exists \varepsilon > 0 \ \text{s.t.} \ y + N_\varepsilon(y) \subset K^*$$

$$\Leftrightarrow \quad \langle y + \delta, x \rangle \geq 0, \quad \forall x \in K, \quad \forall \delta \ \text{s.t.} \ ||\delta|| < \varepsilon$$

$$\Leftrightarrow \quad \langle y, x \rangle > 0, \quad \forall x \in K.$$

Indeed, the existence of $\hat{x} \in K$ s.t. $\langle y, \hat{x} \rangle \leq 0$ would lead to the contradiction:

$$\langle y + \delta, \hat{x} \rangle = \langle y, \hat{x} \rangle + \langle \delta, \hat{x} \rangle < 0,$$

with $\delta = -\alpha \hat{x}$ and $0 < \alpha < \varepsilon \backslash ||\hat{x}||$. **(13i)** Ab absurdo, suppose that $\exists \hat{a} \in \text{int}\, K^*$ and $\exists \hat{x} \in K \backslash \{O\}$ s.t. $\langle \hat{a}, \hat{x} \rangle = 0$. $\hat{a} \in \text{int}\, K^*$ implies the existence of a sphere $N_{\overline{\varepsilon}}$, with centre at O and radius $\bar{\varepsilon} > 0$, s.t. $\hat{a} + y \in \text{int } K^*$, $\forall y \in N_{\bar{\varepsilon}}$. Choose $\hat{y} := -(\bar{\varepsilon}/2||\hat{x}||)\hat{x}$. Since $\hat{x} \in K$, from (2.2.12b) we have:

$$\langle \hat{a} + \hat{y}, \hat{x} \rangle \geq 0,$$

or

$$0 = \langle \hat{a}, \hat{x} \rangle \geq -\langle \hat{y}, \hat{x} \rangle = \frac{\bar{\varepsilon}}{2}||\hat{x}|| > 0.$$

Now, again ab absurdo, suppose that $\exists \hat{a} \notin \text{int } K^*$ s.t.

$$\langle \hat{a}, x \rangle > 0, \quad \forall x \in K\backslash\{O\},$$

so that, of course, $\hat{a} \in \text{frt } K^*$ and $\exists \delta \in \mathbb{R}^n$, with $||\delta|| > 0$ and arbitrarily small, s.t. $\hat{a} + \delta \notin K^*$. Therefore, $\exists \hat{x} \in K$ s.t. $\langle \hat{a} + \delta, \hat{x} \rangle < 0$. It is not restrictive to consider the subset, say $\overline{K}$, of K, whose elements have norm 1, so that $\overline{K}$, being a closed subset of a compact set, is compact and hence the minimum of $\langle \hat{a}, x \rangle$, say μ, on $\overline{K}$ is >0. Then, by choosing δ in a such way that the minimum of $\langle \delta, x \rangle$ on $\overline{K}$ be $\geq -\mu$, we obtain the contradiction $0 > \langle \hat{a} + \delta, \hat{x} \rangle \geq 0$. $\qquad\qquad\square$

Some of the properties of Theorem 2.2.6 do not require the convexity; they can be verified on the Figs. 2.2.1-2.2.6; this is of help in understanding such a theorem; in particular, Fig. 2.2.2 allows us to check that $K^{**} \neq K$ may occur. The last part of (i) is due to Farkas. Now, let us prove one of the many consequences of polarity.

Corollary 2.2.4. Let $K \subset \mathbb{R}^n$ be a nonempty and convex cone (with apex at the origin), such that:

$$K = (\text{cl } K)\backslash\{O\}. \tag{2.2.31}$$

There exists a halfspace H^+ of $\mathbb{R}^n$ with $O \in \text{frt } H^+$, such that:

$$K \subset \text{int } H^+, \tag{2.2.32}$$

if and only if K is pointed.

Proof. Only if. $(-\text{int } H^+) \cap \text{int } H^+ = \varnothing$ and $-K \subset -\text{int } H^+$ imply $K \cap (-K) = \varnothing$. This and (2.2.31) imply (2.1.8). **If.** Theorem 2.2.6 (9i) $\Rightarrow \dim K^* = \dim (\text{cl } K)^* = n$, so that $\text{int } K^* \neq \varnothing$. Then Theorem 2.2.6 (12i) (where the inclusion holds as equality) gives (2.2.32), since every element of $\text{int } K^*$ is the gradient of a halfspace H^+, which fulfils (2.2.32). $\qquad\qquad\square$

The next theorem and lemma — which are a slight generalization of Theorem 2.1 of [IV14]; see also [41] — give a property of separating hyperplanes.

Theorem 2.2.7. Let $K \subset \mathbb{R}^n$ be a nonempty and convex cone with apex at $O \notin K$, such that (2.1.14) holds, namely

$$K + \text{cl } K = K, \tag{2.2.33}$$

and F be any face of K. Let $S \subset \mathbb{R}^n$ be nonempty with $O \in \text{cl } S$ and such that $S - \text{cl } K$ is convex. F is contained in every hyperplane which separates K and S, if and only if

$$F \subseteq TC(S - \text{cl } K),\tag{2.2.34}$$

where $TC(S - \text{cl } K)$ is the tangent cone to $S - \text{cl } K$ at O.

Before proving Theorem 2.2.7, let us state some preliminary properties. See Examples 3.4.9, 3.4.12 and 3.4.13 for some illustrations of (2.2.34).

Lemma 2.2.1. Under the same assumptions of Theorem 2.2.7, we have:

$$K \cap S = \varnothing \ \Leftrightarrow \ K \cap (S - \text{cl } K) = \varnothing.\tag{2.2.35}$$

$$\left\{ \begin{array}{c} \text{a hyperplane separates} \\ K \text{ and } S \end{array} \right\} \Leftrightarrow \left\{ \begin{array}{c} \text{the same hyperplane} \\ \text{separates } K \text{ and } S - \text{cl } K \end{array} \right\}\tag{2.2.36}$$

$$S \subseteq TC(S), \ \ S - \text{cl } K \subseteq TC(S - \text{cl } K), \ \ TC(S) \subseteq TC(S - \text{cl } K).\tag{2.2.37}$$

Proof. Let us start with the proof of (2.2.35). ($\Leftarrow$) Since $O \in \text{cl } K$ implies $S \subset S - \text{cl } K$, then obvioulsy the latter of (2.2.35) implies the former. ($\Rightarrow$) Ab absurdo, suppose that the latter of (2.2.35) be false, so that $\exists x^1 \in S$ and $\exists x^2 \in \text{cl } K$ s.t. $x^1 - x^2 \in K$. Because of (2.2.33), $(x^1 - x^2) + x^2 \in K$, or $x^1 \in K$; therefore $x^1 \in K \cap S$, which contradicts the assumption. Hence (2.2.35) follows. Now, let us prove (2.2.36). ($\Leftarrow$) Since $O \in \text{cl } K$ implies $S \subset S - \text{cl } K$, then obviously a hyperplane, which separates K and $S - \text{cl } K$, separates K and S too. ($\Rightarrow$) With the notation of the beginning of this section, let H^0 be any hyperplane s.t. $K \subset H^+, S \subseteq H^-$. Ab absurdo, suppose that $\exists \hat{x} \in S - \text{cl } K$ s.t. $\hat{x} \notin H^-$. $\hat{x} \in S - \text{cl } K \ \Rightarrow \ \exists x^1 \in S, \ \exists x^2 \in \text{cl } K$, s.t. $\hat{x} = x^1 - x^2$. $\hat{x} \notin H^- \Rightarrow \langle a, \hat{x} \rangle > 0$, or

$$0 \geq \langle a, x^1 \rangle > \langle a, x^2 \rangle \geq 0,$$

where the 1st inequality is implied by $x^1 \in S \subseteq H^-$, and the 3rd by $x^2 \in \text{cl } K \subseteq H^+$. Hence (2.2.36) follows. (2.2.37) is a straightforward consequence of Theorem 2.1.4. $\square$

Proof of Theorem 2.2.7. Only if. Since $O \in \text{cl } S$, $TC(S)$ and $TC(S - \text{cl } K)$ exists. Now, ab absurdo, suppose that $\exists \bar{x} \in F \backslash TC(S - \text{cl } K)$. Because of (2.2.37), Corollary 2.2.3 (ii) gives the existence of a hyperplane, say H^0 and let $\langle a, x \rangle = b$ with $a \in \mathbb{R}^n \backslash \{O\}$. and $b \in \mathbb{R}$ be its equation, such that:

$$\langle a, x \rangle \leq b < \langle a, \bar{x} \rangle, \ \ \forall x \in TC(S - \text{cl } K).\tag{2.2.38}$$

$O \in TC(S - \text{cl } K) \ \Rightarrow \ b \geq 0$, so that we can set $b = 0$ in (2.2.38), which becomes:

$$\langle a, x \rangle \leq 0 < \langle a, \bar{x} \rangle, \ \ \forall x \in TC(S - \text{cl } K).\tag{2.2.39}$$

By exploiting again (2.2.37), from the first of inequalities (2.2.39) we draw:

$$\langle a, x \rangle \leq 0, \ \ \forall x \in S - \text{cl } K.\tag{2.2.40}$$

Now, we prove that

$$\langle a, x \rangle \geq 0, \ \ \forall x \in K.\tag{2.2.41}$$

Ab absurdo, suppose that $\exists \hat{x} \in K$ s.t. $\langle a, \hat{x} \rangle < 0$. Then, whatever $\tilde{x} \in S - \mathrm{cl}\, K$ we may choose, we have $\tilde{x} - \alpha\hat{x} \in S - \mathrm{cl}\, K$, $\forall \alpha \in \mathbb{R}_+$, so that:

$$\lim_{\alpha \to +\infty} \langle a, \tilde{x} - \alpha\hat{x} \rangle = +\infty,$$

which contradicts (2.2.40). Therefore, (2.2.41) follows, and then H^0 separates K and $S - \mathrm{cl}\, K$. Because of (2.2.36), H^0 separates also K and S; then, due to the assumption, $F \subseteq H^0$ so that $\langle a, \overline{x} \rangle = 0$, which contradicts (2.2.39). **If.** Suppose that $\exists a \in \mathbb{R}^n \setminus \{O\}$ s.t. the hyperplane H^0, whose equation is $\langle a, x \rangle = 0$, separates K and S. Because of (2.2.36), H^0 separates also K and $S - \mathrm{cl}\, K$, or

$$\langle a, x \rangle \leq 0 \leq \langle a, y \rangle, \quad \forall x \in S - \mathrm{cl}\, K, \quad \forall y \in K. \tag{2.2.42}$$

These inequalities imply $TC(S - \mathrm{cl}\, K) \subseteq H^-$, where H^- is the halfspace identified by $\langle a, x \rangle \leq 0$. Hence $F \subset H^-$. Besides, since $F \subseteq \mathrm{cl}\, K$, (2.2.42) $\Rightarrow F \subset H^+$, where H^+ is the halfspace identified by $\langle a, x \rangle \geq 0$. It follows $F \subseteq H^- \cap H^+ = H^0$. $\qquad \square$

The class of cones (2.2.33) has been discussed briefly after (2.1.14); it will be considered again in Sect. 3.2, where its role for developing the theory will appear clearly. From the proof, we note that, in Theorem 2.2.7, F need not be a face of K; it can be merely a subset of $\mathrm{cl}\, K$. If $O \notin \mathrm{cl}\, S$, then, according to Definition 2.1.9, the considered tangent cones are not defined; indeed, to define the TC at an exterior point makes no sense. Note that the convexity of $S - \mathrm{cl}\, K$ (which has been assumed in Theorem 2.2.7) does not require that of S; this will be fundamental for the applications in Sect. 3.2 and in the following chapters. The thesis of Theorem 2.2.7 becomes self-evident, if S is affine and hence, due to $O \in \mathrm{cl}\, S$, linear; in fact, in this case $TC(S - \mathrm{cl}\, K) = S - \mathrm{cl}\, K$.

Example 2.2.4. Set $K = \{x \in \mathbb{R}^3 : x_3 - x_2 > 0,\, x_3 + x_2 > 0\}$ and $S = \{x \in \mathbb{R}^3 : x_1 = x_3 = 0\}$, so that $S - \mathrm{cl}\, K = \{x \in \mathbb{R}^3 : x_3 \leq 0\}$. The assumptions of Theorem 2.2.7 are evidently satisfied. Set $F = \{x \in \mathbb{R}^3 : x_1 \geq 0,\, x_2 = x_3 = 0\}$. The only plane which separates K and S is identified by $x_3 = 0$; it contains F, which is contained in $TC(S - \mathrm{cl}\, K) = S - \mathrm{cl}\, K$. Note that K is not pointed, even if it fulfils (2.2.33). $\qquad \square$

Example 2.2.5. Set $K = \{x \in \mathbb{R}^3_+ : x_2 = 0\}$ and $S = \{x \in \mathbb{R}^3 : x_1 \leq 0,\, x_2 = 0,\, x_3 = \sqrt{-x_1}\}$, so that K is convex, S is not convex while $S - \mathrm{cl}\, K = S - K = \{x \in \mathbb{R}^3 : x_1 \leq 0,\, x_2 = 0,\, x_3 \leq \sqrt{-x_1}$ is convex. Since K is closed, according to the remarks which follow (2.1.14), (2.2.33) is fulfilled. The planes which separate K and S are those of equation $ax_1 + bx_2 = 0$ with $a, b \in \mathbb{R}$, $a^2 + b^2 > 0$, and they all contain the face $F = \{x \in \mathbb{R}^3 : x_1 = x_2 = 0,\, x_3 \geq 0\}$ of K. In agreement with Theorem 2.2.7, we have

$$F \subset TC(S - \mathrm{cl}\, K) = TC(S - K) = \{x \in \mathbb{R}^3 : x_1 \leq 0,\, x_2 = 0\}. \qquad \square$$

With the same notation of the beginning of this section, we state the following theorems.

Theorem 2.2.8. Let $K \subset \mathbb{R}^n$ be a nonempty and convex cone with apex at $O \notin K$, such that (2.2.33) holds, and $S \subset \mathbb{R}^n$ be a nonempty affine manifold. If $K \cap S = \varnothing$, then there exists a hyperplane $H^0(b)$, such that:

$$S \subseteq H^0(b), \quad K \subset \text{int } H^+(b), \tag{2.2.43}$$

where $H^+(b)$ is the halfspace identified by $\langle a, x \rangle \geq b$.

Proof. Because of Theorem 2.2.1, there exists a hyperplane, which contains S and disjoint from $\text{ri } K$ (possibly empty). Hence, the former of (2.2.43) is achieved; while, instead of the latter of (2.2.43), we have obtained $\text{ri } K \subset \text{int } H^+(b)$. This inclusion implies $\mu := \inf_{x \in K} \langle a, x \rangle \geq b$ and then, being $O \in \text{cl } K$, $\mu = 0$ and $b \leq 0$. If $b < 0$, then the latter of (2.2.43) is obvious. Let $b = 0$. Because of Proposition 2.1.5 (i), $\text{cl } K$ is convex, so that also $S - \text{cl } K$ is convex. Since S is affine and $\text{cl } K$ a closed cone, we have $TC(S - \text{cl } K) = S - \text{cl } K$. Because of (2.2.35)(which can be exploited since (2.2.33) holds), for any face F of K we have:

$$F \nsubseteq S \;\Rightarrow\; F \nsubseteq S - \text{cl } K \;\Rightarrow\; F \nsubseteq TC(S - \text{cl } K).$$

Consider any face F of K with $F \subseteq K$. Since $F \nsubseteq S$, and hence $F \nsubseteq TC(S - \text{cl } K)$, from the "only if part" of Theorem 2.2.7 we draw that at least one of the hyperplanes , which separate K and S — their existence has been assured before — call it H_F^0 and denote by $\langle a_F, x \rangle = 0$ its equation, cannot contain F (otherwise, $F \subseteq TC(S - \text{cl } K)$ and then $F \subset S$). Let $\mathcal{F}$ denote the family of faces of K which belong to K, and $\tilde{a}$ denote any proper convex combination of the gradients a_F, $F \in \mathcal{F}$. We have $\langle \tilde{a}, x \rangle > 0$, $\forall x \in K$, and the latter of (2.2.43) follows. $\qquad\square$

Example 2.2.4 can be used to illustrate also Theorem 2.2.8, since S is a line disjoint from K. Theorem 2.2.8 expresses disjunctive separation.

Example 2.2.6. Set $K = \{x \in \mathbb{R}^3 : x_3 - x_2 > 0, \; x_3 + x_2 > 0\} \cup \{x \in \mathbb{R}^3 : x_2 = x_3 = 0\}$ and $S = \{x \in \mathbb{R}^3 : x_1 = x_3 = 0\}$. It is immediate to see that the only plane which separates K and S has $x_3 = 0$ as equation. It does not satisfy the latter of (2.2.43). However, not all the assumptions of Theorem 2.2.8 are fulfilled. Indeed, we have $x^1 = (1, 0, 0) \in K$, $x^2 = (-2, 0, 0) \in \text{cl } K$, and $x^1 + x^2 \notin K$, so that (2.2.33) does not hold. In passing, note that K is not pointed.

Theorem 2.2.9. Let $U \subset \mathbb{R}^\ell$, $V \subset \mathbb{R}^\ell$ be nonempty, closed, convex and (properly) pointed cones with apex at the origin, and with $\text{int } U \neq \varnothing$. Set $U_0 := (U \backslash \{O_\ell\})$, $\overset{o}{U} := \text{int } U$, and $n = \ell + m$; let F_U denote any face of U different from O_ℓ. Let $S \subset \mathbb{R}^n$ be nonempty with $O \in \text{cl } S$ and such that $S - \text{cl } K$ be convex. **(i)** Both $C_0 := U_0 \times V$ and $\overset{o}{C} := \overset{o}{U} \times V$ fulfil (2.2.33). **(ii)** Whether we set $K = C_0$ or $K = \overset{o}{C}$, if $K \cap S = \varnothing$, then there exists a hyperplane H^0 through the origin, such that:

$$S - \text{cl } K \subseteq H^-, \quad K \subseteq H^+, \tag{2.2.44}$$

where H^- and H^+ are the halfspaces identified by H^0. **(iii)** Set $K := U_0 \times V$, and $F := F_U \times \{O_m\}$. If and only if

$$K \cap S = \varnothing \tag{2.2.45a}$$

and

$$F \nsubseteq TC(S - \operatorname{cl} K), \tag{2.2.45b}$$

there exists a hyperplane H^0, such that (2.2.44) holds, and

$$F \cap H^0 = \{O\}. \tag{2.2.46}$$

Proof. **(i)** Because of (2.1.8) and of the convexity of U, we have that $u^1 \in U_0$ and $u^2 \in U$ imply $u^1 + u^2 \in U_0$. From the convexity of V, we draw that $v^1, v^2 \in V$ imply $v^1 + v^2 \in V$ (Proposition 2.1.6). Therefore, $(u^1 + v^1) \in C_0$, $(u^2 + v^2) \in \operatorname{cl} C_0 = U \times V$ imply $(u^1, v^1) + (u^2, v^2) \in C_0$. Analogously, $u^1 \in \overset{o}{U}$ and $u^2 \in U$ imply that $\exists \varepsilon > 0$ s.t. $(u^1 + \varepsilon d) + u^2 \in U$, $\forall d \in N_\varepsilon(u^1)$ (N_ε denoting hypersphere with centre at u^1 and radius ε), which shows that $u^1 + u^2 \in \overset{o}{U}$; as before, $v^1, v^2 \in V$ imply $v^1 + v^2 \in V$. Thus, $(u^1, v^1) \in \overset{o}{C}$, $(u^2, v^2) \in \operatorname{cl} \overset{o}{C} = U \times V$ imply $(u^1, v^1) + (u^2, v^2) \in \overset{o}{C}$. **(ii)** It is an obvious consequence of Theorem 2.2.4(i). **(iii) If.** Because of (2.2.45a) and of the previous (ii), (2.2.44) hold here too. Ab absurdo, suppose that (2.2.46) does not hold, so that, for every hyperplane which fulfils (2.2.44), (2.2.46) is violated. Then, by applying Theorem 2.2.7 (whose assumptions are satisfied because of the previous (i)), (2.2.45b) is contradicted. **Only if.** It follows, by applying again Theorem 2.2.7. $\qquad \square$

Theorem 2.2.10. If $K \subset \mathbb{R}^n$ is nonempty, convex and compact, then:

$$K = \operatorname{conv} \operatorname{vert} K. \tag{2.2.47}$$

Proof. $\operatorname{vert} K \subseteq K \Rightarrow \operatorname{conv} \operatorname{vert} K \subseteq K$. Ab absurdo, let us suppose that there exists $\bar{x} \in K \setminus \operatorname{conv} \operatorname{vert} K$. Because of Corollary 2.2.2, $\exists a \in \mathbb{R}^n \setminus \{O\}$ and $\exists b \in \mathbb{R}$, s.t.

$$\langle a, \bar{x} \rangle = 0, \quad \langle a, x \rangle \geq b > 0, \quad \forall x \in \operatorname{conv} \operatorname{vert} K. \tag{2.2.48}$$

Because of the compactness of K and of the continuity of $\langle a, x \rangle$, $\exists x^0 \in K$ s.t.

$$\langle a, x^0 \rangle = \min_{x \in K} \langle a, x \rangle,$$

and $F := \{x \in K : \langle a, x - x^0 \rangle = 0\}$ is a compact face (a closed subset of a compact set is compact) of K and admits (at least) one extreme point, say $\hat{x}$. Hence $\hat{x} \in \operatorname{vert} K$ and then $\hat{x} \in \operatorname{conv} \operatorname{vert} K$, so that $\langle a, \hat{x} \rangle \geq b$. This inequality, $\hat{x} \in K$, and (2.2.48) lead to the contradiction:

$$\langle a, \hat{x} \rangle = \langle a, x^0 \rangle \leq \langle a, \bar{x} \rangle = 0 < b \leq \langle a, \hat{x} \rangle. \qquad \square$$

The above theorem is due to Krein and Milman [30].

2.3. Convex Functions

The concept of convex function is fundamental for the theory of constrained extrema. Even if it might be carried out as a special case of the theory of convex sets, it is useful to develop it in a functional language, as usual.

Definition 2.3.1. Let $K \subseteq \mathbb{R}^n$ be nonempty and convex. $f : K \to \mathbb{R}$ is called *convex*, iff

$$(1 - \alpha)f(x^1) + \alpha f(x^2) - f(x(\alpha)) \geq 0, \quad \forall x^1, x^2 \in K, \quad \forall \alpha \in [0, 1], \tag{2.3.1}$$

where $x(\alpha) := (1 - \alpha)x^1 + \alpha x^2$. Iff the above inequality is verified as strict inequality $\forall \alpha \in\,]0, 1[$ and $\forall x^1, x^2 \in K$ with $x^1 \neq x^2$, f is called *strictly convex*. f is (*strictly*) *concave* iff $-f$ is (*strictly*) convex. f is *affine*, iff is both convex and concave or, equivalently, iff (2.3.1) holds as equality or iff differs from a linear function because of a constant.

In the above definition, by convex function is meant what often is called *proper convex function*: a function whose epigraph is nonempty and does not contain vertical lines, or $f(x) < +\infty$ for at least one $x \in K$ and $f(x) > -\infty$, $\forall x \in K$, or $\text{dom}\, f \neq \varnothing$ and f is finite on K. The graph of a strictly convex function does not contain any (nondegenerate) segment (of $K \times \mathbb{R}$). If $n \geq 2$, f is strictly convex, and $\text{card lev}_{\leq \alpha} f > 1$ with $\alpha \in \text{Im} f$, then $\text{frt lev}_{\leq \alpha} f$ does not contain any (nondegenerate) segment (of K). The contrary happens, for $n \geq 1$, if f is convex, but not strictly.

Theorem 2.3.1. Let $K \subseteq \mathbb{R}^n$ be nonempty and convex, and $f : K \to \mathbb{R}$. **(i)** f is convex if and only if $\text{epi}\, f$ is convex. **(ii)** If, $\forall \overline{x} \in \text{ri}\, K$, there exists $\sigma \in \mathbb{R}^n$ (depending on $\overline{x}$), such that:

$$E(\overline{x}, x, \sigma) := f(x) - f(\overline{x}) - \langle \sigma, x - \overline{x} \rangle \geq 0, \quad \forall x \in K, \tag{2.3.2a}$$

then f is convex on $\text{ri}\, K$; if f is convex on K, then (2.3.2a) holds, $\forall \overline{x} \in \text{ri}\, K$; if f is differentiable, then (2.3.2a) becomes:

$$E(\overline{x}, x, f'(\overline{x})) = f(x) - f(\overline{x}) - \langle f'(\overline{x}), x - \overline{x} \rangle \geq 0, \quad \forall x \in K. \tag{2.3.2b}$$

If f is continuous on K, then (2.3.2a) with $\overline{x} \in K$ is necessary and sufficient for the convexity of f on K. **(3i)** f is convex, if and only if

$$\rho(f) := \sup \left[f\left(\sum_{i=1}^{r} \alpha_i x^i\right) - \sum_{i=1}^{r} \alpha_i f(x^i) \right] = 0, \tag{2.3.3}$$

where the supremum is performed with respect to all possible vectors $(\alpha_1, ..., \alpha_r)$ with $\alpha_i \in [0, 1], i = 1, ..., r, \sum_{i=1}^{r} \alpha_i = 1$, and all possible sets of r vectors $x^1, ..., x^r \in K$, and whatever the positive integer r may be. **(4i)** Let $K = \mathbb{R}^n$. f is convex if and only if, $\forall d \in \mathbb{R}^n$ with $||d|| = 1$, $\forall x \in \mathbb{R}^n$, we have:

$$[Q(t_2) - Q(t_1)](t_2 - t_1) \geq 0, \quad \forall t_1, t_2 \in \mathbb{R}_+ \setminus \{0\}, \tag{2.3.4}$$

where $Q(t) := \frac{1}{t}[f(x+td) - f(x)]$. **(5i)** Let $K = \mathbb{R}^n$. f is convex if and only if, $\forall x, x^1, x^2 \in \mathbb{R}^n$ with $x \in]x^1, x^2[$, we have:

$$\frac{f(x) - f(x^1)}{||x - x^1||} \leq \frac{f(x^2) - f(x^1)}{||x^2 - x^1||} \leq \frac{f(x^2) - f(x)}{||x^2 - x||}. \tag{2.3.5}$$

Proof. Set $y^1 := f(x^1)$, $y^2 := f(x^2)$, $x(\alpha) := (1-\alpha)x^1 + \alpha x^2$, $y(\alpha) := (1-\alpha)y^1 + \alpha y^2$, $z(\alpha) := (1-\alpha)z^1 + \alpha z^2$ (see Fig. 2.3.1). **(i)** (2.3.1) holds iff

$$z(\alpha) \geq f(x(\alpha)), \quad \forall \alpha \in [0,1], \quad \forall x^1, x^2 \in K, \quad \forall z^1 \geq f(x^1), \quad \forall z^2 \geq f(x^2), \tag{2.3.6}$$

or iff

$$\text{conv } \{(x^1, z^1), (x^2, z^2)\} = [(x^1, z^1), (x^2, z^2)] \subset \text{epi } f, \tag{2.3.7}$$

which allows us to conclude by noting that (x^1, z^1) and (x^2, z^2) are any elements of epi f. **(ii)** Let (2.3.2a) hold, and consider any $x^1, x^2 \in \text{ri } K$, so that, $\forall \alpha \in [0,1]$, $x(\alpha) \in \text{ri } K$. Consider the two inequalities obtained from (2.3.2a) for $x = x^1$ and for $x = x^2$; multiply them, respectively, by $1 - \alpha$ and by α, and sum them side by side; set $\bar{x} = x(\alpha)$. We obtain $y(\alpha) - f(x(\alpha)) \geq 0$, and then (2.3.6), with ri K in place of K, follows (or $\forall \alpha \in [0,1]$, $z(\alpha) \in \text{epi } f_{/\text{ri } K}$) and (i) leads to the conclusion. Now, let f be convex on K. Because of (i), epi f is convex. $\forall \hat{x} \in K$, epi f and the singleton $\{(\hat{x}, f(\hat{x}))\}$ are identified with K_1 and K_2 of Theorem 2.2.4 (ii); it gives the existence of a separating

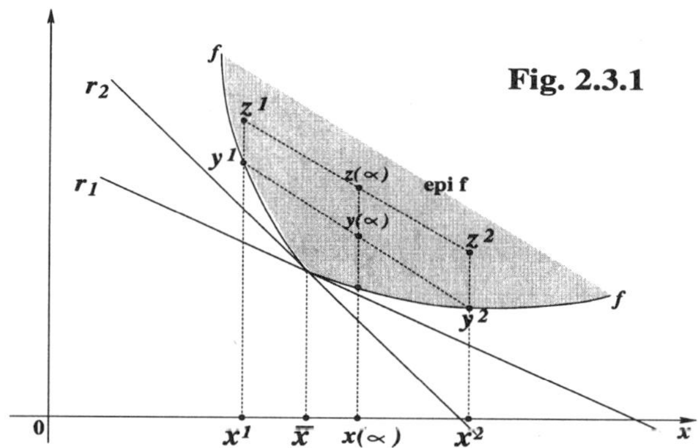

hyperplane (the lines r_1 and r_2 in Fig. 2.3.1 and their convex combination), which, being $(\hat{x}, f(\hat{x})) \in \text{epi } f$, is also a supporting hyperplane for epi f. Its equation is $y = f(\hat{x}) + \langle \sigma, x - \hat{x} \rangle$ with $\sigma \in \mathbb{R}^n$. (2.3.2) follows. When f is differentiable, $\sigma = f'(\hat{x})$ comes from the very definition of differentiability. The last part is achieved by noting that now epi f is closed and repeating the above reasoning under this additional property. **(3i)** It is enough to note that (2.3.1) holds iff, $\forall (\alpha_1, ..., \alpha_r)$ with $\alpha_i \in [0,1]$, $i = 1, ..., r$ and $\sum_{i=1}^{r} \alpha_i = 1$, and for each set of r vectors $x^1, ..., x^r \in K$, we have:

$$f(\sum_{i=1}^{r} \alpha_i x^i) \leq \sum_{i=1}^{r} \alpha_i f(x^i). \tag{2.3.8}$$

The fact that (2.3.8), called *Jensen Inequality* [28], be equivalent to (2.3.1) is elementary. **(4i) Only if.** Let $t, u \in \mathbb{R}_+\backslash\{0\}$. Since $x + td = \frac{u}{t+u}x + \frac{t}{t+u}[x + (t + u)d]$, we have:

$$Q(t + u) - Q(t) = \frac{f(x + (t + u)d) - f(x)}{t + u} - \frac{f(x + td) - f(x)}{t} \geq$$

$$\geq \frac{f(x + (t + u)d) - f(x)}{t + u} - \frac{\frac{u}{t+u}f(x) + \frac{t}{t+u}f(x + (t + u)d) - f(x)}{t} = 0$$

where the inequality is a consequence of (2.3.1). Since $(t + u) - t = u > 0$, (2.3.4) follows. **If.** Ab absurdo, suppose that $\exists x^1, x^2 \in \mathbb{R}^n$, $\exists \alpha \in]0, 1[$, s.t.

$$f(x(\alpha)) > (1 - \alpha)f(x^1) + \alpha f(x^2) \quad \text{or} \quad \alpha[f(x^2) - f(x^1)] < f(x^1 + \alpha(x^2 - x^1)) - f(x^1).$$

From the last inequality, by setting $d := x^2 - x^1$ so that $x^2 = x^1 + d$, we draw:

$$Q(1) = f(x^1 + d) - f(x^1) < \frac{1}{\alpha}[f(x^1 + \alpha d) - f(x^1)] = Q(\alpha),$$

and then $[Q(1) - Q(\alpha)](1 - \alpha) < 0$, which contradicts (2.3.4). **(5i) Only if.** Consider any $t_1, t_2 \in \mathbb{R}_+\backslash\{0\}$ with $t_1 < t_2$. From (2.3.4) we have $Q(t_1) < Q(t_2)$ or, $\forall d \in \mathbb{R}^n$ with $||d|| = 1$,

$$\frac{f(y + t_1 d) - f(y)}{t_1} \leq \frac{f(y + t_2 d) - f(y)}{t_2}, \quad \forall y \in \mathbb{R}^n. \tag{2.3.9}$$

By setting $y = x^1$, $y + t_1 d = x$ and $y + t_2 d = x^2$, so that $t_1 = ||x - x^1||$ and $t_2 = ||x^2 - x^1||$, (2.3.9) becomes the former of (2.3.5). By setting $y = x^2$, $y + t_1 d = x$ and $y + t_2 d = x^1$, so that $t_1 = ||x^2 - x||$ and $t_2 = ||x^2 - x^1||$, (2.3.9) becomes the latter of (3.2.5). From the above positions, we draw:

$$0 < ||x - x^1|| < ||x^2 - x^1||, \qquad 0 < ||x^2 - x|| < ||x^2 - x^1||.$$

Because of the Triangle Inequality, the above inequalities imply $x \in]x^1, x^2[$. **If.** It is enough to follow the above deductions in reverse order. $\square$

In the last part of Theorem 2.3.1(ii), there is the assumption of continuity. This means that the mere convexity on K does not guarantee the continuity as the following Example 2.3.1 shows. (2.3.4) expresses the isotonicity of $Q(t)$; see Definition 2.3.2.

Without any loss of generality, in (2.3.3) we can assume $r \leq n + 2$. In fact,

$$\left(\sum_{i=1}^{r} \alpha_i x^i, \ \sum_{i=1}^{r} \alpha_i f(x^i) \right)$$

is a convex combination of elements of graph f and then, by Proposition 2.1.2, belongs to conv gr f. Therefore, because of Theorem 2.1.1, it can be expressed as convex combination of at most $n + 2$ elements of gr f.

The inequalities (2.3.5) are intuitively obvious, if interpreted in terms of quotient ratios.

Example 2.3.1. Set $K = \{x \in \mathbb{R}^2 : 0 \leq x_i \leq 1,\ i = 1, 2\}$, $f(x) = 0$ if $0 \leq x_1 \leq 1$ and $0 < x_2 \leq 1$, and $f(x) = 1 - x_1^2$ if $0 \leq x_1 \leq 1$ and $x_2 = 0$ (see Fig. 2.3.2). Observe that, $\forall \overline{x} \in \operatorname{ri} K$, (2.3.2) is satisfied for $t = (0, 0)$, but f is not convex on K. This shows that the 2nd part of (ii) of Theorem 2.3.1 cannot be inverted. It shows also that the validity of (2.3.2) on $\operatorname{ri} K$ does not give a meaningful information of the behaviour of f on $K \backslash \operatorname{ri} K$. $\qquad\qquad\Box$

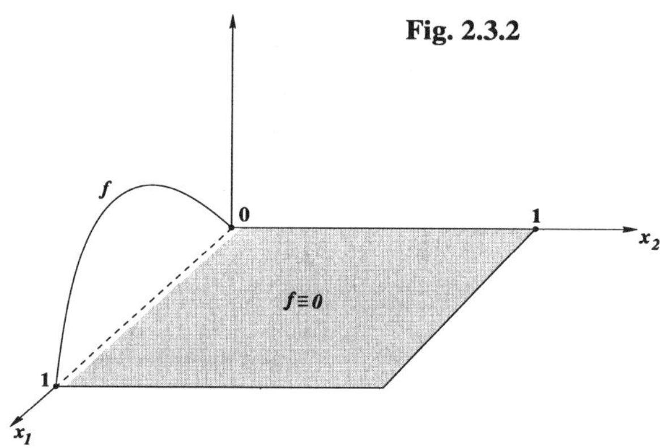

Note that, in the proof of Theorem 2.3.1 (ii), we can exploit also (ii) of Theorem 2.2.4 to obtain even proper separation, since the relative interiors of epi f and a finite element of graph f are obviously disjoint.

The function E which appears in (2.3.2) expresses support of K at $\overline{x}$; it is called *Weierstrass Excess Function* (for short, E-*function*) since it was introduced by Weierstrass; it has shown to be useful. If σ exists and is unique, then E is infinitesimal of higher order (see the comment after Definition 2.3.3). (2.3.8) is called *Jensen Inequality*.

Among the several consequences which can be derived from Theorem 2.3.1, its (5i) implies that a not constant convex function, having the entire space as domain, cannot be bounded from above.

Theorem 2.3.2. Let $K \subseteq \mathbb{R}^n$ be nonempty and convex, and $f : K \to \mathbb{R}$. **(i)** A necessary condition for f to be (strictly) convex on K is that $\operatorname{lev}_{\leq \alpha} f$ be (strictly) convex $\forall \alpha \in \mathbb{R}$. **(ii)** A sufficient (but, obviously, not necessary) condition for f to be (strictly) convex on K is that $\operatorname{lev}_{\leq \alpha} f$ be (strictly) convex $\forall \alpha \in \mathbb{R}$, and $\exists a \in \mathbb{R}^n$ s.t., up to a translation, the graph of f be a cone with apex at the origin and $f(x) \geq \langle a, x \rangle, \forall x \in K$.

Proof. **(i)** $\forall x^1, x^2 \in \operatorname{lev}_{\leq \alpha} f$, the convexity of f implies:

$$f(x(\alpha)) \leq (1 - \alpha) f(x^1) + \alpha f(x^2) \leq \alpha,$$

where $x(\alpha) := (1-\alpha)x^1 + \alpha x^2$. Hence $x(\alpha) \in \mathrm{lev}_{\leq\alpha}f$, $\forall \alpha \in [0,1]$. **(ii)** It is enough to prove the thesis for $a = O$; the case $a \neq O$ can be, obviously, reduced to the previous one by replacing $f(x)$ with $f(x) - \langle a, x \rangle$. Then let $a = O$. Ab absurdo, suppose that $\exists x^1, x^2 \in K$ and $\exists \alpha \in]0,1[$ s.t.

$$f(x(\alpha)) > (1-\alpha)f(x^1) + \alpha f(x^2),$$

where $x(\alpha) := (1-\alpha)x^1 + \alpha x^2$. It is not restrictive to assume $f(x^1) \geq f(x^2) \geq 0$. Since graph f is a non-negative cone, then $\exists \beta \in [0,1]$ s.t. $f(\beta x^1) = f(x^2)$. If $f(x^1) = 0$, set $\beta = 1$; otherwise, set $\beta = f(x^2)/f(x^1)$. We have $\alpha\beta + 1 - \alpha > 0$ and thus:

$$\frac{1-\alpha}{1-\alpha+\alpha\beta}(\beta x^1) + \frac{\alpha\beta}{1-\alpha+\alpha\beta}(x^2) = \frac{\beta}{1-\alpha+\alpha\beta}x(\alpha),$$

so that

$$x(\alpha,\beta) := \frac{\beta}{1-\alpha+\alpha\beta}x(\alpha) \in [\beta x^1, x^2].$$

By multiplying both sides of the absurd inequality by $\beta/(1-\alpha+\alpha\beta)$, we obtain the inequality:

$$f(x(\alpha,\beta)) > \frac{1-\alpha}{1-\alpha+\alpha\beta}f(\beta x^1) + \frac{\alpha\beta}{1-\alpha+\alpha\beta}f(x^2) = f(x^2),$$

which, taking into account that $x(\alpha,\beta) \in [\beta x^1, x^2]$, contradicts the convexity of the lower sets of f if $\beta x^1 \neq x^2$, or becomes $f(x^2) > f(x^2)$ if $\beta x^1 = x^2$. Obvious changes are required for the strict case. $\qquad\square$

If the assumption of non-negativity (or nonpositivity) is removed in the above theorem, then the thesis may not hold, as simple examples show; take, for instance, $f : \mathbb{R} \to \mathbb{R}$ with $f(x) = 2x$ if $x \leq 0$, $f(x) = x$ if $x > 0$. The fact that the necessary condition, expressed by (i) of the above theorem, be not sufficient is trivial: every nondecreasing or nonincreasing $f : \mathbb{R} \to \mathbb{R}$ proves it; see also next Example 2.3.2. Analogous remark holds for the strict case.

Example 2.3.2. Set $K = \mathbb{R}$ and $f(x) = x^2/(1+x^2)$. We have:

$$\mathrm{lev}_{\leq\alpha}f = \left\{ x \in \mathbb{R} : \; -\sqrt{\frac{\alpha}{1-\alpha}} \leq x \leq \sqrt{\frac{\alpha}{1-\alpha}} \right\}, \quad \alpha \in [0,1[,$$

which is obviously convex. Note that f has one and only one m.p.; see Fig. 2.3.3. $\square$

Theorem 2.3.3. Let $K \subseteq \mathbb{R}^n$ be nonempty and convex, and $f : K \to \mathbb{R}$. A necessary condition for f to be convex is that:

$$(1-\alpha)f(x^1) + \alpha f(x^2) \geq \inf_{x\in K\cap N} f(x), \quad \forall x^1, x^2 \in K, \quad \forall \alpha \in [0,1], \qquad (2.3.10)$$

where N denotes any neighbourhood of $x(\alpha) := (1-\alpha)x^1 + \alpha x^2$.

Proof. Since obviously $f(x(\alpha)) \geq \inf_{x\in K\cap N} f(x)$, (2.3.1) implies (2.3.10). The instance $K = \mathbb{R}$, $f(x) = x^2$ if $x \neq 0$, $f(x) = 1$ if $x = 0$, shows that (2.3.10) is not sufficient. $\square$

Trivial examples show that the condition expressed by Theorem 2.3.2 is not sufficient.

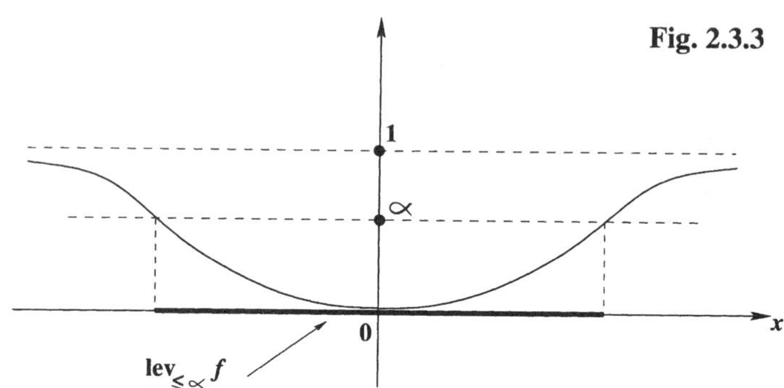

Definition 2.3.2. Let $K \subseteq \mathbb{R}^n$ be nonempty and convex. $F : K \to \mathbb{R}^n$ is called *isotone* (or *monotone nondecreasing*) iff

$$\langle F(x^1) - F(x^2), x^1 - x^2 \rangle \geq 0, \quad \forall x^1, x^2 \in K. \tag{2.3.11a}$$

It is called *antitone* (or *monotone nonincreasing*) iff

$$\langle F(x^1) - F(x^2), x^1 - x^2 \rangle \leq 0, \quad \forall x^1, x^2 \in K. \tag{2.3.11b}$$

It is called *strictly isotone* or *strictly antitone* iff, respectively, in (2.3.11) the inequalities are strictly verified when $x^1 \neq x^2$. When $n = 1$, the terms isotone, strictly isotone, antitone and strictly antitone are synonyms of *nondecreasing, increasing, nonincreasing* and *decreasing*, respectively.

Theorem 2.3.4. Let $K \subseteq \mathbb{R}^n$ be nonempty and convex. We have:

(i) If $f_1, ..., f_r : K \to \mathbb{R}$ are (strictly) convex, then $f_1 + ... + f_r$ is (strictly) convex.

(ii) If $f : K \to \mathbb{R}$ is convex and $\varphi : \mathbb{R} \to \mathbb{R}$ is convex and isotone, then $F := \varphi \circ f$ is convex. If, furthermore, f is strictly convex and φ is increasing, then F is strictly convex.

(3i) If $f : K \to \mathbb{R}$ is convex, $A \in \mathbb{R}^{n \times m}$ and $b \in \mathbb{R}^n$, then $F(x) := f(Ax + b), x \in \mathbb{R}^n$, is convex. If, furthermore, f is strictly convex, $n = m$ and $\det A \neq 0$, then F is strictly convex.

(4i) Let Ξ be any set of parameters and $\{f_\xi(x), \xi \in \Xi\}$ be any family of convex functions $f_\xi : K \to \mathbb{R}$. $F(x) := \sup_{\xi \in \Xi} f_\xi(x)$ is convex, possibly with empty epigraph.

Proof. (i) It is an obvious consequence of Definition 2.3.1. The case of convexity comes also from Theorem 2.3.1(i), and from the fact that the convexity of $\mathrm{epi} f_1, ..., \mathrm{epi} f_r$

implies the convexity of $\sum_{i=1}^{r} \text{epi} \, f_i$ and then the convexity of $\text{epi} \sum_{i=1}^{r} f_i$. **(ii)** Let us set $x(\alpha) := (1 - \alpha)x^1 + \alpha x^2$. $\forall x^1, x^2 \in K$, and, $\forall \alpha \in [0, 1]$, we have:

$$F(x(\alpha)) = \varphi(f(x(\alpha))) \leq \varphi((1 - \alpha)f(x^1) + \alpha f(x^2)) \leq$$

$$\leq (1 - \alpha)\varphi(f(x^1)) + \alpha\varphi(f(x^2)) = (1 - \alpha)F(x^1) + \alpha F(x^2),$$

where the 1st inequality is due to the convexity of f and the isotonicity of φ (which allows one to use (2.3.11a)), and the 2nd inequality is a consequence of the convexity of φ. If φ is increasing and f strictly convex, then the above inequalities are both strict (of course, only one of the two assumptions is not enough, as shown by trivial examples: $\varphi(y) = y$, $f(x) = x$). **(3i)** Set $x(\alpha) := (1 - \alpha)x^1 + \alpha x^2$. $\forall x^1, x^2 \in K$, $\forall \alpha \in [0, 1]$, we have:

$$F(x(\alpha)) = f((1 - \alpha)(Ax^1 + b) + \alpha(Ax^2 + b)) \leq (1 - \alpha)F(x^1) + \alpha F(x^2).$$

Under the additional assumptions, we have $Ax^1 + b = Ax^2 + b$ iff $x^1 = x^2$; then, if $x^1 \neq x^2$ the above inequality is strict verified. **(4i)** Because of Proposition 2.1.1, $\bigcap_{\xi \in \Xi} \text{epi} \, f_\xi(x)$ is convex (but not necessarily $\neq \varnothing$). It is easy to show that $\bigcap_{\xi \in \Xi} \text{epi} \, f_\xi(x) = \text{epi} \, F(x)$. Then F is convex due to Theorem 2.3.1 (i). $\qquad\qquad\qquad\qquad\qquad\qquad\qquad\qquad\qquad\qquad\qquad\quad\square$

Among the several applications of Theorem 2.3.4, let us consider a few special cases. *If f is (strictly) convex on K, then $\exp f(x)$ is (strictly) convex*; it comes from (ii). *If $K \subset \mathbb{R}^n$ is convex and bounded, and $f : K \to]0, +\infty[$ is (strictly) concave, then $-\log f(x)$ is (strictly) convex*; in fact, $-\log(-y)$ with $y < 0$ is increasing and $y = -f(x)$ is (strictly) convex; then apply again (ii). *If $f(x) \geq 0$ and (strictly) convex on K, and $\alpha \in [1, +\infty[$, then $f(x)^\alpha$ is strictly convex*; apply again (ii) with $\varphi(y) = y^\alpha$.

If f is convex, then $\max\{0, f(x)\}$ is convex; in fact, its epigraph is the intersection of that of f and a closed halfspace; then apply Proposition 2.1.1 and Theorem 2.3.1 (i).

A special but very important class of convex functions is that of differentiable ones, for which we recall the following results.

Theorem 2.3.5. Let $K \subseteq \mathbb{R}^n$ be nonempty, open and convex, and $f : K \to \mathbb{R}$ be twice differentiable on K. f is convex (or strictly convex) on K, if and only if its Hessian matrix $f''(x)$ is positive semidefinite (or definite) on K.

Proof. Only if. Consider any $x^1, x^2 \in K$, $\alpha \in \mathbb{R}$, and set $y = x^1 + \alpha x^2$. Because of the double differentiability, we have:

$$f(y) = f(x^1) + \alpha\langle f'(x^1), x^2 \rangle + \tfrac{1}{2}\alpha^2 \langle x^2, f''(x^1)\, x^2 \rangle + \alpha^2 \varepsilon(\alpha),$$

where $\varepsilon(\alpha)$ is infinitesimal with respect to α. Because of the convexity of f, from (2.3.2) we draw:

$$f(y) - f(x^1) - \alpha\langle f'(x^1), x^2 \rangle \geq 0.$$

A comparison of the 2 above relations leads to:

$$\tfrac{1}{2}\alpha^2 \langle x^2, f''(x^1)\,x^2 \rangle + \alpha^2 \varepsilon(\alpha) \geq 0.$$

Dividing both sides of the above inequality by $\alpha^2/2$ and letting α go to zero, we obtain $\langle x^2, f''(x^1)x^2 \rangle \geq 0$, and therefore ($x^2$ being arbitrary) $f''(x^1)$ turns out to be positive semidefinite. **If.** Because of Taylor Theorem, $\forall x^1, x^2 \in K$, we have:

$$f(x^2) = f(x^1) + \langle f'(x^1), x^2 - x^1 \rangle + \tfrac{1}{2}\langle x^2 - x^1, f''(z(\alpha))(x^2 - x^1) \rangle,$$

where $z(\alpha) := (1-\alpha)x^1 + \alpha x^2$ and $\alpha \in\,]0, 1[$ is suitable. Since f'' is positive semidefinite, from the above equality we deduce:

$$f(x^2) \geq f(x^1) + \langle f'(x^1), x^2 - x^1 \rangle,$$

which expresses the non-negativity of the Weierstrass E-function, so that the convexity of f comes from Theorem 2.3.1 (ii). With regard to the strict convexity, it is enough to note that, if $x^1 \neq x^2$, the above inequalities hold in strict sense. $\qquad\square$

Theorem 2.3.6. Let $K \subseteq \mathbb{R}^n$ be nonempty, open and convex, and $f : K \to \mathbb{R}$ be differentiable on K. f is convex on K, if and only if f' is isotone, or:

$$\langle f'(x^2) - f'(x^1), x^2 - x^1 \rangle \geq 0, \quad \forall x^1, x^2 \in K. \tag{2.3.12}$$

f is strictly convex, if and only if the above inequality holds in strict sense when $x^1 \neq x^2$.

Proof. Only if. We can apply Theorem 2.3.1 (ii) with $\sigma = f'(\overline{x})$; from (2.3.2) we have:

$$E(x^2, x^1, f'(x^2)) + E(x^1, x^2, f'(x^1)) = \langle f'(x^2) - f'(x^1), x^2 - x^1 \rangle \geq 0.$$

If. Set $x(\alpha) := (1 - \alpha)x^1 + \alpha x^2$. By Lagrange Mean-Value Theorem, $\exists \alpha \in\,]0, 1[$ s.t.

$$f(x^2) - f(x^1) = \langle f'(x(\alpha)), x^2 - x^1 \rangle, \quad \forall x^1, x^2 \in K. \tag{2.3.12$'$}$$

From (2.3.12) we have:

$$\langle f'(x(\alpha)) - f'(x^1), x(\alpha) - x^1 \rangle = \alpha \langle f'(x(\alpha)) - f'(x^1), x^2 - x^1 \rangle \geq 0$$

and hence $\langle f'(x(\alpha)), x^2 - x^1 \rangle \geq \langle f'(x^1), x^2 - x^1 \rangle$. This inequality and (2.3.12$'$) imply $E(x^1, x^2, f'(x^1)) \geq 0$. Therefore, K being open, Theorem 2.3.1 (ii) gives the convexity of f. With regard to the strict convexity, it is enough to note that the above inequalities hold in strict sense, when $x^1 \neq x^2$. $\qquad\square$

If (2.3.12) is replaced by (2.3.11b), then concavity is characterized, instead of convexity. Note that, when f is derivable, if we set $H(x, y) := \langle f'(x), y \rangle - f(y)$, then inequality (2.3.2a) holds iff

$$H(\overline{x}, x) \leq H(\overline{x}, \overline{x}), \quad \forall x \in K. \tag{2.3.2a$'$}$$

In fact, we have $E(\overline{x}, x, f'(\overline{x})) = H(\overline{x}, \overline{x}) - H(\overline{x}, x)$. This allows another geometric interpretation of a convex function: consider the hyperplane through the origin and with gradient $f'(\overline{x})$; call it π. Let $z_{\overline{x}}$ and z_x be the ordinates of the points of π having $\overline{x}$ and x as abscissas, respectively. Then, $H(\overline{x}, \overline{x}) = z_{\overline{x}} - f(\overline{x})$ and $H(\overline{x}, x) = z_x - f(x)$, and (2.3.2a)$'$ says that, at every x, such a difference (between the ordinate of π and that of f) must have its maximum at $\overline{x}$. This will receive an interpretation in terms of conjugate function (see (2.3.21) and subsequent remarks).

Example 2.3.3. An important special case is that of a quadratic function:

$$f(x) = a_o + \langle a, x \rangle + \frac{1}{2}\langle x, Ax \rangle, \quad x \in \mathbb{R}^n, \tag{2.3.13}$$

where $a_o \in \mathbb{R}$ and the matrix A is symmetric. (2.3.1) is equivalent to:

$$\langle x^1 - x^2, A(x^1 - x^2) \rangle \geq 0, \quad \forall x^1, x^2 \in \mathbb{R}^n, \tag{2.3.14}$$

and agrees with Theorem 2.3.5. Also (2.3.12) is equivalent to (2.3.14) in agreement with Theorem 2.3.6.

Example 2.3.4. Set $K = \mathbb{R}$ and, as function f of Definition 2.3.1, consider the integrand in (1.2.8) as function of x' or $\psi_0(x') = \sqrt{1 + x'^2}$ (see also (1.2.3a)). Because of Theorem 2.3.5, being $\psi_0'' = 1/(1 + x'^2)^{3/2} > 0$, ψ_0 is strictly convex. Consider also the functional to be minimized in (1.2.8), namely (without any fear of confusion, next f does not act as that of Definition 2.3.1, which is replaced here by ψ_0):

$$f(x) = \int_T \sqrt{1 + x'^2} \ dt, \quad x \in C^1(T).$$

The above functional is l.s.c., even if not continuous. Indeed, if $\{x_r(t), \, t \in T\}_1^\infty$ is a sequence of functions in $C^1(T)$ s.t. $\exists x = \lim_{x \to +\infty} x_r$, the equality $\lim_{r \to +\infty} f(x_r) = f(x)$ does not hold necessarily. This is shown by simple instances. Take, for instance,

$$x_r(t) = \frac{\sin rt}{\sqrt{r}}, \quad t \in T = [0, \pi], \quad r = 1, 2, \dots \ .$$

From one side, since $0 \leq |\sin rt| \leq 1$, $x = \lim_{r \to +\infty} x_r = 0$, $\forall t \in T$, and then $f(x) = \pi$. From the other side, since

$$f(x_r) = \int_T \sqrt{1 + r\cos^2 rt} \ dt > \int_T \sqrt{r \cos^2 rt} \ dt = \sqrt{r} \int_0^\pi |\cos rt| \ dt =$$

$$= \frac{\sqrt{r}}{r} \int_0^{r\pi} |\cos \tau| \ d\tau = 2\sqrt{r},$$

we have $\lim_{r \to +\infty} f(x_r) = +\infty$. Now, let us show that f is l.s.c. Because of the convexity of ψ_0, from (2.3.2a) we have:

$$\sqrt{1 + x_r'^2} \geq \sqrt{1 + x'^2} + \frac{x'}{\sqrt{1 + x'^2}}(x_r' - x'), \quad \forall x \in C^1(T).$$

By integrating both sides on T, we find:

$$f(x_r) \geq f(x) + \int_T \frac{x'}{\sqrt{1 + x'^2}}(x'_r - x')\, dt, \quad \forall x \in C^1(T) =$$

$$= f(x) + \int_T (x'_r - x')\frac{df}{dx'}\, dt = f(x) + [x_r - x]_{t_0}^{t_1} - \int_T (x_r - x)\frac{d^2 f}{dx'^2}\, dt \geq$$

$$\geq f(x) - 2\delta - f^0 \delta(t_1 - t_0),$$

where

$$f^0 := \max_{t \in T}\left|\frac{d^2 f}{dx'^2}\right|; \qquad |x_r(t) - x(t)| < \delta, \quad \forall t \in T. \tag{2.3.14'}$$

Therefore, if $\delta < \varepsilon/[2 + f^0(t_1 - t_0)]$, then we have that, $\forall \varepsilon > 0$, $\exists \delta > 0$ s.t. for each $x_r \in C^1(T)$ which fulfils the latter of (2.3.14)', the inequality

$$f(x_r) > f(x) - \varepsilon$$

holds and shows that f is l.s.c. (the assumptions on $x_r(t)$ and $x(t)$ are more than necessary). Note that f cannot be u.s.c.; in fact, from $f(x_r) < f(x) + \varepsilon$ we would draw $|f(x) - f(x_r)| < \varepsilon$, so that $\lim_{r \to +\infty} f(x_r) = f(x)$ and f would be continuous. $\qquad \square$

Example 2.3.4 points out the importance of connections between convexity and continuity of a function. Unfortunately, we cannot claim that the two classes are ordered. Obviously, a continuous function may be not convex, and a convex function may be discontinuous (take, for instance, $K =]-1, +\infty[$, $f(-1) = 2$ and $f(x) = x^2$ if $x > -1$). However, with a suitable limitation, the latter claim becomes true, as shown by Theorem 2.3.7.

Lemma 2.3.1. Let $K \subseteq \mathbb{R}^n$ be nonempty and convex and $f : K \to \mathbb{R}$ be convex. **(i)** $\forall \overline{x} \in \operatorname{ri} K$, there exists an (open) neighbourhood of $\overline{x}$, where f is bounded from above by a finite constant. **(ii)** f is upper semicontinuous.

Proof. Without any loss of generality, we can suppose that:

$$d := \dim \operatorname{aff} K > 0.$$

(i) Let $\overline{x} \in \operatorname{ri} K$, so that $\exists \delta > 0$ s.t.

$$C_\delta(\overline{x}) := \{x \in \mathbb{R}^n : |x_i - \overline{x}_i| \leq \delta, \ i = 1, ..., n\} \cap \operatorname{aff} K \subseteq K,$$

$$N_\delta(\overline{x}) := \{x \in \mathbb{R}^n : ||x - \overline{x}|| < \delta\} \cap \operatorname{aff} K \subset C_\delta(\overline{x}).$$

Let x^i, $i = 1, ..., 2^d$ be the vertices of the hypercube $C_\delta(\overline{x})$. Because of (2.3.8), we have:

$$f(x) \leq k := \max\{f(x^i), \ i = 1, ..., 2^d\}, \quad \forall x \in C_\delta(\overline{x}).$$

Hence, f is bounded in the open hypersphere $N_\delta(\overline{x})$. **(ii)** Ab absurdo, suppose that f be not u.s.c., so that $\operatorname{hypo} f$ is not closed. Hence, $\exists \hat{x} \in K$ and $\exists \hat{y} \in \mathbb{R}$, s.t. $\hat{y} > f(\hat{x})$

and $(\hat{x}, \hat{y})$ is an accumulation point of gr f. Therefore, $\exists \{x^i\}_1^\infty \subset K$, with $x^i \neq \hat{x}$ and $\lim\limits_{i \to +\infty} x^i = \hat{x}$, s.t. $\hat{y} = \lim\limits_{i \to +\infty} f(x^i)$. Since these conditions are satisfied by the restriction of f to any ray of K with apex at $\hat{x}$, it is not restrictive to assume that $x^i \in [\hat{x}, x^1]$, $\forall i$. Since $\hat{y} > f(\hat{x})$, $\exists r, s \in \mathbb{N}$, s.t.

$$x^s \in]\hat{x}, x^r[, \quad f(\hat{x}) < f(x^r) \leq f(x^s).$$

Thus, $\exists \alpha \in]0, 1[$, s.t. $x^s = (1 - \alpha)\hat{x} + \alpha x^r$. Because of the convexity of f, we reach a contradiction:

$$f(x^s) = f((1 - \alpha)\hat{x} + \alpha x^r) = (1 - \alpha)f(\hat{x}) + \alpha f(x^r) < f(x^s). \qquad \square$$

Theorem 2.3.7. Let $K \subseteq \mathbb{R}^n$ be nonempty and convex, and $f : K \to \mathbb{R}$ be convex. Then f is continuous on ri K.

Proof. Without any loss of generality, we can suppose that dim aff $K > 0$, that $O \in$ ri K, and that $f(O) = 0$. Because of (i) of Lemma 2.3.1, $\exists \delta \in \mathbb{R}_+ \setminus \{0\}$, $\exists k \in \mathbb{R}$, s.t. f is bounded from above by k in $N_\delta := \{x \in \mathbb{R}^n : ||x|| < \delta\} \cap$ aff K. Consider any $x \in N_\delta \setminus \{O\}$, and set $\hat{x} := (\delta/||x||)x$, so that $\hat{x} \in$ frt N_δ and $||\hat{x}|| = \delta$. Because of (2.3.1),

$$x \in]0, \hat{x}] \;\; \Rightarrow \;\; f(x) \leq \frac{||\hat{x} - x||}{\delta} f(O) + \frac{||x||}{\delta} f(\hat{x}) \leq \frac{k}{\delta}||x||;$$

$$O \in] - \hat{x}, x[\;\; \Rightarrow \;\; f(O) \leq \frac{||x||}{\delta + ||x||} f(-\hat{x}) + \frac{\delta}{\delta + ||x||} f(x)$$

or

$$f(x) \geq -\frac{||x||}{\delta} f(-\hat{x}) \geq -\frac{k}{\delta}||x||.$$

From the above relations, we draw the inequalities:

$$-\frac{k}{\delta}||x|| \leq f(x) \leq \frac{k}{\delta}||x||,$$

which show the continuity of f at O. $\qquad \square$

With the above proof we have achieved also that *a convex function is Lipschtzian on any closed and bounded subset of the relative interior of its domain.*

If the domain of f is not finite dimensional, then Lemma 2.3.1 is no longer true, and its thesis is taken as hypothesis, if we want to guarantee the continuity of a convex function. Even if the domain is finite dimensional and f is closed — in the sense that cl epi $f =$ epi f — its convexity does not guarantee its continuity on the frontier of the domain (unless it is a subset of $\mathbb{R}$), as Example 2.3.5 shows. Besides this, let us observe also that cl epi $f =$ epi $f \not\Rightarrow$ cl dom $f =$ dom f; take, for instance, $K = \mathbb{R}$, $f(x) = 1/x$ if $x > 0$ and $f(x) = +\infty$ if $x \leq 0$.

Example 2.3.5. Let us set $K = \{x \in \mathbb{R}^2 : x_1 \geq 0\}$, $f(x) = x_2^2/x_1$ if $x_1 > 0$,

$f(0,0) = 0$, and $f(0,x_2) = +\infty$ if $x_2 \neq 0$. Consider any $x = (x_1, x_2)$ and $y = (y_1, y_2)$. The convexity of f is consequence of the obvious identity:

$$[(1-\alpha)x_2 + \alpha y_2]^2 = [(1-\alpha)x_1 + \alpha y_1]\left[(1-\alpha)\frac{x_2^2}{x_1} + \alpha\frac{y_2^2}{y_1}\right] - \alpha(1-\alpha)\frac{(x_1 y_2 - x_2 y_1)^2}{x_1 y_1}.$$

f is l.s.c., but it is discontinuous at $(0,0)$. $\qquad\qquad\square$

The fact that the convexity of a function (with domain in $\mathbb{R}^n$) does not imply its continuity on the boundary of its domain prevents us from claiming that a convex function on a compact set can have minimum; this is indeed false, as trivial examples show (e.g., $X = [0,1]$, $f(x) = x$ if $x \in \,]0,1]$ and $f(0) = 1$). While, because of (ii) of Lemma 2.3.1, the restriction of a convex function to a compact set, contained in $\mathrm{dom}f$, has maximum.

The convex functions are heartily welcome to the Optimization Community, since they enjoy many nice properties. A few have been described previously; the following theorem is, perhaps, the most important fact for extremum problems. However, contrary to the general belief, convex functions may behave very badly, as we will see in Sect. 2.5 (see (2.5.9)).

Theorem 2.3.8. Let $K \subseteq \mathbb{R}^n$ be nonempty and convex, and $f : K \to \mathbb{R}$ be convex. **(i)** Every $\overline{x} \in K$, which is either a global minimum point, or a local minimum point, or a lower semistationary point of f, enjoys also the other properties. **(ii)** The set K^0 of global minimum points (or local minimum points or lower semistationary ones) is convex; if f is strictly convex, then $\mathrm{card}\,K^0 \leq 1$; if K is closed and f is linear, then K^0 is a face of K; if f is linear and K is a polyhedron, then K^0 is polyhedral.

Proof. If $\mathrm{card}\,K \leq 1$ or $K^0 = \varnothing$, then the thesis is trivial. Then, suppose $\mathrm{card}\,K > 1$ and $K^0 \neq \varnothing$. **(i)** Let $\overline{x}$ be local m.p., so that there exists a neighbourhood $N(\overline{x})$ s.t. $f(x) \geq f(\overline{x})$, $\forall x \in K \cap N(\overline{x})\backslash\{\overline{x}\}$. Suppose, ab absurdo, that $\overline{x}$ be not a global m.p., so that $\exists \hat{x} \in K\backslash\{\overline{x}\}$ s.t. $f(\hat{x}) < f(\overline{x})$. Because of the convexity of f, $\exists\alpha \in \,]0,1[$ s.t. $f(x) \leq (1-\alpha)f(\hat{x}) + \alpha f(\overline{x}) < f(\overline{x})$, which contradicts the assumption. If $\overline{x}$ is a global m.p., then $f(x) - f(\overline{x}) \geq 0$, $\forall x \in K$. Therefore, the lower limit, as $x \to \overline{x}$, of $[f(x) - f(\overline{x})]/\|x - \overline{x}\|$ is non-negative and, according to Definition 1.1.1, $\overline{x}$ is a lower semistationary point. Now, suppose that $\overline{x}$ be a lower semistationary point, and, ab absurdo, that be not a local m.p., so that $\exists \hat{x} \in K\backslash\{\overline{x}\}$ s.t. $f(\hat{x}) < f(\overline{x})$. Because of the convexity of K, $x(\alpha) := (1-\alpha)\overline{x} + \alpha\hat{x} \in K$, $\forall\alpha \in [0,1]$, and, because of the convexity of f, we have (see (2.3.5)):

$$\frac{f(x(\alpha)) - f(\overline{x})}{\|x(\alpha) - \overline{x}\|} \leq \frac{f(\hat{x}) - f(\overline{x})}{\|\hat{x} - \overline{x}\|} < 0, \quad \forall\alpha \in \,]0,1],$$

which, according to Definition 1.1.1, contradicts the lower semistationarity of $\overline{x}$. **(ii)** Let $x^1, x^2 \in K^0$, so that $f(x^1) = f(x^2) = \min_{x \in K} f(x)$. The convexity of K implies $x(\alpha) := (1-\alpha)x^1 + \alpha x^2 \in K$, $\forall\alpha \in [0,1]$. Moreover, the convexity of f implies that:

$$f(x(\alpha)) \leq (1 - \alpha)f(x^1) + \alpha f(x^2) = \min_{x \in K} f(x),$$

and thus $f(x(\alpha)) = \min_{x \in K} f(x)$, $\forall \alpha \in [0, 1]$. Hence $x(\alpha) \in K^0$ (because of (i), this extends to the case of local m.p. or to lower semistationary ones). If f is strictly convex, then, $\forall \alpha \in]0, 1[$, the last inequality must hold in strict sense and this implies $x^1 = x^2$. Let K be closed and f be linear or $f(x) = \langle a, x \rangle$ with $a \in \mathbb{R}^n \backslash \{O\}$ (the thesis is trivial for $a = 0$). $\forall \overline{x} \in K^0$, we have $\langle a, x - \overline{x} \rangle \geq 0$, $\forall x \in K$. Therefore $\langle a, x - \overline{x} \rangle = 0$ is, according to Definition 2.2.1, the equation of a supporting hyperplane for K. Hence, being $\langle a, x - \overline{x} \rangle = 0$, $\forall x \in K^0$, and $\langle a, x - \overline{x} \rangle > 0$, $\forall x \in K \backslash K^0$, according to Definition 2.2.2, K^0 is a face of K. According to Definition 2.1.6, the last part is obvious. $\square$

It is easy to see that (i) and the 1st part of (ii) of Theorem 2.3.8 hold also if $K \subseteq B$ (recall that B denotes a Banach space) with the same proof. If, in the 3rd part of (ii) of the above theorem, K is open, then $K^0 = \varnothing$; however, the same statement can be achieved, by defining K^0 as the set of infimum (or Weierstrass) points; if K is neither closed nor open, then K^0 may not be a face of K, as next example shows.

Example 2.3.6. Set $K = \{x \in \mathbb{R}^2_+ : x_1 \leq 2, \ x_2 \leq 2 \ \text{if} \ x_1 \leq 1 \ \text{and} \ x_2 < 2 \ \text{if} \ x_1 > 1\}$ and $f(x) = -x_2$. We find $K^0 = \{x \in \mathbb{R}^2_+ : x_1 \leq 1, \ x_2 = 2\}$, which is not a face of K; indeed, the only support line of K, which contains K^0, has equation $x_2 = 2$, and the intersection of such a line with cl K (facet of K; see Definition 2.2.2) strictly contains K^0. $\square$

Theorem 2.3.8 shows a tight connection between convexity and minimum points. However, meaningful failures exist in such a connection, as next example shows.

Example 2.3.7. Set $K = \mathbb{R}$, $\alpha(x) = \frac{1}{2} + \frac{1}{2}\sin\frac{1}{x}$ if $x \neq O$, $\alpha(0) = 0$, and $f(x) = [1 + \alpha(x)]x^2$. f is derivable. Elementary calculations show that $\overline{x} = 0$ is the unique global m.p. of f on K. Notwithstanding this, f is not convex in every neighbourhood of $\overline{x}$ (see Example 3.5.5). $\square$

An interesting class of functions is that of *positively homogeneous functions of degree* α, defined by:

$$f(tx) = t^\alpha \, f(x), \quad \forall t \in \mathbb{R}_+, \quad x \in \mathbb{R}^n. \tag{2.3.15a}$$

Particularly important are those of degree one (which appear in Theorem 2.3.2):

$$f(tx) = tf(x), \quad \forall t \in \mathbb{R}_+, \quad x \in \mathbb{R}^n. \tag{2.3.15b}$$

When the degree is not mentioned, the degree one is understood. The graph and epigraph of (2.3.15b) are obviously cones with apex at the origin. (2.3.15b), iff its epigraph (ipograph) is convex, is called *sublinear (superlinear)*. *A positively homogeneous function is convex (and hence sublinear) if and only if it is subadditive* i.e. $f(x + y) \leq f(x) + f(y)$, $\forall x, y \in \mathbb{R}^n$, as it easy to prove by exploiting Proposition 2.1.6 and noting that the subadditivity of f is equivalent to epi f being closed under vector addition and non-negative scalar multiplication. In fact, assuming subadditivity, we have that $(x^i, y_i) \in \text{epi } f$, $i = 1, 2$, mean $y_i \geq f(x^i)$, $i = 1, 2$, and imply:

$$y_1 + y_2 \geq f(x^1) + f(x^2) \geq f(x^1 + x^2),$$

so that we have:

$$(x^1, y_1) + (x^2, y_2) = (x^1 + x^2, y_1 + y_2) \in \text{epi } f,$$

or closure under vector addition; closure under non-negative scalar multiplication is a trivial consequence of positive homogeneity; vice versa, assuming convexity, we have:

$$(x^i, f(x^i)) \in \text{epi } f, \ i = 1, 2 \quad \Rightarrow \quad (x^1 + x^2, f(x^1) + f(x^2)) \in \text{epi } f$$

$$\Rightarrow \quad f(x^1 + x^2) \leq f(x^1) + f(x^2),$$

where the former implication is due to the closure of epi f under vector addition (besides the obvious inclusion graph $f \subset \text{epi } f$), and the latter implication is due to the fact that $(x^1 + x^2, f(x^1 + x^2)) \in \text{graph } f$.

A positively homogeneous function is not necessarily convex, even if all its lower level sets are convex (i.e., quasiconvex; see Definition 2.4.1), as simple examples show (see that which follows Theorem 2.3.2).

If f is convex and subadditive, then $\exists \overline{x}$ s.t. $f(\overline{x} + x) - f(\overline{x})$ is positively homogeneous of 1st degree. This is easily shown by taking into account Proposition 2.1.6 and the fact that a function is subadditive iff its epigraph is closed under addition (in fact, let $y_i \geq f(x^i)$, $i = 1, 2$, so that $y_1 + y_2 \geq f(x^1) + f(x^2)$; then, taking into account these inequalities, we draw:

$$f(x^1 + x^2) \leq f(x^1) + f(x^2),$$

$$f(x^1 + x^2) \leq y_1 + y_2,$$

$$(x^1 + x^2, y_1 + y_2) \in \text{epi } f,$$

$$(x^1, y_1) + (x^2, y_2) \in \text{epi } f,$$

and prove the claim). In passing, the above statement suggests a nice geometrical rule to detect whether or not a function is subadditive: assuming, without any loss of generality, that $f(O) = 0$, f is subadditive iff the epigraph of f contains the shifting of graph f from the origin to any point $(x, f(x))$, or

$$\text{epi } [f(x) + f(y - x)] \subseteq \text{epi } f(y), \quad \forall x \in \mathbb{R}^n.$$

Another useful property is that *a continuous and positively homogeneous of degree $\alpha > 0$ function admits the existence of x^1, x^2, such that:*

$$f(x^1) \cdot ||x||^\alpha \leq f(x) \leq f(x^2) \cdot ||x||^\alpha, \quad \forall x \in \mathbb{R}^n.$$

In fact, because of a well known theorem of Weierstrass (see Theorem 1.1.1 and Corollary 3.2.1), f admits minimum and maximum on $S := \{x \in \mathbb{R}^n : ||x|| \leq 1\}$. Hence, $\exists x^1, x^2 \in S$ s.t.

$$f(x^1) \leq f\left(\frac{1}{||x||}x\right) \leq f(x^2), \quad \forall x \in \mathbb{R}^n \backslash \{O\}.$$

From these inequality, taking into account (2.3.15a), we draw the thesis.

A convex function is not necessarily differentiable; however, it enjoys the nice property expressed by next:

Theorem 2.3.9. Let $K \subseteq \mathbb{R}^n$ be nonempty, convex, and such that dim aff $K \geq 1$, and let $f : K \to \mathbb{R}$ be convex. If $\overline{x} \in K$, then f is directionally derivable, and its directional derivative at $\overline{x}$ in the direction $d \in K - \overline{x}$, say $f'(\overline{x}; d)$, is a positively homogeneous, convex, and u.s.c. function with respect to d. Furthermore, we have:

$$f'(x^2; x^1 - x^2) \leq f(x^1) - f(x^2), \quad \forall x^1, x^2 \in K; f'(\overline{x}; -d) \geq -f'(\overline{x}; d).$$

If f is sublinear, then

$$f'(\overline{x}; d) \leq f(d), \quad \forall d \in \mathbb{R}^n \setminus \{O\},$$

where the equality holds if $\overline{x} + d = \alpha\overline{x}$ with $\alpha \in \mathbb{R}_+\backslash\{0\}$.

Proof. The directional derivative of f at $\overline{x} \in K$ in the direction $d \in K - \overline{x}$ is (see Definition 3.1.1):

$$f'(\overline{x}; d) := \lim_{t\downarrow 0} \frac{f(\overline{x} + td) - f(\overline{x})}{t}.$$

The above limit exists (finite or infinite); in fact, taking into account (4i) of Theorem 2.3.1, the argument of the previous limit does not increase as $t \downarrow 0$; this implies also that $\exists k \in \mathbb{R}$ s.t.

$$f'(\overline{x}; d) \leq \frac{f(\overline{x} + td) - f(\overline{x})}{t} < k,$$

which leads to show that f' is u.s.c. in $\overline{x}$. The following relations, which prove, respectively, the positive homogeneity (see (2.3.15b)), the convexity and the first 2 inequalities of f', are immediate (use (2.3.1) for the 2nd):

$$f'(\overline{x}; \alpha d) = \alpha f'(\overline{x}; d), \quad \forall \alpha > 0, \quad \forall d \in K - \overline{x};$$

$$f'(\overline{x}; (1-\alpha)d^1 + \alpha d^2) \leq (1-\alpha)f'(\overline{x}; d^1) + \alpha f'(\overline{x}; d^2), \quad \forall d^1, d^2 \in K - \overline{x}, \quad \forall \alpha \in [0,1];$$

$$f'(\overline{x}; -d) = -\lim_{t\uparrow 0}\frac{1}{t}[f(\overline{x}+td)-f(\overline{x})] \geq f'(\overline{x}; d), \quad \forall d \in K - \overline{x}.$$

The last but one claim is proved by the following obvious relations:

$$f'(x^2; x^1 - x^2) = \lim_{t\downarrow 0}\frac{f((1-t)x^2 + tx^1) - f(x^2)}{t}$$

$$\leq \lim_{t\downarrow 0}\frac{(1-t)f(x^2) + tf(x^1) - f(x^2)}{t} = f(x^1) - f(x^2).$$

Being obviously $f'(\overline{x}; O) = 0$, due to the convexity of f', we have:

$$0 = f'(\overline{x}; O) = f'\left(\overline{x}; \frac{1}{2}d + \frac{1}{2}(-d)\right) \le \frac{1}{2}f'(\overline{x}; d) + \frac{1}{2}f'(\overline{x}; -d),$$

which proves the inequality of the last claim. From the last claim, we have that f is differentiable, iff $f'(\overline{x}; -d) = -f'(\overline{x}; d)$, $\forall d \in K - \overline{x}$; iff this equality holds for a given d, f turns out to be differentiable only along the line identified by d. Since $\overline{x} + td = (1-t)\overline{x} + t(\overline{x} + d)$ and it is not restrictive to assume $t \in]0, 1]$ (t must go to zero), because of the convexity of f, we have:

$$f'(\overline{x}; d) \le \lim_{t \downarrow 0} \frac{t[f(\overline{x} + d) - f(\overline{x})]}{t} \le f(\overline{x}) + f(d) - f(\overline{x}) = f(d),$$

where the former inequality is due to the convexity of f and holds as equality if $\overline{x} + d = \alpha\overline{x}$, and the latter inequality is due to the subadditivity of f enjoyed because of its convexity and positive homogeneity. $\qquad\square$

The inequalities (2.3.2) suggest an interesting generalization of the concept of differential with regard to the convex functions. Indeed, the halfspace defined by $y = f(\overline{x}) + \langle \sigma, x - \overline{x} \rangle$ is a supporting halfspace for the epigraph of f at $x = \overline{x}$. Therefore, the epigraph of f may be regarded as the "envelope" of such halfspaces.

Definition 2.3.3. Let $K \subseteq \mathbb{R}^n$ be nonempty and convex, $f : K \to \mathbb{R}$ be convex and $\overline{x} \in \mathrm{dom}\, f$. The set of $\sigma \in \mathbb{R}^n$ which fulfil (2.3.2a) is called *subdifferential of f at $\overline{x}$* and denoted by $\partial f(\overline{x})$; σ is called *subgradient*. Iff $\partial f(\overline{x}) \ne \varnothing$, f is said to be *subdifferentiable at $\overline{x}$*.

It is immediate to prove that (f being convex) $\partial f(\overline{x})$ *is convex*. When card $\partial f(\overline{x}) = 1$, according to (2.3.2), the unique supporting hyperplane is tangent to epi f, so that $E(\overline{x}, x, \sigma)/\|x - \overline{x}\|$ is infinitesimal and therefore, if $\overline{x} \in \mathrm{int}\, K$, f fulfils the classic definition of differentiability of a not necessarily convex function. See also Definition 3.1.2.

Theorem 2.3.10. Let $K \subseteq \mathbb{R}^n$ with card $K > 1$ be convex. **(i)** If $f : K \to \mathbb{R}$ is convex on K, then $\partial f(x) \ne \varnothing$, $\forall x \in \mathrm{ri}\, K$. **(ii)** If $\partial f(x) \ne \varnothing$, $\forall x \in \mathrm{ri}\, K$, then f is convex on ri K. **(iii)** If $f_1, ..., f_r : K \to \mathbb{R}$ are convex on K, then

$$\partial \sum_{i=1}^{r} f_i(x) = \sum_{i=1}^{r} \partial f_i(x), \quad \forall x \in K.$$

(4i) Let f be convex on K. $\sigma \in \partial f(\overline{x})$ with $\overline{x} \in \mathrm{dom}\, f$, if and only if

$$f'(\overline{x}; d) \ge \langle \sigma, d \rangle, \quad \forall d \in K - \overline{x}.$$

(5i) If f is convex on K, then, $\forall d \in K - \overline{x}$, we have:

$$f'(\overline{x}; d) = \sup\left\{\langle \sigma, d \rangle : \sigma \in \partial f(\overline{x})\right\},$$

which shows that the directional derivative is the support function (see (2.2.10)) of the subdifferential. If f is finite at $\overline{x}$, then the above supremum is a maximum. **(6i)** If f is finite at $\bar{x}$, then $\partial f(\bar{x})$ is closed and convex; if, furthermore, $\bar{x} \in$ ri dom f, then $\partial f(\bar{x})$ is non-empty and bounded. **(7i)** If $K \subset$ int dom f is compact, and f is convex and closed, then

$$\partial f(K) := \bigcup_{x \in K} \partial f(x)$$

is non-empty and compact. Moreover, f is Lipschitz on K with

$$L := \sup \{\|x\| : x \in \partial f(K)\} < +\infty$$

as Lipschitz constant, and

$$f'(x; d) \leq L, \qquad \forall x \in K, \qquad \forall d \in \mathbb{R}^n.$$

Proof. **(i)** It is an obvious consequence of Theorem 2.3.1 (ii). **(ii)** Let $x^1, x^2 \in$ ri K, and set $x(\alpha) := (1 - \alpha)x^1 + \alpha x^2, \quad \alpha \in [0, 1]$. Let $\sigma(\alpha) \in \partial f(x(\alpha)) \neq \varnothing$. From (2.3.2a) we have:

$$f(x^1) \geq f(x(\alpha)) + \langle \sigma(\alpha), \alpha(x^1 - x^2) \rangle, \quad f(x^2) \geq f(x(\alpha)) + \langle \sigma(\alpha), (1 - \alpha)(x^2 - x^1) \rangle.$$

By multiplying these inequalities, respectively, by $1 - \alpha$ and α, and summing up them side by side, we achieve (2.3.1). **(iii)** It is a straightforward consequence of Definition 2.3.3, since we are assuming that dom $f_i = K, \ i = 1, ..., r$. **(4i)** With the position $x = \bar{x} + td$, (2.3.2a) becomes:

$$\frac{f(\bar{x} + td) - f(\bar{x})}{t} \geq \langle \sigma, d \rangle, \quad \forall d \in K - \bar{x}.$$

Because of Theorem 2.3.1(4i), the difference quotient does not increase as $t \downarrow 0$. This proves the claim. **(5i)** Ab absurdo, suppose that the thesis does not hold. Because of (4i), we have:

$$\lim_{t \downarrow 0} \frac{f(\bar{x} + td) - f(\bar{x})}{t} > \sup \{\langle \sigma, d \rangle : \sigma \in \partial f(\bar{x})\}.$$

Hence, $\exists k \in \mathbb{R} \backslash \{0\}$ s.t.

$$f(\bar{x} + td) - f(\bar{x}) > k > \langle \sigma, td \rangle, \quad \forall t > 0, \quad \forall \sigma \in \partial f(\bar{x}).$$

Passing to the limit as $t \downarrow 0$, we reach the contradiction $0 > k > 0$. The last claim comes from the fact that, if f is finite at $\bar{x}$, then $\partial f(\bar{x})$ is compact (as shown by next (6i)). **(6i)** Closedness and convexity are obvious consequences of the fact that the elements σ of $\partial f(\bar{x})$ are solutions of a system of infinite linear inequalities obtained from (2.3.2a) when x varies. Taking into account that we are considering proper convex functions, if $\bar{x} \in$ ri dom f, then the point $(\bar{x}, f(\bar{x}))$ is finite and int epi f is nonempty (besides convex), so that Theorem 2.2.2(ii) can be applied to obtain their proper separation by a non-vertical hyperplane, whose gradient turns out to be an element of $\partial f(\bar{x})$. This is bounded since its support function (which, by (5i), is $f'(\bar{x}; d)$) is finite on ri dom f and this is necessary and sufficient for a set to be bounded. **(7i)** Let $\hat{\sigma} \in \operatorname{cl} \partial f(K)$, so that $\exists \{\sigma^i\}_1^\infty \subset \partial f(K)$, s.t.

$$\hat{\sigma} = \lim_{i \to +\infty} \sigma^i.$$

Then $\exists \{x^i\}_1^\infty \subset K$ s.t. $\sigma^i \in \partial f(x^i)$. Because of the compactness of K, $\{x^i\}_1^\infty$ (or one of its subsequences) has a limit point, say $\hat{x} \in K$. Because of the closedness of $\partial f(\hat{x})$ (see (6i)), $\hat{\sigma} \in \partial f(\hat{x})$ and thus $\hat{\sigma} \in \partial f(K)$, so that the closedness of $\partial f(K)$ follows. $\forall x \in K$, $\partial f(x)$ is bounded (see (6i)), and $f'(x;d)$ is its support function (see (5i)). Therefore, thanks to the equality in (5i), $f'(x;d)$ is finite, since a set (here $\partial f(x)$) is bounded iff its support function is finite (see (2.2.10) and the comments which follow (2.3.17)). Hence we have:

$$L := \sup_{||d||=1} f'_{\mathrm{sup}}(d), \qquad f'_{\mathrm{sup}}(d) := \sup_{x \in K} f'(x;d).$$

The compactness of K and the u.s.c. of $f'(x;d)$ with respect to x (see Theorem 2.3.9), show that f'_{sup} is finite. Being the pointwise supremum of a family of convex functions, f'_{sup} is convex. Thus, f'_{sup} is a convex and finite and hence continuous function (Theorem 2.3.7). It follows that $L < +\infty$. The boundedness of $\partial f(K)$ is achieved. From Theorem 2.3.9, we draw that, $\forall x^1, x^2 \in K$,

$$f(x^1) - f(x^2) \geq f'(x^2; x^1 - x^2) = -f'(x^2; x^2 - x^1).$$

Set $d := (1 \backslash ||x^2 - x^1||) \cdot (x^2 - x^1)$. Since $||d|| = 1$, we have:

$$f(x^2) - f(x^1) \leq f'(x^2; x^2 - x^1) = ||x^2 - x^1|| f'(x;d) \leq L \cdot ||x^2 - x^1||,$$

or

$$f(x^2) - f(x^1) \leq L \cdot ||x^2 - x^1||. \qquad \Box$$

In the above theorem, (ii) cannot be improved by achieving the thesis on the entire K, as shown by Example 2.3.1, where $\partial f(x) \neq \varnothing$, $\forall x \in \operatorname{ri} K$, but f is not convex on K. (i) can be improved, if $K = \mathbb{R}^n$; in such a case f is convex iff has a subgradient $\forall x \in \mathbb{R}^n$.

Theorem 2.3.11. Let $K \subseteq \mathbb{R}^n$ be nonempty and convex, and let $f : K \to \mathbb{R}$ be convex. **(i)** $\overline{x} \in \operatorname{ri} K$ is a global minimum point of f if and only if

$$O \in \partial f(\overline{x}).$$

(ii) $O \in \operatorname{int} \partial f(\overline{x}), \overline{x} \in \operatorname{ri} K$, implies that $\overline{x}$ is the unique global minimum point of f. **(3i)** If x^0 is an isolated minimum point of f, then it is the unique global minimum point of f. **(4i)** If K is compact and f continuous, then there exists an affine function, say $\ell(x)$, such that:

$$\min_{x \in K} \ell(x) = \min_{x \in K} f(x).$$

Proof. **(i)** It is an immediate consequence of (2.3.2a) for $\sigma = O$. **(ii)** Ab absurdo, suppose that $\exists \hat{x} \in K \backslash \{\overline{x}\}$ s.t. $f(\hat{x}) = f(\overline{x})$. The assumption implies that $\exists \overline{\varepsilon} > 0$ s.t. $\overline{\varepsilon}(\hat{x} - \overline{x}) \in \partial f(\overline{x})$ so that, for $\sigma = \overline{\varepsilon}(\hat{x} - \overline{x})$ and $x = \hat{x}$, from (2.3.2a) we draw the inequalities:

$$f(\hat{x}) \geq f(\overline{x}) + \langle \overline{\varepsilon}(\hat{x} - \overline{x}), \hat{x} - \overline{x} \rangle = f(\overline{x}) + \overline{\varepsilon} \|\hat{x} - \overline{x}\|^2 > f(\overline{x}),$$

which contradict the absurd assumption. **(3i)** Because of Theorem 2.3.8(i), x^0 is a global m.p. of f. Ab absurdo, suppose that $\hat{x} \in K \backslash \{x^0\}$ be another global m.p. of f. Hence, from (2.3.1) we draw:

$$f(x) = f(x^0) = f(\hat{x}), \quad \forall x \in]x^0, \hat{x}[\neq \varnothing,$$

which contradicts the isolation of x^0. **(4i)** Because of Definition 2.3.3 and of Theorems 1.1.1 and 3.2.1(ii), we have:

$$f(\overline{x}) := \min_{x \in K} f(x) \geq \min_{x \in K} \left[f(\overline{x}) + \langle \sigma, x - \overline{x} \rangle \right], \quad \forall \sigma \in \partial f(\overline{x}).$$

Ab absurdo, suppose that, $\forall \sigma \in \partial f(\overline{x})$, in the above inequality the equality never holds, or, $\forall \sigma \in \partial f(\overline{x})$, $\exists x_\sigma \in K$ such that:

$$f(\overline{x}) > \min_{x \in K} \left[f(\overline{x}) + \langle \sigma, x - \overline{x} \rangle \right] = f(\overline{x}) + \langle \sigma, x_\sigma - \overline{x} \rangle.$$

Then we draw:

$$\langle \sigma, x_\sigma - \overline{x} \rangle < 0, \quad \forall \sigma \in \partial f(\overline{x}),$$

which implies:

$$\sigma \neq O, \quad \forall \sigma \in \partial f(\overline{x}).$$

If $\overline{x} \in \mathrm{ri}\, K$, then (i) is contradicted. If $\overline{x} \in r\partial K$, then, because of Theorem 2.3.10 (5i), we have:

$$f'(\overline{x}; x_\sigma - \overline{x}) = \max_{\sigma \in \partial f(\overline{x})} \langle \sigma, x_\sigma - \overline{x} \rangle < 0,$$

or

$$\lim_{t \downarrow 0} \frac{f(\overline{x} + t(x_\sigma - \overline{x})) - f(\overline{x})}{t} < 0,$$

which, for t small enough, implies:

$$f(\overline{x} + t(x_\sigma - \overline{x})) < f(\overline{x}).$$

Since $\overline{x} + t(x_\sigma - \overline{x}) \in K$, the minimality of $\overline{x}$ is contradicted. It follows that $\exists \overline{\sigma} \in \partial f(\overline{x})$, s.t. $\ell(x) := f(\overline{x}) + \langle \overline{\sigma}, x - \overline{x} \rangle$ fulfils the thesis. $\qquad \square$

Trivial examples show that statement (i) of the above theorem may not hold at $\overline{x} \in K \backslash \mathrm{ri}\, K$ (take, for instance, $K = \{x \in \mathbb{R} : 1 \leq x \leq 2\}$, $f : K \to \mathbb{R}_+$ given by $f(x) = x^2$, and $\overline{x} = 1$), and that (ii) cannot be inverted (take, for instance, $K = \mathbb{R}$, $f(x) = x^2$, and $\overline{x} = 0$; $\partial f(0) = \{0\}$), even if $\mathrm{int}\, \partial f(\overline{x}) \neq \varnothing$ (take, for instance, $K = \mathbb{R}$, $\overline{x} = 0$, $f(x) = x$ if $x \geq 0$ and $f(x) = x^2$ if $x < 0$; $\mathrm{int}\, \partial f(0) =]0, 1[$).

Compactness, continuity and finite dimensionality of the space can be overcome, in (4i) above, without substantial difficulties by replacing minimum with infimum. This is necessary at least because in (4i) the minimum of f may not exist as trivial examples show. Take, for instance, in $\mathbb{R}$, $K = [0,1]$ and $f(x) = x$ if $x > 0$, $f(0) = 1$; or $K =]0,1]$ and $f(x) = x$; or $K = [0,+\infty[$ and $f(x) = e^{-x}$; where compactness, closure, boundedness have been alternatively removed.

Definition 2.3.3 suggests the following useful strengthening of the concept of convexity.

Definition 2.3.4. Let $K \subseteq \mathbb{R}^n$ be nonempty and convex. $f : K \to \mathbb{R}$ is *strongly convex*, iff $\exists \sigma \in \mathbb{R}^n$ and $\exists \alpha \in \mathbb{R}_+\backslash\{0\}$, such that $E(\overline{x},x,\sigma) \geq \alpha||x - \overline{x}||^2$, $\quad \forall \overline{x}, x \in K$.

Of course, for $\alpha = 0$ the above concept decays, and we have merely convexity or (2.3.2a). It is easy to show that *a derivable f is strongly convex on K, iff ∇f is strongly isotone on K*. As from Definition 2.3.1 we have been led to Definition 2.3.2, so from Definition 2.3.4 we are led to:

Definition 2.3.5. Let $K \subseteq \mathbb{R}^n$ be nonempty and convex. $f : K \to \mathbb{R}^n$ is called *strongly isotone*, iff $\exists \alpha \in \mathbb{R}_+\backslash\{0\}$, such that:

$$\langle f(x^1) - f(x^2), x^1 - x^2 \rangle \geq \alpha||x^1 - x^2||, \quad \forall x^1, x^2 \in K,$$

and *strongly antitone*, iff $\exists \alpha \in \mathbb{R}_+\backslash\{0\}$, such that

$$\langle f(x^1) - f(x^2), x^1 - x^2 \rangle \leq -\alpha||x^1 - x^2||, \quad \forall x^1, x^2 \in K.$$

Let us now consider a few classic examples of convex functions. The *norm function* $f(x) = ||x||$ is convex, as it is easy to prove by exploiting the inequalities $||x + y|| \leq ||x|| + ||y||$, $||\alpha x|| = \alpha||x||$, $\forall x, y$, $\forall \alpha \in \mathbb{R}_+$. Theorem 2.2.5 gives an important property of norm function. Another classic function, which plays a role in Convex Analysis, is the *indicator function*:

$$\delta(x;K) := \begin{cases} 0, & \text{if } x \in K, \\ +\infty, & \text{if } x \notin K. \end{cases} \tag{2.3.16}$$

By using Definition 2.3.1 in extended way allowing $\pm\infty$, it is immediate to see that K *is convex iff* $\delta(x;K)$ is. Another formal role of (2.3.16) is offered by the fact that the subdifferential $\partial\delta(\overline{x};K)$ of *the indicator function of a nonempty, closed and convex set K for $x = \overline{x} \in \mathrm{frt}\, K$ is the normal cone to K at $\overline{x}$*. In fact, according to Definition 2.3.3, $\sigma \in \partial\delta(\overline{x};K)$ iff

$$\delta(x;K) \geq \delta(\overline{x};K) + \langle \sigma, x - \overline{x} \rangle, \quad \forall x \in \mathbb{R}^n,$$

which implies $\overline{x} \in K$ (otherwise, $\delta(x;K) = +\infty$, $\forall x \in \mathbb{R}^n$, and then $K = \varnothing$) and $\langle \sigma, x - \overline{x} \rangle \leq 0$, $\forall x \in K$. This proves the claim.

Consider again the support function defined in (2.2.10). Let $K \subset \mathbb{R}^n$ be nonempty and convex. $\forall x^1, x^2 \in K$, $\quad \forall \alpha \in [0,1]$, set $x(\alpha) := (1-\alpha)x^1 + \alpha x^2$, and note that:

$$\delta^*(x(\alpha);K) = \sup_{y \in K}\langle x(\alpha);y \rangle \leq (1-\alpha)\sup_{y \in K}\langle x^1;y \rangle + \alpha\sup_{y \in K}\langle x^2;y \rangle =$$

$$= (1 - \alpha)\delta^*(x^1; K) + \alpha\delta^*(x^2; K).$$

This shows that *the convexity of K implies that of its indicator function, which turns out to be also closed* (again considering Definition 2.3.1 in extended sense). If K is a cone, then the support function of K is the indicator function of the negative polar of K:

$$\delta^*(x; K) = \delta(x; K^*), \quad x \in \mathbb{R}^n. \tag{2.3.17}$$

If f is convex, then the closure of its directional derivative (as a convex function of the direction) is the support function of the closed and convex subdifferential, or

$$\operatorname{cl} f'(\overline{x}; d) = \delta^*(\overline{x}; \partial f(\overline{x})).$$

There is a tight relationship between (2.2.10) and (2.2.1): *the equation $\langle a, x \rangle = \delta^*(a; K)$, $x \in \mathbb{R}^n$, identifies a supporting hyperplane of K, if $\delta^*(a; K) < +\infty$.* Note also that $\overline{x} \in \operatorname{int} K$, iff $\langle x, \overline{x} \rangle < \delta^*(x; K)$, $\forall x \in \mathbb{R}^n$. It is immediate to note that *K is bounded iff $\delta^*(x; K) < +\infty$, $\forall x \in \mathbb{R}^n$.*

Problem (1.1.4) is equivalent to (see Sect. 5.9):

$$\min y, \quad \text{s.t.} \quad x \in \mathbb{R}, \quad -f(x) + y \geq 0. \tag{2.3.18}$$

Hence, setting $\hat{R} := (R \times \mathbb{R}) \cap \{(x, y) \in \mathbb{R}^n \times \mathbb{R} : -f(x) + y \geq 0\}$ and $a := (0, -1)$, (1.1.4) is equivalent to find $\delta^*(a; \hat{R})$ (of course, this is immediately extended to (1.1.5)).

Example 2.3.8. Set $K = \{x \in \mathbb{R}_+^2 : x_1 + x_2 \leq 1\}$. (2.3.17) becomes:

$$\delta^*(x; K) = \begin{cases} x_1, & \text{if } x \in K_1, \\ x_2, & \text{if } x \in K_2, \\ 0, & \text{if } x \in K_3, \end{cases}$$

where K_1, K_2 and K_3 are given in Fig. 2.3.4. $\qquad\qquad\qquad\square$

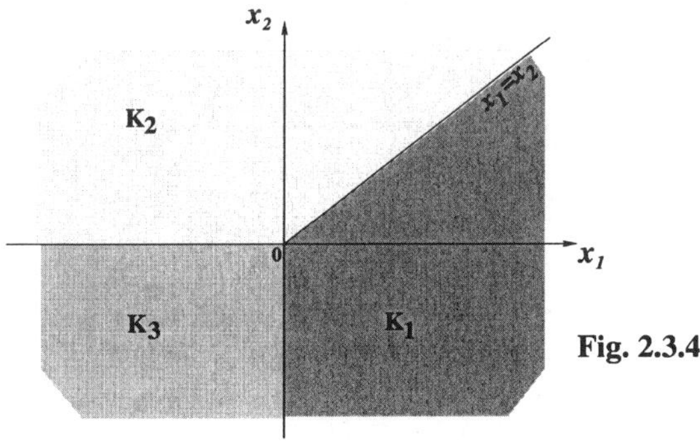

Fig. 2.3.4

Given a nonempty set $K \subseteq \mathbb{R}^n$, the function, defined by

$$M(x; K) := \inf \{t \in \mathbb{R}_+ : x \in tK\}, \tag{2.3.19}$$

is called *Minkowski* (or *gauge*) *function* of K. In Example 2.3.7, (2.3.19) exists only in $\mathbb{R}^2_+$, where $M(x; K) = x_1 + x_2$. $M(0; K) = 0$, whatever K may be. Obviously, the Minkowski function *is positively homogeneous; it is a norm if everywhere finite and symmetric (if K is bounded and $O \in K$, then the support function (2.2.10) is a gauge, but not a norm); it is convex if it is subadditive and vice versa with K convex.* With regard to the last claim, $\forall x^1, x^2, \quad \forall \alpha \in [0, 1]$, we note that:

$$M((1-\alpha)x^1 + \alpha x^2; K) \le M((1-\alpha)x^1; K) + M(\alpha x^2; K) = (1-\alpha)M(x^1; K) + \alpha M(x^2; K);$$

$$M(x^1 + x^2; K) = 2M(\tfrac{1}{2}x^1 + \tfrac{1}{2}x^2; K) \le 2[\tfrac{1}{2}M(x^1; K + \tfrac{1}{2}M(x^2; K)] = M(x^1; K) + M(x^2; K).$$

Given a finite family $\{f_i(x),\ i \in I\}$ of functions, the *max-function* of this family is defined by:

$$\phi(x) := \max_{i \in I} f_i(x). \tag{2.3.20a}$$

By observing that $\operatorname{epi} \phi = \underset{i \in I}{\cap} \operatorname{epi} f_i$, and using Theorem 2.3.1 (i), we see that (2.3.20a) *is convex if all f_ξ are*. Note that Theorem 2.1.3, where we set $\mathcal{F} = \{\operatorname{epi} f_i,\ i \in I\}$, gives a sufficient condition for (2.3.20a) to be finite. Of course, the above definition extends to any family of functions; in this case, in general, sup replaces max:

$$\phi(x) := \sup_{\xi \in \Xi} f(x; \xi), \tag{2.3.20b}$$

where Ξ is a set of parameters ξ; if Ξ is compact and f continuous with respect to ξ, then, of course, sup becomes max as in (2.3.20a).

Given a function $f : \mathbb{R}^n \to \mathbb{R}$, the *conjugate* of f is defined by

$$f^*(x^*) := \sup_{x \in \mathbb{R}^n} [\langle x^*, x \rangle - f(x)], \quad x^* \in \mathbb{R}^n. \tag{2.3.21}$$

Theorem 2.3.14. f^* is convex, and

$$\operatorname{epi} f^* = \{(x^*, y^*) \in \mathbb{R}^n \times \mathbb{R} : \langle x^*, x \rangle - y^* \le f(x), \quad \forall x \in \mathbb{R}^n\}. \tag{2.3.22}$$

Proof. (2.3.22) is a trivial consequence of (2.3.21). Taking into account (2.3.22), the convexity of $\operatorname{epi} f^*$ is immediately obtained by using Definition 2.1.1. Then, the convexity of f^* comes from Theorem 2.3.1(i). $\qquad \square$

The conjugate of a function enjoys several important properties. For instance, it is easy to show that, if f is convex, then $(f^*)^* = \operatorname{cl} f$ or $\operatorname{cl} \operatorname{epi} (f^*)^* = \operatorname{cl} \operatorname{epi} f$. Another property is expressed by the *Fenchel inequality*:

$$\langle x^*, x \rangle \le f^*(x^*) + f(x), \quad \forall x, x^* \in \mathbb{R}^n, \tag{2.3.23}$$

which is easily proved by noting that (2.3.21) holds iff

$$\langle x^*, x \rangle - f^*(x^*) \le f(x), \quad \forall x, x^* \in \mathbb{R}^n. \tag{2.3.24}$$

The conjugate of a function receives a very important interpretation in terms of support. Let $f^*(x^*) < +\infty$, and $\overline{x}$ be s.t. $f^*(x^*) = \langle x^*, \overline{x} \rangle - f(\overline{x})$; this and (2.3.24) show that *the hyperplane $H^0 \subseteq \mathbb{R}^{n+1}$, defined by $z = \langle x^*, x \rangle - f^*(x^*)$, fulfils* (2.2.1) *for $K = epi\, f$ and $(\overline{x}, f(\overline{x}))$ is a supporting point;* if $\overline{x}$ does not exist finite, then H^0 is still a supporting hyperplane, but there is no supporting point (take, for instance, $f(x) = e^x$ and $x^* = 0$). $f^*(x^*) = +\infty$ *means that no supporting hyperplane exists having gradient equal to x^*.* It is easy also to see that *the support function* (2.2.10) *is the conjugate of the indicator function* (2.3.16). In fact, if $K \neq \varnothing$, we have:

$$\sup_{x \in \mathbb{R}^n} \left[\langle x^*, x \rangle - \delta(x; K) \right] = \sup_{x \in K} \langle x^*, x \rangle.$$

The conjugate of a function has many applications. Let us mention, in a simplified version, an application to Economics. *f(x)* is the cost for producing n items at the levels indicated by the elements of x (which now is supposed to be non-negative). x^* is the vector of the corresponding selling prices. Then $\langle x^*, x \rangle - f(x)$ is the profit due to the production and sale of the n items at the quantities denoted by x. $f^*(x^*)$ gives the maximum (or supremum) level that the profit can achieve as a function of the vector of selling prices.

Note that, when f is derivable, $f^*(f'(\overline{x})) = H(\overline{x}, \overline{x})$, where H is the function in (2.3.2a)′.

2.4. Some Extensions of Convexity

In the previous sections we have met some properties of convex sets and functions. Even if they are only a few, they are enough for understanding the importance of convexity. Due to this, in the last three decades there has been an impressive growth of definitions of generalized convexity, both for sets and functions. The way of obtaining them is very simple: if we remove one of the many properties enjoyed by convexity, or we extend one of the terms of the definition, then we obtain a generalized concept; now, the same can be done with the concept just obtained, and so on in a practically endless process. Some of such generalizations are of fundamental importance; unfortunately, many generalizations look like mere formal mathematics without any motivation and contribute to drive mathematics away from the real world. Neglecting the fact that definition is the cornerstone of mathematics and hence is the most difficult task, new generalized concepts of convexity sprout like mushrooms (even 30 meaningless generalizations of convexity can be found in a same recent paper! while E.De Giorgi, in his entire mathematical life, gave only one concept: (p,q)-convexity; and G.Stampacchia dealt with coerciveness (see (1.3.20)); both such extensions of convexity have been introduced and used under strong motivations [157]). Here, we will consider shortly only a few generalizations, which have shown to be useful. The first is suggested by Theorem 2.3.2.

Definition 2.4.1. Let $K \subseteq \mathbb{R}^n$ be convex. $f : K \to \mathbb{R}$ is called *quasiconvex* iff $\mathrm{lev}_{\leq \alpha} f$

is convex $\forall \alpha \in \mathbb{R}$. f is called *quasiconcave* iff $-f$ is quasiconvex. f is called *strictly quasiconvex* iff, $\forall x^1, x^2 \in K$ such that $x^1 \neq x^2$ and $f(x^1) \leq f(x^2)$, we have:

$$f((1-\alpha)x^1 + \alpha x^2) < f(x^2), \quad \forall \alpha \in]0,1[.$$

f is called *strictly quasiconcave* iff $-f$ is strictly quasiconvex.

Of course, a convex function is quasiconvex. It is easy to see that *strict quasiconvexity implies quasiconvexity*. In fact, let $\alpha \in \mathbb{R}$, $x^1, x^2 \in \operatorname{lev}_{\leq \alpha} f$ with $x^1 \neq x^2$ and set $x(\alpha) := (1-\alpha)x^1 + \alpha x^2$. Since it is not restrictive to assume $f(x^1) \leq f(x^2)$, we have $f(x(\alpha)) < f(x^2)$, $\forall \alpha \in]0,1[$. Hence $x(\alpha) \in \operatorname{lev}_{\leq \alpha} f$, $\forall \alpha \in]0,1[$. The inverse claim is not true: take for instance $f(x) = x^2$ if $x \leq 1$ and $f(x) = 1$ if $x > 1$, with $x \in \mathbb{R}$.

It is also easy to show that, *if f is strictly quasiconvex, then* $\operatorname{lev}_{\leq \beta} f$ *is strictly convex* $\forall \beta \in \mathbb{R}$. In fact, the existence of $\overline{\beta} \in \mathbb{R}$, of $x^1, x^2 \in \operatorname{lev}_{\leq \overline{\beta}} f$ with $x^1 \neq x^2$ and of $\overline{\alpha} \in]0,1[$, such that $f(x(\alpha)) \in \operatorname{lev}_{=\overline{\beta}} f$ contradicts the very definition of strict quasiconvexity ($x(\alpha)$ being as above). As a consequence of this property, we have that, if f is quasiconvex, but not strictly quasiconvex, then there must exists a (nondegenerate) segment of K, where f is constant.

It is equally easy to show that, *if f is strictly quasiconvex, then a local m.p., say $\overline{x}$, is also a global one*. In fact, the assumption implies that $\exists N_\rho(\overline{x})$, s.t.

$$f(\overline{x}) \leq f(x), \quad \forall x \in K \cap N_\rho(\overline{x}).$$

Ab asurdo, suppose that $\exists \hat{x} \in K$ s.t. $f(\hat{x}) < f(\overline{x})$. This implies:

$$f(x(\alpha)) < f(\overline{x}), \quad \forall \alpha \in [0,1[,$$

where now $x(\alpha) := (1-\alpha)\hat{x} + \alpha \overline{x}$. If $\alpha > 1 - \rho/\|\hat{x} - \overline{x}\|$, then $x(\alpha) \in K \cap N_\rho(\overline{x})$ and the local optimality of $\overline{x}$ is contradicted. *Uniqueness* is also easy to be proved: two global m.p. $x^1 \neq x^2$ (being $f(x) \geq f(x^1) = f(x^2)$, $\forall x \in K$) imply the absurd $f(x(\alpha)) < f(x^2)$, $\forall \alpha \in]0,1[$, now being $x(\alpha) := (1-\alpha)x^1 + \alpha x^2$.

The concept of quasiconvexity, which comes in a natural way from the property expressed by Theorem 2.3.2, has been the basis of an early study on the extensions of convexity under the name "convex stratifications" [16]; indeed, the now commonly used term quasiconvexity is not much suitable because of the use of "quasi" in the Theory of Misure.

The importance of quasiconvexity is pointed out by the following property, which shows also a relaxation of the properties of convexity; indeed, it mantains only a part of Theorem 2.3.8 (i).

Theorem 2.4.1. Let $K \subseteq \mathbb{R}^n$ be nonempty and convex. **(i)** If $f : K \to \mathbb{R}$ is quasiconvex, then every isolated local minimum point of f is also a global one. **(ii)** Let K be open and f differentiable. f is quasiconvex, if and only if

$$\left. \begin{array}{c} x^1, x^2 \in K \\ f(x^1) \leq f(x^2) \end{array} \right\} \Rightarrow \langle f'(x^2), x^1 - x^2 \rangle \leq 0.$$

Proof. (i) Ab absurdo, suppose that $\bar{x} \in K$ be an isolated (see Definition 1.1.1) local m.p. of f, but not a global one. Then, $\exists \hat{x} \in K \setminus \{\bar{x}\}$ s.t. $f(\hat{x}) < f(\bar{x}) := \bar{y}$. Since $\bar{x}$ is isolated, $\mathrm{lev}_{\leq \bar{y}} f$ contains $\bar{x}$ and $\hat{x}$, but not the entire segment $]\bar{x}, \hat{x}[$. According to Definition 2.4.1, this contradicts the assumption of quasiconvexity. (ii) **Only if.** Let $x^1, x^2 \in K$ be s.t. $f(x^1) \leq f(x^2)$, and set $x(\alpha) := (1 - \alpha)x^1 + \alpha x^2$. The quasiconvexity of f implies $f(x(\alpha)) \leq f(x^2)$, $\forall \alpha \in [0, 1]$. This and the differentiability of f imply:

$$0 \geq f(x(\alpha)) - f(x^2) = (1 - \alpha)\langle f'(x^2), x^1 - x^2 \rangle + \mathbf{o}((1 - \alpha)\|x^1 - x^2\|), \quad \forall \alpha \in [0, 1],$$

where $\mathbf{o}$ denotes an infinitesimal of higher order with respect to $1 - \alpha$. By considering the inequality between the 1st and 3rd sides, subtracting $\mathbf{o}(\bullet)$ from them, dividing them by $1 - \alpha > 0$, we find:

$$\langle f'(x^2), x^1 - x^2 \rangle \leq \mathbf{o}_1(1 - \alpha), \quad \forall \alpha \in [0, 1[,$$

where $\mathbf{o}_1$ denotes an infinitesimal with respect to $1 - \alpha$. Passing to the limit as $\alpha \uparrow 1$, we achieve the thesis. **If.** Consider any $\bar{x} \in K$ and any $x^1, x^2 \in \mathrm{lev}_{\leq f(\bar{x})} f(x)$. Set $x(\alpha) := (1 - \alpha)x^1 + \alpha x^2$. We must show that, $\forall \alpha \in [0, 1]$, we have:

$$f(x(\alpha)) \leq f(\bar{x}), \quad \forall \alpha \in [0, 1].$$

Ab absurdo, suppose that $\exists \bar{\alpha} \in]0, 1[$, s.t.

$$f(x(\bar{\alpha})) > f(\bar{x}),$$

so that:

$$f(x(\bar{\alpha})) > f(x^i), \ i = 1, 2.$$

These inequalities and the assumption imply:

$$\langle f'(x(\bar{\alpha})), x^i - x(\bar{\alpha}) \rangle \leq 0, \quad i = 1, 2.$$

Thus, because of the derivability of $f(x(\alpha))$, we draw:

$$\frac{d}{d\alpha} f(x(\alpha)) /_{\alpha = \bar{\alpha}} = 0.$$

Therefore, $f(x(\alpha))$ turns out to be constant where is greater than $f(\bar{x})$ and hence, being $\leq f(\bar{x})$ for $\alpha = 0, 1$, discontinuous somewhere, contradicting its continuity. $\qquad\square$

Examples 2.4.1. Set $K = \mathbb{R}$ and $f(x) = 3x^4 - 4x^3$. Since $f'(x) \leq 0$ if $x \leq 1$ and $f'(x) \geq 0$ if $x \geq 1$, then f is quasiconvex, even if not convex. $\bar{x} = 0$ is a stationary point, but not a local m.p.; this shows that a part of Theorem 2.3.8 (i) does not hold for quasiconvex functions. Set again $K = \mathbb{R}$, and $f(x) = x^2$ if $x \leq 0$, $f(x) = 0$ if $x \in]0, 1[$, and $f(x) = (x - 1)(x - 2)$ if $x \geq 1$. f is evidently quasiconvex, $\bar{x} = 1/2$ is a local m.p.; being not isolated, $\bar{x}$ is not a global m.p. of f. Now set $K = [0, 1]$ and $f(x) = x(1 - x)$. $\bar{x} = 0$ and $\tilde{x} = 1$ are isolated global m.p. of f on K, f is not quasiconvex; this shows that (i) of Theorem 2.4.1 cannot be inverted. □

Theorem 2.4.2. Let $K \subseteq \mathbb{R}^n$ be nonempty and convex, and $f : K \to \mathbb{R}$ be convex, and $F : f(K) \to \mathbb{R}$ be isotone. Then $F \circ f$ is quasiconvex.

Proof. $\forall \beta \in \mathbb{R}, \quad \forall x^1, x^2 \in \mathrm{lev}_{\leq \beta} F(f(x))$, and $\forall \alpha \in [0, 1]$, we have:

$$F(f((1 - \alpha)x^1 + \alpha x^2)) \leq F((1 - \alpha)f(x^1) + \alpha f(x^2)) \leq F(f(x^2)) \leq \beta,$$

where the 1st inequality is due to the convexity of f and to the isotonicity of F, the 2nd is due to the isotonicity of F and to the not restrictive assumption $f(x^1) \leq f(x^2)$, and the 3rd to the assumption on x^2. □

If in Theorem 2.4.2 we set $n = 1$ and alternatively $f(x) = x$ and $f(x) = -x$, then we obtain the following obvious:

Corollary 2.4.1. Let $K \subseteq \mathbb{R}$ be convex. If $f : K \to \mathbb{R}$ is either isotone or antitone, then it is quasiconvex.

An important generalization of convexity is offered by the following concept; see Sect. 2.5 (the comment about E.E. Levi).

Definition 2.4.2. Let $K \subseteq \mathbb{R}^n$ be nonempty, convex and open. $f : K \to \mathbb{R}$ is called *pseudoconvex at* $\overline{x} \in K$, iff it is directionally derivable at $\overline{x}$ in every direction with convex (and hence sublinear) directional derivative, and

$$x \in K, \quad f'(\overline{x}; x - \overline{x}) \geq 0 \quad \Rightarrow \quad f(x) \geq f(\overline{x}). \tag{2.4.1}$$

f is called *pseudoconvex on* K iff it is pseudoconvex at every $\overline{x} \in K$. f is *strictly pseudoconvex at* $\overline{x} \in K$ (or *on* K), iff (2.4.1) holds with the 2nd inequality in strict sense (for every $\overline{x} \in K$). f is *(strictly) pseudoconcave*, iff $-f$ is (strictly) pseudoconvex. If f is derivable, then (2.4.1) shrinks to:

$$x \in K, \quad \langle f'(\overline{x}), x - \overline{x} \rangle \geq 0 \quad \Rightarrow \quad f(x) \geq f(\overline{x}). \tag{2.4.2}$$

Note that (2.4.1) is not satisfied by the function of next Example 2.4.1 and that of Examples 3.1.4 and 3.1.8. Therefore, pseudoconvexity looks like a condition which guarantees to detect local minima by means of directional derivatives.

Proposition 2.4.1. Let $K \subseteq \mathbb{R}^n$ be nonempty, convex and open. If $f : K \to \mathbb{R}$ is strictly pseudoconvex on K, then it is strictly quasiconvex on K.

Proof. Ab absurdo, suppose that f be not strictly quasiconvex. According to Definition 2.4.1, this implies that $\exists x^1, x^2 \in K$ with $x^1 \neq x^2$ and $\exists \hat{\alpha} \in]0, 1[$ s.t.

$$f(x^1) \leq f(x^2), \quad f(x(\hat{\alpha})) \geq f(x^2), \tag{2.4.3}$$

where $x(\alpha) := (1 - \alpha)x^1 + \alpha x^2$. Then, $f(x(\alpha))$ being continuous (a directionally derivable function at a point is, of course, continuous there), $\exists \overline{\alpha} \in \,]0,1[$ s.t.

$$f(x(\overline{\alpha})) = \max_{x \in [x^1, x^2]} f(x). \tag{2.4.4}$$

Then, because of well known facts, we have ($\overline{x} := x(\overline{\alpha})$):

$$f'(\overline{x}; x^1 - \overline{x}) \leq 0, \quad f'(\overline{x}; x^2 - \overline{x}) \leq 0. \tag{2.4.5}$$

The positive homogeneity and the convexity of f', allow us to state (see Theorem 2.3.9):

$$f'(\overline{x}; x^1 - \overline{x}) = \overline{\alpha} f'(\overline{x}; -(x^2 - x^1)) \geq -\overline{\alpha} f'(\overline{x}; x^2 - x^1),$$

$$f'(\overline{x}; x^2 - \overline{x}) = (1 - \overline{\alpha}) f'(\overline{x}; x^2 - x^1).$$

These relations and (2.4.5) imply $f'(\overline{x}; x^2 - x^1) = 0$, and then (see Theorem 2.3.9):

$$f'(\overline{x}; x^1 - \overline{x}) = \overline{\alpha} f'(\overline{x}; x^1 - x^2) \geq 0.$$

This inequality, because of (2.4.1), implies $f(x^1) \geq f(\overline{x})$, so that, being $\overline{\alpha} \in \,]0,1[$, taking into account the former of (2.4.3) and (2.4.4), we obtain: $f(x^1) = f(\overline{x}) = f(x^2)$, $\overline{x} \in \,]x^1, x^2[$, which contradicts the strict pseudoconvexity of f. $\qquad\square$

Of course, there exist functions, which are strictly quasiconvex, but not pseudoconvex; take, for instance, $f : \mathbb{R} \to \mathbb{R}$ with $f(x) = -x^3$. The sole pseudoconvexity does not imply strict quasiconvexity, as a constant function shows. A finite convex function f is obviously pseudoconvex. In fact, if $f'(\overline{x}; x - \overline{x}) \geq 0$, being $d = x - \overline{x}$, and taking into account Theorem 2.3.10(5i), then (2.3.2a) implies $f(x) \geq f(\overline{x})$. However, note that, if pseudoconvexity is defined assuming differentiability and adopting (2.4.2) instead of (2.4.1) − as often happens −, then we cannot claim that every convex function is pseudoconvex.

Theorem 2.4.3. Let $K \subseteq \mathbb{R}^n$ be nonempty, convex and open, and $f : K \to \mathbb{R}$ be pesudoconvex at $\overline{x} \in K$. If

$$f'(\overline{x}; d) \geq 0, \quad \forall d \in \mathbb{R}^n, \tag{2.4.6}$$

then $\overline{x}$ is a global minimum point of f.

Proof. Obvious consequence of (2.4.1) and of Definition 1.1.1. $\qquad\square$

Without the assumption of pseudoconvexity, (2.4.6) by itself does not guarantee the thesis as next example shows.

Example 2.4.1. Set $n = 2$, $K = \mathbb{R}^2$, $f(x_1, x_2) = (x_2 - \alpha x_1^2)(x_2 - \beta x_1^2)$ with $0 < \alpha < \beta$, and $\overline{x} = (0,0)$. It is immediate to verify that (2.4.6) holds. Notwithstanding this,

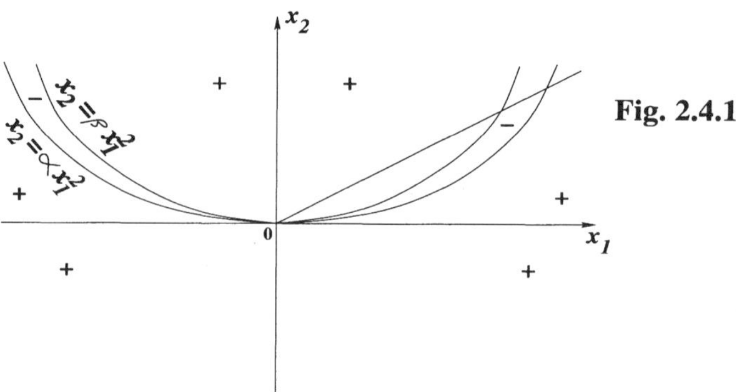

$\overline{x} = (0,0)$ is not even local minimum point, since in every neighbourhood of $\overline{x}$, f takes negative values; in Fig. 2.4.1, the parts of the plane where f is positive or negative have been indicated by $+$ or $-$, respectively. See Example 3.1.8 for an extension: $\overline{x}$ may not be local m.p. even if all the directional derivatives are positive. Note that pseudoconvexity prevents from such a situation.   □

The function of the above example is well known as Peano Function [I37, p.34], and has been fundamental for the development of the theory of the extrema; see Sect. 2.5.

From Definition 2.4.2 and Theorem 2.4.3, we immediately see that a directionally derivable function f is pseudoconvex, iff it does not possess stationary points (in the sense of (2.4.6)), which are not global m.p. Therefore, the drawback of Theorem 2.4.3 is that the assumption is too "close" to the thesis, in the sense that, in general, to verify the assumption or to verify the thesis are not very different activities. However, some tests can be conceived. For instance, *if f is twice differentiable, then a necessary condition for f to be pseudoconvex is that there is no point where its gradient be zero and its Hessian matrix be negative definite.* This condition $-$ which is an obvious consequence of the above remark and expresses the absence of isolated maxima $-$ is not sufficient, as the following example shows.

Example 2.4.2. Set $K = \mathbb{R}$ and $f(x) = x^4$ if $x \leq 0$, $f(x) = 0$ if $0 < x < 1$, $f(x) = -(x-1)^3$ if $x \geq 1$. f fulfils the above condition, but is not pseudoconvex.   □

Definition 2.4.3. Let $K \subseteq \mathbb{R}^n$ be nonempty and convex. $F : \mathbb{R}^n \to \mathbb{R}^n$ is *pseudoisotone* on K, iff $\forall x^1, x^2 \in K$ we have:

$$\langle F(x^1), x^2 - x^1 \rangle \geq 0 \quad \Rightarrow \quad \langle F(x^2), x^1 - x^2 \rangle \leq 0. \qquad (2.4.7)$$

F is *pseudoantitone* on K, iff $\forall x^1, x^2 \in K$ we have:

$$\langle F(x^1), x^2 - x^1 \rangle \leq 0 \quad \Rightarrow \quad \langle F(x^2), x^1 - x^2 \rangle \geq 0. \qquad (2.4.8)$$

It is immediate to show that *F isotone (antitone) implies F pseudoisotone (pseudoantitone)*. In fact, (2.3.11a) is equivalent to $\langle F(x^1), x^2 - x^1 \rangle \leq -\langle F(x^2), x^1 - x^2 \rangle$, which implies (2.4.7). The converse is not true, as easy examples show. Furthermore, *a differentiable f is pseudoconvex, iff ∇f is pseudoisotone* [145].

Proposition 2.4.2. Let $K \subseteq \mathbb{R}^n$ be convex, and $F : K \to \mathbb{R}^n$. **(i)** If F is pseudoisotone, then (1.3.5) implies (1.3.6) for $\phi \equiv 0$. **(ii)** If F is continuous, then (1.3.6) for $\phi \equiv 0$ implies (1.3.5).

Proof. (i) (1.3.6) for $\phi \equiv 0$ is an obvious consequence of (2.4.7). **(ii)** Set $x(\alpha) := (1-\alpha)\overline{x} + \alpha y$. Since $x(\alpha) \in K$, $\forall \alpha \in [0,1]$, for $\phi \equiv 0$, (1.3.6) is equivalent to:

$$\langle F(x(\alpha)), y - \overline{x} \rangle \geq 0, \quad \forall y \in K, \quad \forall \alpha \in [0,1].$$

Because of the continuity of F, for $\alpha \downarrow 0$, we obtain (1.3.5). $\qquad \square$

Note that pseudoisotonicity is "close" to MVI.

The functions which are not necessarily convex, but have convex restrictions to affine manifolds, are useful. Unlike other extensions of convex functions, they admit a characterization which is meaningful from the numerical calculus point of view. A first obvious case is that where the above manifold is a subspace. Let the argument of $f : \mathbb{R}^n \to \mathbb{R}$ be partitioned into 2 subvectors, say x and y. Then, according to Definition 2.3.1, f is convex with respect to x, iff $\forall y$ we have:

$$f(x,y) - f(\overline{x},y) - \langle \sigma, x - \overline{x} \rangle \geq 0, \quad \forall x.$$

Therefore, all the theorems for convex functions can be applied to $f(\bullet, y)$. We can have the convexity of $f(\bullet, y)$ and of $f(x, \bullet)$, but not that of $f(x,y)$.

Example 2.4.3. $f : \mathbb{R}^2 \to \mathbb{R}$, given by $x^2 + y^2 + kxy$ is convex with respect to x and with respect to y whatever $k \in \mathbb{R}$ may be, but is convex with respect to (x,y) iff $-2 \leq k \leq 2$. $\qquad \square$

Indeed, if f has the second derivatives, according to Theorem 2.3.5, the convexity of f with respect to (x,y) holds iff the matrix

$$\begin{pmatrix} f''_{xx}(x,y) & f''_{xy}(x,y) \\ f''_{yx}(x,y) & f''_{yy}(x,y) \end{pmatrix}$$

is positive semidefinite, while the convexity of f with respect to x and with respect to y requires the positive semidefiniteness only of $f''_{xx}(x,y)$ and of $f''_{yy}(x,y)$. Such an analysis can be extended to the case where the subspace is replaced by a linear or affine manifold or by a polyhedron (see Vol. 2).

In some developments of the theory of both scalar and vector problems, we are faced with vector-valued functions. In this case, some concepts of convexity can be extended by introducing a "cono-concept" or a "curvilinear concept". In other words, in some of the preceding definitions, we have implicitly used a special type of cone, namely a halfline, and a special type of curve, namely a segment, without obviously calling them cone and curve. More precisely, the inequality of (2.3.1) can be equivalently written as

$$(1-\alpha)f(x^1) + \alpha f(x^2) - f(x(\alpha)) \in \mathbb{R}_+.$$

When $f : \mathbb{R}^n \to \mathbb{R}^\nu$ with $\nu \geq 1$, then it is immediate to replace the special cone $\mathbb{R}_+$ with a cone of $\mathbb{R}^\nu$.

Definition 2.4.4. Let $X \subseteq \mathbb{R}^n$ be convex and $H \subset \mathbb{R}^\nu$ be a closed and convex cone with apex at the origin. $f : X \to \mathbb{R}^\nu$ is called *H-function* iff

$$(1 - \alpha)f(x^1) + \alpha f(x^2) - f(x(\alpha)) \in H, \quad \forall x^1, x^2 \in X, \quad \forall \alpha \in [0, 1],$$

where $x(\alpha) := (1 - \alpha)x^1 + \alpha x^2$. When $H \subseteq \mathbb{R}^\nu_+$ or $H \supseteq \mathbb{R}^\nu_+$ (respectively, $H \subseteq \mathbb{R}^\nu_-$ or $H \supseteq \mathbb{R}^\nu_-$), then f is called *H-convex* (respectively, *H-concave*). If $X \subseteq \mathbb{C}^n$ and $H \subseteq \mathbb{C}^\nu$, then $f : X \times X \to \mathbb{C}^\nu$ is called *complex* H-*function,* iff

$$(1 - \alpha)f(\zeta, \zeta^*) + \alpha f(\eta, \eta^*) - f((1 - \alpha)\zeta + \alpha\eta, (1 - \alpha)\zeta^* + \alpha\eta^*) \in H,$$

$$\forall \zeta, \eta \in X, \forall \alpha \in [0, 1] \subset \mathbb{R}.$$

Note that a $(\mathbb{R}^\nu_+)$-function (or $(\mathbb{R}^\nu_-)$-function) has all the components convex (concave), and a *(O)*-function has all the components affine. Of course, the above definition can be given without requiring the convexity of H, but this case is of little interest. In a quite similar way, we can extend Definition 2.3.2 and give the concept of cone-operator and, in particular, of cone-isotone and cone-antitone operators.

Definition 2.4.5. Let $X \subseteq \mathbb{R}^n$ be convex and $H \subset \mathbb{R}^\nu$ be a closed and convex cone with apex at the origin. $f : X \to \mathbb{R}^\nu$ is called H-*convexlike* iff

$$\forall x^1, x^2 \in X, \quad \forall \alpha \in [0, 1], \quad \exists \hat{x} \in X \text{ s.t. } (1 - \alpha)f(x^1) + \alpha f(x^2) - f(\hat{x}) \in H.$$

f is called H-*concavelike* iff $-f$ is H-convexlike. When $H = \mathbb{R}^\nu_+$, they are called simply *convexlike* and *concavelike*, respectively. Iff $X \subseteq \mathbb{C}^n$ and $H \subseteq \mathbb{C}^\nu$, then $f : X \times X \to \mathbb{C}^\nu$ is called *complex* H-*convexlike on* X, iff $\forall \zeta, \eta \in X, \forall \alpha \in [0, 1] \subset \mathbb{R}, \exists \tau \in X$, s.t.

$$(1 - \alpha)f(\zeta, \zeta^*) + \alpha f(\eta, \eta^*) - f(\tau, \tau^*) \in H.$$

Note that, for $\nu = 1$ and $\mathbb{C} = \mathbb{R}$, the above definition is fulfilled by every continuous function. For details about convexlike functions see [IV23].

With regard to the "curvilinear concept", note that the 1st condition of Definition 2.1.1 can be written as $x(\alpha) \in K, \quad \forall \alpha \in [0, 1]$, where $x(\alpha) := (1 - \alpha)x^1 + \alpha x^2$. Then, it is obvious to think of $x(\alpha)$ as a not necessarily affine function; of course, any nonlinear function would lead us to a more general (but, often, too loose and useless) concept. Therefore, we must ask $x(\alpha)$ to enjoy some properties. A nice way consists in requiring that $x(\alpha)$ be a geodesic. An excellent treatise on this topic is [I56], where it is shown that such an extension of convexity leads to an effective numerical calculus.

Another type of curvilinear extension is obtained by noting that (2.3.2a) can be equivalently written as:

$$f(x) - f(\bar{x}) - \langle \sigma, \eta(x, \bar{x}) \rangle \geq 0,$$

where $\eta(x,\overline{x}) := x - \overline{x}$. Instead of this special linear function, we can of course think of η as a nonlinear function; iff there exists η s.t. the above inequality be satisfied, f is called *invex*. Again, in the absence of any property for η, invexity is a nice concept with doubtful possibility of numerical applications.

Ideas, different from the above ones, for extending the class of convex functions have been given in [3]. The basic idea is very simple: a convex function $f : [a,b] \subseteq \mathbb{R} \to \mathbb{R}$ fulfils, $\forall [x^1, x^2] \subset]a,b[$, the inequality:

$$f(x) \le y(x), \quad x \in [x^1, x^2], \tag{2.4.9}$$

where $y(x)$ is the solution of the (trivial) differential equation:

$$y'' = 0, \quad y(x^i) = f(x^i), \quad i = 1, 2. \tag{2.4.10}$$

Starting from this remark, generalizations of convex functions have been obtained [7] by replacing (2.4.10) with the differential equation:

$$y'' + a(x)y' + b(x)y = 0, \quad y(x^i) = f(x^i), \quad i = 1, 2, \tag{2.4.11}$$

or with the more general differential equation:

$$y'' = \varphi(x, y, y'), \quad y(x^i) = f(x^i), \quad i = 1, 2, \tag{2.4.12}$$

where φ must fulfil suitable assumptions [39]; therefore, functions f, which fulfil (2.4.9), extend the convex ones. Such an approach has been extended to functions with more than 1 variable, by replacing the above ordinary differential operator with a partial differential one. This has led to study subharmonic and superharmonic functions [8], which, among other things, are of interest for the theory of problems (1.1.7).

2.5. Comments

1.The concept of convexity was already clear to Euclid in his "Elements". However, interesting elaborations of this concept can be found in Archimedes. In the treatise "On the sphere and cylinder" he gives the definition of a convex arc: "In the plane there are (finite) arcs of curves, which lie completely on one side of the lines joining their endpoints, or (however) have no point on the other side"; in this sense, finite means that the curves must have distinct endpoints, so that a closed curve is excluded. In an analogous way, he gives the definition of a convex surface bounded by a plane curve. In the treatise "On the equilibrium of planes; or on centres of gravity of planes" Archimedes states several postulates which are fundamental; one of them says that a convex set contains its centroid: "for each figure, whose boundary be concave from the same side, the centre of gravity must lie in the interior of the figure". At least 17 centuries were to elapse before Archimedes' ideas and results were taken and developed.

Indeed, Archimede's achievements on convexity were fully understood by Fermat and Galilei. However, the development of the Infinitesimal Calculus threw convexity in the shade. The following discover by Augustin Louis Cauchy in 1841 was one of the main reasons for renewing the interest in convexity: the perimeter of a closed and convex curve (i.e., a curve encircling a convex set of the plane) equals π times the mean value of the lengths of the orthogonal projections of the curve onto the lines through a point; the measure of a closed and convex surface (i.e.,enclosing a convex set of the space) equals 4 times the mean value of the measures of the orthogonal projections of the surface onto the planes through a point.

2.The investigation of the properties of the intersection graph associated to a family of convex sets, as well as the other topics related to Helly Theorem, still represent a fruitful and interesting field of research, which can contribute to weaken the hurtful separation between Combinatorial and Continuous Optimization. Such a separation is incomprehensible, since in the dawn of Convex Analysis there have been fundamental achievements, which have shown the tight connections between continuous and discrete spaces. Think of the classic result of H. Minkowski in 1891: "every compact and convex set of $\mathbb{R}^n$, with centre at the origin and volume greater than 2^n, contains at least one lattice point different from the origin".

3.The definition of a pointed cone is generally done by asking:

$$K \cap (-K) = \{O\} \tag{2.5.1}$$

and is less restrictive than (2.1.8); for instance, the cone of Fig. 2.1.6(a) (which contains full lines) fulfils (2.5.1), but not (2.1.8). However, we must observe that (2.5.1) requires $O \in K$, which is not acceptable for some applications (see $\mathcal{H}$ of Sect. 3.2). If (2.5.1) is replaced by $K \cap (-K) \subseteq \{O\}$, then such a drawback disappears, but a convex pointed cone may have more than 1 apex (take, for instance, an open halfspace). Furthermore, for the important properties, like (9i) of Theorem 2.2.6, K is assumed to be closed, so that (2.1.8) and (2.5.1) are equivalent. If (2.1.8) is adopted, then, as already noted, a convex and pointed cone can be called acute in agreement with the classic Geometry.

4.The characterization of the sets which have no supporting hyperplane in no point of their boundaries may have both theoretical interest and numerical utility to set up test problems. The former of Examples 2.2.1 suggests the class of sets $K = \mathbb{R}^n \backslash S$ where S is a convex and bounded set; the latter shows that such a class is not exhaustive.

5.The concept of the tangent cone was introduced by Bouligand in [9], Chapter X, Sect. 68, page 65, under the title "La notion de contingent". He started with the analysis of the existence and uniqueness of "demi-tangente"; he carried on both smooth and nonsmooth cases, and connected the contingent to functions which are "positivement homogène du premier degré".

6.Theorem 2.2.2 has influenced greatly the development of the Convex Analysis and of the Theory of Extrema, and many consequences are still to come, especially for the nonconvex case. In characterizing a convex set by means of its "tangents", it has introduced a "dual" way of looking at a set as "envelope" of its tangents. Such aspects would deserve to be deepened. This of course requires the development of the theory of envelopes; such a theory was created in the smooth case to detect the singular solutions of a differential equation and, strangely, seems to have received no further development. In this context, the connections with Fenchel theory of conjugate functions look interesting; this theory, even if formally independent, has much to do with Theorem 2.2.2.

7.The first meaningful results on polyhedra are due to Euler; in 1752 he established the famous equality for polytopes of $\mathbb{R}^3$:

$$v - e + f = 2, \tag{2.5.2}$$

where v, e, f are, respectively, the numbers of vertices, edges, facets. This famous result is a consequence of Euler's approach, which consists in finding a classification of polyhedra. The "kind" of a polyhedron is identified by a "table", where the vertices of each face are listed in a cyclical order. Two polyhedra are of the same kind (isomorphic) iff, up to renumbering of vertices and faces, their tables are identical. Another interesting consequence of this approach is that *the number of faces with an odd number of vertices is even*. The result (2.5.2) was extended to a polytope $K \subset \mathbb{R}^n$ by H. Poincaré and L. Schläfli at the end of the 19th century [49]: let $\nu_r(K)$ denote the number of r-dimensional faces of K; we have:

$$\sum_{r=0}^{n=1}(-1)^r\nu_r(K) = 1 - (-1)^n. \tag{2.5.3}$$

For $n=3$, this equality collapses to the Euler one.

8.As already noted, the polar of a cone is fundamental for the development of the theory of constrained extrema. The polar of a set, besides an undoubted theoretical importance, has shown to be very useful to solve problem (1.1.7) [1]. Since the vector formulation of (1.1.7) is now appearing in the applications, it should be interesting to extend the theory of the polar of a set to the vector case and to apply it to the vector version of (1.1.7).

9.A convex set is associated with several points, which represent it in a sense to be specified. One of them is the centre. It should be suitable to have a general definition, which contains the several existing concepts of centre. A hint might be the following one. Let $K \subset \mathbb{R}^n$ be a nonempty, compact and convex body, and let $r \in \mathbb{R}_+$. Consider the problem:

$$\min_{y \in K} \min_{r \geq 0} \int_{\partial K} |\,\|x - y\| - r|\, dx,$$

where we have a hypersurface integral (the classic curvilinear integral, if $n = 2$). Let (y^0, r_0) be a solution of the above problem; y^0 may be called *centre* of K. The meaning of the above definition is: to find, among all the hyperspheres with centre in y and radius r, one s.t. the "sum of the deviations" between its boundary and ∂K be minimum; in other words, we search for the "best approximation" of K by means of a hypersphere.The extensions to the cases where K is not convex, or not compact, or not a body, or card

$K < \infty$ (the integral is replaced by a summand), or K is a subset of a Banach space are of interest. The integrand in the above problem may be replaced by a suitable, non-negative function of it; e.g., the square. Just to consider an instance of the existing concepts of centre, let us show a definition of centre, which is a slight strengthening of one in the literature. Let $K \subset \mathbb{R}^n$ have a bounded closure and cardinality >1. Consider any $y \in \mathrm{ri}\, K$, and a function $\phi : (\mathrm{cl}\, K) \times \mathrm{ri}\, K \to \mathbb{R}$. s.t.

$$\phi(x; y) = 0, \quad \forall x \in \mathrm{frt}K,$$
$$\phi(x; y) > 0, \quad \forall x \in \mathrm{ri}\, K,$$
$$\phi(y; y) = 1$$
$$\phi((1 - \alpha)x + \alpha y; y) = \alpha, \quad \forall \alpha \in [0, 1], \quad \forall x \in \mathrm{frt}K.$$

The volume of the set of $\mathbb{R}^n \times \mathbb{R}$ (truncation of a convex cone; see Definition 2.1.7), whose boundary is $[(\mathrm{cl}\, K) \times \{0\}] \cup \mathrm{gr}\, \phi(\bullet; y)$, is constant with respect to y. Among all these sets, consider the (unique) one, whose boundary has minimum measure. Let it be given for $y = y^0$. y^0 may be considered the *centre* of K.

Of course, instead of the above ϕ (which can be considered a gauge-type function; see (2.3.19) for a classic gauge function), we might consider more general (or different) functions (see [I35], pp 559-663, where only the 4 above conditions are given to define the centre, so that a not binding concept is obtained; see also [I1], Chapter VIII).

10. The concept of centre of a convex set K is important, at least for the algorithms which are based on the interior points, and deserves to be deepened. The definition of centre of the above comment corresponds to a criterion which, of course, is not the unique one. For instance, when K is a polytope, the point which minimizes the sum of its distances from the vertices of K (see Example 1.2.3 and the previous comment about (1.2.4)-(1.2.5)) might be considered as centre of K (if K is a triangle of the plane, such a point is the Torricelli point). This point, call it *generalized Torricelli point*, and the centre of Sect. 2.1 do not coincide necessarily. Consider, for instance, the triangle:

$$K_T := \{x \in \mathbb{R}^2_+ : 1 - x_1 - x_2 \geq 0\}.$$

The Torricelli point is $x_T = (x_1 = x_2 = 2 - \sqrt{3} \simeq 0.26)$, while the centre of the above comment is $x_C = \left(x_1 = x_2 = 1 - \frac{\sqrt{2}}{2} \simeq 0.29 \right)$, which is the solution of the problem (of type (1.1.3); $x_1 = \rho \cos\theta$, $x_2 = \rho \sin\theta$):

$$\min_{(\theta, \rho) \in R} \varphi(\theta, \rho),$$

where

$$R_T := \left\{ (\theta, \rho) \in \mathbb{R}^2_+ : \theta \leq \frac{\pi}{2} \right\},$$

and

$$\varphi(\theta, \rho) := \frac{1}{2} \left(\sqrt{1 + \rho^2 \sin\theta} + \sqrt{1 + \rho^2 \cos^2\theta} + \right.$$
$$\left. + \sqrt{\rho^2 + 2\rho^2 \sin\theta \cos\theta - 2\rho(\sin\theta + \cos\theta) + 3} \right)$$

gives the measure of the lateral surface of the pyramid (tetrahedron) whose base is the above triangle and whose altitude has length 1 and foot in the point $(\rho\cos\theta,\ \rho\sin\theta,\ 0)$ of the base. With regard to a polytope, any relation between the generalized Torricelli point and the centre of of the above comment would be useful. In particular, it would be interesting to find conditions under which the solution of the weighted generalized Fermat-Torricelli problem (namely, (1.2.4) where the distances are multiplied by weights) and the centre of the above comment coincide.

If K is given in the form (1.1.2), in the general case it is difficult to use the definition of centre of Sect. 2.1, since the boundary of K is not explicitly available. An alternative definition of centre must be considered, as the following one: y^0 is the centre of K, iff

$$y^0 \in \ \operatorname{argmax}\ \left\{\sum_{i\in \mathcal{I}^+}\ln g_i(x) :\ g_i(x) = 0,\ i\in \mathcal{I}^0,\ g_i(x) > 0,\ i\in \mathcal{I}^+\right\}.$$

This criterion, for the case of the above triangle K_T, becomes:

$$y^0 \in \ \operatorname{argmax}\ \{\ln(1 - x_1 - x_2) + \ln x_1 + \ln x_2 :\ (x_1, x_2)\in \ \operatorname{int}\ K_T\},$$

and gives $y^0 = (y_1^0 = y_2^0 = 1/3)$ as centre, which is different from the two above points, but equal to the barycentre of K_T. The comparison of all these points is of interest. To this end, note that, in the last criterion, there is a drawback. If, unlike K_2, the representation of K of type (1.1.2) contains redundant equalities or inequalities (in the sense that the removal of them does not change K), then the centre may change if they are removed; in other words, if a convex set K has 2 different representations of type (1.1.2), say R_1 and R_2, then the same K may be associated with 2 different centres. In this sense, the problem of the uniqueness of the centre is strictly related with the problem of detecting redundant constraints; these two problems may help one another. Through the convex hull, the concept of centre can be extended to the nonconvex case. The unbounded case is, of course, also of interest.

11. Let $\mathcal{F} = \{H^+\}$ be the family of all the halfspaces considered by Theorem 2.2.2, and $\mathcal{F}^0 = \{H^+\}$ the subfamily s.t. H^+ is a supporting halfspace of K; of course, $\mathcal{F}$ can be equivalently replaced by $\mathcal{F}^0$ in the sense that:

$$\bigcap_{K^+\in\mathcal{F}^0} K^+ = \bigcap_{K^+\in\mathcal{F}} K^+ = K.$$

It is usefull to characterize the minimal (in the sense of inclusion) subfamily of $\mathcal{F}^0$, say $\mathcal{F}^m$, s.t.

$$\bigcap_{K^+\in\mathcal{F}^m} K^+ = K.$$

12.Theorems 2.2.7 and 2.2.8 show the importance of the class of cones defined by (2.1.14). This is seen also in Sect. 2.7. The following chapters will further stress the role of such a class. Therefore, a systematic investigation of the properties of (2.1.14), both in $\mathrm{I\!R}^n$ and in B and comparison with other classes of cones, like for instance that of pointed cones, would be useful.

13.Sect. 2.2, notwithstanding the fact that only a few (fundamental) properties have been considered, shows that linear separation has been investigated and now-a-days forms a good theory, even if something more could be done. It would be interesting and useful to develop theories for more general classes of manifolds. For instance, piecewise linear and continuous manifolds or conic ones may play an important role both in Vector Optimization (see Vol. 2) and in the scalar one (see [III11]). Also the classes of parabolic and exponential manifolds are of great interest; a few definitions and results are in Chapters 4 and 5. For each class, we should achieve a set of theorems like those we know for linear separation. This might lead to introduce the concept of *analytical complexity*; some remarks aimed to give a definition will be done in Sects. 3.5 and 4.10.

14.As we will see in Chapter 5, separation between two sets can be considered as the fundamental basis for the development of the Lagrangian theory of constrained extrema. A pioneer work in this sense, even if with the sole purpose of necessary optimality conditions, is due to Dubovitskii and Milyutin [18], who based their result on the concept of *separation among several sets* [III2, I34]. The sets $K_1, ..., K_r \subset \mathbb{R}^n$ are *separable*, iff there exist r hyperplanes, whose equations are denoted by $\langle a^i, x \rangle = b_i$ with $a^i \in \mathbb{R}^n \backslash \{O\}$ and $b_i \in \mathbb{R}$, such that:

$$\sum_{i=1}^{r} a^i = O, \quad \sum_{i=1}^{r} b_i \leq 0, \quad K_i \subseteq \text{lev}_{\leq 0}(\langle a^i, x \rangle - b_i). \tag{2.5.4}$$

When $r=2$, by choosing $a^2 = -a^1$ and $b^2 = -b^1$, the above definition shrinks to Definition 2.2.5. In [IV14, Corollary 1.1], the following property has been proved: let $K_i \subset \mathbb{R}^n$, $i = 1, ..., r+s$, be convex sets, the first r, $r \geq 1$, of which are open. We have:

$$\bigcap_{i=1}^{r+s} K_i = \varnothing, \tag{2.5.5}$$

if and only if there exist (δ^* is defined in (2.2.10); see also (2.3.17)):

$$\bar{a}^i \in \Delta_i := \{a^i \in \mathbb{R}^n : \delta^*(a^i; K_i) < +\infty\}, \quad i = 1, ..., r+s, \tag{2.5.6}$$

such that, for at least one index $i = 1, ..., r + s$, we have:

$$\bar{a}^i \neq O, \quad \langle \bar{a}^i, x^i \rangle < \delta^*(\bar{a}^i; K_i), \quad \forall x^i \in K_i, \tag{2.5.7}$$

and

$$\sum_{i=1}^{r+s} \bar{a}^i = O, \quad \sum_{i=1}^{r+s} \delta^*(\bar{a}^i; K_i) \leq 0 \tag{2.5.8}$$

In the particular case where every K_i is a cone with apex at the origin, the above statement collapses to a lemma proved by Dubovitskii and Milyutin (see Lemma 5.11 of [134]), since $\Delta_i = -K_i^*$, and the inequality in (2.5.8) is redundant. They exploited such a Lemma to state a general necessary optimality condition − known as the Dubovitskii-Milyutin Maximum Principle − from which they derive − perhaps, for the first time − both the Lagrange multiplier rule and the Euler equation (see Chapter 5). The fact that their lemma deals with cones, and not with sets, is due to the fact that their purpose is only a necessary condition. However, even if statement (2.5.5)-(2.5.8) is a slight generalization of that of Dubovitskii and Milyutin, the proof of the former is completely different from that of the latter, since it is based on the reduction of the separation among several sets to the separation of two sets. Indeed, by introducing

$$\mathcal{H} := \underset{i=1}{\overset{r+s}{\times}} K_i, \quad \mathcal{K} := \{(x^1, ..., x^{r+s}) : x^i \in \mathbb{R}^n, \quad i = 1, ..., r+s\},$$

the proof of the former statement is reduced to a scheme of type (3.2.2), which means that, instead of proving separation among $r+s$ sets (as (2.5.5)-(2.5.8) require), we prove separation between $\mathcal{H}$ and $\mathcal{K}$. Due to the importance of the Dubovitskii and Milyutin result, even if the above reduction is a formal one, it would be interesting to investigate the connections between the two concepts of separation, to clarify whether or not it is useful to carry out the theory of separation among several sets, both in $\mathbb{R}^n$ and B.

The above concept of separation among several sets might be generalized by requiring *m-separation among several sets*: given $m, n \in \mathbb{N}, m \leq n$, and the sets $K_1, ..., K_n$, it is required to search for a condition which guarantees that, if Dubovitskii-Milyutin separation holds for each subfamily $\{K_{i_1}, ..., K_{i_m}\}$, then such a separation holds for the family $\{K_1, ..., K_n\}$.

15.Note that Theorem 2.1.3 is a sufficient condition for (2.5.5) to be false. Therefore, when a condition for the separation of several sets is necessary and sufficient (and in Dubovitskii-Milyutin Theorem this happens), then the negation of such a condition guarantees the thesis of Helly Theorem. The connections between such two theorems might be fruitful. For instance, by exploiting the separation between 2 nonconvex sets (see Chapter 4), and using the transformation outlined in the previous comment, the former theorem might be extended to the nonconvex case, and this result might lead to extend the latter theorem. The m-separation among several sets of the previous comment might be related to the assumption of Helly Theorem.

16.Another important extension of the concept of separation is *asymptotic separation*, which will be discussed in Sect 4.10. It has been investigated only a little, and mainly as a side topic; never has it been related to the separation of several sets.

17.Theorem 2.2.8 expresses disjunctive separation; Theorem 2.2.9 expresses separation, but only with respect to a face of the cone K. Due to their importance for achieving optimality conditions, it would be useful to extend them to the case where K is different from C_0 and $\overset{o}{C}$, and to the case where K is not necessarily a Cartesian product. Special kinds of sets S are of interest.

18. The functional (2.3.3) gives a "measure" of the lack of convexity of a function. It has some useful applications, like e.g., in the Duality Theory. However, such a "measure" may turn out to be rough. For instance, let M and ε be positive reals with M very large and ε very small, and consider the functions $f_1, f_2 : [-M, M] \to \mathbb{R}$, given by:

$$f_1(x) := \frac{\varepsilon}{M^2}(M + x)(M - x), \quad x \in [-M, M],$$

$$f_2(x) := \begin{cases} \frac{1}{\varepsilon}(\varepsilon + x)(\varepsilon - x), & \text{if } x \in [-\varepsilon, \varepsilon], \\ x^2, & \text{if } x \in [-M, M]\setminus[-\varepsilon, \varepsilon]. \end{cases}$$

It is easy to see that $\rho(f_1) = \rho(f_2)$, while they are substantially different: f_1 is strictly concave and f_2 is "almost" strictly convex. It would be extremely useful to improve the definition of $\rho(f)$.

19. After Example 2.3.5, it was noted that a convex function may behave very badly (bad convex function). Indeed, the following problem was posed [25]: let $f : \mathbb{R}^n \to \mathbb{R}$ with $n \geq 2$ be convex, and set $x(t) = (t, 0, ..., 0) \in \mathbb{R}^n$ with $t \in \mathbb{R}$. Assume that $\nabla f(x(t))$ exists $\forall t > 0$, and consider the following limit:

$$\lim_{t \downarrow 0} \nabla f(x(t)). \tag{2.5.9}$$

It was conjectured that (2.5.9) could not exist. If (2.5.9) does not exist, is ∇f norm-bounded? The existence of functions for which (2.5.9) does not exist was proved, independently of each other, in [40, 44, 54]. Let us consider the function constructed in [44], which required the following interesting theorem [15] and its corollary (whose extension to a Hilbert space would be interesting).

Theorem 2.5.1. Let Ξ be a compact set, $f : \mathbb{R}^n \times \Xi \to \mathbb{R}$, and consider the max-function (see (2.3.20b)):

$$\phi(x) := \max_{\xi \in \Xi} f(x; \xi), \tag{2.5.10}$$

Suppose that $f(\bullet; \xi)$ be differentiable and that $f(\bullet; \xi)$ and its gradient (with respect to x) $\nabla_x f(\bullet; \xi)$ depend continuously on $(x; \xi)$. Then, ϕ is directionally derivable, and its directional derivative in the direction d is given by:

$$\phi'(x; d) = \max_{\xi \in \Xi(x)} \langle \nabla_x f(x; \xi), d \rangle, \tag{2.5.11}$$

where $\Xi(x) := \operatorname*{argmax}_{\xi \in \Xi} f(x; \xi)$.

Corollary 2.5.1. Under the assumptions of the above theorem, if in addition $f(\bullet; \xi)$ is convex, and if $\overline{x}$ is such that $\Xi(\overline{x}) = \{\overline{\xi}\}$, then ϕ is differentiable at $\overline{x}$ and

$$\nabla \phi(\overline{x}) = \nabla f(\overline{x}; \overline{\xi}). \tag{2.5.12}$$

Now, let us consider the function $\phi : \mathbb{R}^2 \to \mathbb{R}$, given by:

$$\phi(x_1, x_2) := \max_{(\xi_1, \xi_2) \in \Xi} \{\xi_1 x_1 + \xi_2 x_2 - \frac{1}{2}\xi_1^2\}, \quad (x_1, x_2) \in \mathbb{R}^2, \tag{2.5.13}$$

where

$$\Xi := \left\{(\xi_1, \xi_2) \in \mathbb{R}^2 : 0 < \xi_1 \leq 1, \xi_2 = \sin\frac{1}{2\xi_1}\right\} \cup \{(\xi_1, \xi_2) \in \mathbb{R}^2 : \xi_1 = 0, |\xi_2| \leq 1\}.$$

The set Ξ is obviously compact. The function ϕ is convex with respect to (x_1, x_2), since it is the pointwise maximum of affine functions. In the above corollary, we set $n = 2$, consider the particular Ξ just introduced, and set $f(x; \xi) = \xi_1 x_1 + \xi_2 x_2 - \frac{1}{2}\xi_1^2$, so that we have now:

$$\nabla_x f(x; \xi) = (\xi_1, \xi_2).$$

Then, from Corollary 2.5.1, if $x = (x_1, 0)$ and $0 < |x_1| < 1$, we draw:

$$\Xi(x) = \operatorname*{argmax}_{\xi \in \Xi} f(x; \xi) = \operatorname*{argmax}_{(\xi_1, \xi_2) \in \Xi} \left(\xi_1 x_1 - \frac{1}{2}\xi_1^2 \right) = \left\{ \left(x_1, \sin \frac{1}{2x_1} \right) \right\},$$

so that card $\Xi(x) = 1$. Hence, we obtain (see Definition 2.3.3):

$$\nabla \phi(x_1, 0) = \left(x_1, \sin \frac{1}{2x_1} \right), \quad 0 < |x_1| < 1,$$

$$\partial \phi(0, 0) = \{O\} \times [-1, 1],$$

which show that $\nabla \phi(x_1, 0)$ exists $\forall (x_1, 0)$ s.t. $0 < |x_1| < 1$, and that $\lim_{x_1 \downarrow 0} \nabla \phi(x_1, 0)$ does not exist. This proves the claim about (2.5.9). Independently of [44], a function similar to (2.5.13) was proposed by E. De Giorgi: $\phi(x_1, x_2) = \max_{\xi \neq 0} [x_2 \sin(\log |\xi|) + 2x_1 \xi - \xi^2]$. The functions proposed in [40, 54] are quite different from (2.5.13); in [40] the claim about (2.5.9) is proved to hold also in B and if $x(t)$ is a norm-continuous path ending at the origin. The above functions show that the class of convex functions for which (2.5.9) does not exist is nonempty. Since the elements of this class may offer difficulties both to the theoretical analysis and to the algorithms, it would be interesting to characterize such a class, to evaluate its width (for instance, in terms of measure), to consider the same questions for subsets of that of convex functions, and to extend them to other classes of functions (for instance, quasiconvex functions which are not convex; pseudoconvex functions which are not convex).

20. An early trace of quasiconvex functions is in [37]: at page 307 of Chapter III, J. von Neumann proves a theorem (which now-a-days is a classic one), whose thesis is

$$\max_{\xi} \min_{\eta} f(\xi, \eta) = \min_{\eta} \max_{\xi} f(\xi, \eta).$$

To show this, he assumes that the upper (lower) sets of the restriction of f to ξ (to η) be convex; hence, using the present terminology, he assumes that f be quasiconcave with respect to ξ and quasiconvex with respect to η. Quasiconvex functions were treated by B. De Finetti in [16] under the term "convex stratifications", which means convex level sets; such a paper contains several interesting remarks which may lead to further investigation.

21. In Definition 2.4.1, we have considered, a definition of strict quasiconvexity, which guarantees its inclusion within quasiconvexity. This seems the least requirement for a correct extension of the theory of convex functions. We will quote it as SQC1. In the literature, there is a definition, which is obtained from SQC1 by replacing the assumption $f(x^1) \leq f(x^2)$ with $f(x^1) < f(x^2)$; we will quote it as SQC2. Unlike SQC1, a function SQC2 is not necessarily quasiconvex: take, for instance, $f : \mathbb{R} \to \mathbb{R}$ with

$f(x) = 0$ if $x \neq 0$ and $f(0) = 1$; it is SQC2 but not all its lower level sets are convex. Therefore, it is improper to assign to the functions SQC2 a name which recalls to us quasiconvexity; this may be misleading. Furthermore, a function of type SQC2 does not guarantee uniqueness of m.p.; take, for instance, the function $f : \mathbb{R} \to \mathbb{R}$, with $f(x) = -(2 + x)(1 + x)$ if $x < -1$, $f(x) = 0$ if $x \in [-1, 1]$, $f(x) = -(x - 1)(x - 2)$ if $x > 1$. It is of class SQC2 and even quasiconvex, but uniqueness of m.p. does not happen, since every $x \in [-1, 1]$ is a global minimum point. If correctly named and located, the class SQC2 might be useful. Indeed, it enjoys the property that a local m.p. is also a global one (the proof is quite similar to that given for SQC1), and continues to make true Proposition 2.4.1. Unfortunately, as the above instance shows, SQC2 does not enjoy uniqueness; this reinforces the previous remark. Now, let us make an additional observation. For some applications, the class of strictly convex functions is too feeble, and the class of strong convex functions has been introduced and defined by replacing the inequality (2.3.2a) with

$$E(\overline{x}, x, \sigma) \geq \alpha||x - \overline{x}||^2, \quad \forall x \in K,$$

where $\alpha > 0$ is suitable. Due to the importance of such a concept — for instance, in the analysis of equilibrium problems —, it would be interesting to extend it to quasiconvex functions. To this end, perhaps, we need a strengthening of SQC1. As a hint for this task, let us consider the following definition, which will be quoted as SQC3: a function $f : K \subseteq \mathbb{R}^n \to \mathbb{R}$, with K convex, will be called SQC3, iff it is quasiconvex and

$$\pi_1, \pi_2 \in \operatorname{gr} f \quad \Rightarrow \quad]\pi_1, \pi_2[\; \subset \; \sim \operatorname{gr} f.$$

It is possible to show that SQC3$\Rightarrow$SQC1. Ab absurdo, suppose that $\exists x^1, x^2 \in K$ with $x^1 \neq x^2$ and $f(x^1) \leq f(x^2)$ and that $\exists \overline{\alpha} \in]0, 1[$ s.t. $f(x(\overline{\alpha})) \geq f(x^2)$, where $x(\alpha) := (1 - \alpha)x^1 + \alpha x^2$. Consider any $\alpha \in [\overline{\alpha}, 1[$. Being $f(x(0)) = f(x^1) \leq f(x^2) = f(x(1))$ and $x(\alpha) \in]x^1, x^2[$, because of the quasiconvexity of f, we must have:

$$f(x(\alpha)) \leq f(x^2), \quad \forall \alpha \in [\overline{\alpha}, 1],$$

and hence $f(x(\overline{\alpha})) = f(x^2)$. Besides this, we prove now that:

$$f(x(\alpha)) = f(x^2), \quad \forall \alpha \in [\overline{\alpha}, 1].$$

In fact, if $\exists \hat{\alpha} \in]\overline{\alpha}, 1[$ s.t. $f(x(\hat{\alpha})) < f(x^2)$, then, being $x(\overline{\alpha}) \in]x^1, x(\hat{\alpha})[$, because of the quasiconvexity of f, we must have:

$$\text{either} \quad f(x(\overline{\alpha})) \leq f(x(\hat{\alpha})) \quad \text{or} \quad f(x(\overline{\alpha})) \leq f(x^1),$$

according to, respectively,

$$f(x^1) \leq f(x(\hat{\alpha})) \quad \text{or} \quad f(x(\hat{\alpha})) \leq f(x^1).$$

In both cases we meet a contradiction; in the former:

$$f(x(\overline{\alpha})) \leq f(x(\hat{\alpha})) < f(x^2) = f(x(\overline{\alpha})),$$

and in the latter:

$$f(x(\overline{\alpha})) \leq f(x^1) < f(x^2) = f(x(\overline{\alpha})).$$

Having achieved that the segment $[(x(\overline{\alpha}), f(x(\overline{\alpha}))), (x^2, f(x^2))]$ is contained into the graph of f, the 2nd part of the assumption is contradicted. $\square$

It is easy to show that SQC1 $\nRightarrow$ SQC2. Take, e.g., $f : \mathbb{R} \to \mathbb{R}$ with $f(x) = x$, or with $f(x) = |x|$. Now, we can show a further property of the class SQC2; namely, uniqueness for the minimum points. In fact, ab absurdo, let x^1 and x^2, with $x^1 \neq x^2$, be global m.p. of f on a convex set $R \subseteq K$. Because of the quasiconvexity of f and the convexity of R, being $f(x^1) = f(x^2) = \min_{x \in R} f(x)$, we must have:

$$x(\alpha) := (1 - \alpha)x^1 + \alpha x^2 \in \mathrm{lev}_{=f(x^2)} f(x), \quad \forall \alpha \in [0, 1].$$

It follows that the segment $[(x^1, f(x^1)), (x^2, f(x^2))]$ is contained into gr f, and this contradicts the 2nd part of the assumption. $\square$

22.By using the propositions of Sect. 2.4, it is not difficult to give the following characterization of differentiable pseudoconvex functions: f is pseudoconvex, iff it is quasiconvex and every x s.t. $f'(x) = 0$ is a local m.p. of f. This is conceptually interesting, but difficult to use for finding extrema. Indeed, in the theory of extrema, the extensions of convexity aim to achieve statements of type of Theorem 2.3.8. Hence, a characterization in terms of the existence of local m.p. is evidently a drawback. To overcome this would be interesting.

23.Among the many attempts to extend the concept of convexity, there are those which give the definition of "convex function" over a nonconvex set or even over a graph (see [V43]). For instance, one of them weakens Definition 2.3.1 by allowing K to be any set and asking (2.3.1) to be verified only at those $\alpha \in [0, 1]$ s.t. $x(\alpha) \in K$; of course, the resulting "convexity" is too feeble: any function f, whose domain K is the boundary of a sphere, would be convex. These definitions are of limited validity and are finalized to specific problems. Investigation in this field is extremely useful for at least combinatorial problems. One of the several definitions is originated by the obvious necessary condition for a function $f : K \to \mathbb{R}$ to be convex:

$$(1 - \alpha)f(x^1) + \alpha f(x^2) \geq \inf_{x \in K \cap N} f(x), \quad \forall x^1, x^2 \in K, \quad \forall \alpha \in [0, 1], \tag{2.5.14}$$

where N is a neighbourhood of $x(\alpha) := (1 - \alpha)x^1 + \alpha x^2 (f(x) = x^2$ if $x \in \mathbb{R}\backslash\{O\}$, $f(0) = 1$, shows it is not sufficient). With $N(x) := \{y \in \mathbb{Z}^n : ||x - y|| < 1\}$ we define the neighbourhood of x in $\mathbb{Z}^n$; $x \in \mathbb{Z}^n \Rightarrow N(x) = \{x\}$;$x \notin \mathbb{Z}^n \Rightarrow \mathrm{card}\, N(x) = 2^r$ where r is the number of elements of the n-tuples x which do not belong to $\mathbb{Z}$. Then f might be called *discretely convex* iff (2.9.14) holds with N replaced by $N(x(\alpha))$ and K by a subset of $\mathbb{Z}^n$. Note that the restriction of a convex function to a discrete interval, namely to $X := \{x \in \mathbb{Z}^n : a \leq x \leq b\}$ with $a, b \in \mathbb{R}^n$, is not necessarily discretely convex, as shown by the following example: $X = \mathbb{Z}^2$, $f(x_1, x_2) = (2x_1 - x_2)^2$; since $f''(x) = \begin{pmatrix} 8 & -4 \\ -4 & 2 \end{pmatrix}$, f is obviously convex (Theorem 2.3.5); for $x^1 = (0,0)$, $x^2 = (1,2)$, $\alpha = 1/2$, (2.9.14) with $K \cap N$ replaced by $N(x(\alpha)) = \{y \in \mathbb{Z}^2 : \|\frac{1}{2}x^1 + \frac{1}{2}x^2 - y\| < 1\} = \{(0,1),(1,1)\}$ becomes $0 \geq 1$. This example, even if trivial, and the above short remark show how difficult it is to transfer to a discrete space a concept which is born for a continuous space.

24. In 1910-11 Eugenio Elia Levi (1883-1917) made an important discovery [31, 32]: an open and connected set $\Omega \subset \mathbb{R}^4$ is not necessarily a holomorphic field for a function $f(z_1, z_2)$ of the complex variables $z_1 = x_1 + iy_1$, $z_2 = x_2 + iy_2$ (while, it was known that $\Omega \subset \mathbb{R}^2$ was always a holomorphic field for some suitable function $f(z)$ of the complex variable $z=x+iy$). Levi gave a necessary condition for Ω to be a holomorphic field for a function $f(z_1, z_2)$. Such a condition (which was soon extended to $\mathbb{R}^{2n}$, $n > 2$; see [24]) contained a function which, just after the appearance of the Levi results, was named *pseudoconvex*. The introduction of pseudoconvexity in the field of Optimization has happened much later with the investigations of Mangasarian [33] and, under the term of semiconvex functions, of Tuy [49]; see also [I45].

25. The problem of finding conditions, under which the restriction of a function to an affine manifold or to a polyhedron is convex, can be posed also for the extensions of convexity. In particular, it would be interesting to find conditions under which the restriction of a function to an affine manifold or polyhedron is geodesic convex or coercive or H-convexlike or H-convex. An analogous question exists for operators.

26. The class of generalized convex functions defined by (2.4.12) has had a development substantially independent of the other generalizations. It would be useful to carry out comparisons. For instance, may a generalized convex function defined by (2.4.12) be geodesic convex in the sense of [I56]?

27. As we have seen (Sect. 2.4), convexlikeness is an important concept. The drawback (common to the other extensions of convexity) is that there are not conditions − which be meaningful from the numerical calculus viewpoint − for testing whether or not a function is convexlike. Indeed, if we want concrete applications to become reality, we cannot remain on Definitions 2.4.4 and 2.4.5. With regard to the Definition 2.4.5, for $H = \mathbb{R}^\nu_+$, set $z(\alpha) := (1 - \alpha)f(x^1) + \alpha f(x^2)$, $z(\alpha) = (z_1(\alpha), ..., z_\nu(\alpha))$; the condition for convexlikeness requires that:

$$\bigcap_{i=1}^{\nu} \operatorname{lev}_{\leq z_i(\alpha)} f_i \neq \varnothing, \quad \forall \alpha \in [0,1]. \tag{2.5.15}$$

If $f_1, ..., f_\nu$ are convex, then (2.5.15) holds, since its left-hand side contains $(1 - \alpha)x^1 + \alpha x^2$. An obvious (but restrictive) sufficient condition for (2.5.15) to hold is that:

$$\bigcap_{i=1}^{\nu} \operatorname{lev}_{\leq \beta_i} f_i \neq \varnothing, \quad \forall \beta_i \in \operatorname{Im} f_i, \ i = 1, ..., \nu.$$

If $X \subseteq \mathbb{R}$ (or $n = 1$), then (2.5.15) holds, if each f is nondecreasing (or nonincreasing), since the left-hand side of (2.5.15) contains $\min\{x^1, x^2\}$ (or $\max\{x^1, x^2\}$).

When in Definition 2.4.4 int $H = \varnothing$, then the class of $H-$functions is, of course, very special. However, it might be of some interest to consider some kinds of cones H, having int $H = \varnothing$, and study the corresponding classes of $H-$functions; in particular, the case where qri $H \neq \varnothing$.

28. The function of Example 2.4.1 was conceived by Peano to show a mistake made by Lagrange; see [I37], page 33; see also Examples 2.4.1 and 3.1.8. Lagrange wrote (Théorie des Fonctions; in Oeuvres de Lagrange, Vol. IX, Gauthier-Villars, Paris, 1881, page 290): "Si, dans une fonction quelconque des variables $x, y, z, ...$, on substitue à la place de ces variables les quantités $x + p, y + q, z + r, ...$, et qu'on développe la fonction suivant les puissances et les produits des quantités $p, q, r, ...$, les termes où ces quantités ne se trouveront qu'à la première dimension, étant égalés chacun séparément à zéro, donneront les équations nécessaire pour que la fonction proposée devienne un maximum ou minimum: ensuite on considérera la quantité composée de tous les termes oú $p, q, r, ...$formeront deux dimensions, et il faudra pour le minimum que cette quantité soit toujours positive, et pour le maximum toujours négative, quelles que puissent être les valeurs de $p, q, r, ...$ Si tous ces termes s'évanouissaient à la fois, il faudrait alors, pour l'existence du maximum ou minimum, que tous les termes où $p, q, r, ...$ formeraient trois dimensions disparussent aussi à la fois, et que la quantité composée des termes où $p, q, r, ...$ formeraient quatre dimensions fùt toujours positive pour le minimum et toujours négative pour le maximum, $p, q, r, ...$ ayant des valeurs quelconques" This claims that, if all the terms of the 1st and 2nd dimensions of Taylor' Expansion (Lagrange was the first to understand the importance of Taylor Theorem after about 60 years since its appearence) of a function of $x, y, z, ...$ (where p, q, r are the variations of the arguments) vanish, then it is necessary, for the existence of a maximum or a minimum, that all the terms of the third dimension in $p, q, r, ...$ shall disappear and that the quantity composed of terms where $p, q, r, ...$ form four dimensions shall be always positive for the minimum and always negative for the maximum when $p, q, r, ...$ have any values whatever.

Following Lagrange, all writers on this subject (for instance, Bertrand and Serret) made the same incorrect deductions, until Peano detected it at the end of 19th century: "The proofs for the criteria by which the maxima and minima of functions of several variables are to be recognized, and which are given in most books, depend upon the theorem that in Taylor development for functions of several variables the ratio of the remainder after an arbitrary term to this term has a limit zero when the increments of the variables approach zero. This theorem is in general false when the term in question is *not* a definite form with respect to the increments of the variables, and when it is a definite form, the theorem needs proof" The function of Example 2.4.1 was found by Peano to show the incorrect deduction. Note that Peano function is not pseudoconvex. Such a mistake, and, of course, its detection, have been extremely important for the development of the theory of extrema and related fields during the first 2 decades of 20th century, which has been the basis for the important achievements of the calculus of variations in the third one. This mathematical event is a lesson for all of us, and should be reported, in mathematical and historical details, in all Mathematical Analysis and Optimization teachings. In my opinion, this does not diminish at all the fact that Lagrange has been one of the greatest mathematicians of all times: every mathematician is able to produce a "true theorem" only a genius can make an important mistake!

References

[1] Balas E., "Disjunctive Programming". Annals of Discrete Mathematics, Vol.5, North-Holland Publ.Co., 1979, pp.3-51.

[2] Bazaraa M.S. and Shetty C.M., "Foundations of Optimization". Springer-Verlag, Berlin, 1976.

[3] Beckenbach E.F., "Generalized Convex Functions". Bull.Amer.Math.Soc., Vol.43, 1937, pp.363-371.

[4] Bellman R., "Dynamic Programming". Princeton Univ. Press, 1957.

[5] Berman A., "Cones, matrices and mathematical programming". Lecture Notes in Ec. and Math.Systs., No.79, Springer-Verlag, Berlin, 1973.

[6] Bonnesen T. and Fenchel W., "Theorie der Konvexen Körper". Springer-Verlag, Berlin, 1934.

[7] Bonsall F.F.,"The characterization of generalized convex functions". Quarterly Jou.of Mathematics, Vol.1, 1950, pp.100-111.

[8] Bonsall F.F.,"On generalized subharmonic functions". Proc.Cambridge Phil.Soc., Vol.46, 1950, pp.387-395.

[9] Bouligand G., "Introduction a la Géométrie Infinitésimale Directe". Libraire Vuibert, Paris, 1932.

[10] Carathéodory C.,"Über der Variabilitäsbereich der Koeffizienten von Potenzreihen die gegebene Werte nicht annehmen". Mathematische Annalen, Vol.64, 1907, pp.95-115.

[11] Cauchy A.L., "Méthode générale pour la résolution des systèmes d'équations simultanées". Comptes Rendus, Acad. Sci., Paris, Vol.25, 1847, pp.536-538.

[12] Clarke F.H., "Optimization and Nonsmooth Analysis". J. Wiley, New York, 1983.

[13] Courant R, "Variational methods for the solution of problems of equilibrium and vibrations". Bull. Amer. Math. Soc., Vol.49, 1943, pp.1-23.

[14] Danzer L., Griinbaum B. and Klee V., "Helly's Theorem and its relatives". In [29], pp.101-180.

[15] Danskin J.M., "The Theory of Max-Min, with Applications". SIAM Jou. on Applied Mathematics, Vol.14, No.4, 1966, pp.641-644.

[16] De Finetti B., "Sulle stratificazioni convesse (On the convex stratifications)" (in Italian). Annali di Matematica Pura ed Applicata, Serie 4°, Tomo XXX, Published by N.Zanichelli, Bologna, 1949, pp.173-183.

[17] Dem'yanov F.V. and Rubinov A. (Eds.), "Quasidifferentiability and related topics". Kluwer Acad.Publ., Dordrecht, 2000.

[18] Dubovitskii A.Ya. and Milyutin A.A., "The extremum problem in the presence of constraints". Dokl. Akad. Nauk SSSR, Vol.149, No.4, 1963, pp.759-762.

[19] Ekeland I. and Temam R., "Convex Analysis and Variational Problems". North Holland, Amsterdam, 1975.

[20] Fan K., "Convex sets and their applications". Lecture Notes, Argonne National Laboratory, Illinois, 1959.

[21] Fan K., "Applications of a theorem concerning sets with convex sections". Mathematische Annalen, Vol.163, 1966, pp.189-203.

[22] Fan K., Glicksburg I. and Hoffman A.J., "Systems of Inequalities involving convex functions". Proc.Am. Math. Soc. , Vol.8, 1957, pp.617-622.

[23] Fenchel W., "Convex cones, sets and functions". Lecture Notes, Princeton Univ. Press, Princeton, N.J., 1953.

[24] Fichera G., "Tre battaglie perdute da tre grandi matematici italiani (Three fights which have been lost by three great Italian mathematicians)" (in Italian). Proceedings of the conference in honour of G.Gemignani (Modena, May 20,1994), Accademia Nazionale di Scienze, Lettere ed Arti, Modena, 1994, pp.9-28.

[25] Giannessi F., "A problem on Convex Functions". Jou. of Optimiz. Theory and Appls., Vol.59, No.3, Dec.1988, page 525.

[26] Grünbaum B., "Convex polytopes". Interscience-Wiley, London, 1967.

[27] Helly E., "Über Mengen konvexer Körper mit gemeinschaftlichen Punkten". Jber. Deutsch. Math. Verein., Vol.32, 1923, pp.175-176.

[28] Jensen J.L.W.V., "Sur les functions convexes et les inegalités entre les valeurs moyennes". Acta Mathematica, Vol.30, 1906, pp.175-193.

[29] Klee V.L., "Convexity". Proceedings of Symposia in Pure Mathem., Vol. VII, Amer. Math. Soc., Providence, RI, 1963.

[30] Krein M. and Milman D., "On extreme points of regularly convex sets". Studia Math., Vol.9, 1940, pp.133-138.

[31] Levi E.E., "Studi sui punti singolari essenziali delle funzioni analitiche di 2 o più variabili complesse (Studies on essential singular points of analytic functions of 2 or more variables)" (in Italian). Annali di Matematica Pura ed Applicata, Zanichelli Publisher, Bologna, 1910, pp.61-68 .

[32] Levi E.E., "Sulle ipersuperficie dello spazio a 4 dimensioni che possono essere frontiera del campo di esistenza di una funzione analitica di due variabili complesse (On the ipersurfaces of 4-dimensional space, which can be boundary of the domain of an analytic function of two complex variables)" (in Italian). Annali di Matematica Pura e Applicata, Zanichelli Publisher, Bologna, 1911, pp.69-80.

[33] Mangasarian D.L., "Pseudo-convex functions". SIAM Jou. on Control, ,Vol.3, 1965, p.281-290.

[34] Mastroeni G. and Rapcsák T., "On convex generalized systems". Jou. of Optimiz. Theory and Appls., Vol.104, No.3, March 2000, pp.605-627.

[35] Mifflin R., "Semismooth and semiconvex functions in constrained optimization". SIAM Jou. on Control and Optimiz., Vol.15, 1977, pp.959-972.

[36] Minkowski H., "Theorie der Konvexen Körper". Insbesondere Begründung ihres Oberflächenbegriffs, Gesammelte Abhandlungen II, Leipzig, 1911.

[37] von Neumann J., "Zur Theorie der Gesellschaftsspiele". Mathematische Annalen, 1928.

[38] Panik M.I., "Fundamentals of Convex Analysis". Kluwer Academic Publ., Dordrecht, 1993.

[39] Peixoto M., "Generalized convex functions and second-order differential inequalities". Bull.Amer.Math.Soc., Vol.55, 1949, pp.563-572.

[40] Pontini C., "Solving in the Affirmative a Conjecture about a limit of Gradients". Jou. of Optimiz. Theory and Appls., Vol. 70, No.3, 1991, pp.623-629.

[41] Quang P.H. and Yen N.D., "New proof of a theorem of F. Giannessi". Jou. of Optimiz. Theory and Appls., Vol.68, No.2, 1991, pp.385-387.

[42] Radon J., "Mengen konvexer Körper, die einen gemeinsamen Punkt enthalten". Mathematische. Annalen, Vol.83, 1921, pp.113-115.

[43] Rockafellar R.T., "Convex Analysis". Princeton Univ. Press, Princeton, N.J., 1970.

[44] Rockafellar R.T., "On a Special Class of Convex Functions". Jou. of Optimiz. Theory and Appls., Vol.70, No.3, 1991, pp.619-621.

[45] Rockafellar R.T. and Wets J.B., "Variational Analysis". Springer-Verlag, Berlin, 1998.

[46] Rubinov A.M. and Gasimov R.N., "Strictly increasing positively homogeneous functions with application to exact penalization". Optimization, Vol.52, No.1, 2003, pp.1-28.

[47] Sacks S., "Theory of The Integral". Warszawa-Lwów, 1937.

[48] Stoer J. and Witzgall C., "Convexity and Optimization in Finite Dimensions I". Springer-Verlag, Berlin, 1970.

[49] Tuy H., "Sur les inégalités linéaires". Colloquium Mathematicum, Vol.13, 1964, pp.107-123.

[50] Tverberg H., "A generalization of Radon's Theorem". Jou. of London Math. Soc., Vol.41, 1966, pp.123-129.

[51] Valentine F.A., "Convex sets". McGraw-Hill, New York, 1964.

[52] de la Vallée Poussin Ch.J., "Sur la méthode de l'approximation minimum". Ann. Soc. Sci. de Bruxelles, Vol.35, 1911, pp.1-16.

[53] Weyl H., "Elementare Theorie der konvexen Polyheder". Commentarii Math. Helvetici, Vol.7, 1935, pp.290-306.

[54] Zagrodny D., "An Example of Bad Convex Function". Jou. of Optimiz. Theory and Appls., Vol.70, No.3, 1991, pp.631-637.

[55] Zalinescu C., "Convex Analysis in General Vector Spaces". World Scientific, Singapore, 2002.

[56] Zangwill W.I., "The piecewise concave function". Management Science, Vol.13, 1967, pp.900-912.

CHAPTER 3. INTRODUCTION TO IMAGE SPACE ANALYSIS

3.1. Semidifferentiability

Due to its importance, the concept of differentiability has been the recipient of many generalizations. Most of them have been conceived independently of each other and for special objectives, often different from those of the theory of extrema. Here, we will consider a generalization of differentiability, which is a sort of "container" of several existing concepts and is suitable for achieving necessary optimality conditions. The symbols of this section are independent of those of the other sections, if they overlap.

Let $X \subseteq \mathbb{R}^n$, with card $X > 1$, be a convex cone with apex at $\overline{x}$ (or at least the intersection of the closed unit ball with a convex cone), and denote by $\mathcal{G}$ the set of all functions $\mathcal{D}f : X \times (X - \overline{x}) \to \mathbb{R}$ which are positively homogeneous with respect to the 2nd argument, i.e. $\forall d \in X - \overline{x}$

$$\mathcal{D}f(\overline{x}; \alpha d) = \alpha \mathcal{D}f(\overline{x}; d), \quad \forall \alpha \in \mathbb{R}_+. \tag{3.1.1}$$

$\mathcal{L}$ and $\mathcal{C}$ denote the subsets of $\mathcal{G}$, whose elements are linear (or continuous linear, if $\mathbb{R}^n$ is replaced by B) and convex (and hence sublinear) with respect to the 2nd argument, respectively; obviously, $\mathcal{C}$ and $\mathcal{L}$ are closed with respect to the sum. G denotes any subset of $\mathcal{G}$.

Definition 3.1.1. Let $G \subseteq \mathcal{G}$. A function $f : X \to \mathbb{R}$ is called *G-differentiable* at $\overline{x}$, iff there exists $\mathcal{D}_G f \in G$, such that, $\forall d \in X - \overline{x}$, we have:

$$\lim_{d \to 0} \frac{1}{||d||} \left[\varepsilon(\overline{x}; d) := f(\overline{x} + d) - f(\overline{x}) - \mathcal{D}_G f(\overline{x}; d) \right] = 0. \tag{3.1.2}$$

$\mathcal{D}_G f(\overline{x}; d/||d||)$ is called *the G-derivative of f at $\overline{x}$*. When $n \geq 2$, f is called G-*derivable*, iff each of the n functions of 1 variable, restrictions of f to $x_1, ..., x_n$, is G-differentiable; the G-derivatives are now called *partial G-derivatives*. f is said to be *directionally derivable at $\overline{x}$ in the direction d*, iff the limit

$$f'(\overline{x}; d) := \lim_{\alpha \downarrow 0} \frac{1}{\alpha} \left[f(\overline{x} + \alpha d) - f(\overline{x}) \right] \tag{3.1.3}$$

exists finite (Theorem 2.3.9 gives some properties of f'). If $f : X \to \mathbb{R}^\nu$ with $\nu \geq 1$, then we say that f is G-differentiable, iff each component of f is G-differentiable.

It is easy to see that a $\mathcal{L}$-differentiable function is a differentiable one, and that a G-differentiable function is directionally derivable (the G-derivative being the directional derivative) but not conversely as Examples 3.1.4 and 3.1.5 show. If $G = \{\alpha d : \alpha \in \mathbb{R}_+\}$, so that d can be considered fixed, then (3.1.2) is equivalent to:

$$\lim_{\alpha \downarrow 0} \mathcal{D}_G f \left(\overline{x}; \frac{1}{||\alpha d||} \alpha d \right) = \lim_{\alpha \downarrow 0} \frac{1}{||d||} \frac{f(\overline{x} + \alpha d) - f(\overline{x})}{\alpha}.$$

Multiplying both sides of this equality by $||d||$ and identifying $\mathcal{D}_G f$ with f', we are led to (3.1.3). Note that a $\mathcal{C}$-differentiable function is also directionally derivable, but not vice versa, as simple examples show; take, for instance, $f : \mathbb{R}^2 \to \mathbb{R}$, with $f(x) = ||x||_2$ if $x_1 + x_2 \in \mathbb{Q}$ and $f(x) = -||x||$ otherwise; the two concepts coincide on $\mathbb{R}$. In general, the $\mathcal{C}$-derivative is not either continuous or convex with respect to the 1st argument; see, for instance, Example 3.1.2. If $|| \bullet ||$ denotes a norm in the considered space; then the above definition can be obviously adapted to any normed space.

Example 3.1.1. Set $X = \mathbb{R}$, $\overline{x} = O$, $\alpha \geq 0$, and $f : \mathbb{R} \to \mathbb{R}_+$ with $f(x) = \beta|x| + x^2$ if $x \in \mathbb{Q}$ and $f(x) = \beta|x| + 2x^2$ if $x \notin \mathbb{Q}$. For $\beta > 0$, it is easy to check that f is $\mathcal{C}$-differentiable at $\overline{x}$ (and only there) with $\mathcal{D}_\mathcal{C} f(0; d) = \beta|d|$. Note that f is continuous at $\overline{x}$ only. For $\beta = 0$, f offers an instance of a function which is defined in the entire space, but is derivable only at one point. $\qquad\square$

Example 3.1.2. Set $X = \mathbb{R}$ and $f : [-1, 1] \to \mathbb{R}_+$ with

$$f(x) := \begin{cases} -x & , \text{ if } -1 \leq x \leq 0, \\ f_n(x) := -x^2 + (3 \cdot 2^{-n} + 1)x - 2^{1-2n}, & \text{ if } 2^{-n} < x \leq 2^{1-n}, \ n = 1, 2, ...; \end{cases}$$

see Fig. 3.1.1. By setting:

$$\mathcal{D}_\mathcal{C} f(\overline{x}; x - \overline{x}) := \begin{cases} -(x - \overline{x}) & , \text{ if } -1 \leq \overline{x} \leq 0, \ \forall x \\ (1 - 2^{-n-1})(x - \overline{x}) & , \text{ if } \overline{x} = 2^{-n}, \ x \leq 2^{-n}, \ n = 1, 2, ..., \\ (1 + 2^{-n})(x - \overline{x}) & , \text{ if } \overline{x} = 2^{-n}, \ x > 2^{-n}, \ n = 1, 2, ..., \\ (-2\overline{x} + 3 \cdot 2^{-n} + 1)(x - \overline{x}) & , \text{ if } 2^{-n} < \overline{x} < 2^{1-n}, \quad n = 1, 2, ..., \\ \frac{1}{2}(x - \overline{x}) & , \text{ if } \overline{x} = 1, \ x \leq 1, \end{cases}$$

we see that (3.1.2) is fulfilled at any $\overline{x} \in [-1, 1]$ with $G = \mathcal{C}$. This is easily seen by noting that $1 - 2^{-n-1}$ and $1 - 2^{-n}$ are, respectively, left and right derivatives of $f_n(x)$ for $x = 2^{-n}$, and that $\lim_{n \to +\infty} (1 - 2^{-n-1})(x - \overline{x}) = x - \overline{x}$. The present example shows that continuity and $\mathcal{C}$-differentiability on a convex set do not imply the convexity of f. f is locally Lipschitz at any $\overline{x} \in [-1, 1]$. $\qquad\square$

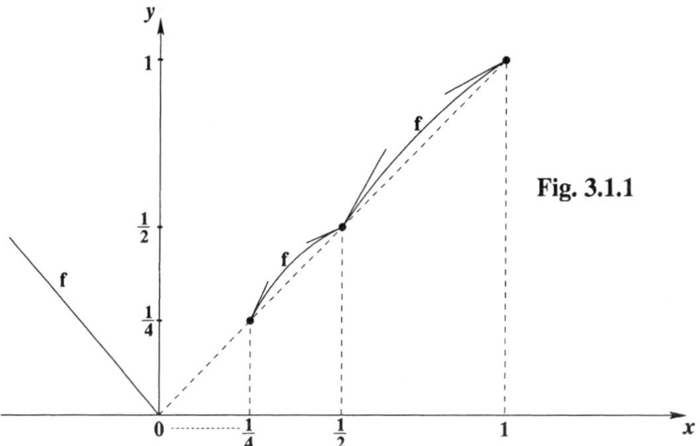

Fig. 3.1.1

Example 3.1.3. Set $X = \mathbb{R}$ and $f : [-1, 1] \to \mathbb{R}_+$ with

$$
f(x) := \begin{cases} -x, & \text{if } -1 \leq x \leq 0, \\ 2^{-n}(-4x^2 + 12 \cdot 2^{-n}x - 8 \cdot 2^{-2n})^{\frac{1}{2}}, & \text{if } 2^{-n} < x \leq 2^{1-n}, \ n = 1, 2, \ldots . \end{cases}
$$

This case is quite analogous to the previous one with the sole exception that $\mathcal{D}_e f = \infty$ at $\overline{x} = 2^{-n}$; now f is not Lipschitz at $\overline{x} = 2^{-n}$. $\square$

Example 3.1.4. Set $X = \mathbb{R}^2$, $\overline{x} = O$ and

$$
\varphi(\alpha) := \begin{cases} 1 & , \text{ if } 0 < \alpha \leq 1 \text{ or } \alpha \geq 3, \\ 3 - 2\alpha & , \text{ if } 1 < \alpha \leq 2, \\ 2\alpha - 5 & , \text{ if } 2 < \alpha < 3, \end{cases}
$$

$$
f(x) := \begin{cases} (x_1^2 + x_2^2)^{\frac{1}{2}} & , \text{ if } x \not> O, \\ \varphi\left(\dfrac{x_2}{x_1^2} \right) (x_1^2 + x_2^2)^{\frac{1}{2}} & , \text{ if } x > O. \end{cases} \tag{3.1.4}
$$

We find that (3.1.3) is fulfilled with $f'(\overline{x}; d) = (d_1^2 + d_2^2)^{\frac{1}{2}}$, while it is not possible to verify (3.1.2) (see Fig. 3.1.2). Therefore, a directionally derivable function is not necessarily G-differentiable, in the sense that a set $G \subseteq \mathcal{G}$ and a $\mathcal{D}_G \in G$ s.t. (3.1.2) be satisfied do not exist necessarily. Let us observe that f is continuous (but not locally Lipschitz) and $f'(0; d) > 0$, $\forall d \in \mathbb{R}^2 \backslash \{O\}$; notwithstanding this, $\overline{x}$ is not a m.p.; f extends the Peano-function (see Example 2.4.1), namely $f : \mathbb{R}^2 \to \mathbb{R}$, given by:

$$
f(x) = (x_2 - \alpha x_1^2)(x_2 - \beta x_1^2) \ , \quad 0 < \alpha < \beta. \tag{3.1.5}
$$

Indeed, note that the function (3.1.4) has been conceived in the light of Peano ideas (see also Example 3.1.8). Note that pseudoconvexity (see Definition 2.4.2) cuts of functions like the present one. $\qquad\square$

Example 3.1.5. Set $X = \mathbb{R}^2,\ \ \overline{x} = O$, and

$$f(x) := \begin{cases} |x_1| + |x_2| & \text{, if } x_2 \leq x_1^2, \\ (5 - 4t)|x_1|^{2\alpha - 1} & \text{, if } x_2 = tx_1^2,\ 1 \leq t \leq \frac{3}{2}, \\ (6t - 10)x_1^2 & \text{, if } x_2 = tx_1^2,\ \frac{3}{2} < t < 2, \\ x_2 & \text{, if } x_2 \geq 2x_1^2. \end{cases}$$

f is directionally derivable at $\overline{x}$ in any direction:

$$f'(O; d) = \lim_{\alpha \downarrow 0} \frac{1}{\alpha} f(\alpha d_1, \alpha d_2) = \left\{ \begin{array}{ll} d_2 & \text{, if } d_2 \geq d_1^2 \\ |d_1| + |d_2|, & \text{if } d_2 < d_1^2 \end{array} \right\} > 0, \quad \forall d \in \mathbb{R}^2 \backslash \{O\}.$$

The same remarks made for Example 3.1.4 hold here too. $\qquad\square$

The class of G-differentiable functions was introduced in [IV14] (under the term generalized differentiable functions, and as a subclass of that of semidifferentiable ones) and developed in [18]. Independently of this, in the same period, the above class (under the term of B-differentiable functions) was introduced in [IV89]. Further developments of semidifferentiability are in [15,16,34,36,37,39,44,48,51,52].

Theorem 3.1.1. Let $X \subseteq \mathbb{R}^n$, $f_i : X \to \mathbb{R}$, $i = 1, ..., r$, be $\mathcal{C}$-differentiable at $\overline{x} \in X$, and $\alpha_i \geq 0$, $i = 1, ..., r$. Then $f := \sum_{i=1}^{r} \alpha_i f_i$ is $\mathcal{C}$-differentiable at $\overline{x}$ with $\sum_{i=1}^{r} \alpha_i \, \mathcal{D}_{\mathcal{C}} \, f_i(\overline{x}; \frac{d}{||d||})$ as $\mathcal{C}$-derivative.

Proof. We have:

$$\lim_{d \to 0} \frac{1}{||d||} \left[\sum_{i=1}^{r} \alpha_i f_i(x) - \sum_{i=1}^{r} \alpha_i f_i(\overline{x}) - \sum_{i=1}^{r} \alpha_i \mathcal{D}_{\mathcal{C}} f_i(\overline{x}; d) \right] =$$

$$= \sum_{i=1}^{r} \alpha_i \lim_{d \to 0} \frac{1}{||d||} \left[f_i(x) - f_i(\overline{x}) - \mathcal{D}_{\mathcal{C}} f_i(\overline{x}; d) \right] = 0 \ ,$$

where the last equality is due to the fact that each f_i fulfils (3.1.2). Because of Theorem 2.3.4 (i, ii with φ linear), α_i being non-negative, we have that $\mathcal{D}_{\mathcal{C}} f_i \in \mathcal{C}$, $i = 1, ..., r$, imply $\sum_{i=1}^{r} \alpha_i \mathcal{D}_{\mathcal{C}} f_i \in \mathcal{C}$. Then we can apply Definition 3.1.1 for $G = \mathcal{C}$. $\qquad\square$

Theorem 3.1.2. Let $X \subseteq \mathbb{R}^n$ and $f : X \to \mathbb{R}$ be $\mathcal{C}$-differentiable at $\overline{x} \in X$. **(i)** The $\mathcal{C}$-derivative is unique. **(ii)** f is continuous at $\overline{x}$.

Proof. **(i)** Ab absurdo, suppose that there exist two distinct $\mathcal{C}$-derivatives of f, say $\mathcal{D}_{\mathcal{C}}^1 f$ and $\mathcal{D}_{\mathcal{C}}^2 f$; this implies the existence of $\hat{d} \in X - \overline{x}$, s.t.

$$\mathcal{D}_{\mathcal{C}}^1 f \left(\overline{x}; \frac{1}{||\hat{d}||} \hat{d} \right) \neq \mathcal{D}_{\mathcal{C}}^2 f \left(\overline{x}; \frac{1}{||\hat{d}||} \hat{d} \right).$$

Set:

$$\varepsilon^1(\overline{x}; d) := f(\overline{x} + d) - f(\overline{x}) - \mathcal{D}_{\mathcal{C}}^1(\overline{x}; d),$$

$$\varepsilon^2(\overline{x}; d) := f(\overline{x} + d) - f(\overline{x}) - \mathcal{D}_{\mathcal{C}}^2(\overline{x}; d).$$

Because of (3.1.1) and of the equality in the square bracket of (3.1.2), $\forall d \in X - \overline{x}$, we have:

$$\mathcal{D}_{\mathcal{C}}^1 f\left(\overline{x}; \frac{1}{||d||}d\right) - \mathcal{D}_{\mathcal{C}}^2 f\left(\overline{x}; \frac{1}{||d||}d\right) = \frac{\varepsilon^2(\overline{x}; d)}{||d||} - \frac{\varepsilon^1(\overline{x}; d)}{||d||}.$$

Passing to the limit as $d \to 0$ in the above equality, because of (3.1.2) the right-hand side tends to zero, unlike the left-hand one (due to the above inequality). **(ii)** Because of (3.1.1), we have:

$$\lim_{x \to \overline{x}} f(x) = \lim_{x \to \overline{x}} [f(\overline{x}) + \mathcal{D}_{\mathcal{C}} f(\overline{x}; x - \overline{x}) + \varepsilon(\overline{x}; x - \overline{x})] =$$

$$= f(\overline{x}) + \lim_{d \to 0} \mathcal{D}_{\mathcal{C}} f(\overline{x}; d) + \lim_{d \to 0} \varepsilon(\overline{x}; d) = f(\overline{x}). \qquad \square$$

Theorem 3.1.3. Let $X \subseteq \mathbb{R}^n$ be open and $f : X \to \mathbb{R}$ be $\mathcal{C}$-differentiable at every $\overline{x} \in X$. Then f is continuous on X.

Proof. Obvious consequence of Theorem 3.1.2 and of the fact that $\mathcal{D}_{\mathcal{C}} f(\overline{x}; d)$ exists $\forall \overline{x} \in X$. $\qquad \square$

Theorem 3.1.4. Let $X \subseteq \mathbb{R}^n$ be open and $f : X \to \mathbb{R}$ be $\mathcal{C}$-differentiable at every $\overline{x} \in X$. Then:

$$\lim_{\alpha \downarrow 0} \mathcal{D}_{\mathcal{C}} f(\overline{x} + \alpha d; d) = \mathcal{D}_{\mathcal{C}} f(\overline{x}; d), \qquad (3.1.6)$$

where $d \in \mathbb{R}^n$ is a fixed direction.

Proof. From Definition 3.1.1 and Theorem 3.1.3, we have:

$$\lim_{\alpha \downarrow 0} \mathcal{D}_{\mathcal{C}} f(\overline{x} + \alpha d; d) = \lim_{\alpha \downarrow 0} [f(\overline{x} + (\alpha + 1)d) - f(\overline{x} + \alpha d) - \varepsilon(\overline{x}; d)] =$$

$$= f(\overline{x} + d) - f(\overline{x}) - \varepsilon(\overline{x}; d) = \mathcal{D}_{\mathcal{C}} f(\overline{x}; d). \qquad \square$$

Note that the proofs of Theorems 3.1.2-3.1.4 do not exploit the fact that $\mathcal{D}f \in \mathcal{C}$; therefore, in the statements, $\mathcal{C}$ might be replaced by any $G \subseteq \mathcal{G}$. Notwithstanding the above properties, unlike the subclass $\mathcal{L}$ of linear functions, and the class $\mathcal{G}$, the class $\mathcal{C}$ is not a linear space, since obviously the opposite of a sublinear function is not necessarily sublinear. As Fig. 3.1.1 shows, in going along a line of X, the $\mathcal{C}$-derivative can change much; for instance, going through $\overline{x} = \frac{1}{2}$, $\mathcal{D}_{\mathcal{C}} f$ passes from a line to an angle (whose edges have 3/4 and 3/2 as slopes) and then again to a line. Hence, at $\overline{x} = 1/2$ we have a sort of break. However, coming from the left side, the line arrives with slope 3/4 and, going to the right, the line starts with slope 3/2. This is in agreement with (3.1.6) which, therefore, expresses a directionally "left continuity" and "right continuity" of the $\mathcal{C}$-derivative.

Definition 3.1.2. Let $X \subseteq \mathbb{R}^n$ and $G \subseteq \mathcal{C}$. The *G-subdifferential* of a *G*-differentiable function $f : X \to \mathbb{R}$ at $\overline{x} \in X$ is given by

$$\underline{\partial}_G f(\overline{x}) := \{\sigma \in \mathbb{R}^n : \mathcal{D}_G f(\overline{x}; x - \overline{x}) \geq \langle \sigma, x - \overline{x} \rangle, \ \forall x \in X\}. \tag{3.1.7a}$$

σ is called the G-*subgradient* of f at $\overline{x}$. If $G \subseteq -\mathcal{C}$, then

$$\overline{\partial}_G f(\overline{x}) := \{\sigma \in \mathbb{R}^n : \mathcal{D}_G f(\overline{x}; x - \overline{x}) \leq \langle \sigma, x - \overline{x} \rangle, \ \forall x \in X\} \tag{3.1.7b}$$

is called G-*superdifferential* and σ the G-*supergradient* of f at $\overline{x}$. The pair $(\underline{\partial}_G f(\overline{x})$, $\overline{\partial}_G f(\overline{x}))$ is called the G-*differential* of f at $\overline{x}$. If f is convex (or concave), then for $G = \mathcal{C}$ (or $G = -\mathcal{C}$), we have $\underline{\partial}_\mathcal{C} f(\overline{x}) = \partial f(\overline{x})$ (or $\overline{\partial} f(\overline{x}) = \partial(-f(\overline{x}))$).

When $X \subseteq B$, then σ must belong to the continuous dual of B. (3.1.7a) is nothing more than the subdifferential (Definition 2.3.3) of $\mathcal{D}_G f(\overline{x}; \cdot)$ or $\underline{\partial}_G f(\overline{x}) = \partial \mathcal{D}_G f(\overline{x}; \cdot)$. Note that (3.1.7a) and (3.1.7b), as well as ∂f, do not collapse — from a formal viewpoint — to the classic differential when f is differentiable; this would happen defining $\underline{\partial}_G f(\overline{x}) = \{\langle \sigma, x - \overline{x} \rangle : \sigma \in \mathbb{R}^n, \ \mathcal{D}_G f(\overline{x}; x - \overline{x}) \geq \langle \sigma, x - \overline{x} \rangle, \ \forall x \in X\}$ (and analogously for (3.1.7b); identifying dx with Δx); however, in this case $\underline{\partial}_G f$ is a singleton and its unique element is the gradient of f. Of course, both $\underline{\partial}_G f(\overline{x}) \neq \varnothing$ and $\overline{\partial}_G f(\overline{x}) \neq \varnothing$ happen iff $\mathcal{D}_G f(\overline{x}; \cdot)$ is linear (so that $G \supseteq \mathcal{L}$). The uniqueness of $\underline{\partial}_G f$ and $\overline{\partial}_G f$ is an obvious consequence of Theorem 3.1.2(i) and subsequent remark. In Examples 3.1.1-3.1.3, we have $\underline{\partial}_\mathcal{C} f(0) = [-1, 1]$. In Example 3.1.2 we have $\underline{\partial}_\mathcal{C} f(2^{-n}) = [1 - 2^{-n-1}, 1 + 2^{-n}], \ n = 1, 2, \ldots$.

Example 3.1.6. Set $X = \mathbb{R}$, $f : \mathbb{R} \to \mathbb{R}$ with $f(x) = x^2 \sin \frac{1}{x}$ if $x \neq 0$ and $f(0) = 0$. We find $D_\mathcal{C} f(0; d) = 0, \ \forall d \in \mathbb{R}$ (indeed, f is differentiable). Thus $\underline{\partial}_\mathcal{C} f(0) = \{O\}$ (see also Example 3.1.9). $\qquad\qquad\square$

Theorem 3.1.5. Let $X \subseteq \mathbb{R}^n$ be nonempty, open and convex, and the functions $f, f_1, \ldots, f_r : X \to \mathbb{R}$ be $\mathcal{C}$-differentiable at $\overline{x} \in X$. Then we have:

$$\underline{\partial}_\mathcal{C}(\alpha f)(\overline{x}) = \alpha \underline{\partial}_\mathcal{C} f(\overline{x}), \ \ \forall \alpha \geq 0. \tag{3.1.8}$$

$$\underline{\partial}_\mathcal{C} \left[\sum_{i=1}^r f_i(\overline{x}) \right] = \sum_{i=1}^r \underline{\partial}_\mathcal{C} f_i(\overline{x}). \tag{3.1.9}$$

Moreover, $\underline{\partial}_\mathcal{C} f(\overline{x})$ is compact and convex and, if f is convex, we have $\underline{\partial}_\mathcal{C} f(\overline{x}) = \partial f(\overline{x})$.

Proof. Since, according to Definition 2.3.1, we are assuming that the $\mathcal{C}$-derivatives be proper convex with the same effective domain, then (3.1.9) comes from Theorem 2.3.10 (iii), and (3.1.8) from Definition 3.1.2. The boundedness of $\underline{\partial}_\mathcal{C} f$ is a consequence of proper convexity; the convexity of $\underline{\partial}_\mathcal{C} f$ is easily drawn from (3.1.7); by applying Theorem 2.3.7 to the $\mathcal{C}$-derivative, also the closedness of $\underline{\partial}_\mathcal{C} f$ comes from (3.1.7); indeed $\underline{\partial}_\mathcal{C} f(\overline{x}) = \bigcap_{d \in X - \overline{x}} S_d^{-1}([0, +\infty[)$, where $S_d(\sigma) := \mathcal{D}_\mathcal{C} f(\overline{x}; d) - \langle \sigma, d \rangle \geq 0$. The last part is obvious. $\qquad\qquad\square$

Theorem 3.1.6. Let $X \subseteq \mathbb{R}^n$ be nonempty, open and convex. **(i)** A convex function

$f : X \to \mathbb{R}$ is $\mathcal{C}$-differentiable at every $\overline{x} \in X$ and its unique $\mathcal{C}$-derivative coincides with the directional derivative of f at $\overline{x}$. **(ii)** If $f_1 : X \to \mathbb{R}$ is convex and $f_2 : X \to \mathbb{R}$ is differentiable, then $f_1 + f_2$ is $\mathcal{C}$-differentiable at every $\overline{x} \in X$ and its $\mathcal{C}$-derivative is the sum of the directional derivative of f_1 and of the derivative of f_2; its $\mathcal{C}$-subdifferential is:

$$\underline{\partial}_{\mathcal{C}}[f_1(\overline{x}) + f_2(\overline{x})] = \partial f_1(\overline{x}) + \{f_2'(\overline{x})\}. \tag{3.1.10}$$

(iii) If $f_1, ..., f_r : X \to \mathbb{R}$ are $\mathcal{C}$-differentiable at $\overline{x} \in X$, then the max-function (see (2.3.20a))

$$f(x) := \max\{f_1(x), ..., f_r(x)\} \tag{3.1.11}$$

is $\mathcal{C}$-differentiable at $\overline{x}$, having

$$\max_{i=1,...,r} \left\{ \mathcal{D}_{\mathcal{C}} f_i \left(\overline{x}; \frac{x - \overline{x}}{||x - \overline{x}||} \right) : f_i(\overline{x}) = f(\overline{x}) \right\}$$

as $\mathcal{C}$-derivative of f at $\overline{x}$.

Proof. **(i)** comes from Theorem 2.3.9. **(ii)** Comparing (3.1.2) for $f = f_i$ with the expansion $f_2(x) = f_2(\overline{x}) + \langle f_2'(\overline{x}), x - \overline{x} \rangle + \varepsilon_2(\overline{x}; x - \overline{x})$ (ε_2 being infinitesimal of higher order), and taking into account (i), we achieve the claim about the $\mathcal{C}$-derivative of $f_1 + f_2$; as a consequence of this fact, by using Definitions 2.3.3 and 3.1.2, we obtain (3.1.10). **(iii)** Without any loss of generality, suppose that $f_i(\overline{x}) = f(\overline{x})$, $i = 1, ..., r$. For each fixed $x \in X$ let i_x be s.t. $f_{i_x}(x) = f(x)$, so that $\max \{\mathcal{D}_{\mathcal{C}} f_i(\overline{x}; x - \overline{x}), \ i = 1, ..., r\} = \mathcal{D}_{\mathcal{C}} f_{i_x}(\overline{x}; x - \overline{x})$. Set:

$$\phi(x) := f_{i_x}(x) - f(\overline{x}) - \mathcal{D}_{\mathcal{C}} f_{i_x}(\overline{x}; x - \overline{x}).$$

Because of (3.1.2), $\forall y \in X - \overline{x}$, $\exists z \in X$, s.t.

$$||z - \overline{x}|| < ||y - \overline{x}||, \quad \phi(z) \le \phi(y).$$

Therefore, being $\phi(O) = 0$, using Theorems 3.1.3 and 2.3.7, we draw:

$$\lim_{x \to \overline{x}} \frac{1}{||x - \overline{x}||} [f(x) - f(\overline{x}) - \max\{\mathcal{D}_{\mathcal{C}} f_i(\overline{x}; x - \overline{x}), i = 1, ..., r\}] =$$

$$= \lim_{x \to \overline{x}} \phi(x) = 0.$$

Because of Theorem 2.3.1(i), $\max\{\mathcal{D}_{\mathcal{C}} f_i(\overline{x}; x - \overline{x}), \ i = 1, ..., r\} \in \mathcal{C}$. $\quad\square$

Theorems quite analogous to Theorems 3.1.5 and 3.1.6 hold if $\mathcal{C}$ is replaced by $-\mathcal{C}$ and $\underline{\partial}$ by $\overline{\partial}$.

Theorem 3.1.7. Let $f_i : X \to \mathbb{R}$, $i = 1, 2$ be $\mathcal{C}$-differentiable at $x = \overline{x}$ and let

$$f_i(x) = f_i(\overline{x}) + \mathcal{D}_{\mathcal{C}} f_i(\overline{x}; z) + \varepsilon_i(\overline{x}; z), \ i = 1, 2,$$

be their expansions, where $\mathcal{D}_{\mathcal{C}} f_i$, $i = 1, 2$, are the $\mathcal{C}$-derivatives. Set $\tilde{f} := f_1 \cdot f_2$ and assume that:

$$\mathcal{D}_{\mathcal{C}}\tilde{f}(\overline{x}; z) := \mathcal{D}_{\mathcal{C}} f_1(\overline{x}; z) \cdot f_2(\overline{x}) + f_1(\overline{x}) \cdot \mathcal{D}_{\mathcal{C}} f_2(\overline{x}; z) \in \mathcal{C}. \tag{3.1.12}$$

Then $\tilde{f}$ is $\mathcal{C}$-differentiable at $\overline{x}$ in the direction z and its expansion is given by

$$\tilde{f}(x) = \tilde{f}(\overline{x}) + \mathcal{D}_{\mathcal{C}}\tilde{f}(\overline{x}; z) + \tilde{\varepsilon}(\overline{x}; z), \tag{3.1.13}$$

where

$$\tilde{\varepsilon}(\overline{x}; z) := \varepsilon_1 \varepsilon_2 + \varepsilon_1[f_2(\overline{x}) + \mathcal{D}_{\mathcal{C}} f_2] + \varepsilon_2[f_1(\overline{x}) + \mathcal{D}_{\mathcal{C}} f_1] + \mathcal{D}_{\mathcal{C}} f_1 \cdot \mathcal{D}_{\mathcal{C}} f_2. \tag{3.1.14}$$

Proof. The expansion of $\tilde{f}$ is trivially obtained from the product of the expansion of f_1 and f_2. Because of assumption (3.1.12) $\mathcal{D}_{\mathcal{C}}\tilde{f}$ is sublinear; hence we have to prove only that $\lim_{z \to 0} \tilde{\varepsilon}/||z|| = 0$. As $z \to 0$, obviously $\varepsilon_1 \varepsilon_2/||z|| \to 0$; the same happens to the 2nd and 3rd terms in the right-hand side of (3.1.14), since the forms in square brackets are bounded. The boundedness of $\mathcal{D}_{\mathcal{C}} f_1/||z||$ and $\lim_{z \to 0} \mathcal{D}_{\mathcal{C}} f_2 = 0$ imply that $\mathcal{D}_{\mathcal{C}} f_1 \cdot \mathcal{D}_{\mathcal{C}} f_2/||z|| \to 0$ as $z \to 0$. $\qquad \square$

Assumption (3.1.12) is fulfilled, when f_1 and f_2 are differentiable, since $\mathcal{D}_{\mathcal{C}} f_i$, $i = 1, 2$ are linear (in this case $\mathcal{D}_{\mathcal{C}}\tilde{f} = \langle f_1'(\overline{x}) f_2(\overline{x}) + f_1(\overline{x}) f_2'(\overline{x}), z \rangle$, which is the classic formula), or when $\mathcal{D}_{\mathcal{C}} f_i$, $i = 1, 2$ are not linear and $f_i(\overline{x}) \geq 0$, $i = 1, 2$. When $f_i(\overline{x}) < 0$, then $\tilde{f}$ may not be $\mathcal{C}$-differentiable; see e.g. the case where $f_1(x) = |x|$, $f_2(x) = |x| - 1$, $x \in \mathbb{R}$.

Definition 3.1.3. Let $X \subseteq \mathbb{R}^n$ and $G \subseteq \mathcal{G}$. $f : X \to \mathbb{R}$ is said to be *upper G-semidifferentiable* at $\overline{x} \in X$, iff there exists a finite $\overline{\mathcal{D}}_G f \in G$ and a function $\overline{\varepsilon} : X \times (X \backslash \{\overline{x}\}) \to \mathbb{R}$, such that

$$\limsup_{x \to \overline{x}} \frac{\overline{\varepsilon}(\overline{x}; x - \overline{x})}{||x - \overline{x}||} \leq 0, \tag{3.1.15a}$$

$$f(x) = f(\overline{x}) + \overline{\mathcal{D}}_G f(\overline{x}; x - \overline{x}) + \overline{\varepsilon}(\overline{x}; x - \overline{x}), \tag{3.1.15b}$$

and, for every pairs of functions $(\overline{\mathcal{D}}_G^{\square}, \overline{\varepsilon}^{\square})$ which fulfil (3.1.15a,b) and with $\overline{\mathcal{D}}_G^{\square} f \in G$, we have:

$$\text{epi } \overline{\mathcal{D}}_G^{\square} f \subseteq \text{epi } \overline{\mathcal{D}}_G f. \tag{3.1.15c}$$

f is said to be *lower G-semidifferentiable* at $\overline{x}$, iff $-f$ is upper G-semidifferentiable at $\overline{x}$. $\overline{\mathcal{D}}_G f(\overline{x}; \frac{x - \overline{x}}{||x - \overline{x}||})$ and $\underline{\mathcal{D}}_G f(\overline{x}; \frac{x - \overline{x}}{||x - \overline{x}||})$ are the *upper* and *lower G-semiderivatives* of f at $\overline{x}$, respectively. If both semiderivatives exist, then f is called *G-semidifferentiable at* $\overline{x}$.

A function, which is upper $\mathcal{C}$-semidifferentiable together with its opposite and with common $\mathcal{C}$-derivative, is not necessarily the restriction of an affine function, as the last of Examples 3.1.7 shows.

If f is both upper and lower G-semidifferentiable with the same G-semiderivative, then obviously f fulfils (3.1.2). The upper and lower semiderivatives, even if they both exists, are not necessarily equal, as the 1st of Examples 3.1.7 shows. Note that (3.1.15c) means that $\overline{\mathcal{D}}_G f$ is *maximal* in the sense of epigraph. The uniqueness of the upper (and, hence, lower) $\mathcal{G}$-semiderivative holds and can be easily proved ab absurdo and by going to contradict (3.1.15c); see also Theorem 3.1.10.

Examples 3.1.7. Set $X = \mathbb{R}$, $\alpha : X \to \mathbb{R}$ defined by $\alpha(x) = \frac{1}{2}(1 + \sin\frac{1}{x})$ with $x \neq 0$, and $f(x) = |x|\,[1 - 2\alpha(x)]$ if $x \neq 0$ and $f(0) = 0$. For $G = \mathcal{C}$ (with $n = 1$), (3.1.15) are easily verified, at $\overline{x} = 0$, by

$$\overline{\mathcal{D}}_{\mathcal{C}}(0; x) = |x| \in \mathcal{C}; \quad \varepsilon(0; x) = -2\alpha(x)|x|, \quad x \neq 0,$$

so that f is upper $\mathcal{C}$-semidifferentiable with $|x|/|x| \equiv 1$ as upper $\mathcal{C}$-semiderivative. In a quite similar way, it is immediate to show that f is lower $(-\mathcal{C})$-semidifferentiable with $-|x|/|x| \equiv -1$ as lower $(-\mathcal{C})$-semiderivative. Now set $X = \mathbb{R}_+$, and $f_1 : X \to \mathbb{R}$ with $f_1(x) = x^2$ if $x \in \mathbb{Q}$ and $f_1(x) = -1$ otherwise. For $G = \mathcal{G}$ (with $n = 1$; we might restrict ourselves to $G = \mathcal{C}$), (3.1.15) are easily verified, at $\overline{x} = 0$, by

$$\overline{\mathcal{D}}_{\mathcal{G}} f_1(0; x) \equiv 0; \quad \varepsilon(0; x) = f_1(x).$$

Now set $X = \mathbb{R}_+$, and $f_2 : X \to \mathbb{R}$ with $f_2(0) = 0$, $f_2(x) = -1$ if $x \in \mathbb{Q}\backslash\{0\}$ and $f_2(x) = x^2$ otherwise. For $G = \mathcal{G}$ (with $n = 1$; also for $G = \mathcal{C}$), (3.1.15) are easily verified, at $\overline{x} = 0$, by:

$$\overline{\mathcal{D}}_{\mathcal{G}} f_2(0; x) \equiv 0; \quad \varepsilon(0; x) = f_2(x).$$

It is easy to check that $f(x) := f_1(x) + f_2(x)$ is upper $\mathcal{G}$-semidifferentiable with upper $\mathcal{G}$-semiderivative $\equiv 0$ and with remainder $\varepsilon = f$. Note that the left-hand side of (3.1.15a) is $-\infty$.

Another example is offered by $f : \mathbb{R}_+ \to \mathbb{R}$ with

$$f(x) = e^{-x} - 1 + 2\alpha(x)(1 - e^{-x}), \text{ if } x > 0,$$

and $f(0) = 0$, where $\alpha(x)$ is as above. It is easy to see that f as well as $-f$ are upper $\mathcal{C}$-differentiable at $\overline{x} = 0$, and have $\overline{x} - x, x > 0$, as common $\mathcal{C}$-derivative; in spite of this, f is not the restriction of an affine function. $\qquad\qquad\square$

When $G = \mathcal{G}$, then the equality holds in (3.1.15a), as it is easy to prove. In fact, if, ab absurdo, the left-hand side is $\ell < 0$, then the pair

$$(\overline{\mathcal{D}}_{\mathcal{G}} f(\overline{x}; d) + \ell||d||, \quad \varepsilon(\overline{x}; d) - \ell||d||)$$

fulfils (3.1.15a,b) and

$$\mathrm{epi}\,[\overline{\mathcal{D}}_{\mathcal{G}} f(\overline{x}; d) + \ell||d||] \not\subseteq \mathrm{epi}\,\overline{\mathcal{D}}_{\mathcal{G}} f(\overline{x}; d),$$

which contradicts the upper $\mathcal{G}$-semidifferentiability of f at $\overline{x}$.

If f is $\mathcal{C}$-differentiable, then obviously most of properties of convex functions can be transferred to $\mathcal{D}_{\mathcal{C}} f(\overline{x}; x - \overline{x})$. For instance, it is immediate to show that, *if f is*

$\mathcal{C}$-*differentiable at* $\overline{x}$, *then* $\underline{\partial}_{\mathcal{C}} f(\overline{x}) \neq \varnothing$. In fact, it is enough to apply Theorem 2.3.1(ii) to $\mathcal{D}_{\mathcal{C}} f$ and take into account (3.1.7a).

The class of semidifferentiable functions has been introduced in [IV14] as a sort of container to embed as many generalizations of differentiability as possible. Indeed, due to the obvious importance of the differentiability, in the last decades, several generalizations have been proposed and not always the connections among them have been investigated. A "container definition" may help in comparing the several definitions. An early idea of container appeared in [49,50]. The main aspects of differentiability and its extensions are the approximation of the graph or the epigraph of f and the control of the error made in replacing f with the approximation. The former aspect consists in choosing the class G, where we want to have the (positively homogeneous) function $\mathcal{D}_G f$ or $\overline{\mathcal{D}}_G f$ which (locally) replaces f; depending on this choice, the derivative may not exist. The latter one consists in asking $f(x) - f(\overline{x}) - g(\overline{x}; x - \overline{x})$ to satisfy a certain criterion; (3.1.15a) is an example; (3.1.2) is another stronger example; of course, several other convergence criteria can be adopted depending on a specific class of problems.

Theorem 3.1.8. Let $X \subseteq \mathbb{R}^n$, $f_i : X \to \mathbb{R}$ $i = 1, ..., r$ be upper $\mathcal{G}$-semidifferentiable at $\overline{x} \in X$ with $\overline{\mathcal{D}}_{\mathcal{G}}^i f_i(\overline{x}; \frac{1}{||d||} d)$, $i = 1, ..., r$ as $\mathcal{G}$-semiderivatives, respectively; and let $\alpha_i \in \mathbb{R}_+$, $i = 1, ..., r$. Then $f := \sum\limits_{i=1}^{r} \alpha_i f_i$ is upper $\mathcal{G}$-semidifferentiable at $\overline{x}$ with $\overline{\mathcal{D}}_{\mathcal{G}} f := \sum\limits_{i=1}^{r} \alpha_i \cdot \overline{\mathcal{D}}_{\mathcal{G}}^i f_i$ as upper $\mathcal{G}$-semiderivative.

Proof. Of course, $\overline{\mathcal{D}}_{\mathcal{G}} f \in \mathcal{G}$. Let us set:

$$\varepsilon_i(\overline{x}; d) := f_i(\overline{x} + d) - f_i(\overline{x}) - \overline{\mathcal{D}}_{\mathcal{G}} f_i(\overline{x}; d), \quad i = 1, ..., r,$$

$$\varepsilon(\overline{x}; d) := \sum_{i=1}^{r} \alpha_i \varepsilon_i(\overline{x}; d).$$

(3.1.15b) at $G = \mathcal{G}$ is trivially satisfied by the pair $(\overline{\mathcal{D}}_{\mathcal{G}} f, \varepsilon)$. Since each ε_i fulfils (3.1.15a) as equality (see the remark after Example 3.1.7), then, $\forall \delta > 0$, $\exists \eta > 0$ s.t.

$$\frac{\varepsilon_i(\overline{x}; d)}{||d||} < \delta, \quad \forall d \in (X - \overline{x}) \cap N_\eta(O),$$

and $\forall d_{\delta,\eta}^i \in (X - \overline{x}) \cap N_\eta(O)$, s.t.

$$-\delta < \frac{\varepsilon_i(\overline{x}; d_{\delta,\eta}^i)}{||d_{\delta,\eta}^i||} < \delta.$$

The α_i being fixed, by multiplying the above inequalities by α_i and summing up with respect to i, we find that ε fulfils (3.1.15a) as equality. Now, ab absurdo, suppose that the pair $(\overline{\mathcal{D}}_{\mathcal{G}} f, \varepsilon)$ does not fulfil (3.1.15c), so that there exists a pair $(\overline{\mathcal{D}}_{\mathcal{G}}^0 f, \varepsilon^0)$, which satisfies (3.1.15a) as equality, besides of course (3.1.15b), s.t.

$$\operatorname{epi} \overline{\mathcal{D}}^0_{\mathcal{G}} f \nsubseteq \operatorname{epi} \overline{\mathcal{D}}_{\mathcal{G}} f.$$

Therefore, $\exists \tilde{d} \in X - \overline{x}$ with $||\tilde{d}|| = 1$ s.t.

$$\overline{\mathcal{D}}_{\mathcal{G}} f(\overline{x}; \tilde{d}) > \overline{\mathcal{D}}^0_{\mathcal{G}} f(\overline{x}; \tilde{d}).$$

The above proof of (3.1.15a) can now be repeated by restricting us to the halfline $\rho := \{x \in \mathbb{R}^n : x = \overline{x} + t\tilde{d}, t \in \mathbb{R}_+\}$. This gives the existence of $d_{\delta,\eta} \in \rho$ (δ, η being arbitrary as above) s.t.

$$-\delta < \frac{\varepsilon(\overline{x}; d_{\delta,\eta})}{||d_{\delta,\eta}||} < \delta,$$

$$\frac{\varepsilon^0(\overline{x}; d_{\delta,\eta})}{||d_{\delta,\eta}||} < \delta.$$

Hence, we draw the inequalities:

$$\overline{\mathcal{D}}_{\mathcal{G}} f(\overline{x}; d_{\delta,\eta}) - \delta||d_{\delta,\eta}|| < f(\overline{x} + d_{\delta,\eta}) - f(\overline{x}) < \overline{\mathcal{D}}^0_{\mathcal{G}} f(\overline{x}; d_{\delta,\eta}) + \delta||d_{\delta,\eta}||,$$

which become self contradictory for δ, η small enough and taking into account the previous inequality between $\overline{\mathcal{D}}_{\mathcal{G}} f$ and $\overline{\mathcal{D}}^0_{\mathcal{G}} f$. $\qquad\square$

Now, let us discuss the role of Definition 3.1.3 as container, by considering the most classic extension of derivability, due to U.Dini [III9, III12, III13], and a recent extension, due to F.H.Clarke [12]. To this end, consider the positively homogeneous functions:

$$\overline{D} f(\overline{x}; d) := \limsup_{t \downarrow 0} \frac{f(\overline{x} + td) - f(\overline{x})}{t}, \tag{3.1.16a}$$

$$\underline{D} f(\overline{x}; d) := \liminf_{t \downarrow 0} \frac{f(\overline{x} + td) - f(\overline{x})}{t}, \tag{3.1.16b}$$

which are the *upper and lower (generalized) Dini directional derivatives*. When (3.1.16) coincide, then Dini derivatives become the directional derivative (see Theorem 2.3.9). It is easy to show that, if f is $\mathcal{G}$-semidifferentiable at $\overline{x}$, then we have:

$$\underline{\mathcal{D}}_{\mathcal{G}} f(\overline{x}; d) \leq \underline{D} f(\overline{x}; d) \leq \overline{D} f(\overline{x}; d) \leq \overline{\mathcal{D}}_{\mathcal{G}} f(\overline{x}; d), \quad d \in X - \{\overline{x}\}. \tag{3.1.17}$$

Now, let us show an useful property of Dini derivatives. To this end, set $\overline{\gamma} := (\overline{x}, f(\overline{x}))$.

Theorem 3.1.9. If $f : \mathbb{R}^n \to \mathbb{R}$ is locally Lipschitz at $\overline{x}$, then we have:

$$TC(\overline{\gamma}; \operatorname{gr} f) \cap [\{\overline{\gamma}\} + \operatorname{int} \operatorname{epi} \overline{D} f(\overline{x}; d)] = \varnothing, \tag{3.1.18a}$$

$$TC(\overline{\gamma}; \operatorname{gr} f) \cap [\{\overline{\gamma}\} + \operatorname{int} \operatorname{hypo} \underline{D} f(\overline{x}; d)] = \varnothing. \tag{3.1.18b}$$

For $\overline{x} = O$ and $f(\overline{x}) = 0$, (3.1.18) obviously become:

$$TC(\operatorname{gr} f) \cap \operatorname{int} \operatorname{epi} \overline{D} f(\overline{x}; d) = \varnothing, \tag{3.1.19a}$$

$$TC(\operatorname{gr} f) \cap \operatorname{int} \operatorname{hypo} \underline{D} f(\overline{x}; d) = \varnothing. \tag{3.1.19b}$$

Proof. Up to an obvious translation, (3.1.18) equal (3.1.19); thus, without any loss of generality, we can prove (3.1.19) only. Since obviously

$$(\text{int } \text{epi}\,\overline{D}f) \cap \text{hypo }\overline{D}f = (\text{int } \text{hypo}\,\underline{D}f) \cap \text{epi}\,\underline{D}f = \varnothing,$$

the thesis is proved, if we show that

$$TC(\text{gr } f) \subseteq (\text{hypo } \overline{D}f) \cap \text{epi } \underline{D}f. \tag{3.1.20}$$

If $TC(\text{gr } f) = \{O\}$, then (3.1.20) is trivial. Let $\hat{\gamma} \in TC(\text{gr } f)\backslash\{O\}$. Then, according to Definition 2.1.9, $\exists\{\gamma^i\}_1^\infty \subset \text{gr } f$ s.t. $\lim\limits_{i\to+\infty} \gamma^i = O$, and $\exists\{\alpha_i\}_1^\infty \subset \mathbb{R}_+\backslash\{0\}$ s.t.

$$\lim_{i\to+\infty} \alpha_i\gamma^i = \hat{\gamma}. \tag{3.1.21}$$

$\gamma^i \in \text{gr } f \Rightarrow \exists x^i \in \mathbb{R}^n$ s.t. $\gamma^i = (x^i, f(x^i))$; obviously, $\lim\limits_{i\to+\infty} x^i = O$. Let us set $\hat{\gamma} = (\hat{x}, \hat{y})$; since f is locally Lipschitz at O, there exist both a neighbourhood N of O and a positive constant L, s.t.

$$|f(x^1) - f(x^2)| \leq L\|x^1 - x^2\|, \quad \forall x^1, x^2 \in N. \tag{3.1.22}$$

$\hat{\gamma} \neq O$ and (3.1.21) imply $\lim\limits_{i\to+\infty} \alpha_i = +\infty$; therefore, $\exists i_0$ s.t., $\forall i > i_0$, we have $\frac{1}{\alpha_i}\hat{x} \in N$ besides $x^i \in N$. Hence, for $x^1 = x^i$ and $x^2 = \frac{1}{\alpha_i}\hat{x}$, (3.1.22) becomes:

$$|f(x^i) - f(\frac{1}{\alpha_i}\hat{x})| \leq \frac{L}{\alpha_i}\|\alpha_i x^i - \hat{x}\|, \quad \forall i > i_0,$$

or

$$\alpha_i f(\tfrac{1}{\alpha_i}\hat{x}) - L\|\alpha_i x^i - \hat{x}\| \leq \alpha_i f(x^i) \leq \alpha_i f(\tfrac{1}{\alpha_i}\hat{x}) + L\|\alpha_i x^i - \hat{x}\|, \quad \forall i > i_0. \tag{3.1.23}$$

(3.1.21) imply $\lim\limits_{i\to+\infty} \alpha_i x^i = \hat{x}$ and $\lim\limits_{i\to+\infty} \alpha_i f(x^i) = \hat{y}$. f being locally Lipschitz, because of Theorem 2.1 of [36], it is not restrictive to assume that $x^i \in [O, \hat{x}]$, so that we have ($\beta_i := 1/\alpha_i$):

$$\liminf_{i\to+\infty} \tfrac{1}{\beta_i} f(\beta_i\hat{x}) = \underline{D}f(O; \hat{x}), \quad \limsup_{i\to+\infty} \tfrac{1}{\beta_i} f(\beta_i\hat{x}) = \overline{D}f(O; \hat{x}).$$

It follows that, taking in (3.1.23) the lower limit of the 1st side, the limit of the 2nd side and the upper limit of the 3rd side as $i \to +\infty$, the inequalities ($d = \hat{x}$)

$$\underline{D}f(O; d) \leq \hat{y} \leq \overline{D}f(O; d)$$

hold and prove the thesis. $\qquad\square$

Next example shows that, if f is not locally Lipschitz, then the thesis of Theorem 3.1.9 may not hold, even if f is continuous.

Example 3.1.8. Set $X = \mathbb{R}^2$, $\overline{x} = O$, and

$$\varphi(\alpha) := \begin{cases} 1 & \text{,if } 0 < \alpha \leq 1 \text{ or } \alpha \geq 3, \\ \alpha & \text{,if } 1 < \alpha \leq 2, \\ 4 - \alpha & \text{,if } 2 < \alpha < 3, \end{cases}$$

$$f(x) := \begin{cases} (x_1^2 + x_2^2)^{1/2} & \text{,if } x \not> O, \\ \varphi\left(\frac{x_2}{x_2^{\frac{1}{2}}}\right)(x_1^2 + x_2^2)^{1/2} & \text{,if } x > O. \end{cases}$$

See Fig. 3.1.2. The present f — which is in the same light of Peano-function of Example 2.4.1 — has a behaviour quite analogous to f of Example 3.1.4 and is not locally Lipschitz. According to Definition 2.1.9, we find:

$$TC(\mathrm{gr}\, f) = \{(x,y) \in \mathbb{R}^2 \times \mathbb{R} : y = (x_1^2 + x_2^2)^{\frac{1}{2}}\} \cup \{(x,y) \in \mathbb{R}_+^2 \times \mathbb{R} : x_2 = 0,\ x_1 \leq y \leq 2x_1\}.$$

According to (3.1.16), we find:

$$\underline{D}f(O;x) = \overline{D}f(O;x) = (x_1^2 + x_2^2)^{\frac{1}{2}}.$$

Hence, (3.1.19b) is true, but (3.1.19a) is false; the opposite happens, if we consider the function f of Example 3.1.4. The functions f of the present example and of Example 3.1.4 are, respectively, lower and upper $\mathcal{G}$-semidifferentiable at O with $(x_1^2 + x_2^2)^{\frac{1}{2}}/\|x\|$ as common (respectively, lower and upper) semiderivative, as it is easy to check by applying Definition 3.1.3. Note that the present function is not pseudoconvex (see Definition 2.4.2). $\square$

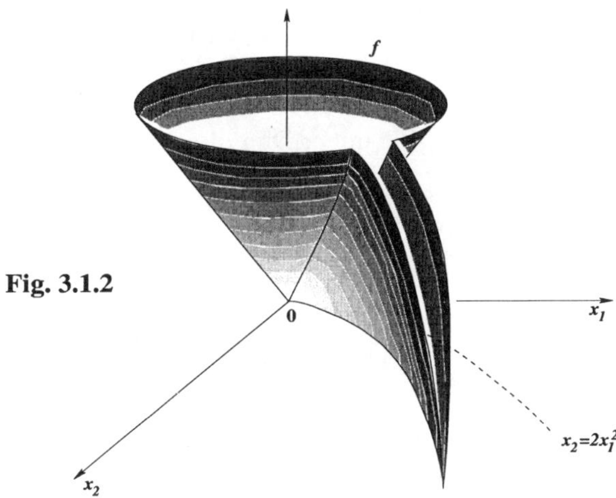

Fig. 3.1.2

In the previous part of this section, nothing has been said about the class G of positively homogeneous functions where the approximation of f must be searched for. If G does not enjoy any property, then the resulting concept of generalized differentiability will turn out to be, perhaps, too loose. An interesting investigation on the conditions, which G must fulfil so that a meaningful theory can be constructed as from Definition 3.1.3, is contained in [III37]. The beginning of such a theory has been developed in the previously quoted papers; in addition, [III34] contains the main rules of a Semidifferentiable Calculus. Apart from the properties of G, in the previous definitions nothing has been said about the dependence of G on f. The desirable situation is, obviously, that where some subsets G of $\mathcal{G}$, like $\mathcal{L}$ and $\mathcal{C}$, are established, and a Calculus is developed for each of them; this identifies classes of functions f, as differentiable and $\mathcal{C}$-(semi)differentiable ones; both the subsets G and the classes of f are expected to be not singletons. However, there is one exception. Consider the following subset of $\mathcal{G}$:

$$\mathcal{C}^0 := \{ g \in \mathcal{G} : \overline{z} + \text{epi } g \subseteq \text{cl } HC(\overline{z}; \text{epi } f) \}, \tag{3.1.24}$$

and the following function:

$$\overline{f}^0(\overline{x}; d) := \lim_{\substack{x \to \overline{x} \\ t \downarrow 0}} \sup \frac{f(x + td) - f(x)}{t}, \tag{3.1.25}$$

where HC denotes hypertangent cone (see Definition 2.1.14). The function $\overline{f}^0(\overline{x}; d/\|d\|)$ is the *Clarke generalized upper directional derivative at $\overline{x}$ in the direction d* [II12].

Theorem 3.1.10. A function $f : \mathbb{R}^n \to \mathbb{R}$, locally Lipschitz at $\overline{x}$, is **(i)** upper (or lower) $\mathcal{G}$-semidifferentiable at $\overline{x}$, having $\overline{D}f(\overline{x}; d/\|d\|)$ (or $\underline{D}f(\overline{x}; d/\|d\|)$) as unique and finite upper (or lower) directional $\mathcal{G}$-semiderivative; **(ii)** upper $\mathcal{C}^0$-semidifferentiable at $\overline{x}$, having f^0 as unique and finite upper directional $\mathcal{C}^0$-semiderivative.

Proof. (i) Let us set $G = \mathcal{G}$ and

$$\mathcal{D}_{\mathcal{G}} f(\overline{x}; d) = \overline{D} f(\overline{x}; d), \quad \varepsilon(\overline{x}; d) = f(\overline{x} + d) - f(\overline{x}) - \overline{D} f(\overline{x}; d)$$

in (3.1.15b). Then, being (3.1.18b) evidently satisfied by $\overline{D}f$, to prove (3.1.15a) we note that, according to (3.1.18a), the upper Dini directional derivative of the function (of x)

$$\varepsilon(\overline{x}; x - \overline{x}) := f(x) - f(\overline{x}) - \overline{D} f(\overline{x}; x - \overline{x})$$

turns out to equal

$$\left[\lim_{t \downarrow 0} \sup \frac{f(\overline{x} + td) - f(\overline{x})}{t} \right] - \overline{D} f(\overline{x}; d) = \overline{D} f(\overline{x}; d) - \overline{D} f(\overline{x}; d) = 0, \quad \forall d \in \mathbb{R}^n.$$

Therefore, epi $\overline{D}\varepsilon(\overline{x}; d) = H^+ := \mathbb{R}^n \times \mathbb{R}_+$. Since ε is locally Lipschitz at $\overline{x}$, Theorem 3.1.9 can be applied and (3.1.19a) becomes here:

$$TC(\text{gr } \varepsilon) \cap \text{int } H^+ = \varnothing.$$

Therefore, ε being obviously continuous, Theorem 2.1.4(iii) (with f replaced by ε) can be invoked to achieve (3.1.15a). The fact that $\mathcal{D}_{\mathcal{G}} f = \overline{D} f$ proves (3.1.15c) (with $G = \mathcal{G}$), so that (3.1.15) — and consequently the uniqueness of $\mathcal{D}_{\mathcal{G}} f$ — are completely proved. $\mathcal{D}_{\mathcal{G}} f$ is finite, since the local Lipschitzianity of f implies that of $\overline{D} f$. In fact, L being the Lipschitz constant of f at $\overline{x}$, we see that:

$$f(\overline{x} + td^1) - f(\overline{x}) \le f(\overline{x} + td^2) - f(\overline{x}) + Lt\|d^1 - d^2\|.$$

Dividing by $t > 0$ both sides of the above inequality and taking upper limits as $t \downarrow 0$, we find:

$$\overline{D}f(\overline{x}; d^1) \leq \overline{D}f(\overline{x}; d^2) + L||d^1 - d^2||.$$

Since d^1 and d^2 can be exchanged, we have that $\overline{D}f$ is locally Lipschitz. The lower part is proved in a quite similar way by replacing $\overline{D}f$ with $\underline{D}f$. **(ii)** In (3.1.15c) we set:

$$G = \mathcal{C}^0, \quad \overline{\mathcal{D}}_{\mathcal{C}^0}f(\overline{x}; d) = \overline{f}^0(\overline{x}; d), \quad \varepsilon(\overline{x}; d) = f(x) - f(\overline{x}) - \overline{f}^0(\overline{x}; d).$$

From Proposition 2.1.1 of [12], we know that $\overline{f}^0$ is positively homogeneous, finite and convex; from Proposition 4B of [III45], we see that $\overline{f}^0 \in \mathcal{C}^0$ and that $\overline{f}^0$ is extreme, in the sense of (3.1.15c), in $\mathcal{C}^0$. Hence, $\mathcal{C}^0 \subset \mathcal{G}$, (3.1.18c), uniqueness and finiteness of $\overline{\mathcal{D}}_{\mathcal{C}^0}f$ are proved. To achieve (3.1.15a), note that $\overline{D}f \leq \overline{f}^0$ implies that the upper Dini directional derivative of the function (of x)

$$\varepsilon(\overline{x}; x - \overline{x}) := f(x) - f(\overline{x}) - \overline{f}^0(x; x - \overline{x})$$

turns out to equal

$$\left[\limsup_{t \downarrow 0} \frac{f(\overline{x} + td) - f(\overline{x})}{t}\right] - \overline{f}^0(\overline{x}; d) = \overline{D}f(\overline{x}; d) - \overline{f}^0(\overline{x}; d) \leq 0, \quad \forall d \in \mathbb{R}^n.$$

Therefore, epi $\overline{D}\varepsilon \supseteq H^+ := \mathbb{R}^n \times \mathbb{R}_+$. Since ε is locally Lipschitz at $\overline{x}$, Theorem 3.1.9 can be applied and (3.1.19a) becomes here:

$$TC(\text{gr}\,\varepsilon) \cap \text{int}\,H^+ = \varnothing.$$

Thus, ε being obviously continuous, Theorem (2.1.4)(iiii) (with f replaced by ε) can be invoked to achieve (3.1.15a). $\qquad\square$

If in (3.1.25) lim sup is replaced by lim inf, then obviously the *Clarke generalized lower directional derivative* is obtained, which enjoys a property quite analogous to Theorem 3.1.10(ii).

Example 3.1.9 (continuation of Example 3.1.6). We have obviously $\mathcal{D}_{\mathcal{C}}(0; d) = \overline{D}f(0; d) = \underline{D}f(0; d) = 0$, $\forall d \in \mathbb{R}$; indeed, the Dini derivatives — as well as the $\mathcal{C}$-derivative and the upper or lower $\mathcal{G}$-semiderivatives — collapse to the (classic) derivative, when f is differentiable. Unlike this, $\overline{f}^0$ does not necessarily shrink to $\langle f'(\overline{x}), d \rangle$ (f' being the gradient of f) when f is differentiable; indeed, as it is easy to prove (see [II12], page 33), in the present instance, we have $\overline{f}^0(0; d) = |d| \not\equiv 0$. $\qquad\square$

Example 3.1.10. Set $X = \mathbb{R}$, $\overline{x} = 0$, and $f(x) = \sqrt{|x|}$. At $\overline{x} = 0$, f is not upper $\mathcal{G}$-semidifferentiable, since $\overline{\mathcal{D}}_{\mathcal{G}}f \in \mathcal{G} \Rightarrow \overline{\mathcal{D}}_{\mathcal{G}}f(0; d) = \alpha d$ with $\alpha \in \mathbb{R}\backslash\{0\}$, so that (3.1.12a) is violated; and is not lower $\mathcal{G}$-semidifferentiable, since $-f(x) = -\sqrt{|x|}$ is not upper $\mathcal{G}$-semidifferentiable; indeed, whatever $\overline{\mathcal{D}}_{\mathcal{G}}(-f)(0; d) = \alpha d$ may be, (3.1.15a) is fulfilled, but (3.1.15c) is not, since there is not a maximal (in the sense of epigraph) upper derivative for $-f$. Note that f is continuous, but not locally Lipschitz, at $\overline{x}$. $\qquad\square$

Example 3.1.11. Set $X = \mathbb{R}$, $\overline{x} = 0$, and $f(x) = |x| - x^2$ if $x \in \mathbb{Q}$ and $f(x) = |x| + x^2$

if $x \notin \mathbb{Q}$. We find $\overline{D}f(0; d) = |d|$, so that (3.1.15b) holds with $\overline{\mathcal{D}}_\mathcal{G} f(0; d) = |d|$, $\varepsilon(0; d) = -d^2$ if $d \in \mathbb{Q}$ and $\varepsilon(0; d) = d^2$ if $d \notin \mathbb{Q}$. Since (3.1.15) are evidently verified, f is upper $\mathcal{G}$-semidifferentiable (and even $\mathcal{C}$-semidifferentiable) at $\overline{x}$. Note that f is continuous only at $\overline{x}$, where it is not locally Lipschitz. $\qquad\qquad\square$

Example 3.1.12. Set $X = \mathbb{R}$, $\quad \overline{x} = 0$, and $f(x) = 2x + x^2$ if $x \leq 0$ and $f(x) = x + x^2$ if $x > 0$. We find $\overline{D}f(0; d) = 2d$ if $d \leq 0$ and $\overline{D}f(0; d) = d$ if $d > 0$. (3.1.15b) holds with $\overline{\mathcal{D}}_\mathcal{G} f(0; d) = \overline{D}f(0; d)$, and $\varepsilon(0; d) = d^2$. Since (3.1.15) are satisfied, f is upper $\mathcal{G}$-semidifferentiable (and even $\mathcal{G}$-differentiable). Note that f is locally Lipschitz at $\overline{x}$.$\square$

Example 3.1.13. Set $X = \mathbb{R}^2$, $\quad \overline{x} = O$ and $f(x) = 2x_1$ if $x_1 \geq 0$ and $x_2 = 0$, and $f(x) = (x_1^2 + x_2^2)^{\frac{1}{2}}$ otherwise. It turns out that $\overline{D}f(0; d) = f(d)$, so that f is upper $\mathcal{G}$-semidifferentiable with $\overline{\mathcal{D}}_\mathcal{G} f = \overline{D}f$ and $\varepsilon \equiv 0$. $\qquad\qquad\square$

Example 3.1.14. Set $X = \mathbb{R}$, $\quad \overline{x} = 0$, $\alpha(x) := \frac{1}{2}[1 + \sin(1/x)]$ for $x \neq 0$, and
$$f(x) = \alpha(x) - 1 + \alpha(x) \cdot x^2 \text{ if } x \neq 0, \quad f(0) = 0.$$
It is easy to check that (3.1.15) are satisfied by $G = \mathcal{L}$, $\mathcal{D}_G f \equiv 0$ and $\varepsilon \equiv f$. Therefore, f is upper $\mathcal{L}$-semidifferentiable. This shows that semidifferentiability at a point does not imply continuity. $\qquad\qquad\square$

In some applications, especially when we deal with (1.1.5), we are faced with differentiability or, more generally, with the semidifferentiability of a composition. Next theorem gives an answer to this question [22, 18]. For the sake of simplicity, without any fear of confusion, in next theorem $\mathcal{G}^n := \mathcal{G} \times \ldots \times \mathcal{G}$ (n times cartesian product) denotes the set of all positively homogeneous (of 1st degree) vector functions of type $x : \mathbb{R} \to \mathbb{R}^n$ (like in (1.1.1) and (1.1.5)); the present $\mathcal{G}$ differs from the previous one only in a formal aspect. Let $\mathcal{F}_n$ be a subset of the set of all continuous and positively homogeneous (of 1st degree) functions of type $\varphi : \mathbb{R}^n \to \mathbb{R}$; $\mathcal{F}_n$ is required to contain the linear ones. We assume that semidifferentiability be *proper*, in the sense that the upper limit in (3.1.12a) must be zero and not merely ≤ 0; analogously for the lower limit. In the rest of this section, the notations D, w are independent of those used throughout the book.

Theorem 3.1.11. Let $x : \mathbb{R} \to \mathbb{R}^n$ be directionally differentiable at $\overline{t}$, and $f : \mathbb{R}^n \to \mathbb{R}$ be upper $\mathcal{F}_n$ semidifferentiable at $\overline{y} := x(\overline{t})$ and let $\mathcal{P} \subseteq \mathcal{G}$,

$$\mathcal{P} := \{\phi \in \mathcal{G} : \exists \varphi \in \mathcal{F}_n : \phi = \varphi \circ x'(\overline{t}; \bullet)\}.$$

Assume that:

$$\limsup_{\substack{d \to 0 \\ D^0}} \frac{1}{\|w(d)\|} [f(\overline{y} + w(d)) - f(\overline{y}) - \overline{\mathcal{D}}_{\mathcal{F}_n} f(\overline{y}; \, w(d))] = 0, \qquad (3.1.26)$$

where $w(d) := x(\overline{t} + d) - x(\overline{t})$ (let us observe that w is continuous at $d = 0$), $D^0 := \{d \in \mathbb{R} : w(d) \neq 0\}$ and $\overline{\mathcal{D}}_{\mathcal{F}_n} f(\overline{y}; y - \overline{y})$ is the upper directional $\mathcal{F}_n$-semiderivative of f in the direction $y - \overline{y}$, at the point $\overline{y}$. Then $F := f \circ x$ is upper $\mathcal{P}$-semidifferentiable at $\overline{t}$ and its upper directional $\mathcal{P}$-semiderivative is given by

$$\overline{\mathcal{D}}_{\mathcal{P}}F(\bar{t}; t - \bar{t}) = \overline{\mathcal{D}}_{\mathcal{F}_n}f(x(\bar{t}); x'(\bar{t}; t - \bar{t})), \qquad (3.1.27a)$$

where $x'(\bar{t}; t - \bar{t})$ is the directional derivative of x at $\bar{t}$ in the direction $t - \bar{t}$. In the particular case where f and x are differentiable, the above assumptions are fulfilled and (3.1.27a) collapses to the classic formula:

$$\langle f'(x(\bar{t})),\ x'(\bar{t})\rangle(t - \bar{t}) \qquad (3.1.27b)$$

where now f' and x' denote gradients.

Proof. By assumption we have (see Definition 3.1.3):

$$f(y) = f(\bar{y}) + \overline{\mathcal{D}}_{\mathcal{F}_n}f(\bar{y};\ y - \bar{y}) + \varepsilon_f(\bar{y}; y - \bar{y}), \quad \forall y \in \mathbb{R}^n,$$

$$x(t) = x(\bar{t}) + x'(\bar{t}; t - \bar{t}) + \varepsilon_x(\bar{t}; t - \bar{t}), \quad \forall t \in \mathbb{R},$$

where $(\overline{\mathcal{D}}_{\mathcal{F}_n}f, \varepsilon_f)$ fulfils (3.1.1), (3.1.12a), (3.1.12c), and x' is the directional derivative of x. Let us set (as before) $d := t - \bar{t}$ and $w(d) := x(\bar{t} + d) - x(\bar{t})$. Obviously we have:

$$F(\bar{t} + d) = f(x(\bar{t} + d)) = f(x(\bar{t})) + \overline{\mathcal{D}}_{\mathcal{F}_n}f(x(\bar{t});\ w(d)) + \varepsilon_f(x(\bar{t}); w(d)).$$

This expansion becomes:

$$F(\bar{t} + d) = F(\bar{t}) + \overline{\mathcal{D}}_{\mathcal{F}_n}f(x(\bar{t}); x'(\bar{t}; d)) + \varepsilon_F(x(\bar{t}); d) \qquad (3.1.28)$$

with ε_F having the analogous property to ε_f with respect to d, if we prove that the directional derivative of $\overline{\mathcal{D}}_{\mathcal{F}_n}f(\bar{y}; w(d))$ (as a function of d) at 0 in the direction d exists and equals $\overline{\mathcal{D}}_{\mathcal{F}_n}f(\bar{y}; x'(\bar{t}; d))$. To this end, note that $x'(\bar{t}; 0) = 0 = \varepsilon_x(\bar{t}; 0)$, that $\lim_{d \to 0}(1/|d|)\varepsilon_x(\bar{t}; d) = 0$, and use the continuity of $\overline{\mathcal{D}}_{\mathcal{F}_n}f(\bar{y}; w)$ with respect to w to find:

$$\lim_{\alpha \downarrow 0} \frac{1}{\alpha}\left[\overline{\mathcal{D}}_{\mathcal{F}_n}f(\bar{y}; w(\alpha d)) - \overline{\mathcal{D}}_{\mathcal{F}_n}f(\bar{y}; 0)\right] =$$

$$= \overline{\mathcal{D}}_{\mathcal{F}_n}f\left(\bar{y}; \lim_{\alpha \downarrow 0}\frac{w(\alpha d)}{\alpha}\right) = \overline{\mathcal{D}}_{\mathcal{F}_n}f(\bar{y}; x'(\bar{t}; d)).$$

Since $\varepsilon_x/|d| \to 0$, $\exists\tilde{\varepsilon} : \mathbb{R}^n \times \mathbb{R} \to \mathbb{R}$ such that $\tilde{\varepsilon}(\bar{y}; d)/|d| \to 0$ and

$$\overline{\mathcal{D}}_{\mathcal{F}_n}f(\bar{y}; w(d)) = \overline{\mathcal{D}}_{\mathcal{F}_n}f(\bar{y}; x'(\bar{t}; d)) + \tilde{\varepsilon}(\bar{y}; d).$$

Hence (3.1.28) follows by setting $\varepsilon_F(x(\bar{t}); d) := \tilde{\varepsilon}(x(\bar{t}); d) + \varepsilon_f(x(\bar{t}); w(d))$. Now we will prove that (3.1.28) satisfies (3.1.1), (3.1.12a), (3.1.12c). $\overline{\mathcal{D}}_{\mathcal{F}_n}f(x(\bar{t}); x'(\bar{t}; \bullet))$ obviously fulfils (3.1.1). (3.1.15a) is proved if we show that:

$$\limsup_{d \to 0}\frac{\varepsilon_F(\bar{y}; d)}{|d|} = \limsup_{d \to 0}\frac{\varepsilon_f(\bar{y}; w(d))}{|d|} = \limsup_{\substack{d \to 0 \\ D^0}}\frac{||w(d)||}{|d|}\cdot\frac{\varepsilon_f(\bar{y}; w(d))}{||w(d)||} =$$

$$= \limsup_{\substack{d \to 0 \\ D^0}} \left\| x'\left(\bar{t}; \frac{d}{|d|}\right) + \frac{\varepsilon_x(\bar{t}; d)}{|d|} \right\| \cdot \frac{\varepsilon_f(\bar{y}; w(d))}{\|w(d)\|} = 0.$$

The 1st of the above inequalities holds since $\tilde{\varepsilon}/|d| \to 0$ as $d \to 0$. The 2nd holds since, without any loss of generality, we can restrict to D^0 (if $0 \neq d \in \mathbb{R} \backslash D^0$, then $\varepsilon_f(\bar{y}; w(d)) = 0$ and $\limsup_{d \to 0} \varepsilon_F(\bar{y}; d)/|d| = 0$). The 3rd is obvious. The 4th is a consequence of the boundedness of $x'(\bar{t}; d/|d|)$ and $\varepsilon_x(\bar{t}; d)/|d|$ in a neighbourhood of 0, and of the existence of $\{d_k\}_{k=1}^{\infty} \to 0$ such that $\varepsilon_f(\bar{y}; w(d_k))/\|w(d_k)\| \to 0$ (because of (3.1.23)). Finally, the pair $(\overline{\mathcal{D}}_{\mathcal{F}_n} f(x(\bar{t}); x'(\bar{t}; d)),\ \varepsilon_F(x(\bar{t}); d))$ fulfils (3.1.15a). Indeed, for every pair (H, η), such that:

$$F(\bar{t} + d) = F(\bar{t}) + H(\bar{t}; d) + \eta(\bar{t}; d), \quad \forall d \in \mathbb{R},$$

and satisfying all (3.1.1), (3.1.15a,b), being $H \in \mathcal{P}$, it must exist $g \in \mathcal{F}_n$ such that $H(\bar{t}; d) = g(x'(\bar{t}; d))$. Now, since by assumption f is upper $\mathcal{F}_n$-differentiable at $x(\bar{t})$, the property (3.1.15c) implies

$$g(x'(\bar{t}; d)) \geq \overline{\mathcal{D}}_{\mathcal{F}_n} f(x(\bar{t}); x'(\bar{t}; d)), \quad \forall d \in \mathbb{R},$$

whence

$$\operatorname{epi} H(\bar{t}; \bullet) \ \subseteq \ \operatorname{epi} \overline{\mathcal{D}}_{\mathcal{F}_n} f(x(\bar{t}); x'(\bar{t}; \bullet)).$$

When f and x are differentiable, then all the assumptions are fulfilled, since $\mathcal{F}_n$ becomes the set of linear (and hence continuous) functions; in (3.1.26) the symbol $\limsup$ is replaced by $\lim$, so that $\overline{\mathcal{D}}_{\mathcal{F}_n} f(\bar{y}; y - \bar{y}) = \langle f'(\bar{y}), y - \bar{y} \rangle$ and $x'(\bar{t}; t - \bar{t}) = x'(\bar{t})(t - \bar{t})$, where now f' and x' denote gradients; $\mathcal{P}$ turns out to be the set of linear functions of one variable, if $x'(\bar{t}) \neq 0$ (otherwise the thesis becomes trivial). In fact, if λ is such a function, then it can be expressed in the form $\lambda(d) = cd$, with $c \in \mathbb{R}$. Set $v := x'(\bar{t})$, $v \in \mathbb{R}^n$, choose $g \in \mathcal{F}_n$ so that $g(v) \neq 0$. Being also $(c/g(v))g$ linear, one has:

$$\frac{c}{g(v)} g(x'(\bar{t}; d)) = \frac{c}{g(v)} g(vd) = cd = \lambda(d), \quad \forall d \in \mathbb{R}.$$

Hence (3.1.27a) collapses to (3.1.27b). $\qquad\qquad\square$

If f is lower-semidifferentiable at $\bar{y}$, then we have a statement quite analogous to Theorem 3.1.10, where in (3.1.26) $\limsup$ is replaced by $\liminf$ and $\overline{\mathcal{D}}_{\mathcal{F}_n} f$ with $\underline{\mathcal{D}}_{\mathcal{F}_n} f$ (lower directional $\mathcal{F}_n$-semiderivative). Then we achieve the lower $\mathcal{P}$-semidifferentiability of $F := f \circ x$ at $\bar{t}$, and (3.1.27a) is replaced by:

$$\underline{\mathcal{D}}_{\mathcal{P}} F(\bar{t}; t - \bar{t}) = \underline{\mathcal{D}}_{\mathcal{F}_n} f(x(\bar{t}); x'(\bar{t}; t - \bar{t})).$$

When only x is differentiable, so that $x'(\bar{t}; \bullet)$ is a linear element of $\mathcal{G}^n$, then (3.1.27a) becomes:

$$\overline{\mathcal{D}}_{\mathcal{P}} F(\bar{t}; t - \bar{t}) = \overline{\mathcal{D}}_{\mathcal{F}_n} f(x(\bar{t}); x'(\bar{t})(t - \bar{t})), \tag{3.1.27c}$$

where now x' denotes gradient. When $n = 1$, so that the continuity of $\overline{\mathcal{D}}_{\mathcal{F}_1} f$ need not to be assumed, then (3.1.27c) becomes:

$$\overline{\mathcal{D}}_{\mathcal{P}} F(\bar{t}; t - \bar{t}) = \overline{\mathcal{D}}_{\mathcal{F}_1} f(x(\bar{t}); \operatorname{sgn} x'(\bar{t}) \cdot (t - \bar{t})) \cdot |x'(\bar{t})|, \qquad (3.1.27d)$$

When $\overline{\mathcal{D}}_{\mathcal{F}_n} f$ is separable, i.e. $\overline{\mathcal{D}}_{\mathcal{F}_n} f(\bar{y}; y - \bar{y}) = \sum_{j=1}^{n} \overline{\mathcal{D}}_{\mathcal{F}_1} f_j(\bar{y}_j; y_j - \bar{y}_j)$ with $\overline{\mathcal{D}}_{\mathcal{F}_1} f_j$ positively homogeneous of 1st degree in the 2nd argument and $\mathcal{F}_1$ like $\mathcal{F}_n$ but having 1-dimensional elements, then (3.1.27c) becomes:

$$\overline{\mathcal{D}}_{\mathcal{P}} F(\bar{t}; t - \bar{t}) = \sum_{j=1}^{n} \overline{\mathcal{D}}_{\mathcal{F}_1} f_j(x_j(\bar{t}); \operatorname{sgn} x'_j(\bar{t})(t - \bar{t}))|x'_j(\bar{t})|,$$

which becomes (3.1.27b) if $\overline{\mathcal{D}}_{\mathcal{F}_1} f_1, ..., \overline{\mathcal{D}}_{\mathcal{F}_1} f_n$ are linear.

Corollary 3.1.1. If $x : \mathbb{R} \to \mathbb{R}^n$ is directionally differentiable at $\bar{t}$ and $f : \mathbb{R}^n \to \mathbb{R}$ is $\mathcal{F}_n$-differentiable at $\bar{y}$, then F is $\mathcal{P}$-differentiable at $\bar{t}$ and it results:

$$\mathcal{D}_{\mathcal{P}} F(\bar{t}; t - \bar{t}) = \mathcal{D}_{\mathcal{F}_n} f(x(\bar{t}); x'(\bar{t}; t - \bar{t})).$$

Proof. By the $\mathcal{F}_n$-differentiability of f at $\bar{y}$ one has:

$$\lim_{z \to 0} \frac{f(\bar{y} + z) - f(\bar{y}) - \mathcal{D}_{\mathcal{F}_n} f(\bar{y}; z)}{||z||} = 0,$$

whence it follows that assumption (3.1.26) is fulfilled, in particular, both with $\limsup\limits_{\substack{d \to 0 \\ D^0}}$ and $\liminf\limits_{\substack{d \to 0 \\ D^0}}$. Therefore, by virtue of Theorem 3.1.11, F turns out to be both upper $\mathcal{P}$-semidifferentiable and lower $\mathcal{P}$-semidifferentiable at $\bar{t}$. This fact implies that F is $\mathcal{P}$-differentiable at $\bar{t}$ and the formula in the thesis holds. $\qquad\square$

The fact that the definition of the set $\mathcal{P}$ depends on the derivative of x at $\bar{t}$ is not particularly restrictive. For instance, if the set $\mathcal{F}_n$ is chosen to be

$$\mathcal{C}_n := \{\varphi : \mathbb{R}^n \to \mathbb{R}, \text{ s.t. } \varphi \text{ is sublinear}\}$$

and x has the property that

$$v_+ := x'(\bar{t}; 1) \quad \text{and} \quad v_- := x'(\bar{t}; -1)$$

are not linearly dependent by a positive constant, then one obtains $\mathcal{P} \supseteq \mathcal{C}_1$. In fact, let ϕ be in $\mathcal{C}_1$; such a function can be expressed in the form:

$$\phi(d) = \begin{cases} r_1 d, & \text{if } d \geq 0, \\ r_2 d, & \text{if } d < 0, \end{cases}$$

with $r_1, r_2 \in \mathbb{R}$, provided that $r_1 \geq r_2$. Now, it is possible to define $g \in \mathcal{C}_n$ in such a way that

$$g(v_+) = r_1, \quad g(v_-) = -r_2;$$

hence it results:

$$g(x'(\bar{t}; d)) = \left\{ \begin{array}{ll} g(v_+ d) = r_1 d, & \text{if } d \geq 0, \\ g(v_-(-d)) = -r_2(-d), & \text{if } d < 0, \end{array} \right\} = \phi(d).$$

When, in particular, x is differentiable at $\bar{t}$ and $x'(\bar{t}) \neq 0$ (namely, it is a regular point of the arc whose equation is $x = x(t)$, $t \in \mathbb{R}$), then obviously $v_- = -v_+$ and $\mathcal{P} = \mathcal{C}_1$. In fact, $\forall d_1, d_2 \in \mathbb{R}$, $\forall \alpha \in [0, 1]$, if $\phi \in \mathcal{P}$, it holds:

$$\phi((1 - \alpha)d_1 + \alpha d_2) = \varphi(x'(\bar{t}; (1 - \alpha)d_1 + \alpha d_2)) =$$

$$= \varphi((1-\alpha)x'(\bar{t}; d_1) + \alpha x'(\bar{t}; d_2)) \leq (1-\alpha)\varphi(x'(\bar{t}; d_1)) + \alpha\varphi(x'(\bar{t}; d_2)) =$$

$$= (1 - \alpha)\phi(d_1) + \alpha\phi(d_2),$$

being $\varphi \in \mathcal{C}_n$ because of the definition of $\mathcal{P}$.

Generally, in the Theorem 3.1.11 it is not possible to avoid the dependence of $x'(\bar{t}; \bullet)$ by replacing the set $\{x'(\bar{t}; \bullet)\}$ with the whole set $\mathcal{G}^n$ in the definition of $\mathcal{P}$. Consider the following counterexample: let $n = 2$ and $\mathcal{F}_2 = \mathcal{L}_2 \cup \{\|\bullet\|\}$, where, as usual, $\mathcal{L}_2$ denotes the class of the linear functionals on $\mathbb{R}^2$; let us define $f : \mathbb{R}^2 \to \mathbb{R}$ in the following way:

$$f(y_1, y_2) = \left\{ \begin{array}{ll} \frac{1}{2}\|(y_1, y_2)\|_2, & \text{if } y_1 = y_2, \ y_1, y_2 \geq 0, \\ \|(y_1, y_2)\|_2, & \text{otherwise}, \end{array} \right.$$

and $x(t) = (t, t)$, $t \in \mathbb{R}$. Now, observe that f turns out to be upper $\mathcal{F}_2$-semidifferentiable at $(y_1, y_2) = (0, 0)$ and its upper $\mathcal{F}_2$-semiderivative is given by:

$$\overline{\mathcal{D}}_{\mathcal{F}_2} f(0, w) = \|w\|_2, \quad w \in \mathbb{R}^2.$$

It is obvious that x is differentiable at $\bar{t} = 0$, and it is $x'(0) = (1, 1)$. By composing x and f one obtains:

$$F(t) = \left\{ \begin{array}{ll} \frac{\sqrt{2}}{2}t, & \text{if } t \geq 0, \\ -\sqrt{2}t, & \text{if } t < 0. \end{array} \right.$$

Here, it results $\overline{\mathcal{D}}f_{\mathcal{F}_2}(0; x'(0; d)) = \sqrt{2}|d|$; this function satisfies property (3.1.26) and

$$\limsup_{d \to 0} \frac{F(d) - F(0) - \overline{\mathcal{D}}_{\mathcal{F}_2} f(0; x'(0; d))}{|d|} = 0.$$

If, as in current formulation of Theorem 3.1.11, $\mathcal{P} := \mathcal{F}_2 \circ \{x'(0; \bullet)\}$, one finds that the thesis is completely achieved. On the other hand, by replacing $\{x'(0; \bullet)\}$ with $\mathcal{G}^2$, one obtains $\mathcal{P} := \mathcal{F}_2 \circ \mathcal{G}^2 = \mathcal{G}$. Since $\mathcal{F} \in \mathcal{G}$ and epi $F \supseteq$ epi $\overline{\mathcal{D}}_{\mathcal{F}_2} f(0; x'(0; \bullet))$, in this case one finds:

$$\overline{\mathcal{D}}_{\mathcal{P}} F(0; d) = F(d), \quad d \in \mathbb{R}.$$

This result does not conform with (3.1.27a) of Theorem 3.1.10.

The following example, due to C.Zălinescu, shows that (3.1.26) cannot be eliminated in the general case.

Example 3.1.15. $n = 1$; $x :] - 1, 1[\to \mathbb{R}$,

$$
x(t) = \begin{cases} \frac{1}{r^2} + t, & \text{if } t \in \left[\frac{1}{r+1}, \frac{1}{r}\right[, \ r = 1, 2, ..., \\ \frac{1}{r^2} - t, & \text{if } t \in \left]-\frac{1}{r}, -\frac{1}{r+1}\right], \ r = 1, 2, ..., \\ 0, & \text{if } t = 0; \end{cases}
$$

$f : \mathbb{R} \to \mathbb{R}$, with

$$
f(y) = \begin{cases} 0, & \text{if } y \leq 0 \text{ or } y = \frac{1}{r}, \ r = 1, 2, ..., \\ -1, & \text{otherwise.} \end{cases}
$$

The function x is directionally differentiable at $\bar{t} = 0$, and we have:

$$
x'(0; t) = |t|;
$$

while the function f turns out to be upper $\mathcal{F}_1$-semidifferentiable at $\bar{y} = x(\bar{t}) = 0$, even with the elements of $\mathcal{F}_1$ convex, and we have:

$$
\overline{\mathcal{D}}_{\mathcal{F}_1} f(0; y) \equiv 0.
$$

Inasmuch as it is $(1/r) \notin x(]-1, 1[\backslash\{0\}) = \underset{r \in \mathbb{N}}{\cup}[1/r^2 + 1/(r + 1), 1/r^2 + 1/r[$, the composition $F(t) = f(x(t))$ turns out to be

$$
F(t) = \begin{cases} 0, & \text{if } t = 0, \\ -1, & \text{otherwise.} \end{cases} \qquad \qquad \square
$$

In (3.1.26) the left-hand side is $-\infty$. F is not upper $\mathcal{P}$-semidifferentiable at $\bar{t} = 0$, where $\mathcal{P} = \{\zeta \in \mathcal{G} : \exists \varphi \in \mathcal{F}_1 : \zeta = \varphi \circ x'(0; \bullet)\}$, even with the elements of $\mathcal{F}_1$ only positively homogeneous, because

$$
\limsup_{t \to 0} \frac{1}{|t|}[F(t) - F(0) - \zeta(t)] = -\infty, \quad \forall \zeta \in \mathcal{G}.
$$

The fact that a function be upper $\mathcal{F}_n$-semidifferentiable does not imply assumption (3.1.26), as shown by the present example.

Example 3.1.16. $n = 2$; $f(y_1, y_2) = (y_1/2)(1 + \sin(1/y_1))$ if $y_1 > 0$, $y_2 > 0$ and $f(y_1, y_2) = (y_1^2 + y_2^2)^{1/2}$ otherwise; $\bar{y} = (0, 0)$. f is upper $\mathcal{F}_2$-semidifferentiable with $\overline{\mathcal{D}}_{\mathcal{F}_2} f = (y_1^2 + y_2^2)^{1/2}$, where $\mathcal{F}_2 \subseteq \mathcal{G}_2$ has only continuous or convex elements, but it is not Gâteaux derivable. However, if $x(t) = (t, t)$ and $\bar{t} = 0$, we have $F(t) = f(t, t)$; F is upper $\mathcal{P}$-semidifferentiable at 0. This shows that the semidifferentiability of the composition can be achieved even if f is not Gâteaux derivable. $\qquad \square$

3.2. Image Problem

The study of the properties of the image of a real−valued function is an old one; recently, it has been extended to multifunctions and to vector-valued functions. However, in most cases the properties of the image have not been the purpose of study and their investigation has occurred as an auxiliary step toward other achievements; see, e.g., [II12, V80, V81].

Traces of the idea of studying the images of functions involved in a constrained extremum problem go back to the work of Carathéodory ([5], Ch.5). In the 1950s, R.Bellman [II4], with his celebrated maximum principle, proposed − for the first time in the field of Optimization − to replace the given unknown by a new one which runs in the image ; however, also here the image is not the main purpose. Only in the late 1960s and 1970s some Authors, independently from each other, brought explicitly such a study into the field of Optimization [4, 19, 24, 25, IV14, V36].

The approach consists in introducing the space, call it Image Space (for short, IS), where the images of the functions of the given optimization problem (or Variational Inequality, or generalized system, as will be seen in Vol. 2) run. Then, a new problem is defined in the IS, which is equivalent to the given one. In a certain sense, such an approach has some analogies with what happens in the Theory of Measure when one goes from Mengoli-Cauchy-Riemann measure to the Lebesgue one.

The analysis in IS must be viewed as a preliminary and auxiliary step − and not as a concurrent of the analysis in the given space − for studying an extremum problem. When a statement has been achieved in the IS, then, of course, we have to write the corresponding (equivalent) statement in terms of the given space B. The latter is, in general, difficult to be conceived without having at disposal the former. If this aspect is understood, then the IS analysis may be highly fruitful. In fact, in the IS we may have a sort of "regularization": the conic extension (see Definition 3.2.1) of the image may be convex or continuous or smooth when the given extremum problem and its image do not enjoy the same property, so that convex or continuous or smooth analysis can be developed in the IS, but not in the given space. If the image of a problem is finite dimensional, then it can be analysed, in IS, by means of the same mathematical concepts which are used for the finite dimensional case, even if X is not finite dimensional (see Fig. 1.1.1). If the image is infinite dimensional, then it is possible to postpone such an infinite dimensionality to the introduction of the IS, which, therefore, can be held finite dimensional.

First of all, we will consider problem (1.1.6) in the case where the images of the constraining functions are finite dimensional, namely $\mathcal{B} = \mathbb{R}$; such a case embraces (1.1.1) and (1.1.4); then, some indications will be given on how to extend the approach to (1.1.5). Throughout this section, we assume that $\mathcal{H}$ fulfils (2.1.14), even if some propositions do not require it. The IS approach arises naturally in as much as an optimality condition for (1.1.6) is achieved through the impossibility of a system. More precisely, by paraphrasing the very Definition 1.1.1 we can say that $\bar{x} \in R$ is a *global minimum point* of (1.1.6), iff the system (in the unknown x):

$$f_{\bar{x}}(x) := f(\bar{x}) - f(x) > 0, \quad g(x) \in D, \quad x \in X \tag{3.2.1}$$

is impossible, or

$$\mathcal{H} \cap \mathcal{K}_{\bar{x}} = \emptyset, \tag{3.2.2}$$

where $\mathcal{H} := \{(u, v) \in \mathbb{R} \times \mathbb{R}^m : u > 0, \ v \in D\}$ and $\mathcal{K}_{\bar{x}} := \{(u, v) \in \mathbb{R} \times \mathbb{R}^m : u = f_{\bar{x}}(x), \ v = g(x), \ x \in X\} = A_{\bar{x}}(X)$ where $A_{\bar{x}}(x) := (f_{\bar{x}}(x), g(x))$. $\mathcal{K}_{\bar{x}}$ is called the *image* of (1.1.6). $A_{\bar{x}}$ is the map which sends the elements of $X \subseteq B$ into elements of the IS; in particular $\bar{x}$ is sent into

$$\bar{z} := (\bar{u}; \bar{v}) := A_{\bar{x}}(\bar{x}) = (0, g(\bar{x})) \in \{O\} \times D.$$

For $\xi = \bar{x}$, $\mathcal{K}_{\bar{x}}$ turns out to be a special case of the map $A(x; \xi)$ of Sect. 1.3, and (3.2.2) is equivalent to the impossibility of (1.3.16).

We might have considered the map $A(x) := (f(x), g(x))$; $A_{\bar{x}}$ has the advantage that, to show optimality, the image faces a set $\mathcal{H}$ which, up to closure, is the non-negative orthant or suborthant. Note that, if $x \in R$, then, because of (3.2.1), $\mathcal{K}_{\bar{x}}$ intersects $\mathrm{cl}\,\mathcal{H}$, while it intersects merely the set $\{(u, v) \in \mathbb{R}^{1+m} : u = 0\}$ if $\bar{x} \notin R$; see Example 3.4.14. It is easy to see that (1.1.6) is equivalent to the following problem :

$$u_{\bar{x}} := \max (u), \quad s.t. \quad (u, v) \in \mathcal{K}_{\bar{x}}, \ v \in D, \tag{3.2.3}$$

which is called *image problem* associated with (1.1.6). The space $\mathbb{R}^{1+m}$, where both $\mathcal{H}$ and $\mathcal{K}_{\bar{x}}$ lay, is called *image space* (IS). $\mathcal{K}_{\bar{x}}^0$ denotes the set of maximum points of (3.2.3).

Most of the analysis carried out in the IS has its root in the system (3.2.1), independently of the fact that it comes from (1.1.6) and of the special form of $f_{\bar{x}}(x)$. This shows, once more, that the mathematically hard topic is the study of a system, like (3.2.1), or the disjunction of two sets, while extremization is a useful language between real problems and the mathematical core.

Let $\xi \in \mathbb{R}^m$, and consider the problems:

$$f^{\downarrow}(\xi) := \min f(x), \quad s.t. \quad x \in R(\xi) := \{x \in X : g(x) \in \{\xi\} + D\}, \tag{3.2.4}$$

and

$$u_{\bar{x}}(\xi, D) := \max(u), \quad s.t. \quad (u, v) \in \mathcal{K}_{\bar{x}}, \ v \in \{\xi\} + D. \tag{3.2.5}$$

Of course, $u_{\bar{x}}(O_m; D) = u_{\bar{x}}$, and $u_{\bar{x}}(\xi; O_m)$ gives the maximum in (3.2.5) under the additional constraint $v = \xi$; with a little abuse of notation, the same functional symbol $u_{\bar{x}}$ has been adopted both in (3.2.3) and (3.2.5), and in (3.2.3) the dependence on D has been understood.

Problem (3.2.4) is a perturbation of (1.1.6), and thus is called *perturbed problem* of (1.1.6); (3.2.5), which is called *perturbed image problem* associated with (1.1.6), is the image problem of (3.2.4). In fact, the image set of (3.2.4) is a translation of $\mathcal{K}_{\overline{x}}$ obtained by considering (O, ξ) as new origin of the IS.

If in (1.1.6) minimization is replaced by maximization, then obviously the entire IS Analysis remains unchanged provided the 1st inequality in (3.2.1) is replaced by $f(x) - f(\overline{x}) > 0$.

Proposition 3.2.1. If the maximum in (3.2.3) exists, then, whatever $\overline{x} \in X$ may be, $f(\overline{x})$ is the minimum of (1.1.6) if and only if (3.2.2) holds.

Proof. If. Ab absurdo, suppose that $\exists \hat{x} \in R$ s.t. $f(\hat{x}) < f(\overline{x})$. Then (3.2.1) is satisfied for $x = \hat{x}$, and (3.2.2) is false. This contradicts the assumption. **Only if.** Trivial. $\quad\square$

In the definition of $f_{\overline{x}}$, $\overline{x}$ is a feasible point of (1.1.6) or $\overline{x} \in R$. Next two examples show that, if merely $x \in X$ and no other assumption is made, then, even if (3.2.2) holds, $\overline{x}$ may not be m.p. and $f(\overline{x})$ may not be minimum of (1.1.6); of course, $f(\overline{x})$ is the minimum of (1.1.6), if that of (3.2.3) exists, as Proposition 3.2.1 shows.

Example 3.2.1. Consider (1.1.1) at $n = m = 1$, $p = 0$, $X = \mathbb{R}$, $g(x) = x$, $f(x) = (x + 1)^2(x - 1)^2$. At $\overline{x} = -1$, we find:

$$\mathcal{K}_{-1} = \{(u, v) \in \mathbb{R}^2 : u = -(v + 1)^2(v - 1)^2\}.$$

Since $\mathcal{H} = \{(u, v) \in \mathbb{R}^2 : u > 0, v \geq 0\}$, we have $\mathcal{H} \cap \mathcal{K}_{-1} = \emptyset$ and thus (3.2.1) is impossible, but $\overline{x}$ is not m.p. of (1.1.1). $f(\overline{x}) = 0$ is the minimum of (1.1.1). $\quad\square$

Example 3.2.2. Consider (1.1.1) at $n = m = 1$, $p = 0$, $X = \mathbb{R}$, $f(x) = (x+1)^2(x-1)^2$ if $x \neq 1$ and $f(1) = 1$, $g(x) = x$. At $\overline{x} = -1$, we find $\mathcal{K}_{-1} = \{(u, v) \in \mathbb{R}^2 : u = -(v + 1)^2(v - 1)^2$ if $v \neq 1$ and $u = -1$ if $v = 1\}$. $\mathcal{H} = \{(u, v) \in \mathbb{R}^2 : u > 0, v \geq 0\}$. We have $\mathcal{H} \cap \mathcal{K}_{-1} = \emptyset$ and thus (3.2.1) is impossible, but neither $\overline{x}$ is m.p. nor $f(\overline{x})$ is minimum of (1.1.4); indeed the minimum does not exist. $\quad\square$

Proposition 3.2.2. $(\hat{u}(\xi), \hat{v}(\xi))$ is a maximum point of (3.2.5) if and only if it is the image — through the pair $(f(\overline{x}) - f(x), g(x))$ — of a minimum point, say $\hat{x}(\xi)$, of (3.2.4), and we have $f(\overline{x}) - \hat{u}(\xi) = f(\hat{x}(\xi))$.

Proof. Only if. $(\hat{u}(\xi), \hat{v}(\xi)) \in \mathcal{K}_{\overline{x}} \cap [\mathbb{R} \times (\{\xi\} \times D)] \Rightarrow \exists \hat{x}(\xi) \in X$ such that $\hat{u}(\xi) = f(\overline{x}) - f(\hat{x}(\xi))$, $\hat{v}(\xi) = g(\hat{x}(\xi)) \in \{\xi\} + D$, and each $(u, v) \in \mathcal{K}_{\overline{x}} \cap [\mathbb{R} \times (\{\xi\} + D)]$ implies that $\exists x \in X$ s.t. $u = f(\overline{x}) - f(x)$, $v = g(x) \in \{\xi\} + D$. Taking into account these relations (the 1st of which proves the last claim), the assumption

$$\hat{u}(\xi) \geq u, \quad \forall(u, v) \in \mathcal{K}_{\overline{x}} \cap [\mathbb{R} \times (\{\xi\} \times D)],$$

implies $f(\overline{x}) - f(\hat{x}(\xi)) \geq f(\overline{x}) - f(x)$ or $f(\hat{x}(\xi)) \leq f(x)$, $\forall x \in R(\xi)$. **If.** Set $\hat{u}(\xi) := f(\overline{x}) - f(\hat{x}(\xi)), \hat{v}(\xi) := g(\hat{x}(\xi))$, so that $(\hat{u}(\xi), \hat{v}(\xi)) \in \mathcal{K}_{\overline{x}} \cap [\mathbb{R} \times (\{\xi\} \times D)]$. From the assumption $f(\hat{x}(\xi)) \leq f(x)$, $\forall x \in R(\xi)$, by setting $u := f(\overline{x}) - f(x)$ and $v := g(x)$, we have $f(\overline{x}) - f(\hat{x}(\xi)) \geq f(\overline{x}) - f(x)$, $\forall x \in R(\xi)$, and hence:

$$\hat{u}(\xi) \geq u, \quad \forall(u, v) \in \mathcal{K}_{\overline{x}} \cap [\mathbb{R} \times (\{\xi\} \times D)]. \quad\square$$

At $\xi = O$, (3.2.4) and (3.2.5) coincide with (1.1.6) and (3.2.3), respectively; in such a case, Proposition 2.6.2 gives a relation between (1.1.6) and its image problem. The function $f(\hat{x}(\xi))$ is called *perturbation function* (called also *optimal value function*) of (1.1.6).

One of the main scopes of the IS Analysis is the study of properties of the image set, whose knowledge represents the foundations of a general theory of constrained extrema. Such a theory is the basis for the construction of methods of solution which aim to find $\bar{x}$ as defined in (3.2.1). This might seem a vicious circle in as much as the image set has been constructed by starting a minimum point $\bar{x}$. It is trivial to note that, for the definition of $\mathcal{K}_{\bar{x}}$, $\bar{x}$ can be any element of X; moreover, a change of $\bar{x}$ gives rise merely to a translation of $\mathcal{K}_{\bar{x}}$ with respect to the u-axis, so that no matter how $\bar{x}$ is chosen in X (see Example 3.2.1). Furthermore, note that a perturbation $-$ of type (3.2.4) $-$ of the constraining functions produces merely a translation of $\mathcal{K}_{\bar{x}}$ in the direction of v-axis.

Before going on with the analysis of the properties of the image of a constrained extremum problem, let us consider some instances of the kind of the regularizations announced at the beginning of this section. Consider the perturbation function:

$$\Phi_{\bar{x}}(\xi) := u_{\bar{x}}(\xi; O_m) = \max_{\substack{(u,v)\in\mathcal{K}_{\bar{x}} \\ v=\xi}} (u).$$

By applying the (weak) Ekeland Theorem [II19, 7, 33], we immediately obtain a more general statement.

Proposition 3.2.3. Let $\Phi_{\bar{x}}$ be upper semicontinuous and bounded from above. $\forall \varepsilon > 0, \ \exists v_\varepsilon \in \text{dom } \Phi_{\bar{x}}(\bullet)$, such that:

$$\Phi_{\bar{x}}(v_\varepsilon) \ \geq \ \max_{v\in D} \Phi_{\bar{x}}(v) - \varepsilon, \tag{3.2.6a}$$

$$\Phi_{\bar{x}}(v_\varepsilon) > \Phi_{\bar{x}}(v) - \varepsilon\|v - v_\varepsilon\|, \quad \forall v \in D, \quad v \neq v_\varepsilon. \tag{3.2.6b}$$

The above proposition is nothing more than the so-called weak Ekeland variational principle applied to the image of (1.1.6). Example 3.4.5 shows that Proposition 3.2.3 is more general than the corresponding theorem in B, even if in the "difference" there are problems which, perhaps, are of theoretical interest only. However, the purpose is here to show that the analysis in the IS can lead to generalize theorems which seem impossible to be extended.

Another instance is offered by coercive functions. The idea which underlies such functions is as simple as it is important: to identify a class of functions (or operators) which "behaves like" a positive definite form at least in a neighbourhood of ∞. A kind of coerciveness for operators is defined by (1.3.20). Let us now consider another type of coerciveness; it will be given in terms of the IS [41]: the image $\mathcal{K}_{\bar{x}}$ of (1.1.1) or (1.1.4) is called *coercive* at $\hat{z} \in \mathcal{K}_{\bar{x}}$, iff there exists a non-empty, (cl $\mathcal{H}$)-compact and strictly convex set $\mathcal{Z} \subset \mathbb{R}^{1+m}$, with int $\mathcal{Z} \neq \phi$, such that:

$$\mathcal{K}_{\bar{x}} \subseteq \mathcal{Z}, \quad \hat{z} \in \text{frt } \mathcal{Z}, \tag{3.2.7}$$

where $(\mathrm{cl}\,\mathcal{H})$-compact means $\mathcal{Z}_z := \mathcal{Z} \cap (z + \mathrm{cl}\,\mathcal{H})$ compact $\forall z \in \mathcal{Z}$. $\mathcal{K}_{\bar{x}}$ is called *regular coercive*, iff there exists a supporting hyperplane (Sect. 2.2) $u + \langle \lambda, v \rangle = 0$, with $\lambda \in D^*$, of $\mathcal{Z}$ at $\hat{z}$.

Proposition 3.2.4. Let $\mathcal{Z}$ be $(\mathrm{cl}\,\mathcal{H})$-compact and strictly convex. If $\mathcal{Z}_z \neq \varnothing$, then

$$\exists!\,\max(u), \quad \text{s.t.} \quad (u, v) \in \mathcal{Z}_z. \tag{3.2.8}$$

Proof. The existence of the (global) maximum is an obvious consequence of the fact that the compactness of $\mathcal{Z}_z \neq \varnothing$ implies the existence of a supporting hyperplane (Sect. 2.2) of type u=cost. which intersects $\mathcal{Z}_z$; call (u_z, v_z) a maximum point. Ab absurdo, suppose that there exists another maximum point; say (u_z, v'_z) with, of course, $v'_z \neq v_z$. Because of the strict convexity of $\mathcal{Z}$, $\exists \varepsilon > 0$ such that

$$z(\varepsilon) := (u_z + \varepsilon, \tfrac{1}{2}(v_z + v'_z)) \in \mathcal{Z}. \tag{3.2.9}$$

Since $z(\varepsilon) \in \mathrm{cl}\,\mathcal{H}$, the maximality of u_z is contradicted. $\qquad\square$

The above proposition shows the meaning of the locution "$\mathcal{Z}$ behaves like" a positive definite form. The coerciveness (3.2.7) is a light generalization of the following classic definition [I5, I42]: a function $\varphi : \mathbb{R}^n \to \mathbb{R}$ is said *coercive* at $\hat{x}$, iff $\exists \alpha > 0$, such that:

$$\varphi(x) \geq \varphi(\hat{x}) + \alpha \|x - \hat{x}\|^2, \quad \forall x. \tag{3.2.9}$$

The opposite of the maximum (if exists) of u over $(u, v) \in \mathcal{K}_{\bar{x}}$ plays the same role of φ in (3.2.9). All the previous definitions remain unchanged if $x \in B$.

Proposition 3.2.5. In (3.2.8) set $z = z(\xi) := (0, \xi)$. Let $x(\xi) \in R(\xi)$ and $(u_{z(\xi)}, v_{z(\xi)})$ be the maximum point of (3.2.8). We have: **(i)** the 1st of (3.2.7) implies that $f(\bar{x}) - u_{z(\xi)}$ is a lower bound of the minimum of (3.2.4); **(ii)** if $\mathcal{K}_{\bar{x}}$ is coercive at $A_{\bar{x}}(x(\xi))$, and $A_{\bar{x}}(x(\xi)) = (u_{z(\xi)}, v_{z(\xi)})$, then $f(\bar{x}) - u_{z(\xi)}$ is the minimum of (3.2.4) and $x(\xi)$ a corresponding minimum point.

Proof. **(i)** $u \leq u_{z(\xi)}, \quad \forall(u, v) \in \mathcal{Z}_{z(\xi)}$ and the 1st of (3.2.7) imply

$$u \leq u_{z(\xi)}, \quad \forall(u, v) \in \mathcal{K}_{\bar{x}} \cap \{(u, v) \in \mathbb{R}^{1+m} : v \in \{\xi\} + D\}.$$

Therefore, we have $f(\bar{x}) - f(x) = u \leq u_{z(\xi)}, \quad \forall x \in R(\xi)$. **(ii)** From (i) we have $f(\bar{x}) - f(x) \leq u_{z(\xi)}, \forall x \in R(\xi)$, and from the equality $A_{\bar{x}}(x(\xi)) = (u_{z(\xi)}, v_{z(\xi)})$ we have $f(\bar{x}) - f(x(\xi)) = u_{z(\xi)}$. Therefore, we draw $f(x(\xi)) \leq f(x), \forall x \in R(\xi)$. Hence, $x(\xi)$ and $f(x(\xi)) - u_{z(\xi)}$ are the m.p. and the minimum of (3.2.4), respectively. $\qquad\square$

The previous two propositions are light generalizations of the results of [41], where some investigation is done about the connections between regular coerciveness and disjunctive separation (see $(2.2.22)'$). Note that the proof of (i) of Proposition 3.2.5 is still valid even if the maximum of (3.2.8) does not exist; in this case, the inclusion of $\mathcal{K}_{\bar{x}}$ in $\mathcal{Z}$ implies the existence of finite supremum.

Now, let us come to the instance announced. In the problem (1.1.1) of Example 3.4.7 (as well as in the problem (1.1.4) of Example 3.4.8), f is not coercive, either in the sense of (3.2.9) or in that of (1.3.20). The image set fulfils (3.2.7) with $\mathcal{Z}$ given by (3.4.6) and $\hat{z}$ any point of its frontier, in particular $(u = 4\sqrt{2},\ v = 0)$. Of course, the existence of an envelope makes it easier to check (3.2.7).

The theory of constrained extrema is full of proposals for changing the data of (1.1.6), without losing minimum and minimum points, and with the purpose of adding a desired property to (1.1.6). Such proposals have been made essentially with reference to the given space, namely B and especially $\mathbb{R}^n$. Here, a different approach will be followed.

Definition 3.2.1. Let $\mathcal{Z} \subset \mathbb{R}^{1+m}$ denote a generic set of the IS associated with problems (1.1.1) or (1.1.4). $\mathcal{E}$ will denote the map which sends $\mathcal{Z}$ into the set $\mathcal{Z} - \mathrm{cl}\,\mathcal{H} \subset \mathbb{R}^{1+m}$; it is called *conic extension* of $\mathcal{Z}$.

Of course, $\mathcal{Z} \subseteq \mathcal{E}(\mathcal{Z})$. In the sequel, our attention will be devoted to the conic extension of the image set, namely $\mathcal{E}(\mathcal{K}_{\overline{x}})$, even if several other kinds of conic extensions could be introduced.

Besides extension, approximation is an important tool. Often the feasible region (1.1.2) or its intersection with a level set of the objective function are very difficult to be analysed; therefore, for special scopes — as necessary optimality condition (Chapter 5) — the above set is replaced with a cone, which can be considered an approximation of it at least in a neighbourhood of a given point; the term approximation must be defined according to the pursued scope. Sect. 2.1 contains the main cones, which are adopted to approximate a set. Such an approach is very popular and classic in the theory of constrained extrema; among the several excellent developments, we can quote the theory of tents [16], that of differential cones [24, 25], and that carried out within Quasidifferential Calculus [8-10]. Here we will adopt the same conceptual line, but it will be exploited in a different way.

Definition 3.2.2. In Problem (1.1.1), suppose that f be $\mathcal{C}$-differentiable and that g be $(-\mathcal{C})$-differentiable on X (see Definition 3.1.1). Consider any $\overline{x} \in X$ and set $d := x - \overline{x}$. The set:

$$\mathcal{K}_{\overline{x}}^h := \{(u, v) \in \mathbb{R}^{1+m} :\ u = -\mathcal{D}_{\mathcal{C}}f(\overline{x}; d);\ v_i = g_i(\overline{x}) + \mathcal{D}_{-\mathcal{C}}\,g_i(\overline{x}; d),\ i \in \mathcal{I};\ d \in X - \overline{x}\}$$

is called *homogenization* of the image set $\mathcal{K}_{\overline{x}}$. It is called *linearization*, if f and g are differentiable.

Indeed, in strict sense, with Definition 3.2.2 we homogenize f and g; more precisely, we replace (1.1.1) with its homogenized form:

$$\min\ [f(\overline{x}) + \mathcal{D}_{\mathcal{C}}f(\overline{x}; d)] \tag{3.2.10a}$$

$$\text{s.t.}\quad g_i(\overline{x}) + \mathcal{D}_{-\mathcal{C}}g_i(\overline{x}; d) = 0,\ \ i \in \mathcal{I}^0, \tag{3.2.10b}$$

$$g_i(\overline{x}) + \mathcal{D}_{-\mathcal{C}}g_i(\overline{x}; d) \geq 0,\ \ i \in \mathcal{I}^+, \tag{3.2.10c}$$

$$d \in X - \overline{x}, \tag{3.2.10d}$$

or, when f and g are differentiable, with its linearized form:

$$\min[f(\overline{x}) + \langle f'(\overline{x}), d \rangle] \tag{3.2.10a}'$$

$$\text{s.t.} \quad g_i(\overline{x}) + \langle g_i'(\overline{x}), d \rangle = 0, \quad i \in \mathcal{I}^0, \tag{3.2.10b}'$$

$$g_i(\overline{x}) \langle g_i'(\overline{x}), d \rangle \geq 0, \quad i \in \mathcal{I}^+, \tag{3.2.10c}'$$

$$d \in X - \overline{x}. \tag{3.2.10d}'$$

When (1.1.1) has only unilateral constraints ($p = 0$, $m \geq 1$), then problems (3.2.10) and (3.2.10)$'$ gain special importance. To show this in the next proposition, let us associate (3.2.10) with the following system (as well as (1.1.1) has beeen associated with system (3.2.1); $\overline{x}$ in (3.2.1) corresponds, here, to $\overline{d} = O$):

$$\mathcal{D}_{\mathcal{C}}f(\overline{x}; d) < 0, g_i(\overline{x}) + \mathcal{D}_{-\mathcal{C}}\, g_i(\overline{x}; d) \begin{cases} > 0, & \text{if } i \in \mathcal{I}_N^+, \\ \geq 0, & \text{if } i \in \mathcal{I}_L^+, \end{cases} d \in X - \overline{x}, \tag{3.2.10}''$$

where $\mathcal{I}_N^+ := \{i \in \mathcal{I}^+ : g_i(\overline{x}) = 0, \varepsilon_i(\overline{x}; d) \nleq 0\}$, $\mathcal{I}_L^+ := \mathcal{I}^+ \backslash \mathcal{I}_N^+$, and ε_i being given by (3.1.2) for $G = \mathcal{C}$. When f and g are differentiable, then (3.2.10)$''$ becomes:

$$\langle f'(\overline{x}), d \rangle < 0, \; g_i(\overline{x}) + \langle g_i'(\overline{x}), d \rangle > 0, i \in \mathcal{I}_N^+, g_i(\overline{x}) + \langle g_i'(\overline{x}), d \rangle \geq 0,$$

$$\tag{3.2.10}'''$$

$$i \in \mathcal{I}_L^+, d \in X - \overline{x}.$$

Note that, if we take the image set of (3.2.10) (or (3.2.10)$'$) at $d = 0$, then the outcome is just $\mathcal{K}_{\overline{x}}^h$ (we put in evidence $\overline{x}$ instead of $d = 0$), as shown by (i) of next Proposition.

Proposition 3.2.6. (i) We have:

$$\mathcal{K}_{\overline{x}}^h \subseteq TC\,(\overline{z}; \mathcal{K}_{\overline{x}}), \tag{3.2.11}$$

and, in a neighbourhood of $\overline{z}$, $\mathcal{K}_{\overline{x}}^h$ is a truncated cone with apex at $\overline{z}$. If, furthermore, X is a cone with apex at $\overline{x}$ (in particular, $X = \mathbb{R}^n$), then $\mathcal{K}_{\overline{x}}^h$ is a cone with apex at $\overline{z}$. **(ii)** The conic extension $\mathcal{E}(\mathcal{K}_{\overline{x}}^h)$ − of the homogenization of the image set − is convex. **(iii)** (Homogenization Lemma) Let $f, -g_i, i \in \mathcal{I}^+$, be $\mathcal{C}$-differentiable at $\overline{x} \in X$. If $\overline{x}$ is a minimum point of (1.1.1a,c,d), then the system (3.2.10)$''$ is impossible.

Proof. (i) Let $\tilde{z} = (\tilde{u}, \tilde{v}) \in \mathcal{K}_{\overline{x}}^h$. Because of Definition 3.2.2 and of (3.1.1), $\exists \tilde{d} \in X - \overline{x}$, s.t.

$$(\tilde{u}^r, \tilde{v}^r) \in \mathcal{K}_{\overline{x}}^h, \quad \forall r = 1, 2, ..., \tag{3.2.12}$$

where

$$\tilde{u}^r := -\mathcal{D}_{\mathcal{C}}f(\overline{x}; \tfrac{1}{r}\tilde{d}), \quad \tilde{v}_i^r := g_i(\overline{x}) + \mathcal{D}_{-\mathcal{C}}g_i(\overline{x}; \tfrac{1}{r}\tilde{d}), \quad i \in \mathcal{I}. \tag{3.2.13}$$

The remainder, which appears in (3.1.2), is here denoted by ε_0 or by ε_i according to it deals with f or g_i, respectively. Then, by using the 1st equality of (3.1.2), we have:

$$(u^r, v^r) \in \mathcal{K}_{\overline{x}}, \quad \forall r = 1, 2, ...,$$

where

$$\begin{cases} u^r := f(\overline{x}) - [f(\overline{x}) + \mathcal{D}_{\mathcal{C}}f(\overline{x}; \tfrac{1}{r}\tilde{d}) + \varepsilon_0(\overline{x}; \tfrac{1}{r}\tilde{d})], \\ v_i^r := g_i(\overline{x}) + \mathcal{D}_{-\mathcal{C}}g_i(\overline{x}; \tfrac{1}{r}\tilde{d}) + \varepsilon_i(\overline{x}; \tfrac{1}{r}\tilde{d}), \quad i \in \mathcal{J}. \end{cases}$$

Set $\beta_r := \tfrac{1}{r}$; $\quad \forall i \in \{0\} \cup \mathcal{J}$, because of the 2nd equality of (3.1.2), we have:

$$\lim_{r \to +\infty} r\varepsilon_i(\overline{x}; \tfrac{1}{r}\tilde{d}) = \lim_{r \to +\infty} \frac{\varepsilon_i(\overline{x}; \beta_r \tilde{d})}{||\beta_r \tilde{d}||} \cdot ||\tilde{d}|| = 0.$$

Therefore, setting $\alpha_r = r$ and using (3.1.1), we find:

$$\lim_{r \to +\infty} \alpha_r[(u^r, v^r) - (\overline{u}, \overline{v})] = (\tilde{u}, \tilde{v}),$$

which, account taken of Definition 2.1.9, gives (3.2.11). Since (3.2.12) remains valid, if in (3.2.13) $\tfrac{1}{r}$ is replaced by $t \in [0,1]$, then $\mathcal{K}_{\overline{x}}^h$ turns out to be a truncated cone in a neighbourhood of $\overline{z}$. If, furthermore, X is a cone with apex at $\overline{x}$, then from $\tilde{d} \in X - \overline{x}$ and (2.1.7b) we have $t\tilde{d} \in X - \overline{x}$, $\forall t \geq 0$. Thus, being $(\tilde{u}, \tilde{v}) \in \mathcal{K}_{\overline{x}}^h$, by using (3.1.1) we have:

$$t[(\tilde{u}, \tilde{v}) - (\overline{u}, \overline{v})] \in \mathcal{K}_{\overline{x}}^h - \overline{z}, \quad \forall t \geq 0,$$

which shows the thesis. **(ii)** The map

$$A_{\overline{x}}^h(d) := (-\mathcal{D}_{\mathcal{C}}f(\overline{x}; d), \quad g_i(\overline{x}) + \mathcal{D}_{-\mathcal{C}}g_i(\overline{x}; d), \quad i \in \mathcal{J}),$$

which is the homogenization of the map $A_{\overline{x}}$, is concave (in the sense that each of its components is concave). The set cl $\mathcal{H}$ is obviously closed, convex, and $-$ as it is easy to show $-$ pointed (see (2.1.8)). Then, Proposition 4.4.1, at $\nu = 1 + m$, $H = \mathcal{H}$, and $A = A_{\overline{x}}^h$ $-$ concavity obviously implies H-concavelikeness (Definition 2.4.5) $-$, can be applied to achieve the convexity of $\mathcal{E}(\mathcal{K}_{\overline{x}}^h)$. **(iii)** Ab absurdo, suppose that (3.2.10)$''$ be possible; let $d = \hat{d}$ be a solution. Then $\alpha\hat{d}$ is a solution of (3.2.10)$''$ $\forall \alpha \in [0,1]$, since $g_i(\overline{x}) \geq 0$ and $\mathcal{D}_{\mathcal{C}}f, \mathcal{D}_{\mathcal{C}}g_i$ satisfy (3.1.1). The assumption implies that the remainders:

$$\varepsilon_0(\overline{x}; d) := f(x) - f(\overline{x}) - \mathcal{D}_{\mathcal{C}}f(\overline{x}; d),$$

$$\varepsilon_i(\overline{x}; d) := g_i(x) - g_i(\overline{x}) - \mathcal{D}_{\mathcal{C}}g_i(\overline{x}; d), i \in \mathcal{J}^+$$

fulfil (3.1.2) for $G = \mathcal{C}$, so that $\exists \hat{\alpha} \in]0,1]$ s.t.

$$\frac{\varepsilon_0(\overline{x}; \hat{\alpha}\hat{d})}{||\hat{\alpha}\hat{d}||} < \frac{-\mathcal{D}_{\mathcal{C}}f(\overline{x}; \hat{d})}{||\hat{d}||}; \quad \frac{\varepsilon_i(\overline{x}; \hat{\alpha}\hat{d})}{||\hat{\alpha}\hat{d}||} > \frac{\mathcal{D}_{\mathcal{C}}g_i(\overline{x}; \hat{d})}{||\hat{d}||}, i \in \mathcal{J}_N^+.$$

From these inequalities, by setting $d^* = \hat{\alpha}\hat{d}$ and noting that $g_i(\overline{x}) = 0$, $\forall i \in \mathcal{J}_N^+$, we have:

$$\mathcal{D}_{\mathcal{C}}f(\overline{x}; d^*) + \varepsilon_0(\overline{x}; d^*) < 0, \quad g_i(\overline{x}) + \mathcal{D}_{\mathcal{C}}g_i(\overline{x}; d^*) + \varepsilon_i(\overline{x}; d^*) > 0, \ i \in \mathcal{J}_N^+. \qquad (*)$$

$\forall i \in \mathcal{J}^+ \backslash \mathcal{J}_N^+$, either $g_i(\overline{x}) = 0$ and $\varepsilon_i \equiv 0$ or $g_i(\overline{x}) > 0$. In the former case, by setting $\tilde{d} := \hat{d}$, we have:

$$g_i(\overline{x}) + \mathcal{D}_{\mathcal{C}}g_i(\overline{x}; \tilde{d}) + \varepsilon_i(\overline{x}; \tilde{d}) \geq 0. \qquad (**)$$

In the latter case, $\exists \alpha^0 \in {]}0, 1]$ s.t.:

$$g_i(\overline{x}) + \mathcal{D}_{\mathcal{C}}g_i(\overline{x}; \alpha\hat{d}) > 0, \quad \forall \alpha \in {]}0, \alpha^0],$$

and thus $\exists \tilde{\alpha} \in {]}0, \alpha^0]$ s.t.:

$$\frac{\varepsilon_i(\overline{x}; \tilde{\alpha}\hat{d})}{||\tilde{\alpha}\hat{d}||} \geq -\frac{g_i(\overline{x}) + \alpha^0\mathcal{D}_{\mathcal{C}}g_i(\overline{x}; \hat{d})}{\alpha^0||\hat{d}||} \geq -\frac{g_i(\overline{x}) + \alpha\mathcal{D}_{\mathcal{C}}g_i(\overline{x}; \hat{d})}{\alpha||\hat{d}||}, \; \forall \alpha \in {]}0, \alpha^0],$$

where the 1st inequality holds since again the remainder ε_i fulfils (3.1.2) and the 2nd side is fixed and negative, the 2nd inequality holds since the 2nd side is obviously the maximum of the 3rd on ${]}0, \alpha^0]$. By setting $\tilde{d} := \tilde{\alpha}\hat{d}$, it follows that:

$$g_i(\overline{x}) + \mathcal{D}_{\mathcal{C}}g_i(\overline{x}; \tilde{d}) + \varepsilon_i(\overline{x}; \tilde{d}) \geq 0. \qquad (***)$$

Collecting the above (*), (**), (***), recalling that $g_i(\overline{x}) = 0$, $i \in \mathcal{I}_N^+$, and using the definition of the remainders ε_i, we obtain the possibility of system (3.2.1) (for $p = 0$), which contradicts the minimality of $\overline{x}$. $\qquad \square$

The Homogenization Lemma, expressed by Proposition 3.2.6(iii), will be extended to problems of type (1.1.5); see Theorem 3.3.4. When system (3.2.10)″ is replaced by (3.2.10)‴ (namely, when the linearization (3.2.10)′ is possible), then Proposition 3.2.6(iii) shrinks to the Linearization Lemma [I1]. Due to the importance of the homogenization of the image set, we will give a different proof of the convexity of $\mathcal{E}(\mathcal{K}_{\overline{x}}^h)$, which does not exploit Proposition 3.4.1; see Proposition 3.2.8(iii). Some instances of $\mathcal{K}_{\overline{x}}$, $\mathcal{K}_{\overline{x}}^h$, their conic extensions and tangent cones can be found in the examples of Sect. 3.4. The inclusion (3.2.11) is illustrated in Fig. 3.4.5a; of course, if the constraints $0 \leq x_i \leq 1$, $i = 1, 2$ of Example 3.4.3 are considered as 4 inequalities of type (1.1.1c), instead of sending them to define X, then $\mathcal{K}_{\overline{x}}^h$ becomes a cone, but of a 6-dimensional IS.

The above concepts enjoy several properties, some of which are now stated. The examples of Sect. 3.4 are useful to follow them.

Proposition 3.2.7. (3.2.2) holds if and only if

$$\mathcal{H} \cap \mathcal{E}(\mathcal{K}_{\overline{x}}) = \varnothing. \qquad (3.2.14)$$

Proof. **If.** It is an obvious consequence of the inclusion $\mathcal{K}_{\overline{x}} \subseteq \mathcal{E}(\mathcal{K}_{\overline{x}})$. **Only if.** Ab absurdo, suppose that $\exists z^1 \in \mathcal{K}_{\overline{x}}$, $\exists z^2 \in \mathrm{cl}\,\mathcal{H}$ — so that $z^1 - z^2 \in \mathcal{E}(\mathcal{K}_{\overline{x}})$ —, and that $z^1 - z^2 \in \mathcal{H}$. Then, because of (2.1.14), $z^1 = (z^1 - z^2) + z^2 \in \mathcal{H} + \mathrm{cl}\,\mathcal{H} = \mathcal{H}$, and hence (3.2.2) is contradicted. $\qquad \square$

As a consequence of the above proposition, we have that problem (3.2.3) is equivalent to:

$$u_{\overline{x}}^e := \max\,(u), \quad \text{s.t.} \quad (u, v) \in \mathcal{E}(\mathcal{K}_{\overline{x}}), \; v \in D, \qquad (3.2.15)$$

and (3.2.5) is equivalent to:

$$u_{\overline{x}}^e(\xi; D) := \max\,(u), \quad \text{s.t.} \quad (u, v) \in \mathcal{E}(\mathcal{K}_{\overline{x}}), \; v \in \{\xi\} + D. \tag{3.2.16}$$

Some of examples of Sect. 3.4 show $\mathcal{E}(\mathcal{K}_{\overline{x}})$. In Example 3.4.2, $\mathcal{K}_{\overline{x}}$ is not convex (Fig. 3.4.3), while $\mathcal{E}(\mathcal{K}_{\overline{x}})$ is convex (Fig. 3.4.4). In a pair of Examples 3.4.4, the image sets are different, while their conic extensions coincide.

An obvious consequence of Proposition 3.2.7 is that (3.2.14) is necessary and sufficient for $\overline{x}$ to be (global) m.p. of (1.1.1).

Proposition 3.2.8. (i) Let X be convex. The conic extension $\mathcal{E}(\mathcal{K}_{\overline{x}})$ is convex, if and only if the map $(f(x), -g(x))$ is $\mathcal{H}$-convexlike. **(ii)** If $\mathcal{K}_{\overline{x}}$ is compact, then $\mathcal{E}(\mathcal{K}_{\overline{x}})$ is closed. **(iii)** If f and $-g_i, i \in \mathcal{J}$ are $\mathcal{C}$-differentiable at $\overline{x}$, then $\mathcal{E}(\mathcal{K}_{\overline{x}}^h)$ and $\mathcal{E}(\mathcal{K}_{\overline{x}}^h - \overline{z})$ are convex.

Proof. (i) By using Definition 2.4.5, it is enough to apply Proposition 3.4.1, where we set $\nu = 1 + m$, $H = \mathcal{H}$, and $A = A_{\overline{x}} = (f(\overline{x}) - f, g)$. **(ii)** Let $\hat{z} \in \mathbb{R}^{1+m}$ be an accumulation point of $\mathcal{E}(\mathcal{K}_{\overline{x}})$. Then $\exists\{\hat{z}^i\}_1^\infty \subset \mathcal{E}(\mathcal{K}_{\overline{x}})$ s.t. $\lim\limits_{i\to+\infty} \hat{z}^i = \hat{z}$. $\hat{z}^i \in \mathcal{E}(\mathcal{K}_{\overline{x}})$ implies that $\exists\{\tilde{z}^i\}_1^\infty \subset \mathcal{K}_{\overline{x}}$ and $\exists\{h^i\}_1^\infty \subset \text{cl}\,\mathcal{H}$ s.t.

$$\hat{z}^i = \tilde{z}^i - h^i, \quad i = 1, 2, \dots .$$

Due to the compactness of $\mathcal{K}_{\overline{x}}$, it is not restrictive to assume that $\exists\tilde{z} := \lim\limits_{i\to+\infty} \tilde{z}^i$ and that $\tilde{z} \in \mathcal{K}_{\overline{x}}$. Hence, due to the above equalities, $\exists\tilde{h} := \lim\limits_{i\to+\infty} h^i$ and, of course, $\tilde{h} \in \text{cl}\,\mathcal{H}$. Passing to the limit in $\hat{z}^i = \tilde{z}^i - h^i$, as $i \to +\infty$, we find $\hat{z} = \tilde{z} - \tilde{h}$ with $\tilde{z} \in \mathcal{K}_{\overline{x}}$ and $\tilde{h} \in \text{cl}\,\mathcal{H}$; thus $\hat{z} \in \mathcal{E}(\mathcal{K}_{\overline{x}})$. **(iii)** It is enough to prove the convexity of $\mathcal{E}(\mathcal{K}_{\overline{x}}^h - \overline{z})$, since $\mathcal{E}(\mathcal{K}_{\overline{x}}^h)$ is a translation of it. If $\mathcal{K}_{\overline{x}}^h = \{\overline{z}\}$, then the thesis is trivial since $\mathcal{E}(\mathcal{K}_{\overline{x}}^h) = \overline{z} - \text{cl}\,\mathcal{H}$. Otherwise, consider the vector

$$\triangle(d) := (-\mathcal{D}_\mathcal{C}f(\overline{x}; d), \;\; \mathcal{D}_\mathcal{C}g_i(\overline{x}; d), i \in \mathcal{J}^0) \in \mathbb{R}^{1+m},$$

and any two distinct elements of $\mathcal{E}(\mathcal{K}_{\overline{x}}^h)$, which can be written as $\triangle(d^i) - h^i$, with $h^i \in \text{cl}\,\mathcal{H}$, $d^i \in X - \overline{x}, i = 1, 2$; $\forall \alpha \in [0, 1]$, their convex combination, say $c(\alpha)$, is such that:

$$c(\alpha) = (1 - \alpha)\triangle(d^1) + \alpha\triangle(d^2) - h(\alpha) \leq \triangle(d(\alpha)) = h(\alpha),$$

where $h(\alpha) := (1 - \alpha)h^1 + \alpha h^2 \in \mathcal{H}$, and $d(\alpha) := (1 - \alpha)d^1 + \alpha d^2 \in X - \overline{x}$. Hence, we have $c(\alpha) = \triangle(d(\alpha)) - \hat{h}(\alpha)$, when $\triangle(d(\alpha)) \in \mathcal{K}_{\overline{x}}^h$ and

$$\hat{h}(\alpha) := [\triangle(d(\alpha)) - (1 - \alpha)\triangle(d^1) - \alpha\triangle(d^2)] + h(\alpha) \in \text{cl}\,\mathcal{H},$$

since the square bracket is non-negative. This shows $c(\alpha) \in \mathcal{E}(\mathcal{K}_{\overline{x}}^h - \overline{z})$. $\qquad\square$

The assumption of Proposition 3.2.8 is obviously satisfied, if f is convex and g concave. If X is a polyhedron or $\mathbb{R}^n$ and f, g affine, then $\mathcal{K}_{\overline{x}}$ is a polyhedron (a polytope, if such is X; or a flat, if such is X), and $\mathcal{E}(\mathcal{K}_{\overline{x}})$ is an unbounded polyhedron. If f and g are differentiable, then $\mathcal{E}(\mathcal{K}_{\overline{x}}^h)$ is the sum of an affine variety and $\mathbb{R}_-^{1+m}$.

Problem (3.2.10) and its image set $\mathcal{K}_{\bar{x}}^{h}$ play a crucial role at least for achieving necessary optimality conditions. To this end, the extremely important aspect would be to be able to claim that, if (3.2.2) holds, then also

$$\mathcal{H} \cap \mathcal{K}_{\bar{x}}^{h} = \varnothing \qquad (3.2.17)$$

holds. Unfortunately, in the general case, such a claim is false, as Example 3.4.9 shows; (3.2.17) is satisfied in Examples 3.4.1-3.4.3. Theorem 2.2.7 is instrumental for deepening such an aspect. Even if for such an analysis $\mathcal{H}$ can lose some faces, now we consider a property quite analogous to Property 3.2.7.

Proposition 3.2.9. Let $\mathcal{H}$ fulfil (2.1.14). (3.2.17) holds if and only if

$$\mathcal{H} \cap \mathcal{E}(\mathcal{K}_{\bar{x}}^{h}) = \varnothing. \qquad (3.2.18)$$

Proof. If. it is an obvious consequence of the inclusion $\mathcal{K}_{\bar{x}}^{h} \subseteq \mathcal{E}(\mathcal{K}_{\bar{x}}^{h})$. **Only if.** Ab absurdo, suppose that $\exists z^{1} \in \mathcal{K}_{\bar{x}}^{h}$, $\exists z^{2} \in \mathrm{cl}\,\mathcal{H}$ — so that $z^{1} - z^{2} \in \mathcal{E}(\mathcal{K}_{\bar{x}}^{h})$—, and that $z^{1} - z^{2} \in \mathcal{H}$. Then, because of (2.1.14), $z^{1} = (z^{1} - z^{2}) + z^{2} \in \mathcal{H} + \mathrm{cl}\,\mathcal{H} = \mathcal{H}$, and hence (3.2.2) is contradicted. $\qquad\qquad \square$

The deepening of the properties of the image set is fundamental also for achieving optimality conditions. As an instance, consider the following proposition, which gives a necessary condition for (3.2.3) to have maximum. To this end, let $w : \mathbb{R}^{1+m} \to \mathbb{R}$ be any continuous function, such that (see Chapter 4 for details):

$$\mathrm{lev}_{\geq 0}\, w(z) \supseteq \mathcal{H}. \qquad (3.2.19)$$

This is a special case of separation functions, which will be investigated in Sect. 4.2; (3.2.19) is a specialization of (4.2.3a). For the sake of simplicity and without any fear of confusion, in the next proposition we will write merely $\mathcal{E}$ instead of $\mathcal{E}(\mathcal{K}_{\bar{x}})$.

Theorem 3.2.1. Let $\mathcal{H}$ fulfil (2.1.14) and F be any face of $\mathrm{cl}\,\mathcal{H}$, and suppose that

$$\mathcal{E}^{0} := (\mathrm{cl}\,\mathcal{H}) \cap \mathrm{cl}\,\mathcal{E} \neq \varnothing. \qquad (3.2.20)$$

If (3.2.1) is impossible (or (3.2.2) holds), then we have:

(i) $TC(z; \mathcal{E}) \cap \mathrm{int}\,\mathcal{H} = \varnothing \quad \forall z \in \mathcal{E}^{0}$;

(ii) $O \in \mathcal{E}^{0} \subseteq \mathrm{cl}\,\mathrm{lev}_{\geq 0}\, w(z)$;

(3i) $\mathcal{E}^{0} \cap \mathrm{ri}\, F \neq \varnothing \;\Rightarrow\; F \subseteq TC(O; \mathcal{E})$;

(4i) $TC(\mathcal{E}) \cap \mathrm{ri}\, F \neq \varnothing \;\Rightarrow\; F \subseteq TC(\mathcal{E})$.

Proof. (i). The thesis is trivial if $\mathrm{int}\,\mathcal{H} = \varnothing$ or $\mathrm{card}\,TC(z; \mathcal{E}) \leq 1$. Otherwise, ab absurdo, suppose that (i) does not hold. Then $\exists \hat{z} \in \mathcal{E}^{0}$ and $\exists \tilde{z} \in \mathbb{R}^{1+m}$, with $\tilde{z} \neq \hat{z}$, s.t.

$$\tilde{z} \in TC(\hat{z}; \mathcal{E}) \cap \mathrm{int}\,\mathcal{H}.$$

Then, according to Definition 2.1.9, $\exists \{z^{i}\}_{1}^{\infty} \subset \mathcal{E}$ and $\exists \{\alpha_{i} > 0\}_{1}^{\infty}$ s.t.

$$\lim_{i \to +\infty} z^{i} = \hat{z}, \quad \lim_{i \to +\infty} \alpha_{i}(z^{i} - \hat{z}) = \tilde{z} - \hat{z}.$$

From this, being $\tilde{z} - \hat{z} \neq O$ so that $\lim_{i \to +\infty} \alpha_i = +\infty$, we draw:

$$z^i = \hat{z} + \frac{1}{\alpha_i}(\tilde{z} - \hat{z}) + 0(\alpha_i),$$

where $0(\alpha_i)$ is an infinitesimal of higher order. It follows that:

$$z^i \in \,]\hat{z}, \hat{z} + \alpha_i(z^i - \hat{z})], \quad \forall i > i_1.$$

The last two relations, $\hat{z} \in \text{cl}\,\mathcal{H}$, and (iii) of Proposition 2.1.5 imply that $z^i \in \text{int}\,\mathcal{H}$, $\forall i > i_1$. Hence, according to Proposition 3.2.7, the impossibility of (3.2.1) is contradicted. **(ii)** To prove the 1st part, it is enough to show that $O \in \text{cl}\,\mathcal{E}$ (since, obviously, $O \in \text{cl}\,\mathcal{H}$). According to Proposition 3.2.7, the impossibility of (3.2.1) implies

$$\mathcal{E}^0 = (r\partial\,\mathcal{H}) \cap r\partial\,\mathcal{E}.$$

Let $\hat{z} \in \mathcal{E}^0$, so that $\hat{z} \in r\partial\,\mathcal{E}$; hence $\exists \{\hat{z}^i\}_1^\infty \subset \mathcal{E}$ s.t. $\lim_{i \to +\infty} \hat{z}^i = \hat{z}$. This implies that $\exists z^i \in \mathcal{K}_{\bar{x}}$ and $\exists h^i \in \text{cl}\,\mathcal{H}$ s.t. $\hat{z}^i = z^i - h^i$. Therefore, we have the following consecutive equalities:

$$\lim_{i \to +\infty} d(z^i, r\partial\,\mathcal{H}) = 0,$$

$$\lim_{i \to +\infty} d(z^i - O, \text{cl}\,\mathcal{H}) = 0,$$

$$\lim_{i \to +\infty} d(z^i - \text{cl}\,\mathcal{H}, O) = 0,$$

the last of which implies $O \in \text{cl}\,\mathcal{E}$. To show the 2nd part, let $\hat{z} \in \text{cl}\,\mathcal{H}$, so that, $\forall \varepsilon > 0$, $\exists z_\varepsilon \in \mathcal{H}$ s.t. $\|z_\varepsilon - \hat{z}\| < \varepsilon$. Then (3.2.19) implies $z_\varepsilon \in \text{lev}_{\geq 0} w(z)$. Since ε is arbitrary and w continuous, then

$$\hat{z} \in \text{cl} \ \text{lev}_{\geq 0} w(z).$$

Hence

$$\text{cl}\,\mathcal{H} \subseteq \text{cl} \ \text{lev}_{\geq 0} w(z)$$

and, as obviously $\mathcal{E}^0 \subset \text{cl}\,\mathcal{H}$, the 2nd part follows. **(3i)** Let $\hat{z} \in \mathcal{E}^0 \cap \text{ri}\,F$, and consider a partial conic extension of $\hat{z}$, namely $\{\hat{z}\} - F$; we have:

$$\{\hat{z}\} - F \subseteq \text{cl}\,\mathcal{E},$$

$$\{\alpha z : z \in F \cap (\{\hat{z}\} - F), \ \alpha > 0\} = F.$$

The thesis is a straightforward consequence of these relationship. **(4i)** Let us consider $\hat{z} \in TC(\mathcal{E}) \cap \text{ri}\,F$. Then $\exists \{z^i\}_1^\infty \subset \mathcal{E}$ and $\exists \{\alpha_i > 0\}_1^\infty$ s.t.

$$\lim_{i \to +\infty} z^i = O, \qquad \lim_{i \to +\infty} \alpha_i z^i = \hat{z}.$$

Set $\hat{z}^i := z^i - (1/\alpha_i)\hat{z}$; we have:

$$\{\hat{z}^i\}_1^\infty \subset \text{cl}\,\mathcal{E}, \qquad\qquad \lim_{i\to+\infty} \hat{z}^i = \lim_{i\to+\infty} \alpha_i \hat{z}^i = O.$$

Now consider the sets:

$$S := F \cap (\{\hat{z}\} - F), \qquad S_i := (\{\hat{z}^i\} + F) \cap (\{z^i\} - F), \quad i = 1, 2, ...,$$

and note that:

$$\lim_{i\to+\infty} S_i = \{O\}, \qquad \lim_{i\to+\infty} \alpha_i S_i = S, \qquad S_i \subseteq \text{cl}\,\xi, \quad i = 1, 2, ...,$$

and that

$$\hat{z} \in \text{ri}\,F \quad\Rightarrow\quad \{\alpha z : z \in S, \alpha > 0\} = F.$$

The fact that $F \subseteq TC(\mathcal{E})$ is now obvious. $\qquad\qquad\qquad\qquad\qquad\qquad \square$

According to the remark which follows (3.2.2), if the feasible region $R \neq \varnothing$, then we have $\mathcal{K}_{\overline{x}} \cap \text{cl}\,\mathcal{H} \neq \varnothing$, so that (3.2.20) is satisfied. In the above theorem, (i) becomes meaningless if $p > 0$, since $\text{int}\,\mathcal{H} = \varnothing$; on the contrary, when $p = 0$, (i) offers a very general necessary condition for $\overline{z}$ to be a maximum point of (3.2.3) and hence for $\overline{x}$ to be m.p. of (1.1.1) or (1.1.4). The right-hand side of (4i) is a special case of (2.2.34).

Connections among (1.1.1) or (1.1.4), (3.2.3) and (3.2.5) have been the subject of some remarks and proofs in this sections; now let us gather them into a unique statement.

Theorem 3.2.2. Let $\hat{x}(\xi) \in R(\xi)$ and set:

$$(\hat{u}(\xi), \hat{v}(\xi)) := (f(\overline{x}) - f(\hat{x}(\xi)),\ g(\hat{x}(\xi)) - \xi). \tag{3.2.21}$$

The following statements are equivalent:

(i) $\hat{x}(\xi)$ is a global minimum point of (3.2.4);

(ii) $(\hat{u}(\xi), \hat{v}(\xi))$ is a global maximum point of (3.2.5);

(iii) $(\hat{u}(\xi), \hat{v}(\xi))$ is a global maximum point of (3.2.16).

Furthermore, we have:

$$f^{\downarrow}(\xi) = f(\overline{x}) - u_{\overline{x}}(\xi; D) = f(\overline{x}) - u_{\overline{x}}^e(\xi; D). \tag{3.2.22}$$

Proof. The equivalence between (i) and (ii), the former of equalities (3.2.22) are given by Proposition 3.2.2. The equivalence between (ii) and (iii), and the latter of (3.2.22) are a straighforward consequence of Proposition 3.2.7, where $\mathcal{H}$ must be replaced by $\{(u, v) \in \mathbb{R}^{1+m} : u > 0, v \in \{\xi\} + D\}$. $\qquad\qquad\qquad\qquad\qquad\qquad \square$

At $\xi = 0$, the above theorem clarifies the connections among problems (1.1.1) or (1.1.4), (3.2.3) and (3.2.15); such equivalences and (3.2.22) can be checked on the examples of Sect. 3.4. Now, let us consider an existence condition, which is based on the results of [47]. To this end, let us introduce the set:

$$U_{\overline{x}} := \{(u, O_m) \in \mathbb{R} \times \mathbb{R}^m : u \geq 0 \text{ and } \exists v \in D \text{ s.t. } (u, v) \in \mathcal{E}(\mathcal{K}_{\overline{x}})\},$$

and consider problems (1.1.1) and (1.1.4) as special cases of (1.1.3). $U_{\bar{x}}$ is the projection, on the non-negative u-semi axis, of the feasible region of (3.2.15).

Theorem 3.2.3. Consider the problem (3.2.3), and suppose that the set

$$\{u \in \mathbb{R} : (u, v) \in \mathcal{K}_{\bar{x}}\} \quad \text{be bounded from above,} \tag{3.2.23a}$$

and that there exists a closed set $S \subset \mathbb{R}^{1+m}$, such that:

$$U_{\bar{x}} \subseteq S \subset \mathcal{E}(\mathcal{K}_{\bar{x}}). \tag{3.2.23b}$$

Then (3.2.3) has maximum and (1.1.3) has minimum.

Proof. It is easy to see that $U_{\bar{x}} \neq \varnothing$, iff the feasible region of (3.2.15) is nonempty, and that $U_{\bar{x}}$ is bounded, iff (3.2.23a) holds. The existence of finite supremum, say $\hat{u}$, for (3.2.15) follows. Ab absurdo, suppose that $\hat{u}$ be not maximum, so that $(\hat{u}, O_m) \notin \mathcal{E}(\mathcal{K}_{\bar{x}})$. Hence $\exists \{(u_i, O_m)\}_1^\infty \subset U_{\bar{x}}$ s.t. $\lim_{i \to +\infty} u_i = \hat{u}$ and, of course, $u_i < \hat{u}$. Thus $(\hat{u}, O_m) \in \text{cl}\, U_{\bar{x}}$, so that $(\hat{u}, O_m) \in S$. Because of (3.2.23b), we meet the contradiction $(\hat{u}, O_m) \in \mathcal{E}(\mathcal{K}_{\bar{x}})$. Therefore, $\hat{u}$ is maximum of (3.2.15). To achieve the thesis, it is enough to appeal to Theorem 3.2.2 for $\xi = 0$, to the equivalence between (3.2.3) and (3.2.15), and to that between (1.1.3) and (3.2.3). $\square$

If $\mathcal{E}(\mathcal{K}_{\bar{x}})$ is closed, then, of course, condition (3.2.23b) is fulfilled; this does not happen necessarily, if $\mathcal{K}_{\bar{x}}$ is closed, as shown by Examples 3.4.14 and 3.4.15: in the former $U_{-1} = [(0,0),(e,0)[$ and $U_{\ln M} = [(0,0),(1/M,0)[$, and in the latter $U_0 = [(0,0),(1,0)[$; in all these cases, $U_{\bar{x}}$ cannot be included in a closed set contained in $\mathcal{E}(\mathcal{K}_{\bar{x}})$, notwithstanding the fact that $\mathcal{K}_{\bar{x}}$ be closed. Obviously, (3.2.23b) is satisfied, if $U_{\bar{x}}$ is closed. Condition (3.2.23a) is equivalent to:

$$f^{\downarrow} := \inf_{x \in R} f(x) > -\infty. \tag{3.2.24}$$

Corollary 3.2.1. If R is nonempty and compact, and f is continuous, then problem (1.1.3) has minimum.

Proof. Straightforward consequence of Proposition 3.2.8(ii) and of Theorem 3.2.3. $\square$

Corollary 3.2.2. If $U_{\bar{x}}$ is unbounded, then the infimum of (1.1.3) is $-\infty$.

Proof. The assumption implies that the supremum of (3.2.15) is $+\infty$. Then Theorem 3.2.2 gives the thesis. $\square$

Corollary 3.2.1 is the well known Weierstrass Theorem, which is a slightly particular case of Theorem 1.1.1. Theorem 3.2.3 can be viewed as a source for deriving existence conditions in the IS and then in the given space. An instance is offered by the next 3 corollaries, which can be stated under more general conditions, like the 3rd, where f can be assumed to be lower semicontinuous.

Corollary 3.2.3. If there exists $\hat{u} \in \mathbb{R}$, such that the set

$$\mathcal{K}_{\bar{x}, \hat{u}} := \{(u, v) \in \mathcal{K}_{\bar{x}} : u \geq \hat{u}, \ v \in D\} \tag{3.2.25}$$

be compact, then the problem (1.1.3) has minimum.

Proof. Set $\mathcal{H}_u := \{(u, v) \in \mathrm{cl}\,\mathcal{H} : v = O_m\}$. If $\hat{u} \leq 0$, then we have the equality

$$U_{\overline{x}} = \mathcal{H}_u \cap \mathrm{proj}_{\mathcal{H}_u} \mathcal{K}_{\overline{x}, \hat{u}},$$

which shows that $U_{\overline{x}}$, being a closed subset of a compact set, is compact, so that (3.2.23) are satisfied. If $\hat{u} > 0$, we reduce ourselves to the previous case, by a suitable translation of $\mathcal{K}_{\overline{x}}$ (or a change of $\overline{x}$): since it is not restrictive to assume that $\exists \hat{x} \in X$ s.t. $\hat{u} = f(\overline{x}) - f(\hat{x})$, we can choose $\overline{x}$ in such a way that $f(\overline{x}) \leq f(\hat{x})$. $\qquad\square$

Corollary 3.2.4. Let $f : X \to \mathrm{I\!R}$ be continuous and the set R, given by (1.1.2), be closed. If there exists $\alpha \in \mathrm{I\!R}$, such that the set $\mathrm{lev}_{\leq \alpha}\, f$ be bounded, then problem (1.1.3) has minimum.

Proof. In order to consider exactly the classic statement, the assumption has been made directly on R and not on g. Since we consider R in the form (1.1.2) — which is, obviously, more general than the classic one — we must make an assumption on g: we suppose that g be continuous on X. That being stated, let us observe that f is continuous on the (compact) set $\mathrm{cl}\,\mathrm{lev}_{\leq \alpha}\, f$; therefore, because of Theorem 1.1.1, f attains its minimum on $\mathrm{cl}\,\mathrm{lev}_{\leq \alpha}\, f$. Hence, its infimum on $\mathrm{lev}_{\leq \alpha}\, f$ exists and is finite (and, of course, the same happens on R). Then (3.2.24) and (3.2.23a) are satisfied. Consider now the set (3.2.25), where we set $\hat{u} = f(\overline{x}) - \alpha$. It is bounded, since it is the image of a bounded set through a couple of continuous functions. It is closed. In fact, ab absurdo, suppose that $\exists (\tilde{u}, \tilde{v}) \in (\mathrm{cl}\,\mathcal{K}_{\overline{x}, \hat{u}}) \backslash \mathcal{K}_{\overline{x}, \hat{u}}$. Then $\exists \{(u_i, v^i)\}_1^\infty \subset \mathcal{K}_{\overline{x}, \hat{u}}$ s.t. $\lim\limits_{i \to +\infty} (u_i, v^i) = (\tilde{u}, \tilde{v})$. This implies that $\exists \{x^i\}_1^\infty \subset X$ s.t. $u_i = f(\overline{x}) - f(x^i)$, $v^i = g(x^i) \geq 0$, $f(x^i) \leq \alpha$. Because of the boundedness of $\mathrm{lev}_{\leq \alpha}\, f$, $\exists \tilde{x} := \lim\limits_{i \to +\infty} x^i$. Then we have:

$$\lim_{i \to +\infty} [f(\overline{x}) - f(x^i)] = f(\overline{x}) - f(\tilde{x}) = \tilde{u}.$$

R being closed and $\tilde{x}$ being an accumulation point, we have $\tilde{x} \in R$. A contradiction has been reached. The compactness of (3.2.25) follows. The thesis is achieved by applying Corollary 3.2.3. $\qquad\square$

In Example 3.4.5, at $\overline{x} = 0$, we have $U_{\overline{x}} = \{(u, v) \in \mathrm{I\!R}^2 : 0 \leq u \leq 1, v = 0\} = \mathcal{E}(\mathcal{K}_{\overline{x}}) \cap \{(u, v) \in \mathrm{I\!R}^2 : u \geq 0, v = 0\}$. Therefore (3.2.23) hold, while Theorem 1.1.1 cannot be applied. Further simple examples show the role of $U_{\overline{x}}$.

Example 3.2.3. In (1.1.1) set $n = m = 1$, $p = 0$, $X = \mathrm{I\!R}$ $f(x) = x(x - 1)$ if $x \neq 1/2$, $f(1/2) = 0$, $g(x) = x$. At $\overline{x} = 1/2$, we find:

$$\mathcal{K}_{\overline{x}} = \left\{ (u, v) \in \mathrm{I\!R}^2 : u = -v(v - 1), \text{ if } v \neq \frac{1}{2}, \text{ and } u = 0, \text{ if } v = \frac{1}{2} \right\},$$

$$\mathcal{E}(\mathcal{K}_{\overline{x}}) = \left\{ (u, v) \in \mathrm{I\!R}^2 : v \leq \frac{1}{2} + \frac{1}{2}\sqrt{1 - 4u},\ u < \frac{1}{4} \right\},$$

$$U_{\overline{x}} = \left\{ (u, v) \in \mathrm{I\!R}^2 : v = 0,\ u \in \left[0, \frac{1}{4}\right[\right\}.$$

Hence $U_{\bar{x}}$ is bounded, but not closed, and there is no set S which fulfils (3.2.23b). The given problem has, of course, finite infimum, but not minimum. □

Example 3.2.4. In (1.1.1) set $n = m = 1, p = 0, X = \mathbb{R}_+, f(x) = \ln x$ if $x > 0$, $f(0) = 0,\ g(x) = x$. At $\bar{x} = 1$, we find:

$$\mathcal{K}_{\bar{x}} = \{(u, v) \in \mathbb{R}^2 : u = -\ln v \text{ if } v > 0,\ u = 0 \text{ if } v = 0\},$$

$$\mathcal{E}(\mathcal{K}_{\bar{x}}) = \{(u, v) \in \mathbb{R}^2 : v \le e^{-u}\},$$

$$U_{\bar{x}} = \{(u, v) \in \mathbb{R}^2 : v = 0,\ u \in [0, +\infty[\}.$$

Hence $U_{\bar{x}}$ is closed, but not bounded; (3.2.23b) is satisfied at $S = U_{\bar{x}}$, but (2.6.23a) is not. □

Example 3.2.5. In (1.1.1) set $n = m = 1,\ p = 0,\ X = \mathbb{R}_+,\ f(x) = \ln(x + 1/2)$ if $x > 0,\ f(0) = 0,\ g(x) = x$. At $\bar{x} = 0$, we find:

$$\mathcal{K}_{\bar{x}} = \left\{(u, v) \in \mathbb{R}^2 : u = -\ln\left(v + \frac{1}{2}\right) \text{ if } v > 0,\ u = 0 \text{ if } v = 0\right\},$$

$$\mathcal{E}(\mathcal{K}_{\bar{x}}) = \left\{(u, v) \in \mathbb{R}^2 : v \le e^{-u} - \frac{1}{2},\ u < \ln 2\right\},$$

$$U_{\bar{x}} = \{(u, v) \in \mathbb{R}^2 : v = 0,\ u \in [0, \ln 2[\}.$$

Hence $U_{\bar{x}}$ is bounded, but not closed, and there is no set S which fulfils (3.2.23b). The given problems has, of course, finite infimum, but not minimum. □

Corollary 3.2.5. Assume that $\exists k \in \mathbb{R}$, such that, $\forall \alpha \in]f^{\downarrow}, k]$, the set $R \cap \text{lev}_{\le \alpha} f$ be nonempty and compact. Then problem (1.1.3) has minimum.

Proof. Of course, $\alpha_1 < \alpha_2 \Rightarrow \text{lev}_{\le \alpha_1} f \subseteq \text{lev}_{\le \alpha_2} f$. It follows that

$$\bigcap_{\alpha \in]f^{\downarrow}, k]} \text{lev}_{\le \alpha} \ne \varnothing$$

and that the set

$$U_{\bar{x}} \cap \{(u, v) \in \mathbb{R}^{1+m} : u \ge f(\bar{x}) - k\}$$

is compact. Hence (3.2.23) are fulfilled and the thesis follows. □

   In the above corollary, the crucial role is played by the projection (on the non-negative u-semi-axis) of the elements of the image set. The assumption of compactness of the sets $R \cap \text{lev}_{=\alpha} f$, instead of $R \cap \text{lev}_{\le \alpha} f$, should be too restrictive; for instance, take $R = \mathbb{R},\ f(x) = 1$ if $x \ne 0,\ f(0) = 0$.

Example 3.2.6. In (1.1.1) set $n = m = 1,\ p = 0,\ X = \mathbb{R},\ f(x) = x^2(x-1)^2$ if $x \ne 1$, $f(1) = 1,\ g(x) = x$. At $\bar{x} = 1$, we find:

$$\mathcal{K}_{\bar{x}} = \{(u, v) \in \mathbb{R}^2 : u = -v^4 + 2v^3 - v^2 + 1 \text{ if } v \ne 1,\ u = 0 \text{ if } v = 1\},$$

$$\mathcal{E}(\mathcal{K}_{\bar{x}}) = \{(u,v) \in \mathbb{R}^2 : u \leq -v^4+2v^3-v^2+1, v > 1\} \cap \{(u,v) \in \mathbb{R}^2 : u < 1, 0 < v \leq 1\} \cap$$

$$\cap \{(u,v) \in \mathbb{R}^2 : u \leq 1, v \leq 0\},$$

$$U_{\bar{x}} = \{(u,v) \in \mathbb{R}^2 : v = 0, u \in [0,1]\}.$$

Therefore, (3.2.23) are satisfied, while the set (3.2.25) is not compact, and $\mathcal{E}(\mathcal{K}_{\bar{x}})$ is not closed. $\qquad\qquad\qquad\qquad\qquad\qquad\qquad\qquad\qquad\qquad\qquad\qquad\qquad\qquad\square$

Theorem 3.2.3, as well as the 3 previous corollaries, are immediately extended to problem (1.1.4).

The previous IS Analysis is valid for constrained extremum problems having a finite dimensional image (see Fig. 1.1.1), typically problems (1.1.1) and (1.1.4). Things change, when the image of a problem is infinite dimensional, as it happens to (1.1.5); in fact, the IS becomes infinite dimensional and the extension of the previous analysis to the case where the IS has infinite dimension is, indeed, a concrete possibility. Of course, in such a case, we are immediately faced with some non-trivial difficulties. Here, following the Lagrange ideas (see Sect. 5.1), we adopt an approach which consists in postponing as long as possible the meeting with the infinite dimensionality [IV15,IV16]. This approach is now shortly outlined; it consists in introducing a multifunction approach, which allows us to circumvent the infinite dimensionality and to be reduced to handling finite dimensional sets in order to study the image of (1.1.4), and thus to exploit the previous scheme. In the sequel, we will use the results of Sect. 4.7.

As already remarked, the image of function x through g_i is again a function defined on T, namely $\tilde{\psi}_i$. The image of $\tilde{\psi}_i$ is a subset of $\mathbb{R}$. Hence, we can introduce the multifunction, which sends x into such a subset of $\mathbb{R}^\nu$, where $\nu := 1 + m$. Let $\bar{x} \in X$ be given, and consider the multifunction $A_{\bar{x}} : X \rightrightarrows Y \subseteq R^\nu$ defined by

$$A_{\bar{x}}(x) := \{(u,v) \in \mathbb{R} \times \mathbb{R}^m : u = f(\bar{x})-f(x) \text{ and } \exists t \in T : v_i = \psi_i(t, x(t), x'(t)), \quad i \in \mathcal{I}\}.$$

$\mathcal{K}_{\bar{x}}(X) := A_{\bar{x}}(X)$ will be called the *image* of (1.1.5). By means of the above definition we reduce ourselves to work in a finite dimensional image space, namely $\mathbb{R}^\nu$, as it has happened for (1.1.1), (1.1.4); however, for these problems $A_{\bar{x}}$ was merely a function, so that the image of x was an element of the IS $\mathbb{R}^\nu$; while now $A_{\bar{x}}$ is a multifunction and the image of x is a set of elements $\mathbb{R}^\nu$. Therefore, we must expect a more complex development than the previous one; however, the present approach has the advantage of postponing the infinite dimensionality to the introduction of the IS.

Let us consider any $\bar{x} \in X$. Obviously, $\bar{x}$ satisfying (1.1.5b) and (1.1.5c) is a global minimum point for (1.1.5) iff the system (in the unknown x):

$$f(\bar{x}) - f(x) > 0; \quad \psi_i(t, x(t), x'(t)) \left\{ \begin{array}{l} = 0, \ i \in \mathcal{I}^0 \\ \geq 0, \ i \in \mathcal{I}^+ \end{array} \right\}, \quad \forall t \in T, \ \forall x \in X \qquad (3.2.26)$$

is impossible; (3.2.26) corresponds to (3.2.1). Let $\mathcal{H}$ be as in (3.2.2). Notwithstanding this, unlike what has happened for (1.1.1) and (1.1.4), the optimality condition (3.2.26) cannot now be expressed in the form of a disjunction, like (3.2.2). In fact, while now

$\mathcal{K}_{\overline{x}}$ continues to be the image of the elements of X, it is a family of sets of the IS and no longer a set of points of the IS. Therefore, the equivalent set-formulation of (3.2.26) is no longer (3.2.2), but

$$A_{\overline{x}}(x) \not\subseteq \mathcal{H}, \quad \forall x \in X. \tag{3.2.27}$$

Since disjunction can be faced by means of separation (Sect. 2.2), to lose the format (3.2.2) and to be obliged to adopt (3.2.27) looks like a severe drawback, which prevents us from exploiting separation and then if undoes the advantage of still having a finite dimensional IS. Such a drawback can be overcome by selection. To this end, consider a vector-valued function $\Theta : 2^Y \to \mathbb{R}^\nu$. As next (3.2.28) shows, Θ allows us to select an element of $A_{\overline{x}}(x)$. This can be done in many ways. We adopt the following one. Consider the functions $\omega_i : T \to \mathbb{R}$, $i \in \mathcal{J}$; denote by Ω the set of vectors $\omega := (\omega_1, ..., \omega_m)$, whose elements are not all zero on T and s.t. $\omega_i \geq 0$, $i \in \mathcal{J}^+$; here Ω represents a class of functional parameters satisfying a suitable condition (depending on the assumptions made on the $\psi_i's$) to be rendered more precise, under which the integral in (3.2.28) makes sense. The selection in this case is obtained by means of the parameter ω, so that Θ is specified to be of type $\Phi : X \times \Omega \to \mathbb{R}^\nu$, defined, $\forall x \in X$, by:

$$\Phi(A_{\overline{x}}(x), \omega) := \int_{A_{\overline{x}}(x)} \omega dt := \left(f(\overline{x}) - f(x), \int_T \omega_i(t)\psi_i(t, x(t), x'(t))dt, i \in \mathcal{J} \right), \tag{3.2.28}$$

where the 1st integral is a short writing to mean selection of an element of $A_{\overline{x}}(x)$ by means of a weighted integration. Φ, as well as any other function Θ, is called *generalized selection function* (for short, GSF) when

$$A_{\overline{x}}(x) \subseteq \mathcal{H} \Leftrightarrow \Phi(A_{\overline{x}}(x), \omega) \in \mathcal{H}, \quad \forall \omega \in \Omega; \tag{3.2.29}$$

ω is called *selection mutiplier* (for short, SM).

It has been shown ([IV16],Theorem 5.1) that, under assumptions which make valid the so-called Fundamental Lemma of Calculus of Variations [5,24,V61], a function of type (3.2.28) is a GSF for (1.1.5a,c,d) with (1.1.5c) independent of x'. The extension of such a proposition to (1.1.5) will now be considered.

Theorem 3.2.4. In problem (1.1.5) let x be twice differentiable. Let $\psi_i \in L_2(R^{1+2n})$, $i \in \mathcal{J}^0$ be G-differentiable (G corresponds to $\mathcal{F}_{2n+1}$ defined before Theorem 3.1.1) with respect to every argument and at each of its values, with continuous G-derivative, and fulfil assumption (3.1.26) where f and x are replaced by ψ_i and (t, x, x'), respectively. Let $\psi_i \in L_2(R^{1+2n})$, $i \in \mathcal{J}^+$ be upper (or lower) G-semidifferentiable with respect to every argument and at each of its values, with continuous G-semiderivative, and fulfil assumption (3.1.26) where f and x are replaced by ψ_i and (t, x, x'), respectively. Then, $\forall x \in X$, Φ given by (3.2.28) with $\omega_i \in L_2(T)$ $i \in \mathcal{J}^+$ is a generalized selection function.

Proof. We have to show that (3.2.28) fulfils (3.2.29). $\forall x \in X$, if $A_{\overline{x}}(x) \subseteq \mathcal{H}$, i.e.

(3.2.26) holds, then, since $\omega_i \geq 0$, $i \in \mathfrak{I}^+$, $\forall \omega \in \Omega$ we have:

$$\int_T \omega_i(t)\psi_i(t,\,x(t),\,x'(t))dt \begin{cases} = 0, & \text{if } i \in \mathfrak{I}^0, \\ \geq 0, & \text{if } i \in \mathfrak{I}^+, \end{cases}$$

and hence $\Phi(A_{\overline{x}};\omega) \in \mathcal{H}$. Conversely, $\forall x \in X$ assume that $\Phi(A_{\overline{x}};\omega) \in \mathcal{H}$, $\forall \omega \in \Omega$. This means that $f(\overline{x}) - f(x) > 0$ and that:

$$\int_T \omega_i(t)\psi_i(t,\,x(t),\,x'(t))dt \begin{cases} = 0, & \text{if } i \in \mathfrak{I}^0, \\ \geq 0, & \text{if } i \in \mathfrak{I}^+, \end{cases} \quad \forall \omega \in \Omega.$$

Ab absurdo, suppose that $A_{\overline{x}}(x) \nsubseteq \mathcal{H}$, so that either $\exists r \in \mathfrak{I}$ and $\exists \tau \in T$ such that $\psi_r(\tau,\,x(\tau),\,x'(\tau)) < 0$, or $\exists r \in I^0$ and $\tau \in T$, such that $\psi_r(\tau,\,x(\tau),\,x'(\tau)) > 0$. In the former case, because of the assumptions we can apply Theorem 3.1.11 to the function $\Psi(t) := \psi_r(t,\,x(t),\,x'(t))$ (Ψ corresponds to $A_{\overline{x}}$; the triplet $(t,\,x,\,x')$ to x; $\psi_r(\bullet,\,\bullet,\,\bullet)$ to $f(\bullet)$), which turns out to be upper $\mathcal{P}$-semidifferentiable; let its expansion be given by:

$$\Psi(t) = \Psi(\tau) + \overline{\mathcal{D}}_{\mathcal{P}}\Psi(\tau; t - \tau) + \varepsilon_\Psi(\tau; t - \tau), \tag{3.2.30}$$

where $\Psi(\tau) < 0$, and where $\overline{\mathcal{D}}_{\mathcal{P}}\Psi$ and ε_Ψ fulfil (3.1.15). Because of (3.1.15a), $\forall \delta > 0$ $\exists t',\,t'' \in T$, with $t' \leq \tau \leq t''$ and $t' < t''$, such that:

$$\varepsilon_\Psi(\tau; t - \tau) < \delta|t - \tau|, \quad \forall t \in [t',t'']. \tag{3.2.31}$$

Let us consider, first of all, the case $t'' > \tau$ (if $t'' = \tau$, we have to consider only the case $t' < \tau$); we find:

$$\int_\tau^{t''} \varepsilon_\Psi(\tau; t - \tau)dt < \frac{\delta}{2}(t'' - \tau)^2. \tag{3.2.32}$$

Because of the positive homogeneity, $\exists \alpha \in \mathbb{R}$ such that:

$$\overline{\mathcal{D}}_{\mathcal{P}}\Psi(\tau;\, t - \tau) = \alpha(t - \tau), \quad \forall t \in [\tau, t''],$$

so that:

$$\int_\tau^{t''} [\Psi(\tau) + \overline{\mathcal{D}}_{\mathcal{P}}\Psi(\tau; t - \tau)]dt = \Psi(\tau)(t'' - \tau) + \frac{\alpha}{2}(t'' - \tau)^2. \tag{3.2.33}$$

Using (3.2.30), (3.2.32), and (3.2.33) we obtain:

$$\int_\tau^{t''} \Psi(t)dt = \int_\tau^{t''} [\Psi(\tau) + \overline{\mathcal{D}}_{\mathcal{P}}\Psi + \varepsilon_\Psi]dt <$$

$$\tag{3.2.34}$$

$$< \Psi(\tau)(t'' - \tau) + \frac{\alpha + \delta}{2}(t'' - \tau)^2 < 0,$$

where the last inequality holds iff

$$\delta < -2\Psi(\tau)/(t'' - \tau) - \alpha. \tag{3.2.35}$$

Of course t'' depends on δ; $\quad \forall \delta > 0 \ \exists t''$ such that (3.2.31) holds; any smaller $t'' > \tau$ makes (3.2.31) valid a fortiori. The right-hand side of (3.2.25) $\to +\infty$ as $t'' \downarrow \tau$; hence (3.2.34) is satisfied if δ and $t'' - \tau$ are small enough.

Let us consider now the case $t' < \tau$; we find:

$$\int_{t'}^{\tau} \varepsilon_\Psi(\tau; \ t - \tau)dt < \frac{\delta}{2}(t' - \tau)^2.$$

Inasmuch as $\exists \beta \in \mathbb{R}$ such that $\overline{\mathcal{D}}_\mathcal{P}\Psi(\tau; \ t - \tau) = \beta(t - \tau)$, $\forall t \in [t', \tau]$, we have:

$$\int_{t'}^{\tau} [\Psi(\tau) + \overline{\mathcal{D}}_\mathcal{P}\Psi(\tau; t - \tau)]dt = \Psi(\tau)(\tau - t') - \frac{\beta}{2}(t' - \tau)^2,$$

so that

$$\int_{t'}^{\tau} \Psi(t)dt = \int_{t'}^{\tau} [\Psi(\tau) + \overline{\mathcal{D}}_\mathcal{P}\Psi + \varepsilon_\Psi]dt < \Psi(\tau)(\tau - t') - \frac{(\beta - \delta)}{2}(t' - \tau)^2 < 0,$$

where the last inequality holds iff

$$\delta < \frac{2\Psi(\tau)}{t' - \tau} + \beta; \tag{3.2.36}$$

the right-hand side of (3.2.36) $\to +\infty$ as $t' \uparrow \tau$. Thus we can conclude that, if $t'' - t'$ is small enough, we have:

$$\int_{t'}^{t''} \Psi(t)dt < 0.$$

Hence, by choosing $\omega_r(t) = 1$ on $[t', t'']$ and zero elsewhere, we obtain:

$$\int_{T} \omega_r(t)\psi_r(t, \ x(t), \ x'(t))dt < 0,$$

which contradicts the non-negativity of $\omega_r(t)$ and the r-th of (1.1.5b,c). It follows that:

$$\psi_r(t, \ x(t), \ x'(t)) \begin{cases} = 0, \ \text{if } i \in \mathcal{J}^0 \\ \geq 0, \ \text{if } i \in \mathcal{J}^+ \end{cases}, \quad \forall t \in T.$$

In the latter case, because of the assumptions, we can apply Corollary 3.1.1 to the composition $\Psi(t) = \psi_r(t, \ x(t), \ x'(t))$ that is now $\mathcal{P}$-differentiable at τ, so that $\forall \delta > 0$ $\exists t', \ t'' \in T$, with $t' < \tau < t''$, such that:

$$\varepsilon_\Psi(\tau; t - \tau) > -\delta|t - \tau|, \quad \forall t \in [t', t''].$$

By operating in the same manner as above, we obtain that, for $t'' - t'$ small enough, it is

$$\int_{t'}^{t''} \Psi(t)dt > 0.$$

This inequality once again leads in an obvious way to the absurdity. $\qquad\square$

Note that, instead of (3.2.29), we can consider the equivalent relationship:

$$A_{\overline{x}}(x) \not\subseteq \mathcal{H} \Leftrightarrow \exists \omega \in \Omega, \quad \text{such that} \quad \Phi(x;\omega) \notin \mathcal{H}. \tag{3.2.37}$$

This shows the dependence of the SM on x, while in (3.2.29) such a dependence is hidden: in (3.2.29) we have to show "something" for every $\omega \in \Omega$, while in (3.2.37) we have to prove the existence of an ω, which of course varies with x. Such a dependence does not appear in Theorem 3.2.4, since it exploits (3.2.29). However the SM depends on x in the general case. This does not appear explicitly in the Calculus of Variations literature.

Theorem 3.2.4 is a general statement; starting from it, several aspects can be carried on and deepened. A function Φ, even if it is a GSF — as stated by Theorem 3.2.4 — may not enjoy desired properties. One of these is continuity. Now, we will shortly show an instance of some results which can be achieved to deepen the analysis. To this end consider the following:

Condition 3.2.1. There exists a continuous function $\alpha : \mathbb{V} \to \mathbb{R}^{\nu}$, such that:

$$\alpha(x) \in A_{\overline{x}}(x) \backslash \mathcal{H}, \quad \forall x \in N(\overline{x}) \tag{3.2.38}$$

where $N(\overline{x})$ is a neighbourhood of $\overline{x} \in \mathbb{V}$.

The above function α will play an important role in the selection approach. Let us introduce the sets:

$$\mathcal{K}_{\overline{x}}(\alpha) := \{(u,v) \in A_{\overline{x}}(X) : (u,v) = \alpha(x), \, x \in X\},$$

$$\mathcal{K}_{\overline{x},N}(\alpha) := \{(u,v) \in A_{\overline{x}}(X) : (u,v) = \alpha(x), \, x \in N(\overline{x})\},$$

where α does not necessarily satisfies condition (3.2.38); the former is a selection of the image set of (1.1.5) and is a set of points of the IS (like the image set $\mathcal{K}_{\overline{x}}$ of (1.1.1) and (1.1.4)); the latter is a local version of the former. Condition (3.2.38) is obviously equivalent to:

$$\mathcal{H} \cap \mathcal{K}_{\overline{x},N}(\alpha) = \varnothing, \tag{3.2.39}$$

which, apart from the local aspect, is of the same type of (3.2.2).

We observe that the existence of a not necessarily continuous function α is necessary and sufficient for $\overline{x}$ to be a minimum point. To suppose the continuity of α allows us to have a further tool in order to develop the analysis. Set $\alpha = (\alpha_0, ... \alpha_m)$. It is simple to prove the following:

Proposition 3.2.10. (3.2.39) holds if and only if $\overline{x}$ is a local minimum point of the problem:

$$\max \alpha_0(x), \quad \text{s.t.} \quad \alpha_i(x) \geq 0, \quad i \in \mathfrak{I}, \quad x \in X. \tag{3.2.40}$$

Proof. $\overline{x}$ is a local m.p. of (3.2.40) iff the system

$$\alpha_0(x) - \alpha_0(\overline{x}) > 0, \quad \alpha_i(x) \geq 0, \quad i \in \mathfrak{I}, \quad x \in N(\overline{x})$$

is impossible. Since $\alpha_0(\overline{x}) = 0$, the impossibility of the previous system is equivalent to (3.2.39). $\square$

Since, given $x \in X$, the first component of the vector $(u, v) \in A_{\overline{x}}(X)$ is uniquely defined, then necessarily it is

$$\alpha_0(x) = f(\overline{x}) - f(x).$$

A direct consequence of Proposition 3.2.10 is the following result.

Proposition 3.2.11. Suppose that Condition 3.2.1 holds. Then, any necessary optimality condition for (3.2.39) is a necessary condition for $\overline{x}$ to be a local minimum point of the given problem (1.1.5a,c,d).

A fundamental aspect of our analysis lies in the possibility of considering well-behaved functions α which fulfil Condition 3.2.1. The function α is a local continuous selection of $A_{\overline{x}}(x)$ in a neighbourhood of $\overline{x}$. Later we will consider suitable assumptions which ensure that Condition 3.2.1 be fulfilled.

We are now in the position to contruct two GSF of type Θ. One, which will allow us to recover classic results of Calculus of Variations, has been announced with (3.2.28). Before considering this, let us construct a GSF, which is interesting even if it does not correspond to classic developments.

Proposition 3.2.11. Let $\overline{x}$ be a minimum point of (1.1.5a,c,d), and

$$\psi_i : C^1(T) \to C^0(T), \quad i \in \{O\} \cup \mathfrak{I},$$

be continuous in $N(\overline{x})$. Then, the function

$$\Phi(x) := \left\{ f(\overline{x}) - f(x), \ \min_{t \in T} \psi_1(t, x(t), x'(t)), \ ..., \ \min_{t \in T} \psi_m(t, x(t), x'(t)) \right\}$$

is a GSF, which is continuous in $N(\overline{x})$.

Proof. We have to show that the functions:

$$f(\overline{x}) - f(x) \quad \text{and} \quad \min_{t \in T} \psi_i(t, x(t), x'(t)), \quad i \in \mathfrak{I},$$

are continuous in $N(\overline{x})$. Let $\tilde{x} \in N(\overline{x})$, and $\varepsilon > 0$; we must prove the inequality:

$$\left| \int_T [\psi_0(t, \tilde{x}(t)) - \psi_0(t, x(t), x'(t))] dt \right| < \varepsilon, \quad \forall x \in N(\tilde{x}),$$

where $N(\tilde{x})$ is a neighbourhood of $\tilde{x}$. Since ψ_0 is continuous at $\tilde{x}$, $\exists \delta > 0$, s.t. the inequality

$$\sup_{t \in T}|\tilde{x}(t) - x(t)| < \delta \tag{3.2.41}$$

implies

$$\sup_{t \in T} |\psi_0(t, \tilde{x}(t), \tilde{x}'(t)) - \psi_0(t, x(t), x'(t))| < \frac{\varepsilon}{b - a}.$$

Hence, for each x which fulfils (3.2.41), we have:

$$\left| \int_T [\psi_o(t, \tilde{x}(t), \tilde{x}'(t)) - \psi_0(t, x(t), x'(t))]dt \right| < \int_T \frac{\varepsilon}{b - a}dt = \varepsilon.$$

Let $i \in \mathfrak{I}$. Consider the inequality:

$$\left| \min_{t \in T} \psi_i(t, \tilde{x}(t), \tilde{x}'(t)) - \min_{t \in T} \psi_i(t, x(t), x'(t)) \right| < \varepsilon. \tag{3.2.42}$$

ψ_i being continuous at $\tilde{x}$, $\exists \overline{\delta} > 0$, s.t. the inequality

$$\sup_{t \in T} |\tilde{x}(t) - x(t)| < \overline{\delta} \tag{3.2.43}$$

implies:

$$\sup_{t \in T} |\psi_i(t, \tilde{x}(t), \tilde{x}'(t)) - \psi_i(t, x(t), x'(t))| < \varepsilon.$$

Let

$$\psi_i(\overline{t}, \tilde{x}(\overline{t}), \tilde{x}'(\overline{t})) = \min_{t \in T} \psi_i(t, \tilde{x}(t), \tilde{x}'(t)),$$

$$\psi_i(t^0, x(t^0), x'(t^0)) = \min_{t \in T} \psi_i(t, x(t), x'(t)).$$

We have, $\forall t \in T$,

$$\psi_i(\overline{t}, \tilde{x}(\overline{t}), \tilde{x}'(\overline{t})) - \psi_i(t, x(t), x'(t)) \leq \psi_i(t, \tilde{x}(t), \tilde{x}'(t)) - \psi_i(t, x(t), x'(t)) < \varepsilon,$$

so that:

$$\psi_i(\overline{t}, \tilde{x}(\overline{t}), \tilde{x}'(\overline{t})) < \psi_i(t^0, x(t^0), x'(t^0)) + \varepsilon.$$

Similarly, $\forall t \in T$,

$$\psi_i(t^0, x(t^0), x'(t^0)) - \psi_i(t, \tilde{x}(t), \tilde{x}'(t)) \leq \psi_i(t, x(t), x'(t)) - \psi_i(t, \tilde{x}(t), \tilde{x}'(t)) < \varepsilon,$$

so that:

$$\psi_i(\overline{t}, \tilde{x}(\overline{t}), \tilde{x}'(\overline{t})) > \psi_i(t^0, x(t^0), x'(t^0)) - \varepsilon.$$

Hence, for each x which fulfils (3.2.43), we have that (3.2.42) is satisfied. $\qquad \square$

Now, we will consider the other particular case of Θ, namely (3.2.28), for which Ω and its element ω will receive here a more detailed form. The function α is now defined as:

$$\alpha_i(x) = \int_T \omega_i(t, x)\psi_i(t, x, x')dt, \quad i \in \mathfrak{I}, \tag{3.2.44}$$

where $\omega_i : T \times X \to \mathbb{R}$ $i \in \mathfrak{I}$, and $\omega := (\omega_1, ..., \omega_m) \in \Omega$, Ω being a given class of parameters. Without any fear of confusion, for the sake of simplicity, the previous set $\mathcal{K}_{\bar{x}}(\alpha)$ is now denoted by $\mathcal{K}(\omega)$; hence we now understand the dependence on $\bar{x}$ and that on α is restricted to ω; this avoids the cumbersome presence of an integral as argument of $\mathcal{K}$. Thus we consider the set:

$$\mathcal{K}(\omega) := \bigcup_{x \in X} \{\Phi(A_{\bar{x}}(x); \omega)\} =$$

$$= \{(u, v_1, ..., v_m) \in \mathbb{R} \times \mathbb{R}^m : u = f(\bar{x}) - f(x), v_i = g_i(x; \omega_i), \quad x \in X, \quad i \in \mathfrak{I}\},$$

where

$$g_i(x; \omega_i) := \int_T \omega_i(t, x)\psi_i(t, x, x')dt.$$

$\mathcal{K}(\omega)$ will be called the *selected image*, and will play the same role as $\mathcal{K}_{\bar{x}}$ in (3.2.2). More precisely, $A_{\bar{x}}$ is now a multifunction and $A_{\bar{x}}(x)$ is a set, not necessarily a singleton. Thus, as previously said, the optimality for (1.1.5a,c,d) cannot be expressed by a disjunction of $\mathcal{H}$ and $\mathcal{K} = A_{\bar{x}}(X)$ as in (3.2.2). However, by selecting an element from $A_{\bar{x}}(x)$ or from its convex hull, say conv $A_{\bar{x}}(x)$ (this means to select $\mathcal{K}(\omega)$ from $\mathcal{K}_{\bar{x}}$), we may hope to reduce ourselves to the scheme of (3.2.2). The infinite dimensionality of the image is overcome by the selection: instead of considering the image of (1.1.5a,c,d), which would lead us to an infinite dimensional image space, we introduce the multifunction $A_{\bar{x}}$, so that we have a finite dimensional image space, where the scheme (3.2.2) can be adopted by replacing $\mathcal{K}_{\bar{x}}$ with $\mathcal{K}(\omega)$. The selected element from $A_{\bar{x}}(x)$ is $A_{\bar{x}}(x; \omega) := (f(\bar{x}) - f(x), g_i(x; \omega_i), \quad i \in \mathfrak{I})$; hence $\bar{z} := A_{\bar{x}}(\bar{x}; \omega)$ is the selected image of $\bar{x}$ and will play a role quite analogous to that played by $\bar{z}$ in (3.2.2) where $A_{\bar{x}}(x)$ was a singleton.

As in (3.2.2), the analysis will be carried out within the class of $\mathcal{C}$-differentiable functions (see Sect. 3.1); in the sequel the $\mathcal{C}$-derivative will be always assumed to be bounded (with respect to the 2nd argument). In the present situation, there is a further difficulty: the $\mathcal{C}$-differentiability must be enjoyed by $f(x)$ and $g_i(x; \omega_i)$, $i \in \mathfrak{I}$, and it should be unsuitable to make an assumption on f, $g_1, ..., g_m$; it is more appropriate that any assumption is made on the given data ψ_i and on the selection multiplier ω_i. To this end, we will need Theorem 3.1.7, where the general case of SM depending on the unknown x is considered. In fact, the following examples show the need of enlarging the class of SM from $\omega_i(t)$ to $\omega_i(t; x)$; X is the set of continuous functions. Next example shows that this may happen also when X is made by C^1 functions.

Example 3.2.7. In (1.1.5) set $T = [0, 3], p = 0, m = 1; \psi_0(t, x, x') = x; \psi_1(t, x, x') = t^2 - 3t + 2 + x(t); \bar{x}(t) = 3t - t^2 + 2, \quad \forall t \in T; \ X = C^1(T)$. Now consider the functions:

$$\hat{x}(t) = \begin{cases} 0, & \text{if } 0 \leq t \leq 1 \\ (t-1)^2, & \text{if } 1 < t \leq \frac{3}{2} \\ -\frac{7}{9}t^2 + \frac{10}{3}t - 3, & \text{if } \frac{3}{2} < t \leq 3 \end{cases} \qquad \tilde{x}(t) = \begin{cases} -\frac{7}{9}t^2 + \frac{4}{3}t, & \text{if } 0 \leq t \leq \frac{3}{2} \\ (t-2)^2, & \text{if } \frac{3}{2} < t \leq 2 \\ 0, & \text{if } 2 < t \leq 3. \end{cases}$$

Neither $\hat{x}$ nor $\tilde{x}$ fulfil (1.1.5c), so that $(\omega \equiv \omega_1, \, \psi \equiv \psi_1)$:

$$A_{\overline{x}}(\hat{x}) \not\subseteq \mathcal{H} = (\mathbb{R}_+ \setminus \{0\}) \times \mathbb{R}_+; \quad A_{\overline{x}}(\tilde{x}) \not\subseteq \mathcal{H}.$$

Hence, $\Phi(A_{\overline{x}}(\hat{x}); \omega) \notin \mathcal{H}$ and $\Phi(A_{\overline{x}}(\tilde{x}); \omega) \notin \mathcal{H}$ must be true for some SM ω. Assume that ω does not depend on $x(t)$. Since $f(\overline{x}) - f(\hat{x})$ and $f(\overline{x}) - f(\tilde{x})$ are obviously positive, $\Phi(F(\hat{x}); \omega) \notin \mathcal{H}$ and $\Phi(A_{\overline{x}}(\tilde{x}); \omega) \notin \mathcal{H}$ are equivalent to $(\psi(t, x, x') \equiv \psi(t, x))$:

$$\int_T \omega(t)\psi(t, \hat{x})dt = \int_0^1 \omega(t)(t^2 - 3t + 2)dt + \int_1^{3/2} \omega(t)(2t^2 - 5t + 3)dt +$$

$$(3.2.45a)$$

$$+ \int_{\frac{3}{2}}^3 \omega(t)(\frac{2}{9}t^2 + \frac{1}{3}t - 1)dt < 0$$

and

$$\int_T \omega(t)\psi(t, \tilde{x})dt = \int_0^{3/2} \omega(t)(\frac{2}{9}t^2 - \frac{5}{3}t + 2)dt +$$

$$(3.2.45b)$$

$$+ \int_{\frac{3}{2}}^2 \omega(t)(2t^2 - 7t + 6)dt + \int_2^3 \omega(t)(t^2 - 3t + 2)dt < 0,$$

respectively. Let $S_1 \subseteq [1, \frac{3}{2}]$, $S_2 \subseteq [\frac{3}{2}, 2]$ be s.t. $\omega(t) = 0$ on $[1, \frac{3}{2}] \setminus S_1$ and on $[\frac{3}{2}, 2] \setminus S_2$. Since $\psi(t, \hat{x}(t))$ and $\psi(t, \tilde{x}(t))$ are continuous and are negative, respectively, only on $]1, \frac{3}{2}[$ and on $]\frac{3}{2}, 2[$, it follows that (3.2.45) hold only if:

$$\int_{S_1} \omega(t)(2t^2 - 5t + 3)dt + \int_{S_2} \omega(t)(\frac{2}{9}t^2 + \frac{1}{3}t - 1)dt < 0, \qquad (3.2.46a)$$

$$\int_{S_1} \omega(t)(\frac{2}{9}t^2 - \frac{5}{3}t + 2)dt + \int_{S_2} \omega(t)(2t^2 - 7t + 6)dt < 0. \qquad (3.2.46b)$$

These inequalities, summing them up side by side, imply: $5 \int_{S_1 \cup S_2} \omega(t)(\frac{2}{3}t - 1)^2 dt < 0$,

which is evidently false, being $\omega(t) \geq 0, \quad \forall t \in T$. Therefore, with a $SM \, \omega$ independent of x, we cannot have, at the same time, both $\Phi(A_{\overline{x}}(\hat{x}); \omega) \notin \mathcal{H}$ and $\Phi(A_{\overline{x}}(\tilde{x}); \omega) \notin \mathcal{H}$. Note that the above $\overline{x}$ is merely a feasible solution of (1.1.5) and not an optimal one. $\square$

Example 3.2.8. In (1.1.5a,c,d), let us set $m = 2, \psi_0 = -x_1(t), \psi_2 \equiv 1$, and

$$\psi_1 = \begin{cases} x_2(t), & \text{if } x_1 \in X_1^- := \{x_1 : \int_T x_1(t)dt \leq 0\}, \\ x_2(t) - x_1(t)^2, & \text{if } x_1 \in X_1^+ := \{x_1 : \int_T x_1 dt > 0\}, \text{ and} \\ & \quad x_2 \in X_2^- := \{x_2 : \int_T x_2 dt \leq 0\} \\ x_2(t) + x_1(t)^2, & \text{if } x_1 \in X_1^+, \, x_2 \in X_2^+ := \{x_2 : \int_T x_2 dt > 0\} \end{cases}$$

with $x_1, x_2 \in C^0(T)$. The selected problem, namely (1.1.5a,c,d) where (1.1.5b) is replaced by:

$$g_i(x; \omega_i) \geq 0, \quad i \in \mathfrak{I}, \qquad (3.2.47)$$

becomes now:

$$\min \int_T - x_1(t)dt, \quad \text{s.t.} \quad g_2(x;\omega_2) = \int_T \omega_2(t)dt \geq 0, \text{and } g_1(x;\omega_1) \geq 0,$$

where:

$$g_1(x;\omega_1) = \begin{cases} \int_T \omega_1(t)x_2(t)dt, & x_1 \in X_1^-, \\ \int_T \omega_1(t)x_2(t)dt - \int_T \omega_1(t)x_1(t)^2 dt, & x_1 \in X_1^+, \ x_2 \in X_2^-, \\ \int_T \omega_1(t)x_2(t)dt + \int_T \omega_1(t)x_1(t)^2 dt, & x_1 \in X_1^+, \ x_2 \in X_2^+. \end{cases}$$

Note that $\psi_1 \equiv 0 \Leftrightarrow x_1 \in X_1^-$ and $x_2 \equiv 0$. In fact:

$$x_1 \in X_1^+ \text{ and } x_2 - x_1^2 \equiv 0 \Rightarrow x_2 \in X_2^+; \quad x_1 \in X_1^+ \text{ and } x_2 + x_1^2 \equiv 0 \Rightarrow x_2 \in X_2^-.$$

It follows that $(x_1 \in X_1^-, \ x_2 \equiv 0)$ are the only admissible pairs, and hence $\bar{x} = (\bar{x}_1 \equiv 0, \bar{x}_2 \equiv 0)$ is the unique m.p. of (1.1.5a,c,d). $\qquad \square$

Example 3.2.9. Let us identify (1.1.5a,c,d) with:

$$\min \int_T \cos x(t)dt, \quad \text{s.t.} \quad x(t) = 0, \quad \forall t \in T := [0, 1], \quad x(t) \in C^0(T).$$

Of course, $x(t) \equiv 0$ is the m.p. The selected problem, namely (1.1.5a,c,d) where (1.1.5b) is replaced by (3.2.47), is

$$\min \int_T \cos x(t)dt, \quad \text{s.t.} \quad \int_T \omega(t)x(t)dt = 0, \ x \in C(T),$$

where $\omega(t) \in C(T)$ is arbitrary. We prove that $x(t) \equiv 0$ is not solution of the selected problem. To this end, it is enough to show that the selected problem admits a not identically zero feasible solution. Note that we have: $\int_T \cos x(t)dt \leq 1$. Therefore, we look for a solution of type

$$x(t) = at + b, \quad a \neq 0, b \neq 0,$$

$$\int_T \omega(t)x(t)dt = [W(t)x(t)]_0^1 - \int_0^1 W(t)x'(t)dt,$$

where $W(t)$ is an antiderivative (or primitive) of $\omega(t)$ ($W(t)$ exists, since $\omega \in C(T)$). Hence, we must have:

$$0 = \int_T \omega(t)x(t)dt = W(1)(a + b) - W(0)b - a\int_T W(t)dt.$$

If a and b are chosen in such a way to have:

$$a\left[W(1) - \int_T W(t)dt\right] + b[W(1) - W(0)] = 0, \quad a, b \neq 0,$$

then we obtain that $x(t) = at + b \not\equiv 0$. Now, let us evaluate the objective function in the found point:

$$\int_T \cos(at+b)dt = \frac{2}{a}\sin\frac{a}{2}\cos\frac{a+2b}{2}.$$

If the minimum were 1, then we should have:

$$\sin\frac{a}{2}\cos\frac{a+2b}{2} = \frac{a}{2}.$$

By choosing $\frac{a}{2} \notin [-1,1]$, the previous equality is false. $\qquad\square$

Example 3.2.10. Let us identify (1.1.5a,c,d) with:

$$\int_T x(t)dt, \quad \text{s.t.} \quad x^2(t) \le 0, \quad \forall t \in T := [0,1], \ x \in C^0(T).$$

Of course, $x(t) \equiv 0$ is the unique feasible (and hence optimal) solution. The selected problem, namely (1.1.5a,c,d) where (1.1.5c) is replaced by (3.2.47), is

$$\min\int_T x(t)dt, \quad \text{s.t.} \quad \int_T \omega(t)x^2(t)dt \le 0, \ x \in C^0(T),$$

where $\omega(t) \in C^0(T)$. If we choose $\omega(t) > 0, \forall t \in T$, then necessarily $x(t) \equiv 0$ to have $x(t)$ admissible for the selected problem; this, therefore, turns out to be equivalent to the given problem. $\qquad\square$

The preceeding examples show that, in the general case, we cannot make true the proposition:

$$A_{\bar{x}}(x) \not\subseteq \mathcal{H} \Leftrightarrow \exists \omega \in \Omega, \quad \text{s.t.} \quad \Phi(A_{\bar{x}}(x);\omega) \notin \mathcal{H}, \qquad (3.2.29)'$$

if the elements of Ω are independent of x.

3.3. Stationarity

Often, the direct search for a m.p. of a function on its (open or not) domain or of its restriction to a given subset of the domain — like in (1.1.6) — is impracticable. In other words, it is difficult, in general, to work only on the set of m.p. of (1.1.6). Hence, it is suitable to introduce a superset of such a set; of course, it should be as small as possible and the search for its elements should be easier than that of m.p. To belong to such a superset is, of course, a necessary condition for being a m.p. To make these rough ideas precise within a rather wide context is one of the aims of the present book; it has been one of the main aims of the theory of extrema from its beginning. We will start with a function on a generic domain, and later (1.1.6) will be considered (recall that B and $\mathcal{B}$ denote Banach spaces).

Definition 3.3.1. Let Y be a subset of B, and $f : B \to \mathcal{B}$. $\bar{x} \in Y$ is called a *lower semistationary point* of f on Y, if and only if

$$\liminf_{\substack{x \to \overline{x} \\ x \in Y \setminus \{\overline{x}\}}} \frac{f(x) - f(\overline{x})}{||x - \overline{x}||} \geq 0, \tag{3.3.1}$$

where $||\cdot||$ is a norm in B. $\overline{x} \in Y$ is an *upper semistationary point* of f on Y if and only if it is a lower semistationary point of $-f$ on Y. $\overline{x} \in Y$ is a *stationary point* of f on Y, if and only if it is both lower and upper semistationary; in this case we have:

$$\lim_{\substack{x \to \overline{x} \\ x \in Y \setminus \{\overline{x}\}}} \frac{f(x) - f(\overline{x})}{||x - \overline{x}||} = 0. \tag{3.3.2}$$

The function which appears in both (3.3.1) and (3.3.2) is called *generalized difference quotient*.

Definition 3.3.1 is motivated by the following property. If the norm in B is not generated by a scalar product (B is not a Hilbert space), then, of course, $\langle f'(\overline{x}), x - \overline{x} \rangle$ is understood as the application of a continuous linear functional (from the continuous dual of B) to the variation $x - \overline{x}$; f' in (3.3.3) and (3.3.4) becomes the Fréchet derivative, and in (3.3.5) becomes the Gâteaux one. Note that (ii) holds in the more general case of $\mathcal{C}$-differentiable; (3.3.3) becomes $\mathcal{D}_{\mathcal{C}}f(\overline{x}; x - \overline{x}) \geq 0$, $\forall x \in R$; in the proof it is enough to replace the "scalar product" with the $\mathcal{C}$-derivative of f.

Theorem 3.3.1. **(i)** If $\overline{x}$ is a local minimum point of (1.1.6), then (3.3.1) with $Y{=}R$ holds. **(ii)** If f is differentiable at $\overline{x}$, then (3.3.1) becomes:

$$\langle f'(\overline{x}), x - \overline{x} \rangle \geq 0, \quad \forall x \in R, \tag{3.3.3}$$

which, if $\overline{x} \in \text{int } R$, collapses to:

$$f'(\overline{x}) = 0, \tag{3.3.4}$$

where, in case of (1.1.4)-(1.1.5), f' denotes the first variation of functional f. **(iii)** If R and f are convex, then (3.3.1) becomes:

$$f'(\overline{x}; x - \overline{x}) \geq 0, \quad x \in R, \tag{3.3.5}$$

f' denoting directional derivative in the direction $x - \overline{x}$, and every lower semistationary point of f on R is a global minimum point.

Proof. **(i)** It is not restrictive to assume that R be not a singleton. By assumption, there exists a neighbourhood of $\overline{x}$, say $N(\overline{x})$, such that $f(x) \geq f(\overline{x})$ $\forall x \in R \cap N(\overline{x})$, so that $[f(x) - f(\overline{x})]/||x - \overline{x}|| \geq 0$ $\forall x \in R \cap N(\overline{x}) \setminus \{\overline{x}\}$, and therefore (3.3.1) at $Y = R$ holds. **(ii)** We have $f(x) = f(\overline{x}) + \langle f'(\overline{x}), x - \overline{x} \rangle + \varepsilon(\overline{x}; x - \overline{x})$, where ε is an infinitesimal of higher order than $||x - \overline{x}||$. Consider the set:

$$S := \{d \in B : ||d|| = 1, \exists \tau \in \mathbb{R}_+ \setminus \{0\} \text{ s.t. } \overline{x} + td \in R, \quad \forall t \in [0, \tau]\}.$$

(3.3.1) at $Y{=}R$ becomes:

$$\sup_{\substack{\tau>0}} \inf_{\substack{d\in S \\ t\in]0,\tau]}} [\langle f'(\overline{x}), d\rangle + \frac{1}{t}\varepsilon(\overline{x}; td)] \geq 0. \tag{3.3.6}$$

(3.3.3) follows, if we show that $(3.3.6) \Rightarrow \ell := \inf_{d\in S} \langle f'(\overline{x}), d\rangle \geq 0$. Ab absurdo, suppose that $\ell < 0$. Then, $\forall \eta > 0 \ \exists d_\eta \in S$, such that $\langle f'(\overline{x}), d_\eta\rangle < \ell + \eta$. Without any loss of generality, assume that $\ell + \eta < 0$. Select $\delta \in]0, -(\ell + \eta)[$. If τ is positive and small enough, because of the differentiability, we have $\frac{1}{t}\varepsilon(\overline{x}; td) < \delta, \quad \forall d \in S, \quad \forall t \in]0,\tau]$, and in particular at $d = d_\eta$. Therefore,

$$\langle f'(\overline{x}), d_\eta\rangle + \frac{1}{t}\varepsilon(\overline{x}; td_\eta) < \ell + \eta + \delta < 0, \quad \forall t \in]0,\tau],$$

which shows that the infimum of the expression within the square brackets of (3.3.6) is less than a negative constant. This condradicts (3.3.6). If $\overline{x} \in \text{int}\,R, \ \exists \varepsilon > 0$ small enough, such that $\overline{x} \pm \varepsilon f'(\overline{x}) \in R$, so that, at $x = \overline{x} \pm f'(\overline{x})$, (3.3.3) becomes $0 \leq \langle f'(\overline{x}), \pm f'(\overline{x})\rangle = \pm ||f'(\overline{x})||^2$ and implies (3.3.4). **(iii)** To achieve (3.3.5), it is enough to repeat the proof of (ii) with the following changes: the expansion of f is now $f(x) = f(\overline{x}) + f'(\overline{x}; x - \overline{x}) + \varepsilon(\overline{x}; x - \overline{x})$; the scalar product which appears in (3.3.6) and consequently in the sequel must be replaced by $f'(\overline{x}; d)$ and later by $f'(\overline{x}; d_\eta)$. Now, to show that $\overline{x}$ is a global minimum point, suppose, ab absurdo, that $\exists \hat{x} \in R$, such that $f(\hat{x}) < f(\overline{x})$. This fact and the convexity of f imply:

$$\frac{1}{||\alpha(\hat{x} - \overline{x})||}[f(\overline{x} + \alpha(\hat{x} - \overline{x})) - f(\overline{x})] \leq \frac{1}{||\hat{x} - \overline{x}||}[f(\hat{x}) - f(\overline{x})] < 0, \quad \forall \alpha \in]0, 1].$$

From this we easily draw that the left-hand side of (3.4.1) is less than or equal to $\frac{f(\hat{x})-f(\overline{x})}{||\hat{x}-\overline{x}||} < 0$; this contradicts (3.3.1) at $Y=R$. $\qquad\qquad\square$

Note that (3.3.3), which obviously implies the non-negativity of the directional derivative like (3.3.5), is a special case of a Variational Inequality (see Vol. 2). Theorem 3.3.1(i) is a quite general necessary optimality condition; however, its application may be impracticable, especially when R contains nonlinear constraining equations and we are unable to prove the existence of the implicit function (by using, e.g., Dini Theorem) defined by them. This is the reason why Lagrange was led to remove the constraining equation from R and add the corresponding functions g_i to the objective function f, after having multiplied each of them by an indeterminate coefficient (called multiplier). Surely, this approach was suggested by mechanical applications in the smooth case, where f,g are potentials and their gradients are forces. In the next chapter, a general approach will be considered, which is based on the Lagrange idea; with the following proposition, we take a step towards it.

Let us consider (1.1.6) at $\mathcal{B} = \mathbb{R}$, namely in the case of the finite dimensional image, and associate it with the function

$$L(x; \theta, \lambda) := \theta f(x) - \langle \lambda, g(x)\rangle, \quad \theta \in \mathbb{R}, \quad \lambda \in \mathbb{R}^m;$$

at $\theta = 1$, L is the classic linear Lagrangian function (the term "linear" will be clear later); to avoid cumbersome notation, in Sect. 5.2 $L(x; 1, \lambda)$ will be replaced merely by $L(x; \lambda)$. As in Sect. 3.2, $\overline{z}$ is the image of $\overline{x}$ through the map A (see after (3.2.2)); the homogenization $\mathcal{K}^h_{\overline{x}}$ of the image set $\mathcal{K}_{\overline{x}}$ has been defined in Sect. 3.2 as well as the polar $(\mathcal{K}^h_{\overline{x}} - \overline{z})^*$.

Theorem 3.3.2. (semistationariness of L). Let f and $-g$ be $\mathcal{C}$-differentiable. **(i)** If and only if

$$-(\theta, \lambda) \in (\mathcal{K}^h_{\overline{x}} - \overline{z})^*, \tag{3.3.7}$$

then

$$\liminf_{\substack{x \to \overline{x} \\ x \in X \setminus \{\overline{x}\}}} \frac{L(x; \theta, \lambda) - L(\overline{x}; \theta, \lambda)}{||x - \overline{x}||} \geq 0. \tag{3.3.8}$$

If, in addition, the $\mathcal{C}$-derivatives of f and g_i, $i \in \mathcal{I}$, are bounded from above by a constant in a neighbourhood of $\overline{x}$, or $B = \mathbb{R}$, then $\liminf$ collapses to $\lim$. **(ii)** If $\overline{x} \in \operatorname{int} X$ and f, g_i $i \in \mathcal{I}$ are differentiable (so that $\mathcal{C}$ can be reduced to the set of linear functionals), then (i) becomes: if and only if

$$-(\theta, \lambda) \in (\mathcal{K}^h_{\overline{x}} - \overline{z})^{\perp}, \tag{3.3.7'}$$

then

$$L'_x(\overline{x}; \theta, \lambda) = 0. \tag{3.3.8'}$$

Proof. **(i)** (3.3.7) is equivalent to:

$$-(\theta, \lambda) \in \{(u^*, v^*) \in \mathbb{R} \times \mathbb{R}^m : u^* u + \langle v^*, v - v^* \rangle \geq 0, \quad \forall (u, v) \in \mathcal{K}^h_{\overline{x}}\},$$

or

$$\theta \mathcal{D}_{\mathcal{C}} f(\overline{x}; d) - \sum_{i \in \mathcal{I}} \lambda_i \mathcal{D}_{\mathcal{C}} g_i(\overline{x}; d) \geq 0, \quad \forall d \in X - \overline{x}.$$

Since $\mathcal{C}$ is closed under addition, the left-hand side of the last inequality is the $\mathcal{C}$-derivative of L; namely, $\mathcal{D}_{\mathcal{C}} L(\overline{x}; d; \theta, \lambda) = \theta \cdot \mathcal{D}_{\mathcal{C}} f(\overline{x}; d) - \sum_{i \in \mathcal{I}} \lambda_i \cdot \mathcal{D}_{\mathcal{C}} g_i(\overline{x}; d)$. It follows that, at $x \neq \overline{x}$, (3.3.7) is equivalent to:

$$\frac{1}{||d||} \mathcal{D}_{\mathcal{C}} L(\overline{x}; d; \theta, \lambda) \geq 0, \quad \forall d \in (X - \overline{x}) \setminus \{O\}.$$

This inequality, by adding to both sides the quantity:

$$\frac{1}{||d||} \overline{\varepsilon}(\overline{x}; d; \theta, \lambda) := \frac{1}{||d||} \left[\theta \varepsilon_f(\overline{x}; d) - \sum_{i \in \mathcal{I}} \lambda_i \varepsilon_{g_i}(\overline{x}; d) \right],$$

ε_f and ε_{g_i} being the infinitesimals of the expansions, respectively, of f and g_i, is equivalent to (recall that $d = x - \overline{x}$):

$$\frac{1}{||d||} [L(x; \theta, \lambda) - L(\overline{x}; \theta, \lambda)] \geq \frac{1}{||d||} \overline{\varepsilon}(\overline{x}; d; \theta, \lambda), \quad \forall x \in X.$$

Since $\bar{\varepsilon}$ is an infinitesimal as $d \to 0$, the above inequality implies (3.3.8); the vice versa follows, since we can prove that (3.3.8) implies the non-negativity of $\mathcal{D}_{\mathcal{C}}L(\bar{x}; d; \theta, \lambda)$ by reasoning in a quite similar way as for (ii) of Theorem 3.3.1. The remaining part is obvious. **(ii)** Since now $\mathcal{K}_{\bar{x}}^{h}$ is an affine manifold, the polar of $\mathcal{K}_{\bar{x}}^{h} - \bar{z}$ becomes its orthogonal complement, and hence $\liminf$ collapses to $\lim$, and this is zero, as both $\geq$ and $\leq$ must hold. $\square$

The concept of stationarity expressed by (3.3.1) is equivalent to that mentioned in [V91, page 58]. In fact, the latter requires the existence of a neighbourhood $N(\bar{x})$ of $\bar{x}$ and a function $\varepsilon : X \times (X - \bar{x}) \to \mathbb{R}$, with $\lim\limits_{x \to \bar{x}} \varepsilon(\bar{x}; x - \bar{x})/\|x - \bar{x}\| = 0$, such that:

$$f(x) \geq f(\bar{x}) + \varepsilon(\bar{x}; x - \bar{x}), \quad \forall x \in X \cap N(\bar{x}).$$

When $x \neq \bar{x}$, this inequality is equivalent to:

$$[f(x) - f(\bar{x})]/\|x - \bar{x}\| \geq \varepsilon(\bar{x}; x - \bar{x})/\|x - \bar{x}\|, \quad \forall x \in X \cap N(\bar{x}),$$

and hence to (3.3.1).

Note that the statement (ii) of Theorem 3.3.1 holds, more generally, for the case where the functions (in particular the Lagrangian function of Theorem 3.3.2) are $\mathcal{C}$-differentiable. Taking into account this fact, it is immediate to recognize that (3.3.8) can be written as:

$$\mathcal{D}_{\mathcal{C}}L(\bar{x}; d; \theta, \lambda) \geq 0, \quad \forall d \in X - \bar{x}; \tag{3.3.8}''$$

in other words: iff (3.3.7) holds, then the $\mathcal{C}$-derivative of the Lagrangian function is non-negative along every d.

If we want to deepen the analysis, it is suitable to distinguish between a finite dimensional image — problems (1.1.1) and (1.1.4) — and an infinite dimensional one — problem (1.1.5) —. The former case will be treated extensively in Chapter 5. Now, we will briefly give an example of how we propose to analyse the latter one; for the sake of simplicity, we will consider the case of unilateral constraints, namely (1.1.5a,c,d).

Examples 3.2.7-3.2.10 show to need, in the general case, to consider a selection multiplier (SM) depending on x. As a consequence of this, the vector of multipliers (θ, λ), introduced in Theorem 3.3.2, must be considered as a function of x, which will turned out to be factorized as a product of a constant and a factor depending on x; the former will correspond to separate the selected image $\mathcal{K}(\omega)$ of (1.1.5a,c,d) (see (3.2.44)) and $\mathcal{H}$; the latter corresponds to the SM (see (3.2.29)). In other words, by means of the selection, (1.1.5a,c,d) is reduced to the case of finite dimensional image. Such an approach allows us to avoid the introduction of an infinite dimensional IS, and postpone the infinite dimensionality, which is limited to the selection (see Sect. 3.2). Such an approach will be furtherly clarified in Chapter 5.

We will assume the $\mathcal{C}$-differentiability of $\psi_0, -\psi_i$, $i \in \mathcal{J}$ with respect to the set of 2nd and 3rd arguments of ω_i with respect to the 2nd argument, and will assume that all the hypotheses of Theorem 3.2.4 be satisfied. As a consequence, we will have the following expansion (for the sake of simplicity, in the sequel $\overline{x}$ will be replaced merely by x):

$$f(x + \delta x) = f(x) + \int_T \mathcal{D}_{\mathcal{C}}\psi_0(t, x, x'; \delta x, \delta x')dt + \int_T \varepsilon_{\psi_0}(t, x, x'; \delta x, \delta x')dt, \quad (3.3.9a)$$

$$g_i(x + \delta x; \omega_i) = g_i(x; \omega_i) + \int_T \mathcal{D}_{\mathcal{C}}\pi_i(t, x, x'; \delta x, \delta x')dt+$$

$$\int_T \varepsilon_i^{\pi}(t, x, x'; \delta x, \delta x')dt, \quad i \in \mathcal{I}, \quad (3.3.9b)$$

where

$$\pi_i := \omega_i \cdot \psi_i; \quad \mathcal{D}_{\mathcal{C}}\pi_i := \mathcal{D}_{\mathcal{C}}\omega_i(t, x; \delta x) \cdot \psi_i(t, x, x') + \omega_i(t, x) \cdot \mathcal{D}_{\mathcal{C}}\psi_i(t, x, x'; \delta x, \delta x');$$

$$\varepsilon_i^{\pi} := \varepsilon_{\omega_i} \cdot \varepsilon\psi_i + \varepsilon_{\omega_i} \cdot [\psi_i(t, x, x') + \mathcal{D}_{\mathcal{C}}\psi_i] + \varepsilon_{\psi_i} \cdot [\omega_i(t, x) + \mathcal{D}_{\mathcal{C}}\omega_i(t, x, \delta x)] + \mathcal{D}_{\mathcal{C}}\omega_i \cdot \mathcal{D}_{\mathcal{C}}\psi_i;$$

and where the pairs $(\mathcal{D}_{\mathcal{C}}\omega_i, \varepsilon_{\omega_i})$, $(\mathcal{D}_{\mathcal{C}}\psi_i, \varepsilon_{\psi_i})$ give the expansions of ω_i, ψ_i, respectively. Since $\mathcal{D}_{\mathcal{C}}$ is an operator which denotes $\mathcal{C}$-derivative, the use of $\mathcal{D}_{\mathcal{C}}\pi_i$ as a symbol would be improper; this does not happen here since π_i is $\mathcal{C}$-differentiable due to Theorem 3.1.7. When ω_i and ψ_i are differentiable $\mathcal{D}_{\mathcal{C}}\pi_i$ collapses to the usual derivative of a product. If ω_i is constant with respect to x, so that can be denoted by $\omega_i(t)$, then $\mathcal{D}_{\mathcal{C}}\pi_i = \omega_i(t) \cdot \mathcal{D}_{\mathcal{C}}\psi_i$ and $\varepsilon_i^{\pi} = \varepsilon_{\psi_i} \cdot [\omega_i(t, x) + \mathcal{D}_{\mathcal{C}}\omega_i(t, x, \delta x)]$.

For the sake of simplicity, we will assume that Condition 3.2.1 be fulfilled by a selection function α, where:

$$\alpha_i(x) = \int_T \omega_i(t)\psi_i(t, x, x')dt, \quad i \in \mathcal{I}, \quad (3.3.10)$$

for $x \in N(\overline{x})$, a neighbourhood of x. We observe that (3.3.10) coincides with (3.2.44) except for the fact that the parameters ω_i do not depend on x, but only on t. Next theorem is a consequence of the above assumptions and of the results stated in Sect. 3.2.

Theorem 3.3.3. Assume that Condition 3.2.1 be fulfilled and that $\alpha_i(x)$ be defined by (3.3.10), $i \in \mathcal{I}$. If the system

$$f(\overline{x}) - f(x) > 0; \quad \psi_i(t, x(t), x'(t)) \geq 0, \quad i \in \mathcal{I}, \quad \forall t \in T, \quad x \in X, \quad (3.3.11a)$$

is impossible (for $x \in X \cap N(\overline{x})$), then the following system is also impossible:

$$f(\overline{x}) - f(x) > 0, \quad g(x, \omega) \geq 0, \quad x \in X \cap N(\overline{x}) \quad (3.3.11b)$$

Proof. It follows from Proposition 3.2.10 taking into account (3.3.10). $\qquad\square$

Next theorem extends Proposition 3.2.6(iii).

Theorem 3.3.4. (Homogenization). Assume that Condition 3.2.1 be fulfilled and that $\alpha_i(x), i \in \mathcal{I}$, be defined by (3.3.10). Let ψ_0 and $-\psi_i$, $i \in \mathcal{I}$ be $\mathcal{C}$-differentiable with respect to the set of the 2nd and 3rd arguments. If $\overline{x}$ is a minimum point of (1.1.5a,c,d), then there exists a non-negative SM $\overline{\omega}(t) = (\overline{\omega}_i(t), \ i \in \mathcal{I}) \in C^0(T)^m$ and a neighbourhood in the sense of closeness of order one (the norms of the difference of any two elements and of that of their derivatives are small enough), say $N^{(1)}(\overline{x})$, such that the system (in the unknown $\delta = x - \overline{x}; \ \delta\overline{x}' = x' - \overline{x}'$):

$$\int_T \mathcal{D}_e \psi_0(t,\overline{x},\overline{x}';\delta\overline{x},\delta\overline{x}')dt < 0; \quad \int_T \overline{\omega}_i(t) \cdot \mathcal{D}_{-e}\psi_i(t,\overline{x},\overline{x}';\delta\overline{x},\delta\overline{x}')dt > 0, \quad i \in \mathfrak{I}^0(\omega),$$

$$(3.3.12)$$

$$g_i(\overline{x};\overline{\omega}_i) + \int_T \overline{\omega}_i(t) \cdot \mathcal{D}_{-e}\psi_i(t,\overline{x},\overline{x}';\delta\overline{x},\delta\overline{x}')dt \geq 0, \quad i \in \mathfrak{I}\backslash\mathfrak{I}^0(\omega); \quad x \in X \cap N^{(1)}(\overline{x}),$$

is impossible, where $\mathfrak{I}^0(\omega) := \{i \in \mathfrak{I} : g_i(\overline{x};\overline{\omega}_i) = 0, \int_T \overline{\omega}_i(t) \cdot \varepsilon_i(t,\overline{x},\overline{x}';\delta\overline{x},\delta\overline{x}')dt \not\equiv 0\}$.

Proof. By applying Theorem 3.3.3, we get the existence of $\overline{\omega}$ such that (3.3.11) is impossible. Now, ab absurdo, suppose that, at the same $\omega = \overline{\omega}$, (3.3.12) be possible, and let $\hat{x} \neq \overline{x}$ be a solution. Then $\alpha\hat{x}$ is a solution of (3.3.12) $\forall\alpha \in]0,1]$, since $g_i(\overline{x};\overline{\omega}_i) \geq 0$ and $\mathcal{D}_e f$, $\mathcal{D}_{-e}\psi_i$, $i \in \mathfrak{I}$ are positively homogeneous of 1st degree (see (3.2.15b)). The assumption implies that the remainders:

$$\int_T \varepsilon_{\psi_0}dt, \quad \int_T \overline{\omega}_i\varepsilon_i dt, \quad i \in \mathfrak{I}$$

are infinitesimal of order > 1 with respect to $||(\delta\overline{x},\delta\overline{x}')||$, so that, setting $\hat{y} = (\hat{x},\hat{x}')$, $\overline{y} = (\overline{x},\overline{x}')$ and $\delta\overline{y} = (\hat{x} - \overline{x}, \hat{x}' - \overline{x}') = (\delta\overline{x},\delta\overline{x}')$, $\exists\hat{\alpha} \in]0,1]$ such that:

$$\frac{1}{||\hat{\alpha}\delta\overline{y}||}\int_T \varepsilon_{\psi_0}(t,\overline{y};\hat{\alpha}\delta\overline{y})dt < -\frac{1}{||\delta\overline{y}||}\int_T \mathcal{D}_e f(t,\overline{y};\delta\overline{y})dt,$$

$$\frac{1}{||\hat{\alpha}\delta\overline{y}||}\int_T \overline{\omega}_i(t) \cdot \varepsilon_i(t,\overline{y};\hat{\alpha}\delta\overline{y})dt > -\frac{1}{||\delta\overline{y}||}\int_T \overline{\omega}_i(t) \cdot \mathcal{D}_{-e}\psi_i(t,\overline{y};\delta\overline{y})dt, \quad i \in \mathfrak{I}^0(\omega).$$

From these inequalities, by noting that $g_i(\overline{x};\overline{\omega}_i) = 0, \quad \forall i \in \mathfrak{I}^0(\omega)$, we have:

$$\int_T [\mathcal{D}_e\psi_0(t,\overline{y};\hat{\alpha}\delta\overline{y}) + \varepsilon_f(t,\overline{y};\hat{\alpha}\delta\overline{y})]dt < 0, \quad (3.3.13a)$$

$$g_i(\overline{x};\overline{\omega}_i) + \int_T \{\overline{\omega}_i(t)[\mathcal{D}_{-e}\psi_i(t,\overline{y};\hat{\alpha}\delta\overline{y}) + \varepsilon_i(t,\overline{y}_i;\hat{\alpha}\delta\overline{y})]\}dt > 0, \quad i \in \mathfrak{I}^0(\omega). \quad (3.3.13b)$$

$\forall i \in \mathfrak{I}\backslash\mathfrak{I}^0(\omega)$ either $g_i(\overline{x};\overline{\omega}_i) = 0$ and $\int_T \overline{\omega}_i\varepsilon_i dt \equiv 0$ or $g_i(\overline{x};\overline{\omega}_i) > 0$. In the former case, with $\hat{\alpha} = 1$, we obviously have:

$$g_i(\overline{x};\overline{\omega}_i) + \int_T \{\overline{\omega}_i(t)[\mathcal{D}_{-e}\psi_i(t,\overline{y};\hat{\alpha}\delta\overline{y}) + \varepsilon_i(t,\overline{y};\hat{\alpha}\delta\overline{y})]\}dt \geq 0. \quad (3.3.13c)$$

In the latter case $\exists\alpha^0 \in]0,1]$ such that:

$$g_i(\overline{x};\overline{\omega}_i) + \int_T \overline{\omega}_i(t) \cdot \mathcal{D}_{-e}\psi_i(t,\overline{y};\alpha\delta\overline{y})dt > 0, \quad \forall\alpha \in]0,\alpha^0],$$

and thus $\exists\tilde{\alpha} \in]0,\alpha^0]$ such that:

$$\frac{1}{||\tilde{\alpha}\delta\overline{y}||}\int_T \overline{\omega}_i(t) \cdot \varepsilon_i(t,\overline{y};\tilde{\alpha}\delta\overline{y})dt \geq$$

$$\geq -\frac{1}{\alpha^0||\delta\overline{y}||}\left[g_i(\overline{x};\overline{\omega}_i) + \alpha^0\int_T \overline{\omega}_i(t) \cdot \mathcal{D}_{-e}\psi_i(t,\overline{y};\delta\overline{y})dt\right] \geq$$

$$\geq -\frac{1}{\alpha\|\delta\overline{y}\|}\left[g_i(\overline{x};\overline{\omega}_i) + \alpha\int_T \overline{\omega}_i(t)\cdot\mathcal{D}_{-\mathcal{C}}\psi_i(t,\overline{y};\delta\overline{y})dt\right], \quad \forall\alpha\in]0,\alpha^0],$$

where the 1st inequality holds since $\int_T \overline{\omega}_i\varepsilon_i\,dt$ is infinitesimal of order > 1 with respect to $\|(\delta\overline{x},\delta\overline{x}')\|$ and the 2nd side is fixed and negative, the 2nd inequality holds since the 2nd side is obviously the maximum of the 3rd on $]0,\alpha^0]$. With $\overline{\alpha} := \tilde{\alpha}$ it follows that:

$$g_i(\overline{x};\overline{\omega}_i) + \int_T \{\overline{\omega}_i(t)[\mathcal{D}_{-\mathcal{C}}\psi_i(t,\overline{y};\hat{\alpha}\delta\overline{y}) + \varepsilon_i(t,\overline{y};\hat{\alpha}\delta\overline{y})]\}dt \geq 0. \tag{3.3.13d}$$

Collecting all (3.3.13), recalling that $g_i(\overline{x};\overline{\omega}_i) = 0$, $i \in \mathcal{J}^0(\omega)$, and using the definition of the remainders ε_i, we obtain the possibility of system (3.3.11), and hence the contradiction. $\qquad\square$

As in Sect. 3.2, the impossibility of system (3.3.13) can be expressed as disjunction of the two sets of the IS associated to (1.1.5a,c,d). To this end introduce the sets:

$$\mathcal{H}(\omega) := \{(u,v) \in \mathbb{R}\times\mathbb{R}^m : u > 0; v_i > 0, i \in \mathcal{J}^0(\omega); v_i \geq 0, i \in \mathcal{J}\backslash\mathcal{J}^0(\omega)\};$$

$$\mathcal{K}^h(\omega) := \{(u,v) \in \mathbb{R}\times\mathbb{R}^m : u = -\int_T \mathcal{D}_{\mathcal{C}}f\,dt;$$

$$v_i = g_i(\overline{x};\omega_i) + \int_T \omega_i\mathcal{D}_{-\mathcal{C}}\psi_i dt, \; i \in \mathcal{J}; \; x \in X\}.$$

It is easily seen that the impossibility of system (3.3.12) holds iff (compare with (3.2.2))

$$\mathcal{H}(\omega) \cap \mathcal{K}^h(\overline{\omega}) = \varnothing.$$

Note that the system (3.3.12) is set up with the homogeneous parts of f and the selections g_i and hence $\mathcal{K}^h(\omega)$ represents the homogenization of the selected image $\mathcal{K}(\omega)$; $\mathcal{H}(\omega)$ simply follows the changes in the types of inequalities in going from (3.3.11a) to (3.3.12).

When ψ_0,ψ_1, $i \in \mathcal{J}$ are differentiable ($\mathcal{C}$ is replaced with its subset $\mathcal{L}$ of linear elements), then (3.3.12) becomes:

$$\int_T [\langle\nabla_x\psi_0, x - \overline{x}\rangle + \langle\nabla_{x'}\psi_0, x' - \overline{x}'\rangle]dt < 0;$$

$$\int_T \overline{\omega}_i[\langle\nabla_x\psi_i, x - \overline{x}\rangle + \langle\nabla_{x'}\psi_i, x' - \overline{x}'\rangle]dt > 0, \quad i \in \mathcal{J}^0(\omega);$$

$$g_i(\overline{x};\omega_i) + \int_T \overline{\omega}_i[\langle\nabla_x\psi_i, x - \overline{x}\rangle +$$

$$+\langle\nabla_{x'}\psi_i, x' - \overline{x}'\rangle]dt \geq 0, \quad i \in \mathcal{J}\backslash\mathcal{J}^0(\omega); \quad x \in X \cap N^{(1)}(\overline{x}),$$

and in this case Theorem 3.3.4 extends to problem (1.1.5a,c,d) a well known linearization lemma (see [V1]; such a lemma, which will be considered, in a more general form, in Chapter 5, consists in the validity of (3.2.17)). Note that Theorem 3.3.4 can be slightly sharpened by requiring differentiability or $\mathcal{C}$-differentiability only for those ψ_i such that $g_i(\overline{x};\overline{\omega}_i) = 0$ and continuity for the remaining ones. Theorem 3.3.4 can be generalized to semidifferentiable functions.

For problem (1.1.5a,c,d), the Lagrangian function, introduced for Theorem 3.3.2, becomes

$$L(x; \theta, \lambda, \omega) = \theta f(x) - \langle \lambda, g(x; \omega) \rangle, \quad (\theta, \lambda) \in \mathbb{R} \times \mathbb{R}^m, \quad \omega \in \Omega.$$

Note that, if we set $\lambda_i(t) := \lambda_i \cdot \omega_i(t)$, then the above L is the classic Lagrangian function associated to (1.1.5a,c,d) [24,V17,V61]. Hence, here the Lagrangian multiplier is splitted into two parts: a selection part, i.e. $\omega_i(t)$ — which, in a wider contest, becomes $\omega_i(t, x)$ — and a separation part, i.e. λ_i (see comments before (3.3.9)).

Now, we can particularize Theorem 3.3.2 to the present case (1.1.5a,c,d). Let $\overline{k}(\omega) := (0, g(\overline{x}; \omega)) := (\overline{u}, \overline{v}(\omega))$ a selection of the image of $\overline{x}$. Unlike before, $y := (x, x')$, $\overline{y} := (\overline{x}, \overline{y})$, $\delta \overline{y} := y - \overline{y}$.

Theorem 3.3.5.(semistationariness). Let ψ_0 be $\mathcal{C}$-differentiable and let ψ_i, $i \in \mathcal{I}$, be $(\text{-}\mathcal{C})$-differentiable with respect to the set of 2nd and 3rd arguments at any value of them. **(i)** If $\exists \overline{\omega} \in \Omega$ such that:

$$-(\theta, \lambda) \in [\mathcal{K}^h(\overline{\omega}) - \overline{k}(\overline{\omega})]^*, \tag{3.3.14}$$

then

$$\liminf_{\substack{x \to \overline{x} \\ x \in X \setminus \{\overline{x}\}}} \frac{L(x; \theta, \lambda, \overline{\omega}) - L(\overline{x}; \theta, \lambda, \overline{\omega})}{\|x - \overline{x}\|} \geq 0. \tag{3.3.15}$$

If $\lim_{\|\delta \overline{y}\| \downarrow 0} \mathcal{D}_{\mathcal{C}} \psi_0(t, \overline{y}; \frac{\delta \overline{y}}{\|\delta \overline{y}\|})$ and $\lim_{\|\delta \overline{y}\| \downarrow 0} \mathcal{D}_{-\mathcal{C}} \psi_i(t, \overline{y}; \frac{\delta \overline{y}}{\|\delta \overline{y}\|})$, $i \in \mathcal{I}$ exist, then the lower limit of (3.3.1) collapses to the ordinary limit. **(ii)** If $\overline{x} \in \text{int } X$ and $\psi_0, \psi_i \in \mathcal{I}$ are differentiable, then (i) becomes: if

$$-(\theta, \lambda) \in [\mathcal{K}(\overline{\omega}) - \overline{k}(\overline{\omega})]^{\perp},$$

then

$$L'_x(\overline{x}; \theta, \lambda, \overline{\omega}) = 0.$$

Proof. (i) (3.3.14) is equivalent to:

$$-(\theta, \lambda) \in \{(u^*, v^*) \in \mathbb{R} \times \mathbb{R}^m : \langle (u^*, v^*), (u - \overline{u}, v(\overline{\omega}) - \overline{v}(\overline{\omega})) \rangle \geq 0, \quad \forall (u, v(\overline{\omega})) \in \mathcal{K}(\omega) \},$$

or

$$\mathcal{D}_{\mathcal{C}} L(x; \delta \overline{x}; \theta, \lambda, \overline{\omega}) - \mathcal{D}_{\mathcal{C}} L(\overline{x}; \delta \overline{x}; \theta, \lambda, \overline{\omega}) \geq 0, \quad \forall x \in X, \tag{3.3.16}$$

where

$$\mathcal{D}_{\mathcal{C}} L = \int_T \left[\theta \mathcal{D}_{\mathcal{C}} \psi_0 - \sum_{i \in \mathcal{I}} \lambda_i \mathcal{D}_{-\mathcal{C}} \psi_i \right] dt.$$

Divide both sides of (3.3.16) by $\|\delta \overline{x}\|$ and add to them:

$$\frac{1}{\|\delta \overline{x}\|} \overline{\varepsilon}(\overline{x}; \delta \overline{x}; \theta, \lambda, \overline{\omega}) := \frac{1}{\|\delta \overline{x}\|} \int_T \left(\theta \varepsilon_{\psi_0} - \sum_{i \in \mathcal{I}} \lambda_i \varepsilon_i \right) dt;$$

then (3.3.16) becomes:

$$\frac{1}{||\delta\overline{x}||}[L(x;\theta,\lambda,\overline{\omega}) - L(\overline{x};\theta,\lambda,\overline{\omega})] \geq \frac{1}{||\delta\overline{x}||}\overline{\varepsilon}(\overline{x};\delta\overline{x};\theta,\lambda,\overline{\omega}), \quad \forall x \in X\backslash\{\overline{x}\}.$$

Now (3.3.15) follows, since $\overline{\varepsilon}/||\delta\overline{x}|| \to 0$ as $x \to \overline{x}$. The remaining part is obvious. **(ii)** Since $\mathcal{K}(\overline{\omega})$ is now affine, the polar becomes the orthogonal complement and therefore lim inf collapses to lim and this is zero since both $\geq$ and $\leq$ must hold. $\qquad\square$

3.4. Some Examples

In this section we develop some simple examples, which serve to illustrate some of the concepts introduced in Sect. 3.2, as well as in other sections. In several parts of the book, these examples will be considered again.

Example 3.4.1. In (1.1.1) set $X = \mathbb{R}^2$, $p = 0$, $m = 1$ (so that $\mathcal{I}^0 = \varnothing$, $\mathcal{I}^+ = \mathcal{I} = \{1\}$, $n = 2$, $x = (x_1, x_2)$) and $f(x) = x_1^2 x_2$, $g_1(x) = g(x) = x_2$. At $\overline{x} = O = (0,0)$ we find $\mathcal{K}_{\overline{x}} = \mathcal{K}_0 = \{(u,v) \in \mathbb{R}^2 : u = -x_1^2 x_2, \ v = x_2, \ (x_1,x_2) \in \mathbb{R}^2\} = \{(u,v) \in \mathbb{R}^2 : u = -\alpha v, \ \alpha \in \mathbb{R}_+\}$. The image of (1.1.1) is now a family of lines which form a double cone of $\mathbb{R}^2$; its closure is the union of 2nd and 4th quadrants (Fig. 3.4.1). Since (3.2.2)

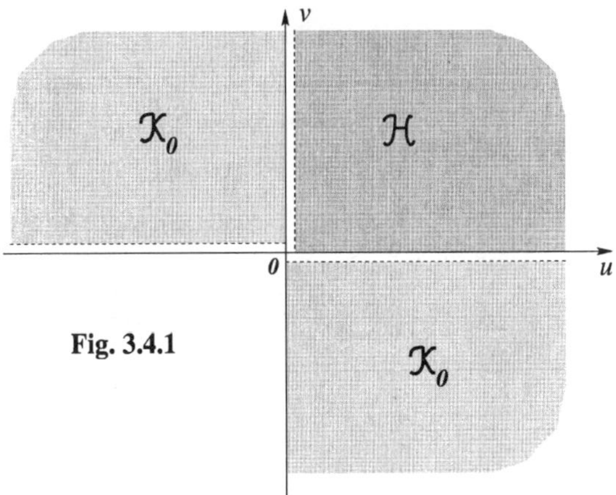

Fig. 3.4.1

is fulfilled, then $\overline{x}$ is m.p. of (1.1.1) (of course, this can be seen immediately by direct inspection). However, in spite of the apparent simplicity of the problem, the image set $\mathcal{K}_0$ is nonconvex and $\mathcal{K}_0$ and $\mathcal{H}$ are not separable, neither by a line nor by any smooth curve. Fig. 3.4.2 shows the conic extension of the image, which turns out to be the complement of $\mathcal{H}$. Now, let us consider the homogenization of $\mathcal{K}_{\overline{x}}$. According to Definition 3.2.2, we have (here $\mathcal{C} = \mathcal{L}$):

$$\mathcal{D}_{\mathcal{C}}f(\overline{x};d) = \mathcal{D}_{\mathcal{L}}f(\overline{x};d) = \langle f'(\overline{x}), d \rangle \equiv 0,$$

$$\mathcal{D}_{-e}g(\overline{x}; d) = \mathcal{D}_{\mathcal{L}}g(\overline{x}; d) = \langle g'(\overline{x}), d \rangle = d_2, \quad d \in \mathbb{R}^2.$$

Therefore we find:

$$\mathcal{K}_{\overline{x}}^h = \{(u, v) \in \mathbb{R}^2 : u = 0\},$$

which means that $\mathcal{K}_{\overline{x}}^h$ is the axis v of Fig. 3.4.1. It is easy to see that the tangent cone to $\mathcal{K}_{\overline{x}}$ at $\overline{x}$ (Definition 2.1.9) is given by the union of 2nd and 3rd quadrants, or

$$TC(\overline{x}; \mathcal{K}_{\overline{x}}) = \operatorname{cl} \mathcal{K}_{\overline{x}}.$$

Hence, we have that the homogenization is strictly contained into the tangent cone and is not a good representative of the image set. $\qquad \square$

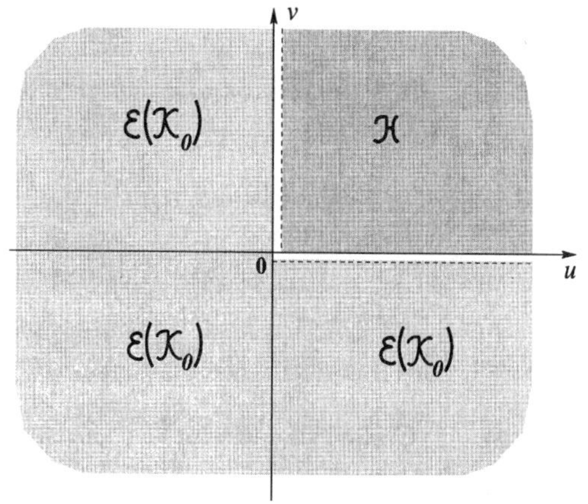

Fig. 3.4.2

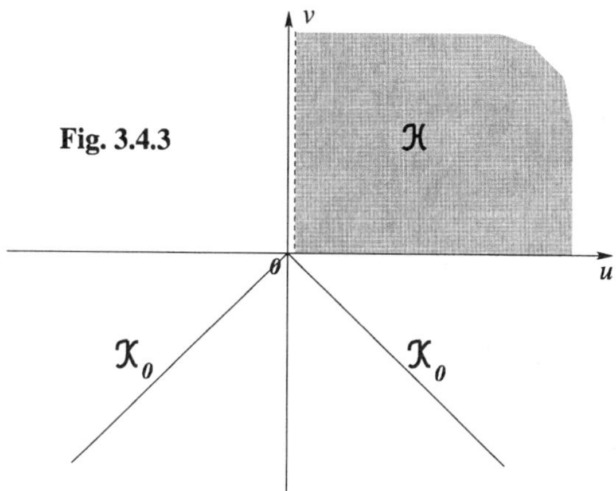

Fig. 3.4.3

Example 3.4.2. In (1.1.1) set $X = \mathbb{R}$, $p = 0$, $m = 1$ (so that $\mathcal{J}^0 = \varnothing$ $\mathcal{J}^+ = \mathcal{J} = \{1\}$, $n = 1$, and $f(x) = x$, $g(x) = -|x|$. At $\overline{x} = 0$ ($R = \{O\}$), we find $\mathcal{K}_0 = \{(u, v) \in \mathbb{R}^2 : v = -|u|\}$, which is shown in Fig. 3.4.3. Since we have chosen $\overline{x}$ as global m.p., then

(3.2.2) obviously holds. $\mathcal{H}$ and $\mathcal{K}_0$ are (linearly) separable; indeed, there are infinitely many separating lines even if $\mathcal{K}_0$ is not convex; however, its conic extension (which is shown in Fig. 3.4.4) is convex and this is the important property with respect to the linear separation. Now, let us consider the homogenization of $\mathcal{K}_{\overline{x}}$. According to Definition 3.2.2, we have:

$$\mathcal{D}_{\mathcal{C}}f(\overline{x};d) = \mathcal{D}_{\mathcal{L}}f(\overline{x};d) = f'(\overline{x}) \cdot d = d, \quad \mathcal{D}_{-\mathcal{C}}g(\overline{x};d) = -|d|, \quad d \in \mathbb{R}.$$

Therefore, we find:
$$\mathcal{K}_{\overline{x}}^h = \{(u,v) \in \mathbb{R}^2 : v = -|u|\}.$$

Since here the application of Definition 2.1.9 is trivial, we have:

$$\mathcal{K}_{\overline{x}}^h = TC(\overline{x}; \mathcal{K}_{\overline{x}}) \qquad\qquad \square$$

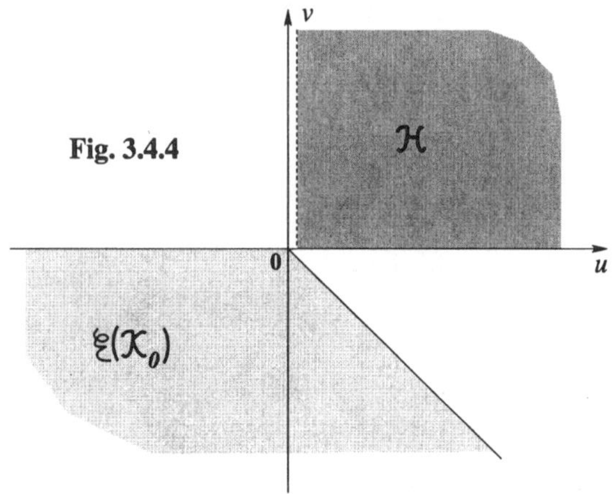

Fig. 3.4.4

Example 3.4.3. In (1.1.1) set $n = 2$, $p = 0$, $m = 1$, $f(x) = x_1 + 2x_2 + 3x_1(1 - x_1) + 3x_2(1-x_2)$, $g(x) = 2x_1+x_2-\xi$ with $\xi \in [0,3]$, and $X = \{x \in \mathbb{R}^2 : 0 \le x_i \le 1, i = 1,2\}$. Therefore, (1.1.1) is now a family of problems described by the parameter ξ. With a slight abuse of notation for introducing the dependence on the parameter ξ, the image of (1.1.1) (Sect. 3.2) is now:

$$\mathcal{K}_{\overline{x}}(\xi) = \{(u,v) \in \mathbb{R}^2 : u = f(\overline{x}) - x_1 - 2x_2 - 3x_1(1 - x_1) - 3x_2(1 - x_2),$$

$$v = 2x_1 + x_2 - \xi, \quad x_i \in [0,1], \ i = 1,2\} =$$

$$= \{(u,v) \in \mathbb{R}^2 : u = 3v^2 - (12x_1 + 5 - 6\xi)v + 15x_1^2 + (6 - 12\xi)x_1 + 3\xi^2 - 5\xi + f(\overline{x}),$$

$$2x_1 - \xi \le v \le 2x_1 - \xi + 1, \quad x_1 \in [0,1]\},$$

and hence it is a family of arcs of parabolae. As noted in Sect. 3.2, a change of $\overline{x}$ (or of ξ) gives rise to a translation of $\mathcal{K}_{\overline{x}}(\xi)$ in the direction of u-axis (or of v-axis). Even if the analysis we will perform is not affected by the choice of $\overline{x}$, due to the simplicity of the example we use $\overline{x}$ as global m.p. of (1.1.1), which is easily found by direct inspection:

$$\overline{x}(\xi) = \begin{cases} (\tfrac{\xi}{2},0) \,, & \text{if } 0 \le \xi \le \tfrac{2}{3}, \\ (\tfrac{1}{3},0),(1,0) \,, & \text{if } \xi = \tfrac{2}{3}, \\ (1,0) & \text{, if } \tfrac{2}{3} < \xi \le 2, \\ (1,\xi-2) & \text{, if } 2 < \xi \le 3, \end{cases} \qquad f(\overline{x}(\xi)) = \begin{cases} -\tfrac{3}{4}\xi^2 + 2\xi & \text{, if } 0 \le \xi < \tfrac{2}{3}, \\ 1 & \text{, if } \tfrac{2}{3} \le \xi \le 2, \\ -3\xi^2 + 17\xi - 21 & \text{, if } 2 < \xi \le 3. \end{cases}$$

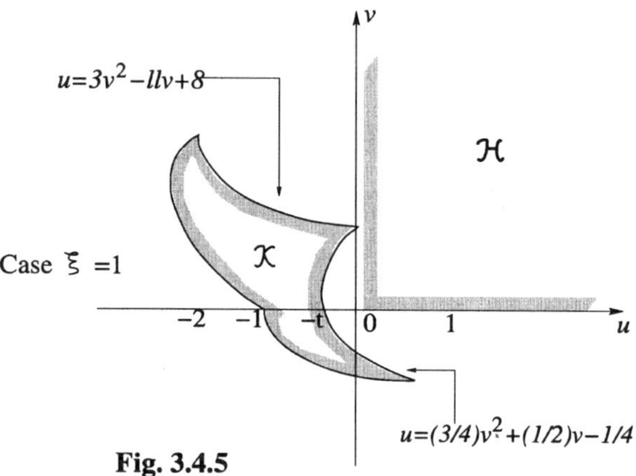

Fig. 3.4.5

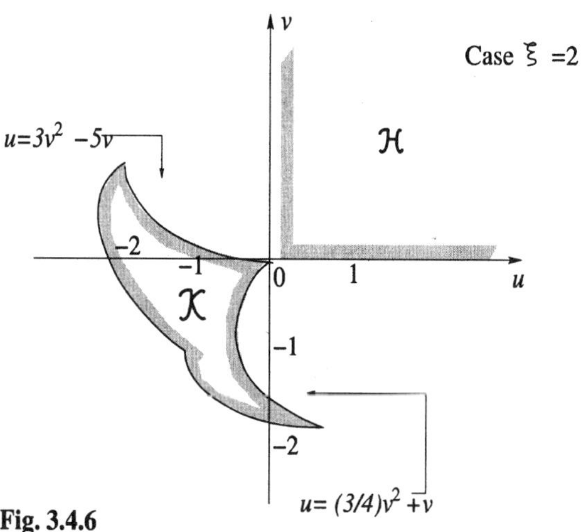

Fig. 3.4.6

Figs. 3.4.5-3.4.7. show $\mathcal{K}_{\overline{x}(\xi)}(\xi)$ at $\xi = 1,2,\tfrac{5}{2}$, respectively; in all three cases, $\overline{x}$ and $f(\overline{x})$ have been chosen from the above formulas, so that (3.2.2) obviously holds. In the 1st case, notwithstanding the nonconvexity of (1.1.1) and of the image, this set and $\mathcal{H}$ are linearly separable by infinitely many lines. The situation drastically changes after a mere translation of the constraint (2nd and 3rd case). Now let us consider the homogenization of the image set at $\xi = 1$, namely of $\mathcal{K}_{\overline{x}}$ at $\overline{x} = (1,0)$. According to

Definition 3.2.2, we have (here $\mathcal{C} = \mathcal{L}$):

$$\mathcal{D}_{\mathcal{C}} f(\overline{x}; d) = \mathcal{D}_{\mathcal{L}} f(\overline{x}; d) = \langle f'(\overline{x}), d \rangle = -2d_1 + 5d_2,$$

$$\mathcal{D}_{\mathcal{C}}\, g(\overline{x}; d) = \mathcal{D}_{\mathcal{L}} g(\overline{x}; d) = \langle g'(\overline{x}), d \rangle = 2d_1 + d_2.$$

Therefore we find:

$$\mathcal{K}^h_{\overline{x}} = \{(u, v) \in \mathbb{R}^2 : u = 2d_1 - 5d_2, \quad v = 1 + 2d_1 + d_2, \quad d_1 \in [-1, 0], \quad d_2 \in [0, 1]\} =$$

$$= \mathrm{conv}\{P_1, P_2, P_3, P_4\},$$

where $P_1 \equiv (0, 1)$, $P_2 \equiv (-5, 2)$, $P_3 \equiv (-27, 4)$, $P_4 \equiv (-22, 3)$; see Fig. 3.4.5bis, where

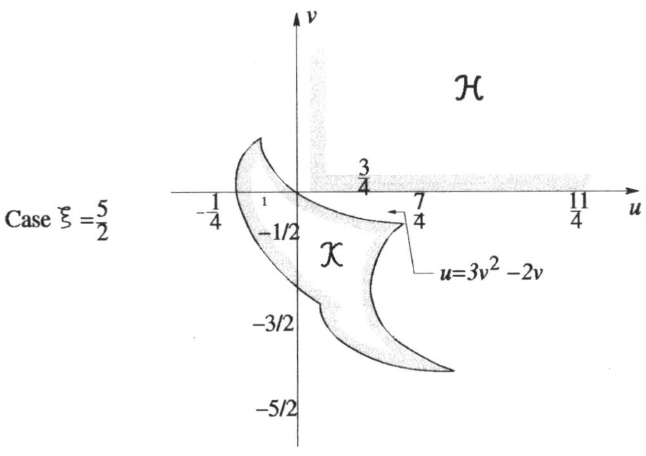

Fig. 3.4.7

the unit on v-axis is 3 times that on u-axis; thus $\mathcal{K}^h_{\overline{x}}$ is the parallelogram whose sides lie on the lines:

$$u = -5v + 5, \quad u = 17 - 11v, \quad u = -5v - 7, \quad u = 11 - 11v.$$

Since here X is a proper subset of $\mathbb{R}^2$, then $\mathcal{K}^h_{\overline{x}}$ is not a cone; however, in a small enough neighbourhood of P_1 it is a truncated cone. Furthermore, $\mathcal{K}^h_{\overline{x}}$ is strictly contained into the tangent cone, which, according to Definition 2.1.9, is given by:

$$TC(\overline{x}; \mathcal{K}_{\overline{x}}) = \{(u, v) \in \mathbb{R}^2 : u \leq -5v + 5, \, u \leq v - 1\},$$

and is represented, in Fig. 3.4.5bis, by the angle with apex P_1 and sides the rays r_1 and r_2. This example continues in Example 5.2.5. $\qquad\qquad\square$

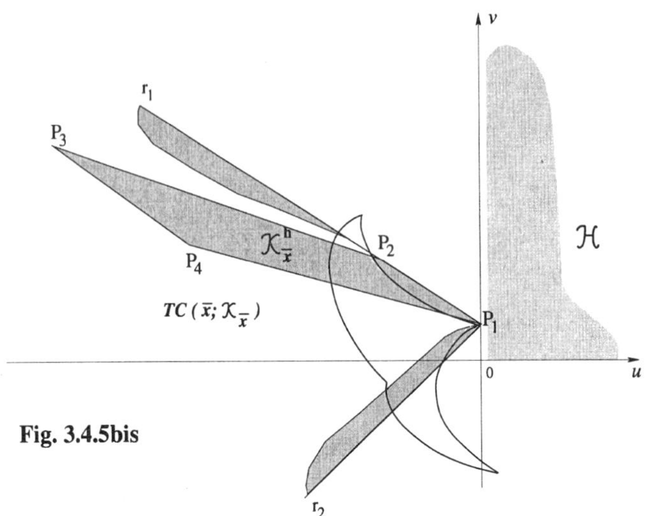

Fig. 3.4.5bis

Example 3.4.4. In (1.1.4) set $n=1$, $p=m=1$, $t_0 = 0$, $t_1 = 1$, $X = \{x \in C^1([0,1]) :$ $x(0) = x(1) = 1/2\}$, $\psi_0 = \sqrt{1 + x'(t)^2}$, $\psi = x(t) - \frac{2+\pi}{8}t$, so that (1.1.4) is a classic isoperimetric problem (like the 3rd of Examples 1.2.2; here, the area being constrained to a small enough value, a parametric representation of the unkown curve is not necessary):

$$\min f(x) = \int_0^1 \sqrt{1 + x'(t)^2}\,\mathrm{d}t, \tag{3.4.1a}$$

subject to

$$g(x) = \int_0^1 x(t)\mathrm{d}t - \frac{2 + \pi}{8} = 0, \quad x \in X. \tag{3.4.1b}$$

Due to the simplicity of the problem, a global m.p. of (3.4.1) is easily recognized to be $\bar{x}(t) = \sqrt{-t^2 + t + 1/4}$, so that

$$f(\bar{x}) = \frac{\sqrt{2}}{2} \int_0^1 \frac{\mathrm{d}t}{\sqrt{-t^2 + t + 1/4}} = \frac{\sqrt{2}}{4}\pi.$$

As noted in Sect. 3.2 and also in Example 3.4.3, in order to construct the image of a problem, the knowledge of a m.p. $\bar{x}(t)$ is unessential, in the sense that the important properties of the image set are not affected by the choice of $\bar{x}$. However, to simplify the analysis, we have chosen the m.p. System (3.2.1) becomes:

$$u = \frac{\sqrt{2}}{4}\pi - \int_0^1 \sqrt{1 + x'(t)^2}\,\mathrm{d}t > 0, \quad v = \int_0^1 x(t)\,\mathrm{d}t - \frac{2 + \pi}{8} = 0, \quad x \in X, \tag{3.4.2}$$

and $\mathcal{K}_{\bar{x}}$ is the set of (u, v) s.t. (3.4.2) hold. To determine $\mathcal{K}_{\bar{x}}$, it is useful the perturbation function (3.2.5); D is now the origin of $\mathbb{R}$, so that $v = \xi \in \mathbb{R}$. Thus, for the sake of simplicity, here $u_{\bar{x}}(\xi, D)$ is denoted merely by $u(v)$. Suitable but elementary calculation leads to the parametric equations of $u(v)$:

$$\begin{cases} u = \varphi_u(\alpha) = \begin{cases} \frac{\sqrt{2}}{4}\pi - \sqrt{4\alpha^2 + 1}\,\arctan\frac{1}{2\alpha}, & \text{if } \alpha > 0, \\ \frac{\sqrt{2}-2}{4}\pi & , \text{if } \alpha = 0, \end{cases} \\[2mm] v = \varphi_v(\alpha) = \begin{cases} \frac{2-4\alpha-\pi}{8} + \frac{4\alpha^2+1}{4}\,\arctan\frac{1}{2\alpha}, & \text{if } \alpha > 0, \\ \frac{1}{4} & , \text{if } \alpha = 0. \end{cases} \end{cases} \tag{3.4.3}$$

The values of the perturbation function $u(v)$ correspond, in the given space, to the arcs of circumference $x(t;\alpha) = \frac{1}{2} - \alpha + \sqrt{-t^2 + t + \alpha^2}$, $t \in [0,1]$. Since the length of a curve including a prescribed area is obviously unbounded from above, it is easy to note that $\mathcal{K}_{\bar{x}}$ can be regarded as the union of the rays $(u \le \varphi_u(\alpha),\ v \le \varphi_v(\alpha))$ $\alpha \ge 0$. Fig. 3.4.8 shows $\mathcal{K}_{\bar{x}}$. $A \equiv (\frac{\sqrt{2}-2}{4}\pi, \frac{1}{4})$, $O \equiv (0,0)$, $B \equiv (\frac{\sqrt{2}}{4}\pi - 1, \frac{2-\pi}{8})$; in fact, we have:

$$\lim_{\alpha \to +\infty} \varphi_u(\alpha) = \frac{\sqrt{2}}{4}\pi - 1, \qquad \lim_{\alpha \to +\infty} \varphi_v(\alpha) = \frac{2-\pi}{8}.$$

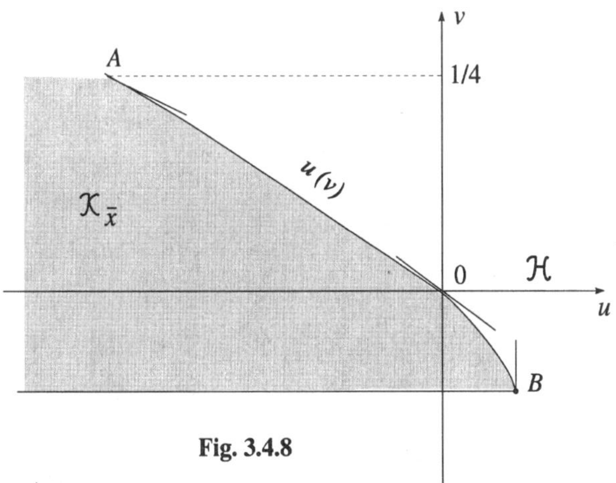

Fig. 3.4.8

Furthermore, we have:

$$u'(v) = \frac{\varphi_u'(\alpha)}{\varphi_v'(\alpha)} = -\frac{2}{\sqrt{4\alpha^2 + 1}}, \qquad \alpha \ge 0,$$

which shows the strict concavity of $u(v)$, so that $\mathcal{K}_{\bar{x}}$ turns out to be convex. In Fig. 3.4.8 there are the tangents to $u(v)$ at A, O, B, whose slopes are $-2, -\sqrt{2}, 0$, respectively. $\mathcal{H}$ is now the positive u-axis, or $\mathcal{H} = \{(u,v) \in \mathbb{R}^2 : u > 0, v = 0\}$. Hence, (3.2.2) is obviously satisfied, since we have chosen $\bar{x}$ as m.p.; the line tangent to $\mathcal{K}_{\bar{x}}$ (and to $u(v)$) at O, whose equation is $u + \sqrt{2}v = 0$, separates (see Sect. 2.2) $\mathcal{H}$ and $\mathcal{K}_{\bar{x}}$. Note that, if in (3.4.1b) the $=$ is replaced by $\ge$, then of course $\mathcal{K}_{\bar{x}}$ does not change while $\mathcal{H} = \{(u,v) \in \mathbb{R}^2 : u > 0, v \ge 0\}$, so that (3.2.2) still holds, and the previous tangent continues to separate $\mathcal{H}$ and $\mathcal{K}_{\bar{x}}$. Of course, the explicit construction of $\mathcal{K}_{\bar{x}}$ has been made possible because of the simplicity of the particular problem; such an achievement is not necessary for the analysis, but useful for illustrating it. Now consider a second example: in (1.1.1) set $n = 2$, $p = m = 1$, $X = \mathbb{R} \times [\frac{2-\pi}{8}, \frac{1}{4}]$, $f(x_1, x_2) = x_1^2 - u(x_2)$, $g(x_1, x_2) = x_2$, where $u(\bullet)$ is the previous function; (1.1.1) becomes:

$$\min[f(x_1, x_2) = x_1^2 - u(x_2)], \tag{3.4.4a}$$

$$\text{s.t.} \quad g(x_1, x_2) = x_2 = 0, \tag{3.4.4b}$$

$$x = (x_1, x_2) \in X = \mathbb{R} \times \left[\tfrac{2-\pi}{8}, \tfrac{1}{4} \right]. \tag{3.4.4c}$$

Note that f is strictly convex, and the feasible region is convex. It is trivial to see that $x^\circ := (0, 0)$ is (the unique global) m.p. of (3.4.4). The image set of (3.4.4) is

$$\mathcal{K}_{x^\circ} = \{(u, v) \in \mathbb{R}^2 : u = -x_1^2 + u(x_2), \ v = x_2, \ x \in X\} =$$

$$= \{(u, v) \in \mathbb{R}^2 : u = u(v) - x_1^2, \ \tfrac{2-\pi}{8} \leq v \leq \tfrac{1}{4}, x_1 \in \mathbb{R}\}.$$

Therefore, the image of (3.4.4) equals that of (3.4.1), or $\mathcal{K}_{x^\circ} = \mathcal{K}_{\bar{x}}$. In fact, $\mathcal{K}_{x^\circ}$ turns out to be a family of arcs, which are translations, in the negative direction of u-axis, of $u(v)$ (the arc AOB of Fig. 3.4.8). Hence, the infinite dimensional problem (3.4.1) and the finite dimensional one (3.4.4) have the same image set. Now consider a third example: in (1.1.1) set $n = 1$, $p = m = 1$, $X = [\frac{2-\pi}{8}, \frac{1}{4}]$, $f(x) = -u(x)$, $g(x) = x$, where $u(\bullet)$ is as above; (1.1.1) becomes:

$$\min [f(x) = -u(x)], \quad \text{s.t.} \quad g(x) = x = 0, \quad x \in \left[\tfrac{2-\pi}{8}; \tfrac{1}{4} \right]. \tag{3.4.5}$$

$\hat{x} := O$ is easily recognized to be (the unique global) m.p. of (3.4.5); its image set is

$$\mathcal{K}_{\hat{x}} = \left\{(u, v) \in \mathbb{R}^2 : u = u(x), \ v = x, \ x \in \left[\tfrac{2-\pi}{8}, \tfrac{1}{4}\right]\right\} =$$

$$= \left\{(u, v) \in \mathbb{R}^2 : u = u(v), \ \tfrac{2-\pi}{8} \leq v \leq \tfrac{1}{4}\right\},$$

which is the arc AOB of Fig. 3.4.8. Thus, (3.4.1) and (3.4.5) have not the same image set. However, two problems, which have different image sets, may have the same conic extension. For instance, in (3.4.1b) replace the $=$ with $\geq$, and perform the same change in (3.4.5). Then, the two image sets do not change (and hence continue to be different), while the conic extensions (see Sect. 3.2) are equal; they are shown by Fig. 3.4.9, where of course $\mathcal{H} = \{(u, v) \in \mathbb{R}^2 : u > 0, v \geq 0\}$. $\qquad\square$

Example 3.4.5. In (1.1.1) let us set $n = 1$, $p = 0$, $m = 1$, $= \mathbb{R}$, and

$$
f(x) = \begin{cases}
4, & \text{if } -\infty < x \le -2, \\
4x + 8, & \text{if } -2 < x \le -1, \\
x^2 + 3, & \text{if } -1 < x < 1,\ x \ne 0, \\
1, & \text{if } x = 0, \\
0, & \text{if } x = 1, \\
x^2 - 2x, & \text{if } 1 < x < 2, \\
x - 3, & \text{if } 2 \le x < 3, \\
0, & \text{if } 3 \le x < +\infty,
\end{cases}
$$

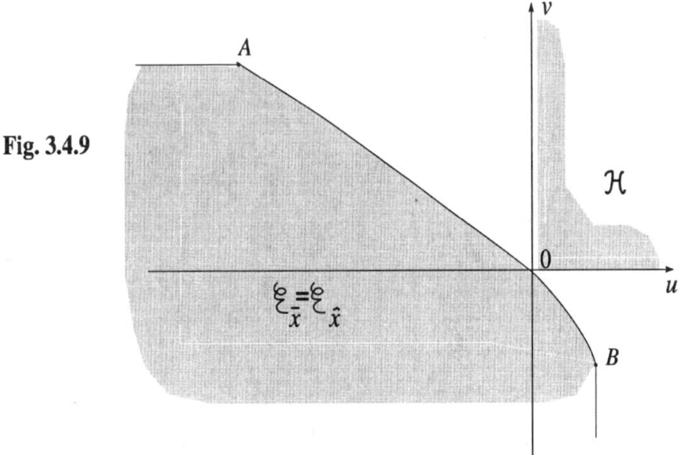

Fig. 3.4.9

$$
g(x) = \begin{cases}
-x + 1, & \text{if } -\infty < x \le -2, \\
x + 2, & \text{if } -2 < x \le -1, \\
-x + 2, & \text{if } -1 < x < 1, \\
0, & \text{if } x = 1, \\
-x - 1, & \text{if } 1 < x < 2, \\
2x - 6, & \text{if } 2 \le x < 3, \\
-x, & \text{if } 3 \le x < +\infty.
\end{cases}
$$

At $\bar{x} = 0$, we find:

$$\mathcal{K}_0 = \left\{ (u,v) \in \mathbb{R}^2 : u = \left\{ \begin{array}{ll} 0 & , \text{if} - \infty < v \leq -3, \\ -v^2 - 4v - 3 & , \text{if} - 3 < v < -2, \\ -\frac{1}{2}v & , \text{if} - 2 \leq v < 0, \\ 0 & , \text{if } v = 0, \\ -4v & , \text{if } 0 < v \leq 1, \\ -v^2 + 4v - 7 & , \text{if } 1 < v < 3, \\ & v \neq 2, \\ -1 & , \text{if } v = 2, \\ -4 & , \text{if } 3 \leq v < +\infty. \end{array} \right. = \Phi_0(v) \right\}.$$

Note that Φ_0 is upper semicontinuous, while f is not lower semicontinuous even in every neighbourhood of $x = 1$, which is the global m.p. of (1.1.1) in the present example. Hence, Proposition 3.2.3 can be applied, while Ekeland Principle cannot. The restriction of f to R is lower semicontinuous, but to take this as assumption would be mathematically meaningless.   □

Example 3.4.6. In (1.1.1) let us set $X = \mathbb{R}$, $p = 0$, $m = 1$, $n = 1$ and

$$f(x) = \left\{ \begin{array}{ll} x, & \text{if } x \leq 0 \\ x + 1, & \text{if } x > 0 \end{array} \right. , \quad g(x) = x.$$

Set $\bar{x} = 0$. We find (see Fig. 3.4.10):

$$\mathcal{K}_{\bar{x}} = \left\{ (u,v) \in \mathbb{R}^2 : u = u(v) := \left\{ \begin{array}{ll} -v, & \text{if } v \leq 0 \\ -v - 1, & \text{if } v > 0 \end{array} \right. \right\}, \quad \mathcal{E}_0 = \text{hypo } u(v).$$

The conic extension removes the disconnection of $\mathcal{K}_0$.   □

Example 3.4.7. In (1.1.1) set $X = \mathbb{R}^2$ $p = 0$, $m = 1$, (so that $\mathcal{J}^0 = \varnothing$, $\mathcal{J}^+ = \mathcal{J} = \{1\}$), $n = 2$, $x = (x_1, x_2)$ and $f(x) = x_1|x_1| + x_2|x_2|$, $g(x) = 16 - x_1^4 - x_2^4$. At $\bar{x} = O = (0,0)$, we find:

$$\mathcal{K}_{\bar{x}} = \mathcal{K}_0 = \{(u,v) \in \mathbb{R}^2 : v = -u^2 - 2x_2|x_2|u + 16 - 2x_2^4, \ x_2 \in \mathbb{R}\}.$$

The image of (1.1.1) is now a family of parabolae depending on the parameter $x_2 \in \mathbb{R}$; Fig. 3.4.11 shows 3 of them. It is easy to see that such a family admits an envelope. This is obtained by eliminating x_2 from the system:

$$\left\{ \begin{array}{l} u^2 + 2x_2|x_2|u + 2x_2^4 - 16 + v = 0, \\ |x_2|(u + 2x_2|x_2|) = 0, \end{array} \right.$$

whose latter equation is obtained by equating to zero the derivative, with respect to x_2, of the left-hand side of the former. The envelope turns out to be $v = 16 - \frac{1}{2}u^2$; see

Fig. 3.4.11. The obvious properties of it are:

$$-u^2 - 2x_2|x_2|u + 16 - 2x_2^4 \leq 16 - \frac{1}{2}u^2, \quad \forall u, \quad \text{or} \quad (u + 2x_2|x_2|)^2 \geq 0, \quad \forall u,$$

and $\forall x_2$, $\exists u$ s.t. equality holds in the above inequality (at $u = -2x_2|x_2|$); such properties imply that, in (3.2.3), $\mathcal{K}_0$ can be replaced by

$$\{(u,v) \in \mathbb{R}^2 : \quad v \leq 16 - \frac{1}{2}u^2\}, \tag{3.4.6}$$

so that (3.2.3) is equivalent to

$$\max(u), \quad \text{s.t.} \ \ 0 \leq v \leq 16 - \frac{1}{2}u^2, \tag{3.4.7}$$

whose solution is $(u = 4\sqrt{2}, \ v = 0)$. As a consequence, the solution of (1.1.1) is obtained as solution of the system:

$$x_1|x_1| + x_2|x_2| = -2\sqrt{8}, \quad x_1^4 + x_2^4 = 16, \tag{3.4.8}$$

and is $(x_1 = x_2 = -\sqrt[4]{8})$. Note that, in the present case, (1.1.1) is not convex, while such a nonconvexity has disappeared in (3.4.7).

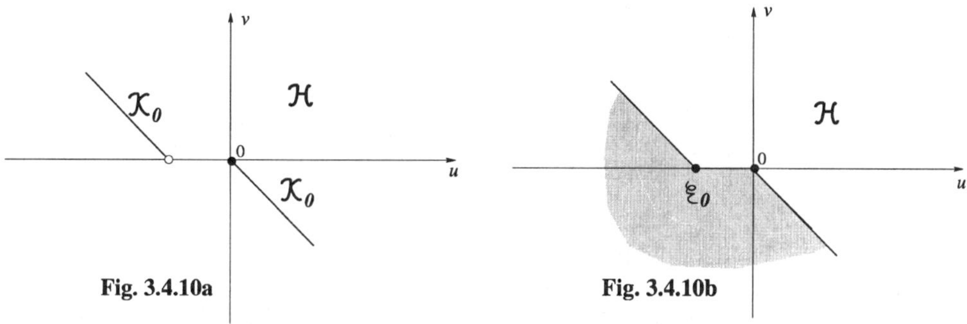

Fig. 3.4.10a   Fig. 3.4.10b

Now, let us consider the homogenization of the image set at $\overline{x}$. According to Definition 3.2.2, we have:

$$\mathcal{D}_e f(\overline{x}; d) = \mathcal{D}_{\mathcal{L}} f(\overline{x}; d) = \langle f'(\overline{x}), d \rangle = \mathcal{D}_{-e} g(\overline{x}; d) = \mathcal{D}_{\mathcal{L}} g(\overline{x}; d) = \langle g'(\overline{x}), d \rangle \equiv 0.$$

Therefore, we find $\mathcal{K}_{\overline{x}}^h = \{(0, 16)\}$. $\qquad\qquad\square$

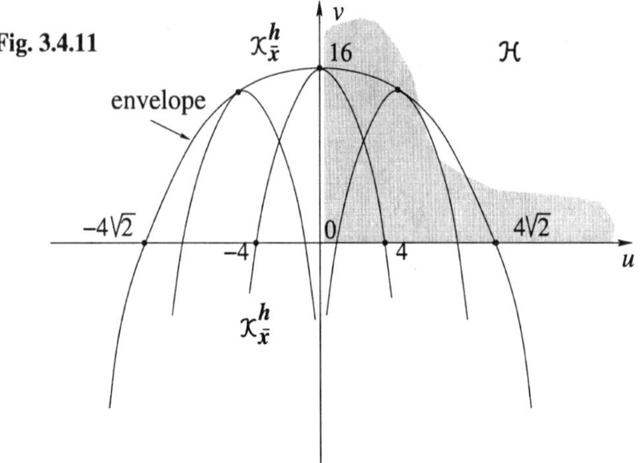

Example 3.4.8. In (1.1.4)   set $n = 2$, $t_0 = 0$, $t_1 = 1$, $p = 0$, $m = 1$ (so that $\mathfrak{I} = \varnothing$, $\mathfrak{I}^+ = \{1\}$, $B = C^0([0,1])^2)$, and

$$X = \{x \in B : x_1(0) = x_2(0) = 0\},$$

$$\psi_0(t, x(t), x'(t)) = 2 \sum_{i=1}^{2} \operatorname{sgn}(y_i) x_i(t) x_i'(t),$$

$$\psi_1(t, x(t), x'(t)) = 16 - 4 \sum_{i=1}^{2} x_i(t)^3 x_i'(t),$$

where $y_1 := x_1(1)$ and $y_2 := x_2(1)$. Therefore, (1.1.4) is now a special boundary value problem: indeed, it has been constructed in order to have the same image set of the problem of Example 3.4.8. By setting $\bar{x}(t) = (\bar{x}_1(t)) \equiv 0$, $\bar{x}_2(t) \equiv 0)$, and by trivial integrations, we find that $\mathcal{K}_{\bar{x}}$ is defined by the system:

$$\begin{cases} u = -\sum_{i=1}^{2} \int_0^1 \operatorname{sgn}(y_i) dx_i^2 = -\sum_{i=1}^{2} y_i |y_i|, \\ v = 16 \int_0^1 dt - \sum_{i=1}^{2} \int_0^1 dx_i^4 = 16 - \sum_{i=1}^{2} y_i^4. \end{cases}$$

Then, the present image set $\mathcal{K}_{\bar{x}}$ coincides with that of Example 3.4.8. Hence, the infinite dimensional problem (1.1.4) can be reduced to the finite dimensional one (3.4.4), whose solution ($u = 4\sqrt{2}$, $v = 0$) leads now to the system (3.4.5) with y_i in place of x_i. We find that any $(x_1(t), x_2(t)) \in X$ and such that $x_1(1) = x_2(1) = -\sqrt[4]{8}$ is a solution of (1.1.4). One might think that the coincidence of the image sets of this example and the previous one is due to the peculiarity of ψ_0 and ψ_1; this is not the case and the choice of such special ψ_0, ψ_1 has been done to make computation trivial. $\qquad\square$

Example 3.4.9. In (1.1.1) set $X = \mathbb{R}$ $p = 0$, $m = 1$, $n = 1$, $f(x) = x$, $g(x) = -x^2$.

At $\bar{x} = 0$, we find:

$$\mathcal{K}_0 = \{(u, v) \in \mathbb{R}^2 : v = -u^2\}.$$

According to Definition 3.2.2, we have:

$$\mathcal{D}_\mathfrak{C} f(\bar{x}; d) = \mathcal{D}_\mathcal{L} f(\bar{x}; d) = f'(\bar{x})d = d, \quad \mathcal{D}_{-\mathfrak{C}} g(\bar{x}; d) = \mathcal{D}_\mathcal{L} g(\bar{x}; d) = g'(\bar{x})d = 0 \cdot d = 0.$$

Therefore, we immediately find: $\mathcal{K}_0^h = \{(u, v) \in \mathbb{R}^2 : v = 0\} = TC(0; \mathcal{K}_0) = RC(0; \mathcal{K}_0)$; see Definitions 2.1.9 and 2.1.10. This is illustrated in Fig. 3.4.12. Note that (3.2.2) is

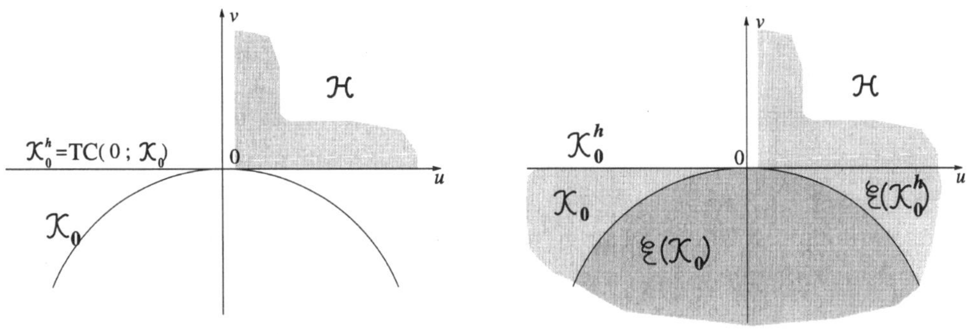

Fig. 3.4.12   Fig. 3.4.13

satisfied, while (3.2.15) does not, notwithstanding the fact that $\bar{x} = 0$ is m.p. of (1.1.1). In other words, $\bar{x} = 0$ is m.p. of (1.1.1), but not of (3.2.10). The conic extension of the image set and of its homogenization are easily found to be:

$$\mathcal{E}(\mathcal{K}_0) = \{(u, v) \in \mathbb{R}^2 : v \leq 0, u \leq \sqrt{-v}\},$$

$$\mathcal{E}(\mathcal{K}_0^h) = \{(u, v) \in \mathbb{R}^2 : v \leq 0\}.$$

They are illustrated in Fig. 3.4.13. While $\mathcal{K}_0$ is not convex, $\mathcal{E}(\mathcal{K}_0)$ does. In spite of this nice fact, (3.2.15) does not hold; in agreement with Theorem 2.2.7 (set $K = \mathcal{H}$, $F =$ positive u-axis, $S = \mathcal{K}_0^h$, so that $S - \text{cl}\, K = \mathcal{E}(\mathcal{K}_0^h)$), the line which separates $\mathcal{H}$ and $\mathcal{K}_0^h$ contains the positive u-axis. $\qquad\square$

Example 3.4.10. In (1.1.1), set $X = \mathbb{R}_-$, $p = 0$, $m = 1$, $n = 1$, $f(x) = x$, $g(x) = -\sqrt{-x}$. At $\bar{x} = 0$, we find:

$$\mathcal{K}_0 = \{(u, v) \in \mathbb{R}^2 : u \geq 0, v = -\sqrt{u}\},$$

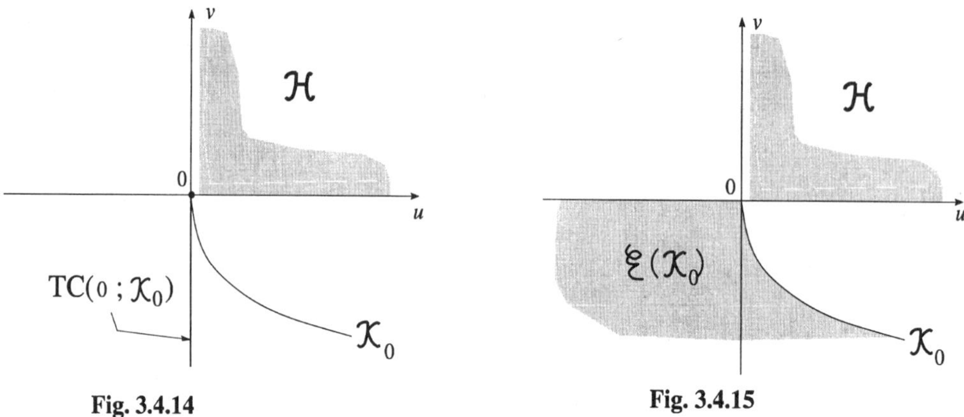

Fig. 3.4.14   **Fig. 3.4.15**

and (3.2.2) is fulfilled in agreement with the obvious fact that we have chosen $\bar{x}$ as m.p. of (1.1.1). Now, $\mathcal{K}_0^h$ does not exist, since g is not $(-\mathcal{C})$-differentiable at $\bar{x}$. Then, we must make a shift and consider the tangent cone (which here equals the reachable one; see Definitions 2.1.9 and 2.1.10), which is easily found to be the non-positive v-axis; see Fig. 2.8.14. The conic extension of $\mathcal{K}_0$ is:

$$\mathcal{E}(\mathcal{K}_0) = \{(u,v) \in \mathbb{R}^2 : v \leq 0; \ u \leq v^2\}.$$

See Fig. 3.4.15. $\hfill\square$

Example 3.4.11. In (1.1.1) set $X = \mathbb{R}$, $p = 0$, $m = 1$, $n = 1$, $\varphi(x) = \frac{1}{2} + \frac{1}{2}\sin\frac{1}{x-1}$, and

$$f(x) = \begin{cases} 4x & \text{,if } x \leq 1, \\[2mm] 4x + (8 - 4x - 4e^{1-x})\varphi(x), & \text{if } x > 1, \end{cases} \qquad g(x) = 1 - x^2.$$

At $\bar{x} = -1$, we find:

$$\mathcal{K}_{-1} = \{(u,v) \in \mathbb{R}^2 : v = -\frac{u^2}{16} - \frac{u}{2}, u \geq -8\} \cup \{(u,v) \in \mathbb{R}^2 : u = -4 - 4\sqrt{1-v} -$$

$$-(8 - 4\sqrt{1-v} - 4e^{1-\sqrt{1-v}})\varphi(\sqrt{1-v})\}.$$

Fig. 3.4.16 (where the unit on v-axis is the double of that on u-axis) shows $\mathcal{K}_{-1}$, which is union of an arc of parabola and a trigonometric curve having $u = 2v - 8$ as common tangent line at $(u = -8, v = 0)$. Now, choose any $\hat{x} \in X$; for instance, $\hat{x} = \frac{1}{2}$; its image

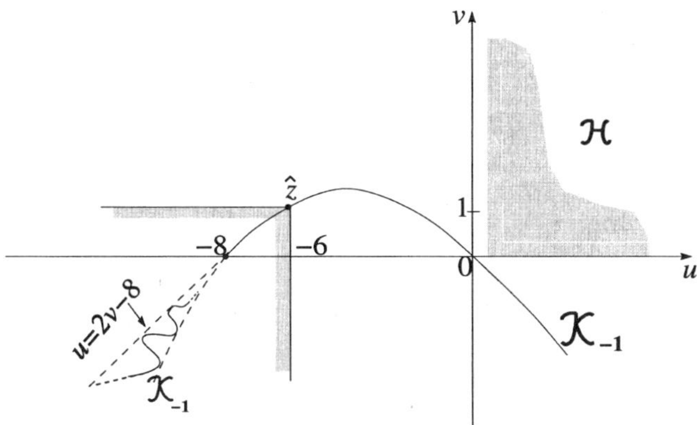

Fig. 3.4.16

is $\hat{z} = A_{-1}(\hat{x}) = (\hat{u} = -6, \hat{v} = \frac{3}{4})$. Instead of performing the entire transformation $\mathcal{E}(\mathcal{K}_{-1})$, we perform only a part of it, namely

$$\hat{z} - \mathrm{cl}\,\mathcal{H} = \left(-6, \tfrac{3}{4}\right) - \mathrm{cl}\,\mathcal{H}.$$

This set, which in Fig. 3.4.16 is the right shaded angle, is the set in the last square brackets of (3.5.25). Such a partial conic extension of the image set, turns $\mathcal{K}_{-1}$ into

$$\hat{\mathcal{K}}_{-1} := \mathcal{K}_{-1} \cup \left\{ (u,v) \in \mathbb{R}^2 : u \leq -6,\ v \leq \tfrac{3}{4} \right\}.$$

Since Proposition 3.2.7 holds — a fortiori — if we perform a part of the conic extension, then the maximum in (3.2.3) does not change, if $\mathcal{K}_{-1}$ is replaced by $\hat{\mathcal{K}}_{-1}$. By means of such an estension, a part of $\mathcal{K}_{-1}$ — which might be undesirable — "disappears". $\square$

Example 3.4.12. In (1.1.1) set $X =]-\infty, 0]$, $p = 0$, $m = 1$, $n = 1$, $f(x) = x$, $g(x) = -\sqrt{-x}e^x$. At $\overline{x} = 0$, we find:

$$\mathcal{K}_0 = \{(u,v) \in \mathbb{R}^2 : v = -\sqrt{u}e^{-u}, u \geq 0\},$$

$$TC(\mathcal{K}_0) = \{(u,v) \in \mathbb{R}^2 : u > 0, v < 0\}, \quad \mathcal{E}(\mathcal{K}_0) = TC(0; \mathcal{K}_0) \cup \mathbb{R}^2_-,$$

$$TC(\mathcal{E}(\mathcal{K}_0)) = \{(u,v) \in \mathbb{R}^2 : v \leq 0\}.$$

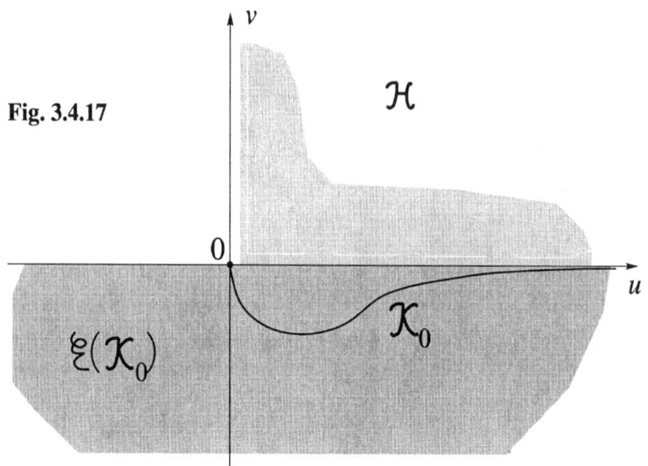

Fig. 3.4.17 shows the image set and its conic extension. The fact that the tangent cone to the conic extension contains the positive u-semi-axis is in agreement with Theorem 2.2.7; indeed, the only way of separating $\mathcal{H}$ and $\mathcal{K}_0$ is to take the u-axis. If we restrict the analysis to a neighbourhood of $\overline{x}$, then $\mathcal{H}$ and the image set (which is only a piece of $\mathcal{K}_0$ exiting from the origin) can be separated by a line different from the u-axis, and the points of the positive u-semi-axis are no longer accumulation points of $\mathcal{E}(\mathcal{K}_0)$. $\qquad\square$

Example 3.4.13. In (1.1.1) set $X = \mathbb{R}$, $p = 0$, $m = 1$, $n = 1$, $f(x) = 0$ if $x = 0$ and $f(x) = -2$ if $x \neq 0$, $g(x) = -x^2$. At $\overline{x} = 0$, we find:

$$\mathcal{K}_0 = \{(u, v) \in \mathbb{R} \times \mathbb{R}_- : u = 0 \text{ if } v = 0, u = 2 \text{ if } v < 0\},$$

$$TC(\mathcal{K}_0) = \{O\}, \quad \mathcal{E}(\mathcal{K}_0) = \mathbb{R}^2_- \cup \{(u, v) \in \mathbb{R}^2 : 0 < u \leq 2, v < 0\},$$

$$TC(\mathcal{E}(\mathcal{K}_0)) = \{(u, v) \in \mathbb{R}^2 : v \leq 0\}.$$

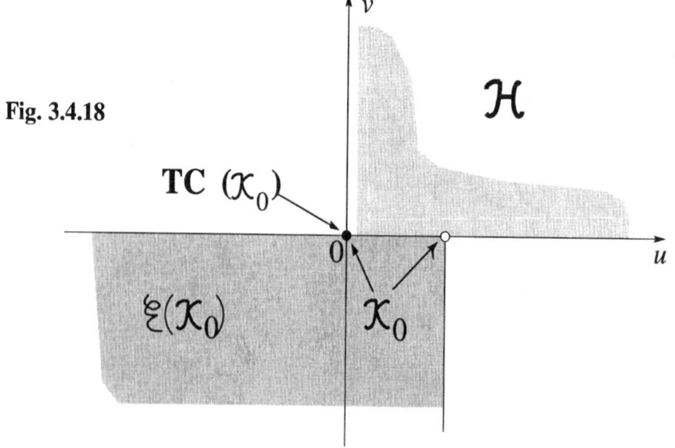

Fig. 3.4.18 shows the image set and its conic extension. The fact that the tangent cone to $\mathcal{E}(\mathcal{K}_0)$ contains the positive u-semi-axis is in agreement with Theorem 2.2.7; indeed, the only way of separating $\mathcal{H}$ and $\mathcal{K}_0$ is to take the u-axis. The example shows also the importance of the conic extension: the tangent cone to the image set (at the origin), being a singleton, would not bring us to any result.

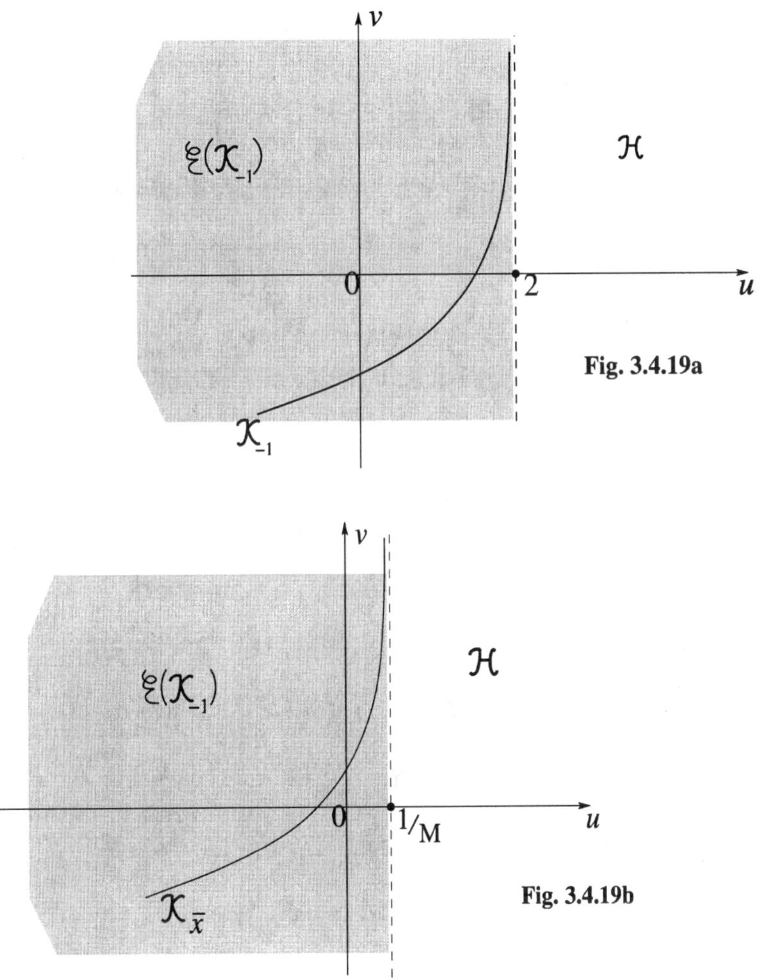

Fig. 3.4.19a

Fig. 3.4.19b

Example 3.4.14. In (1.1.1) set $X = \mathbb{R}$, $p = 0$, $m = 1$, $n = 1$, $f(x) = \exp(-x)$, $g(x) = x$. At $\bar{x} = -1$, we find:

$$\mathcal{K}_{-1} = \{(u,v) \in \mathbb{R}^2 : u = e - e^{-v}\}, \quad \mathcal{E}(\mathcal{K}_{-1}) = \{(u,v) \in \mathbb{R}^2 : u < e\}.$$

Fig, 3.4.19a shows $\mathcal{K}_{-1}$ and its conic extension. According to the remark which comes before (3.2.3), being $\bar{x} \notin R$, $\mathcal{K}_{-1}$ intersects the v-axis, but not the non-negative v-semi-axis. Now, let us change $\bar{x}$ and set $\bar{x} = \log M$ with $M > 0$. We find:

$$\mathcal{K}_{\bar{x}} = \{(u,v) \in \mathbb{R}^2 : u = \frac{1}{M} - e^{-v}\}, \quad \mathcal{E}(\mathcal{K}_{\bar{x}}) = \{(u,v) \in \mathbb{R}^2 : u < \frac{1}{M}\}.$$

Fig. 3.4.19b shows $\mathcal{K}_{\bar{x}}$ and its conic extension under the condition $M \geq 1$, so that, according to the above mentioned remark, being $\bar{x} \in \mathbb{R}$, now $\mathcal{K}_{\bar{x}}$ intersects the non-negative v-semi-axis. Furthermore, we see that, whatever a translation of $\mathcal{K}_{\bar{x}}$ in the direction of u-axis may be, $\mathcal{K}_{\bar{x}}$ intersects $\mathcal{H}$, and

$$\left(\lim_{M \to +\infty} \mathcal{K}_{\ln M}\right) \cap \mathrm{cl}\,\mathcal{H} = \varnothing, \tag{3.4.9}$$

where the limit is in the classic sense: $\forall \varepsilon > 0, \exists M_\varepsilon > 0$, s.t.

$$\mathcal{K}_{\ln M} \subseteq (\lim_{M \to +\infty} \mathcal{K}_{\ln M}) + N_\varepsilon(O), \quad \forall M > M_\varepsilon.$$

Since (3.2.1) or (3.2.2) cannot be satisfied $\forall \bar{x} \in R$, then we conclude that the problem has not minimum. The fact that (3.2.2) holds asymptotically, as (3.4.9) shows, lets us connect (3.4.9) with the existence of finite supremum for (3.2.3) and hence of finite infimum for (1.1.1) or (1.1.4). Note that $\mathcal{K}_{\bar{x}}$ is closed, but its conic extension is not. $\square$

Example 3.4.15. In (1.1.1) set $X =\,]-1,1[\times\mathbb{R}$ $p = 0$, $m = 1$, $n = 2$, $f(x_1, x_2) = x_1$, $g(x_1, x_2) = 1/(1 - x_1^2) + x_2^2$. At $\bar{x} = (0,0)$, we find:

$$\mathcal{K}_{\bar{x}} = \left\{(u,v) \in \mathbb{R}^2 : v \geq \frac{1}{1 - u^2},\ u \in\,]-1,1[\right\}, \quad \mathcal{E}(\mathcal{K}_{\bar{x}}) = \{(u,v) \in \mathbb{R}^2 : u < 1\}.$$

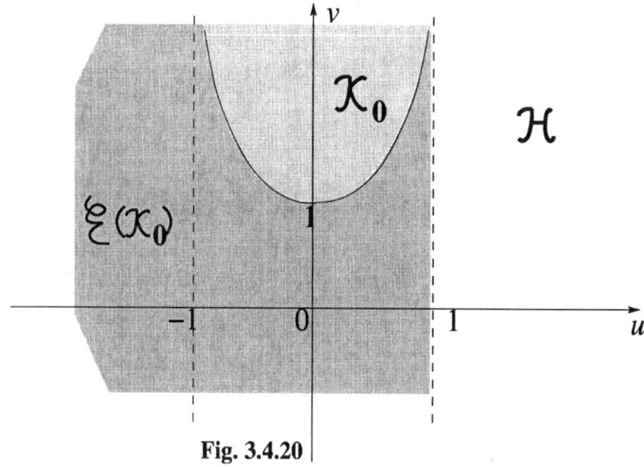

Fig. 3.4.20

Fig. 3.4.20 shows $\mathcal{K}_0$ and its conic extension. As in the previous example, whatever $\bar{x} \in\,]-1,1[$ may be, $\mathcal{H} \cap \mathcal{K}_{\bar{x}} \neq \varnothing$, so that, according to (3.2.2), $\bar{x}$ is not a m.p.; asymptotically we find (the limit is in the same sense as in (3.4.9)):

$$\left(\lim_{\bar{x} \downarrow -1} \mathcal{K}_{\bar{x}}\right) \cap \mathrm{cl}\,\mathcal{H} = \varnothing, \tag{3.4.10}$$

so that (3.2.3) has finite supremum (which is not maximum) and hence (1.1.1) has finite infimum (which is not minimum). $\mathcal{K}_{\bar{x}}$ is closed, but its conic extension is not. $\qquad\square$

Examples 3.4.16. In (1.1.1) set $X = \mathbb{R}^2$, $p = 0$, $m = 1$, $n = 2$, and

$$f(x) = (x_2 + 1)^2(x_2 - 1)^2, \quad g_1(x) = -x_1^2, \quad g_2(x) = x_2.$$

At $\bar{x} = (0,0)$, we find:

$$\mathcal{K}_{\bar{x}} = \{(u, v_1, v_2) \in \mathbb{R}^3 : u = -v_2^4 + 2v_2^2, \ v_1 \le 0\},$$

$$\mathcal{E}(\mathcal{K}_{\bar{x}}) = \left\{(u, v_1, v_2) \in \mathbb{R}^3 : u \le 1, \ v_1 \le 0, \ v_2 \le \sqrt{1 + \sqrt{1 - u}}\right\}.$$

It is easy to see that (3.2.2) — or (3.2.14) — is not satisfied, so that $\bar{x}$ is not m.p. of (1.1.1). Indeed, it is immediate to solve (3.2.3) and obtain $(u = 1, v_1 = 0, v_2 = 1)$ as its maximum point. Now, consider a translation of the above example: $X = \mathbb{R}^2$, $p = 0$, $m = 1$, $n = 2$, $f(x) = x_2(x_2 - 2)^2$, $g_1(x) = -x_1^2$, $g_2(x) = x_2$. At $\bar{x} = (0,1)$, we find:

$$\mathcal{K}_{\bar{x}} = \{(u, v_1, v_2) \in \mathbb{R}^3 : u = -(v_2 - 1)^4 + 2(v_2 - 1)^2, \ v_1 \le 0\},$$

$$\mathcal{E}(\mathcal{K}_{\bar{x}}) = \left\{(u, v_1, v_2) \in \mathbb{R}^3 : u \le 1, \ v_1 \le 0, \ v_2 \le 1 + \sqrt{1 + \sqrt{1 - u}}\right\}.$$

Like before, $\bar{x}$ does not fulfil (3.2.2), so that it is not m.p. of (1.1.1). Now (3.2.3) has 2 maximum points: $(1, 0, 0)$ and $(1, 0, 2)$. $\qquad\square$

Examples 3.4.17. In (1.1.1) set $X = \mathbb{R}$, $p = 0$, $m = n = 1$, and

$$f(x) = x^4 + \frac{1}{3}x^3 - 2x^2 - x + \frac{5}{3}, \quad g(x) = x.$$

At $\bar{x} = 1$, we find (see Fig. 3.4.21a):

$$\mathcal{K}_{\bar{x}} = \{(u, v) \in \mathbb{R}^2 : u = -f(v)\},$$

$$\mathcal{E}(\mathcal{K}_{\bar{x}}) = \{(u, v) \in \mathbb{R}^2 : u \le 0, \ v \le 1\} \cup \{(u, v) \in \mathbb{R}^2 : u \le -f(v), \ v > 1\}.$$

Notwithstanding the nonconvexity of f and of $\mathcal{K}_{\bar{x}}$, $\mathcal{E}(\mathcal{K}_{\bar{x}})$ turns out to be convex. Now, replace the above f with

$$\tilde{f}(x) = (2 - x)^4 + \frac{1}{3}(2 - x)^3 - 2(2 - x)^2 - (2 - x) + \frac{5}{3},$$

which is symmetric of f with respect to the line $x = 1$. Again at $\bar{x} = 1$, we find (see Fig. 3.4.21b):

$$\mathcal{K}_{\bar{x}} = \{(u, v) \in \mathbb{R}^2 : u = -\tilde{f}(v)\},$$

$$\mathcal{E}(\mathcal{K}_{\bar{x}}) = \{(u, v) \in \mathbb{R}^2 : u \le 0, \ v \le 1\} \cup \{(u, v) \in \mathbb{R}^2 : u \le -\tilde{f}(v),$$

$$v \in]1, v^0[\cup]3, +\infty[\} \cup \{(u, v) \in \mathbb{R}^2 : u \le \frac{4}{3}, \ v \in [v^0, 3]\},$$

where v^0 is the unique root of the system $f(2 - v^0) = 4/3$, $v^0 \in]1, 2[$. Now, $\mathcal{E}(\mathcal{K}_{\overline{x}})$ is nonconvex, notwithstanding the fact that $\tilde{f}$ be symmetric of f, so that it has the same lack of convexity in the sense of (2.3.3) or $\rho(\tilde{f}) = \rho(f)$ (see Example 3.5.4). $\square$

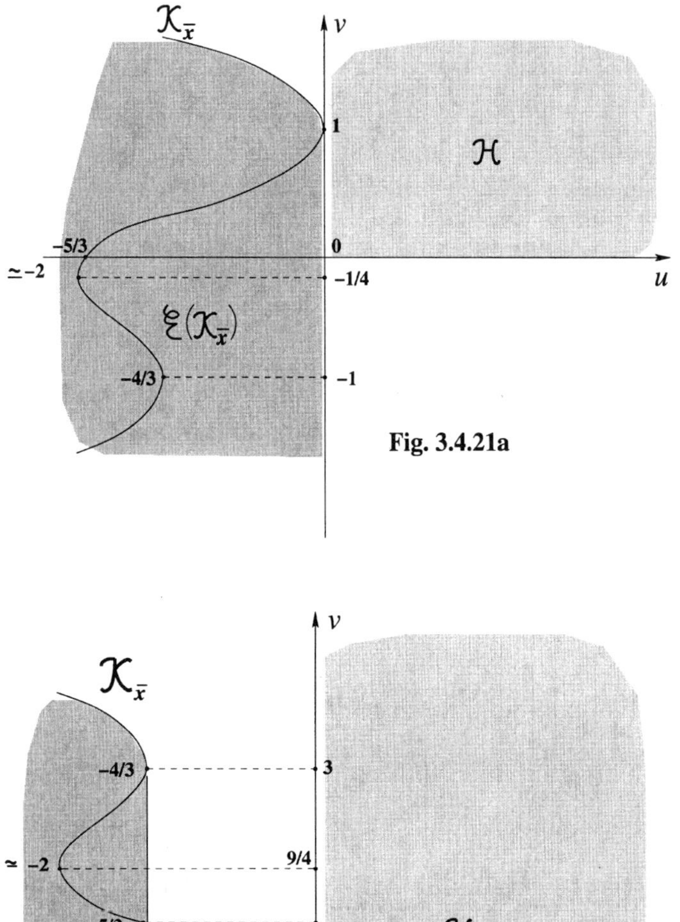

Fig. 3.4.21a

Fig. 3.4.21b

Example 3.4.18. In (1.1.1) set $X = \mathbb{R}^2$, $p = 0$, $m = n = 1$, $f(x) = x_1^2 - 3x_1 x_2 + x_2^2$, $g(x) = -x_1 x_2$. At $\overline{x} = (0, 0)$, we find (see Fig. 3.4.22):

$$\mathcal{K}_{\overline{x}} = \{(u, v) \in \mathbb{R}^2 : u + v \leq 0,\ u + 5v \leq 0\} = \xi(\mathcal{K}_{\overline{x}}).$$

In fact, the map which defines $\mathcal{K}_{\bar{x}}$, namely

$$A_{\bar{x}}(x) = (-x_1^2 + 3x_1 x_2 - x_2^2, \; -x_1 x_2),$$

turns the pencil of lines of $X = \mathbb{R}^2$ through the origin into a pencil of halflines of the IS (u, v) with apices at the origin. More precisely, the axes of X ($x_1 = 0$ and $x_2 = 0$) are turned into the same halfline: $u \leq 0$, $v = 0$. The lines $x_2 = \alpha x_1$, $\alpha \neq 0$, are turned into the halflines

$$\left(u = \frac{\alpha^2 - 3\alpha + 1}{\alpha} v, \; v \leq 0 \right), \quad \text{if } \alpha > 0,$$

$$\left(u = \frac{\alpha^2 - 3\alpha + 1}{\alpha} v, \; v \geq 0 \right), \quad \text{if } \alpha < 0,$$

which form a convex pencil, namely the above $\mathcal{K}_{\bar{x}}$. The fact that $\mathcal{E}(\mathcal{K}_{\bar{x}}) = \mathcal{K}_{\bar{x}}$ is now trivial. Notwithstanding the fact that f be not convex and g not concave in every neighbourhood of $\bar{x}$, the image set and its conic extensions turn out to be convex. $\quad\square$

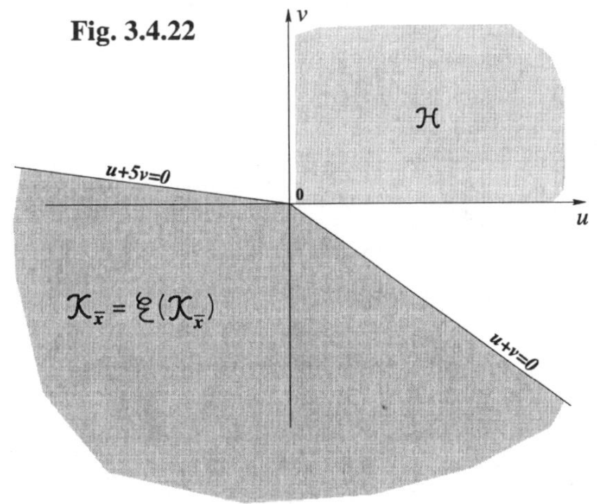

Example 3.4.19. In (1.1.1) set $X = \mathbb{R}$, $p = 0$, $m = n = 1$, $f(x) = x^2$, $g(x) = x^3 - 1$. At $\bar{x} = 1$, we find (see Fig. 3.4.23):

$$\mathcal{K}_{\bar{x}} = \left\{ (u, v) \in \mathbb{R}^2 : u = 1 - (v + 1)^{2/3} \right\}.$$

Since f and g are derivable, so that $\mathcal{D}_e f$ and $\mathcal{D}_{-e} g$ are the ordinary derivatives, the homogenization of $\mathcal{K}_{\bar{x}}$ becomes:

$$\mathcal{K}_{\bar{x}}^h = \left\{ (u, v) \in \mathbb{R}^2 : u = \frac{2}{3} v \right\}$$

and is the tangent to $\mathcal{K}_{\bar{x}}$ at $(u, v) = (0, 0)$.

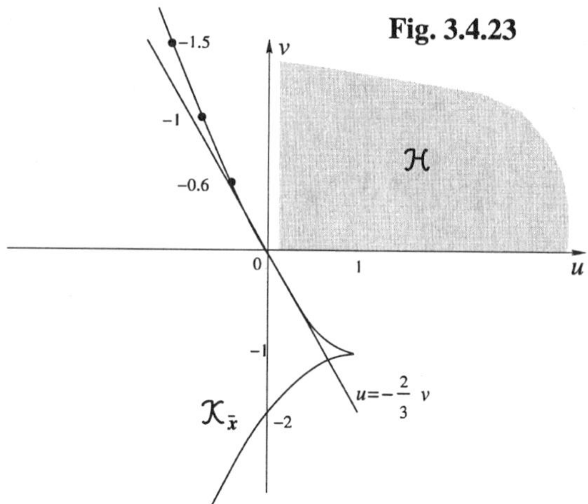

Being derivable, f and g can be replaced by their linear approximation at $\xi \in \mathbb{R}$, namely by

$$\xi^2 + 2\xi(x - \xi) \quad \text{and} \quad \xi^3 - 1 + 3\xi^2(x - \xi),$$

respectively. Now, we consider a slightly more general form of the homogenization of $\mathcal{K}_{\bar{x}}$, by using the above linear approximations at $\mathcal{E}$; this way, instead of $\mathcal{K}_{\bar{x}}^h$, we obtain:

$$\mathcal{K}_{\bar{x}}^h(\xi) := \{(u, v) \in \mathbb{R}^2 : \ u = -2\xi x + \xi^2 + 1, \ v = 3\xi^2 x - 2\xi^3 - 1\}, \qquad (3.4.11)$$

which gives the family of tangents to $\mathcal{K}_{\bar{x}}$, $\forall \xi \neq 0$. Of course, $\mathcal{K}_{\bar{x}}^h(1) = \mathcal{K}_{\bar{x}}^h$. According to the theory of envelopes [A9], by equating to zero the derivative of $u = -2\xi x + \xi^2 + 1$ with respect to ξ, we find $\xi = x$; then $u = -2\xi x + \xi^2 + 1$ becomes $u = 1 - x^2$. By equating to zero the derivative of $v = 3\xi^2 x - 2\xi^3 - 1$ with respect to ξ, we find either $\xi = 0$ or $\xi = x$; then $v = 3\xi^2 x - 2\xi^3 - 1$ becomes, respectively, either $v = -1$ or $v = x^3 - 1$. We have thus obtained precisely the parametric equations which define $\mathcal{K}_{\bar{x}}$. Hence $\mathcal{K}_{\bar{x}}$ is the envelope of its homogenization (linearizations), and (1.1.1) is, in this sense, the envelope of its homogenized (linearized) problems (3.2.10). $\qquad\square$

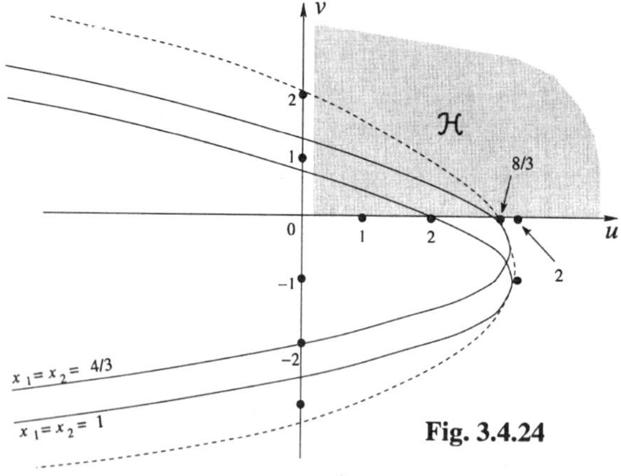

Example 3.4.20. In (1.1.1) set $X = \mathbb{R}^2$, $p = 0$, $m = 1$, $n = 3$, $f(x) = \sum_{j=1}^{3}(x_j^2 - 2x_j)$,

$g(x) = \sum_{j=1}^{3} x_j - 4$. At $\bar{x} = (2, 2, 2)$, which is merely a feasible (and not minimum) point, we find (see Fig. 3.4.24):

$$\mathcal{K}_{\bar{x}} = \{(u, v) \in \mathbb{R}^2 : u = -v^2 + 2(x_1 + x_2 - 3)v - 2(x_1^2 + x_2^2 + x_1 x_2 - 4x_1 - 4x_2 + 4), x_1, x_2 \in \mathbb{R}\}$$

which is the union of parabolas; two of them (for $x_1 = x_2 = 1$ and $x_1 = x_2 = 4/3$) can be seen in Fig. 3.4.24. Such a family of parabolas admit the envelope, whose equation is found to be:

$$u = -\frac{1}{3}v^2 - \frac{2}{3}v + \frac{8}{3};$$

it is the dotted curve of Fig. 3.4.20. The minimum point is easily found to be $x^0 = (6/8, 6/8, 6/8)$ and the minimum $f(x^0) = -8/3$, which is the opposite of the u-coordinate of the envelope at $v = 0$ (see Fig. 3.4.24) in agreement with Proposition 3.2.2 for $\xi = 0$. The conic extension of $\mathcal{K}_{\bar{x}}$ (which is the same as that of the envelope) is:

$$\mathcal{E}_{\bar{x}} = \left\{(u, v) \in \mathbb{R}^2 : u \le -\frac{1}{3}v^2 - \frac{2}{3}v + \frac{8}{3} \text{ if } v \ge -1, u \le 3 \text{ if } v < -1\right\}. \qquad \square$$

Example 3.4.21. In (1.1.1) set $X = \mathbb{R}$, $p = 0$, $m = 1$, $n = 1$, $f(x) = x^4 - \frac{13}{3}x^3 + 5x^2$, $g(x) = x$. At $\bar{x} = 2$ (which is a local, but not global, m.p.), we find (see Fig. 3.4.25a):

$$\mathcal{K}_{\bar{x}} = \left\{(u, v) \in \mathbb{R}^2 : u = -v^4 + \frac{13}{3}v^3 - 5v^2 + \frac{4}{3}\right\}.$$

Now let us perform only a partial conic extension. More precisely, the conic extension (Definition 3.2.1) can be seen as the union of sets of the type $z - \mathrm{cl}\,\mathcal{H}$ with $z \in \mathcal{K}_{\bar{x}}$, or

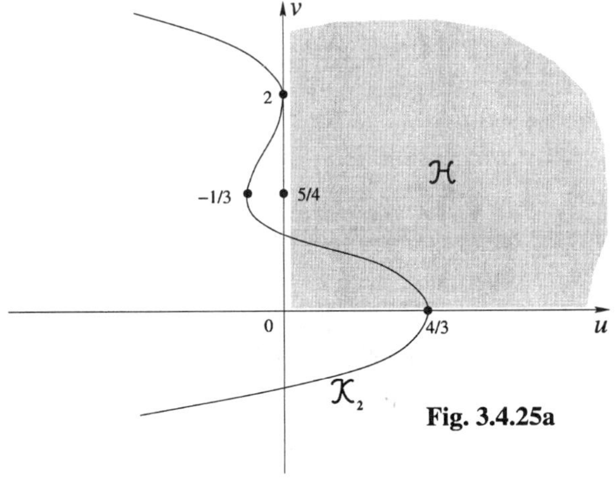

Fig. 3.4.25a

extensions of each element of $\mathcal{K}_{\bar{x}}$ (see 3.5.12). Here we extend only $(u = 0, v = 2)$ and obtain the set (see Fig. 3.4.25b):

$$\mathcal{K}_{\overline{x}} \cup \{(u,v) \in \mathbb{R}^2 : u \le 0,\ v \le 2\}.$$

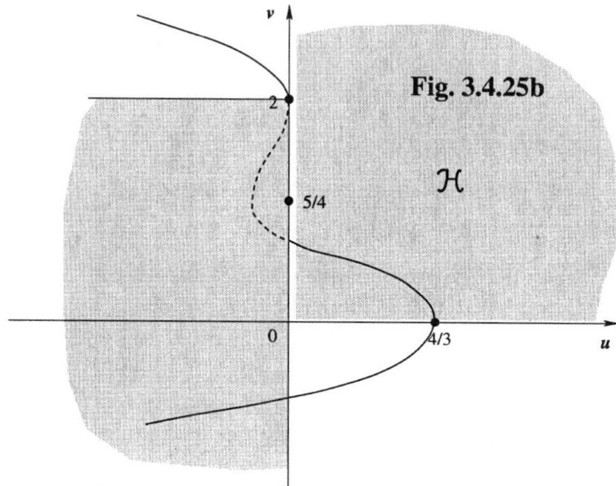

With such a partial conic extension, performed at the image of the current feasible point $\overline{x} = 2$, we do not eliminate all the convexities, but only a part of them. This suggests a "sequential reduction" of the nonconvexities of the problem. See Sect. 3.5 around (3.5.12), (3.5.13), and continuation in Examples 3.5.1 and 3.5.3, where Figs.3.4.25c,d will be discussed. $\qquad\square$

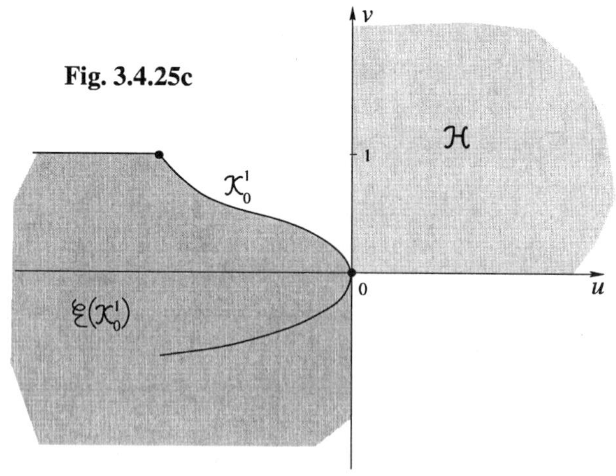

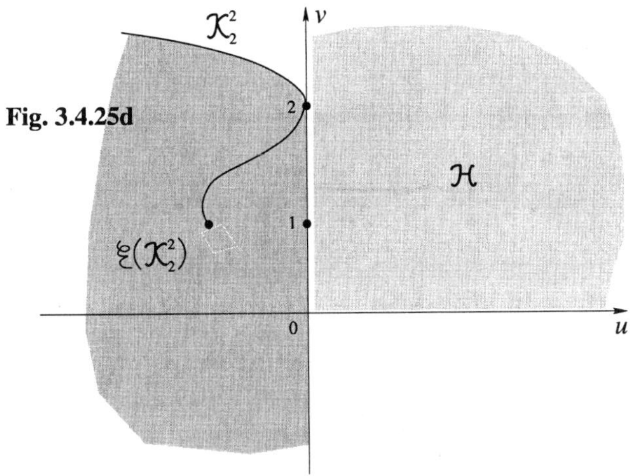

Example 3.4.22. In (1.1.1) set $X = \mathbb{R}$, $p = 0$, $m = 2$, $n = 1$,

$$f(x) = x, \; g_1(x) = x - x^3, \; g_2(x) = x.$$

At $\bar{x} = 0$, we find (see Fig. 3.4.26a)

$$\mathcal{K}_{\bar{x}} = \{(u, v_1, v_2) \in \mathbb{R}^3 : u = -v_2, \; v_1 = v_2 - v_2^3, \; v_2 \in \mathbb{R}\}.$$

Since the curve $\mathcal{K}_{\bar{x}}$ lies on the plane $u + v_2 = 0$ which is disjoint from $\mathcal{H}$, then (3.2.2) holds and shows that $\bar{x}$ is global m.p. , as it was obvious to detect by direct inspection. However, (1.1.1) is here not convex even in every neighbourhood of $\bar{x}$, since g_1 is not concave (see Sect. 3.5 for the continuation). It is easy to see that both $\mathcal{K}_{\bar{x}}$ and its conic extension are not convex. However, it is possible to make a "composition" of the constraints into exactly one constraint, so that the IS becomes a plane; see Example 3.5.2. $\qquad\square$

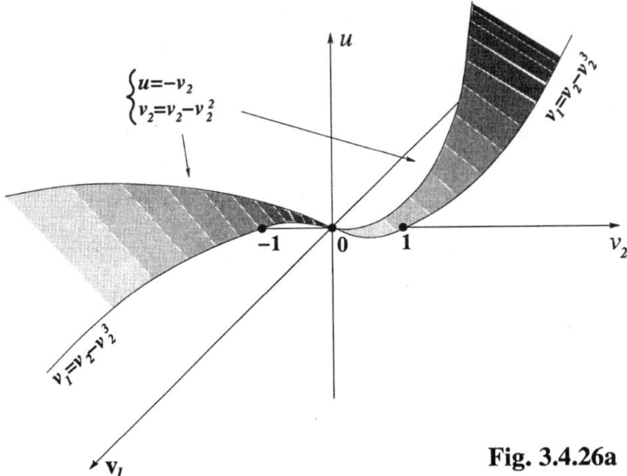

Fig. 3.4.26a

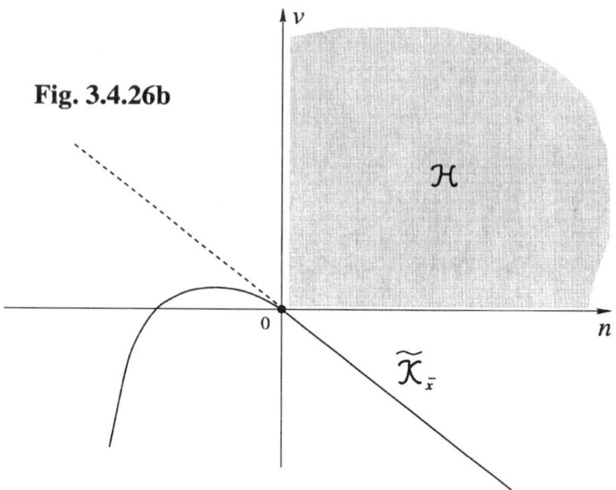

3.5. Comments

1.The concept of semidifferentiability introduced by Definition 3.1.3 is not new. In [45], page 28, a *lower semidifferentiability function* is defined as

$$f(x) \geq f(\overline{x}) + \langle \sigma, d \rangle + \varepsilon(\overline{x}; d), \tag{3.5.1}$$

where

$$\lim_{d \to 0} \frac{\varepsilon(\overline{x}; d)}{||d||} = 0,$$

and where σ is called a *lower semigradient*. In other words, in the classic expansion of differentiabilty, the equality is replaced with an inequality. This concept is different from that of lower semidifferentiability of Definition 3.1.3 (the comparison requires us to set $G = \mathcal{L}$) in as much as the former is the result of a relaxation performed on the expansion of differentiabilty (and means that f must be supported at $\overline{x}$ by a differentiable function), while the latter comes from a relaxation performed on the remainder, which must fulfil merely:

$$\liminf_{d \to 0} \frac{\varepsilon(\overline{x}; d)}{||d||} \geq 0.$$

The concept (3.5.1) seems too feeble for achieving a necessary optimality condition; it accepts functions like that of Example 3.1.10 and excludes functions like that of Example 3.1.12. Note that a $\mathcal{C}$-differentiable function is semidifferentiable in the sense of (3.5.1). In fact, $\mathcal{D}_{\mathcal{C}}f(\overline{x}; \bullet)$ being convex, epi $\mathcal{D}_{\mathcal{C}}f$ admits a supporting hyperplane

(Theorem 2.3.1), say $y = f(\overline{x}) + \langle \overline{\sigma}, d \rangle$; then, from Definition 3.1.1 we draw:

$$f(\overline{x} + d) = f(\overline{x}) + \mathcal{D}_{\mathcal{C}}f(\overline{x}; d) + \varepsilon(\overline{x}; d) \geq f(\overline{x}) + \langle \overline{\sigma}, d \rangle + \varepsilon(\overline{x}; d), \quad \forall d \in \mathbb{R}^n,$$

where ε satisfies (3.1.2); this shows that f fulfils (3.5.1). Clearly, the opposite is false. This fact suggests stating a necessary optimality condition for a problem, like (1.1.1) and subsequent ones, within the class of $\mathcal{C}$-differantiable functions, and extending the results to the class of functions defined by (3.5.1), by considering the former functions as "lower supports" of the latter ones.

2.The concept expressed by Definition 3.1.2 can be extended to semidifferentiable functions, as a set of (linear) supports to the epigraph (or hypograph) of the upper (or lower) G-semiderivative, if it is sublinear (or superlinear). Again in the order of ideas to have a container concept, it would be useful to compare such a definition of upper (or lower) semidifferential with the existing generalizations of differential.

3.Definition 3.1.1 as well as Definition 3.1.3 express a 1st order approximation. It is obviously conceivable to extend it to higher orders. Let $\mathcal{G}^{(r)}$ denote the set of positively homogeneous functions of order $r, r = 1, 2, ...$; so that $\mathcal{G}^{(1)} \equiv \mathcal{G}$ (see Definition 2.3.6). As a hint to define an expansion up to a generic rth order, we might ask the remainder to satisfy the following condition:

$$\lim_{d \to 0} \varepsilon^{(r)}(\overline{x}; d) := \frac{f(\overline{x} + d) - f(\overline{x}) - \sum_{i=1}^{r} \mathcal{D}^{(i)}_{\mathcal{G}^{(i)}} f(\overline{x}; d)}{||d||^r} = 0, \tag{3.5.2}$$

where $\mathcal{D}_{\mathcal{G}^{(i)}} f \in \mathcal{G}^{(i)}$. For $r=1$, (3.5.2) collapses to (3.1.2). Thus, we have the expansion:

$$f(\overline{x} + d) = f(\overline{x}) + \sum_{i=1}^{r} \mathcal{D}^{(i)}_{\mathcal{G}^{(i)}} f(\overline{x}; d) + \varepsilon^{(r)}(\overline{x}; d), \quad d \in \mathbb{R}^n. \tag{3.5.3}$$

If we stop the above expansion at the 2nd order ($r=2$), then an extension of positive semidefiniteness (Theorem 2.3.5) to $\mathcal{C}$-differentiable functions should be achieved. If $n = 1$, $f \in C^\infty(\mathbb{R})$ and, as approximation to the ith order, we choose the particular positively homogeneous function (of degree i) $\mathcal{D}^{(i)}_{\mathcal{G}^{(i)}} f(\overline{x}; d) = \frac{d^i}{i!} f^{(i)}(\overline{x})$, then (3.5.3) becomes the classic Taylor series for functions of one variable. Definition 3.1.2 can be consequently extended. For instance, at $r=2$, the 2nd order version of (3.1.7) might be:

$$\underline{\partial}^2_{\mathcal{G}^{(2)}} f(\overline{x}) := \left\{ \Sigma \in \mathbb{R}^{n \times n} : \; D^{(2)}_{G^{(2)}} f(\overline{x}; d) \geq \langle d, \Sigma d \rangle \right\}.$$

Once the expansion (3.5.3) has been established, a generalization of the class of analytic functions is available; if we use Definition 3.1.3, instead of Definition 3.1.1, then we would define a class of semianalytic functions.

4.In order to investigate the higher orders, it would useful to study some preliminary aspect, such as the extensions to higher orders of the well known relationship among positively homogeneous, convex and subadditive functions.

5.Among the several extensions of Differential Calculus, the Quasidifferential Calculus [I16, II17, 8-10, V24] is a very interesting and effective theory, which has shown to be useful in the applications. Semismooth Analysis [II35], the Theory of Derived Sets [24, 25], and the Theory of Fans and Nonsmooth Analysis [26] are further interesting theories. A comparative study of all these theories with Definitions 3.1.1-3.1.3 might lead to a unifying theory and to further results.

6.Let us denote by $S \subseteq \mathbb{R}^n$ the set of points where a function $f : \mathbb{R}^n \to \mathbb{R}$ is not $\mathcal{C}$-differentiable. If f is locally Lipschitz, then a trivial use of the well known Rademacher Theorem (see, e.g., [V84, II45]) allows one to say that S is of zero (Lebesgue) measure (indeed, S is a subset of the set of points where f is not differentiable). Let f be $\mathcal{C}$-differentiable on X but not everywhere locally Lipschitz; we wonder whether or not S is still of zero measure; namely, whether or not the Rademacher Theorem can be extended to $\mathcal{C}$-differentiable functions. This question might take advantage from finding a function which, at every point, be $\mathcal{C}$-differentiable, but with not linear $\mathcal{C}$-derivative, namely a function which, at every point, be $(\mathcal{C}\backslash\mathcal{L})$-differentiable. Another help might come by keeping in mind the classic result: "the set of points where a convex function $f : \mathbb{R}^n \to \mathbb{R}$ is not differentiable is the union of a denumerable number of compact sets of dimension lower than n" (in other words, a convex function is differentiable almost everywhere), or a more general result related the point of nonsmoothness on a convex manifold (see R.D.Anderson and V.L.Klee, "Convex functions and upper semicontinuous collections". Duke Mathem. Jou., Vol. 19,1952, pp. 349-357. Indeed, much earlier it had been shown that convex functions are even twice differentiable almost everywhere: A.D.Alexandroff, "Almost everywhere existence of the second differential of a convex function and some properties of convex surfaces connected with it". Leningrad State Univ. Annals, Uchenje Zapiski, Mathem. Series, Vol. 6, 1939, pp.3-35).

7.A condition for the integrability (in the sense of Mengoli-Cauchy-Riemann; Pietro Mengoli in 1659 and Augustin Louis Cauchy in 1823 gave the definition, and Georg Friedrich Bernhard Riemann in 1854 gave a necessary and sufficient condition for a function to be integrable) of a $\mathcal{C}$-differentiable function is immediately obtained from Proposition 3.1.3. Of course, not all $\mathcal{C}$-differentiable functions are integrable; for instance, that of Example 3.1.1 (which is based on the Dirichelet function) is not. These facts, the already stated results (which have been quoted in Sect. 3.1), in particular those of [38], form an important part of the $\mathcal{C}$-Differential Calculus. Within this, an interesting aspect would be the introduction of a $\mathcal{C}$-differential equation, the simplest form of which consists in finding a function f, such that (f is the $\mathcal{C}$-antiderivative of g):

$$\mathcal{D}_{\mathcal{C}}f(x; d) = g(x; d), \quad \forall d \in \mathbb{R}^n, \tag{3.5.4}$$

where $g(x; \bullet)$ is a given sublinear function. For instance, if g is identified with the $\mathcal{C}$-derivative of Example 3.1.2, then, in each of the intervals $2^{-n} < x < 2^{1-n}$, $n = 1, 2, ...$, (3.5.4) amounts to solve a differential equation (indeed, a mere integration), and impose, in the points 2^{-n}, $n = 1, 2, ..$, suitable boundary conditions which take into account the $\mathcal{C}$-derivative. The reduction of a $\mathcal{C}$-differential equation to a family of differential equations can be done every time the set of $(\mathcal{C}\backslash\mathcal{L})$-differentiable points is denumerable and they are isolated. Therefore, the properties of the previously set S are important.

8.As well as it is interesting to characterize the class of functions which are derivable, but not differentiable (classic instances are $f : \mathbb{R}^2 \to \mathbb{R}$, given by $f(x_1, x_2) = \sqrt{|x_1 x_2|}$ or by $f(x_1, x_2) = x_1^2 + x_2^2$ if $x_1 \neq x_2$, $f(x_1, x_2) = \sqrt{|x_1|}$ if $x_1 = x_2$), it is interesting to characterize the class of functions which are $(\mathcal{C} \backslash \mathcal{L})$-derivable (i.e., $\mathcal{C}$-derivable but not derivable), but not $(\mathcal{C} \backslash \mathcal{L})$-differentiable (i.e., $\mathcal{C}$-differentiable but not differentiable); an instance is $f : \mathbb{R}^2 \to \mathbb{R}$, given by

$$f(x_1, x_2) = \begin{cases} \sqrt{x_1^2 + x_2^2} + x_1^2 + x_2^2 & \text{if } x_1 \neq x_2, f(x_1, x_2) \\ \sqrt{x_1^2 + x_2^2} + \sqrt{|x_1|} & \text{if } x_1 = x_2. \end{cases}$$

9.$\mathcal{C}$-differentiability of functions $f : \mathbb{R}^n \to \mathbb{R}$ can be extended to functions $f : B \to \mathbb{R}$ and to operators in B. Preliminary results for the former aspect are in [22, V40], and for the latter aspect in [48]; see also Theorems 3.1.1 and 3.2.4.

10.Let f be $\mathcal{C}$-differentiable; fix $\hat{d} \in \mathbb{R}^n$ and consider the ray $\rho := \{d \in \mathbb{R}^n : d = \alpha \hat{d},$ $\alpha \in \mathbb{R}_+ \backslash \{0\}\}$. From Definition 3.1.1, we see that the $\mathcal{C}$-derivative of the restriction of f to ρ equals the restriction to ρ of the $\mathcal{C}$-derivative of f. A minimum requirement to a theory of generalized differentiability should be that the (generalized) derivative of the restriction of the function to any ray be equal to the restriction of the (generalized) derivative to the ray.

11.Starting from the above quoted results and Propositions 3.1.5 and 3.1.6, it should be possible to extend most of the theorems of Differential Calculus to $\mathcal{C}$-(semi)differentiable functions. Some concepts of differential geometry such as curvature, as well as inverse and Dini implicit functions, mean value, composition, other operations like square, square root, logarithm, quotient, sup- and max-functions (see (3.1.11) and [15, 16, 36-39]) would be important topics, as well as the extension of some methods of Numerical Analysis.

12.A question, related to the above one, deals with semiderivability. A function $f : \mathbb{R}^n \to \mathbb{R}$ might be defined to be semiderivable, iff its restrictions to the fundamental directions are semidifferentiable, or iff, $\forall j = 1, ..., n$, the function (of x_j only) $f(x_1, ..., x_{j-1}, \bullet, x_{j+1}, ..., x_n)$ is semidifferentiable. The following example shows that a semiderivable (in the above sense) function is not necessarily semidifferentiable. Set $X = \mathbb{R}^2$, $\bar{x} = (0, 0)$, and

$$\alpha(t) := \begin{cases} \frac{1}{2} + \frac{1}{2} \sin \frac{1}{t}, & \text{if } t \neq 0, \\ 0, & \text{if } t = 0, \end{cases} \quad (t \in \mathbb{R})$$

$$f(x_1, x_2) := \left[1 - \alpha\left(\sqrt{x_1^2 + x_2^2}\right)\right]\left(-\sqrt{x_1^2 + x_2^2}\right) + \alpha\left(\sqrt{x_1^2 + x_2^2}\right)\sqrt{|x_1 x_2|}.$$

If $x_1 = 0$ or $x_2 = 0$, then for $G = \mathcal{L}$ (3.1.12) are fulfilled by $\overline{\mathcal{D}}_G f \equiv 0$ and $\varepsilon = f$. If x is not restricted to the x_1-axis or to the x_2-axis, then f is not upper G-semidifferentiable, even if $G = \mathcal{G}$, since we have a case quite similar to that of Example 3.1.10. Semidifferentiability is not implied even by classic derivability, as shown by the next example. Set $X = \mathbb{R}^2$, $\overline{x} = (0,0)$, and $f(x_1, x_2) = (|x_1 x_2|)^{1/2}$. f is partially derivable (even if not differentiable), but not upper $\mathcal{G}$-semidifferentiable (see, again, Example 3.1.10). Of course, if $n \geq 2$, the derivability of $f : \mathbb{R}^n \to \mathbb{R}$ does not imply its semidifferentiability, since — as is well known — does not imply differentiability; take, for instance, $n = 2$, $f(x_1, x_2) = x_1^2 + x_2^2$ if $x_1 \neq x_2$ and $f(x_1, x_2) = \sqrt{|x_1|}$, if $x_1 = x_2$; at $x_1 = x_2 = 0$ the partial derivatives exist, while f is not differentiable and, with better reason, not semidifferentiable.

13. $C^{1,1}(X)$, with $x \subseteq B$, was introduced to denote the set of functions of the Banach space B having the differential locally Lipschitz on X. A natural extension would be the class of $\mathcal{C}$-differentiable functions whose $\mathcal{C}$-differential is locally Lipschitz.

14. Proposition 3.1.4 and the subsequent comment induce a definition of continuous $\mathcal{C}$-differentiability at $\overline{x}$: $\forall \varepsilon > 0$, $\exists \delta > 0$, s.t.

$$||x - \overline{x}|| < \delta \quad \Rightarrow \quad |\mathcal{D}_{\mathcal{C}} f(x; d) - \mathcal{D}_{\mathcal{C}} f(\overline{x}; d)| < \varepsilon, \tag{3.5.5a}$$

if d is considered fixed; or

$$||x - \overline{x}|| < \delta \quad \Rightarrow \quad \sup_{d \in \mathbb{R}^n} |\mathcal{D}_{\mathcal{C}} f(x; d) - \mathcal{D}_{\mathcal{C}} f(\overline{x}; d)| < \varepsilon, \tag{3.5.5b}$$

otherwise.

15. The class of piece-wise (strictly) concave and continuous functions was introduced in [II56]. In spite of its potential interest, this class has received little attention. Example 3.1.2 shows that it intersects the class of continuous $\mathcal{C}$-differentiable functions. The simultaneous investigations of both classes might have some advantages.

16. The search for the minimum (or the maximum) of a function $f : X \to \mathbb{R}$ by following a piece-wise linear trajectory is an old approach (it goes back to the work of Cauchy [II]) and now-a-days is a commonly used tool. Perhaps, it might be improved within generalized differentiability. To this end, let us make a few remarks. It is easy to show that, if f is lower $\mathcal{C}$-semidifferantiable at $\overline{x} \in X$ and X is convex, then the set $\{d : \underline{\mathcal{D}}_{\mathcal{C}} f(\overline{x}; d) \leq 0\}$ is convex; the same is true if $\underline{\mathcal{D}}_{\mathcal{C}} f$ is replaced by $\overline{\mathcal{D}}_{\mathcal{C}} f$. It is easy to show that, if f is lower $\mathcal{C}$-semidifferentiable at $\overline{x}$, X is convex, and $\underline{\mathcal{D}}_{\mathcal{C}} f(\overline{x}; d) < 0$, then there exists a descent sequence along d or, $\forall \varepsilon > 0$, $\exists t_\varepsilon > 0$, s.t.

$$f(\overline{x} + t_\varepsilon d) < f(\overline{x}). \tag{3.5.6}$$

In fact, we have:

$$f(\overline{x} + td) - f(\overline{x}) = \underline{\mathcal{D}}_{\mathcal{C}} f(\overline{x}; td) + \varepsilon(\overline{x}; td) = t\left[\underline{\mathcal{D}}_{\mathcal{C}}(\overline{x}; d) + \frac{\varepsilon(\overline{x}; td)}{||td||} ||d|| \right]. \tag{3.5.7}$$

Because of Definition 3.1.3, in particular of (3.1.12c) which now becomes

$$\text{hypo } \underline{\mathcal{D}}_{\mathcal{C}}^0 f \subseteq \text{hypo } \underline{\mathcal{D}}_{\mathcal{C}} f,$$

there cannot exist k, $\overline{t} > 0$ s.t. $\varepsilon(\overline{x}; td) \geq k||td||$, $\forall t \in]0, \overline{t}]$, so that for $t \in]0, \overline{t}]$ the form within the square brackets of (3.5.7) is <0 and (3.5.6) follows. It is possible to prove also something about minimum points. Let X be open and convex, f be lower $\mathcal{C}$-semidifferentiable on X, and $\bar{x} \in X$. Consider the sequence $\{x^r\}_1^\infty$ with

$$x^{r+1} := x^r - \frac{t_r}{||\sigma^r||}\sigma^r, \quad t_r > 0, \tag{3.5.8}$$

where σ^r is an element of the lower $\mathcal{C}$-semidifferential of f at $\bar{x}$, defined by an inequality quite analogous to (3.1.7), so that

$$\underline{\mathcal{D}}_\mathcal{C}f(x^r; x - x^r) \geq \langle \sigma^r, x - x^r \rangle, \quad \forall x \in X. \tag{3.5.9}$$

Suppose that, at least for $x = \bar{x}$, one of the following conditions be verified: **(i)** $\varepsilon^r(x^r; x - x^r) \geq 0$ r=1,2,...; **(ii)** $\varepsilon^r(x^r; x - x^r) < 0$ and $\varepsilon^r(x^r; x - x^r) > f(x) - f(x^r)$, r=1,2,.... Then, if $\bar{x}$ is not a minimum point, $\exists \bar{t}_r > 0$, s.t.

$$||x^{r+1} - \bar{x}|| < ||x^r - \bar{x}||, \quad \forall t_r \in]0, \bar{t}_r[. \tag{3.5.10}$$

In fact, by setting $\pi_r := \langle \frac{1}{||\sigma^r||}\sigma^r, \bar{x} - x^r \rangle$ and using (3.5.8), we find:

$$||\bar{x} - x^{r+1}||^2 = ||\bar{x} - (x^r - \frac{t_r}{||\sigma^r||}\sigma^r)||^2 = ||\bar{x} - x^r||^2 + 2t_r\pi_r + t_r^2. \tag{3.5.11}$$

Because of (3.5.9) and of (i) or (ii), we have:

$$\langle \sigma^r, x - x^r \rangle \leq f(x) - f(x^r) - \varepsilon^r(x^r; x - x^r) < 0, \quad \forall x \in X,$$

so that $\pi_r < 0$. Hence,

$$2t_r\pi_r + t_r^2 < 0 \quad \Leftrightarrow \quad 0 < t_r < \bar{t}_r := -2\pi_r.$$

This relation and (3.5.11) prove (3.5.10).

17. Proposition 3.2.3 extends the so-called weak Ekeland variational principle; a similar extension is conceivable for the strong one. The Ekeland variational principles have been applied to several fields and several consequences have been derived [7]. On the basis of the above extensions, it would be interesting to revisit them in terms of the IS Analysis.

18.The analysis carried out in Sect. 3.2 deals with global extrema. Local extrema can be obtained, obviously, by intersecting the domains with suitable neighbourhoods: either X or $\mathcal{K}_{\bar{x}}$ must be replaced, respectively, by $X \cap N(\bar{x})$ or $\mathcal{K}_{\bar{x}} \cap \mathcal{N}(\bar{z})$, where $N(\bar{x})$ and $\mathcal{N}(\bar{z})$ denote neighbourhoods in the given space and in the IS, respectively. These two alternatives are equivalent, when the image, through the map $A_{\bar{x}}(x)$, of $N(\bar{x})$ can be enclosed in an arbitrarily small neighbourhood $\mathcal{N}(\bar{z})$ by suitably choosing $N(\bar{x})$ small enough. Example 3.4.6 (see Fig. 3.4.10a) shows that this may not happen: the diameter of a circle containing the image, through the map

$$A_0(x) = \left(\left\{ \begin{array}{ll} -x - 1, & \text{if } x < 0 \\ 0 & , \text{if } x = 0 \\ -x + 1, & \text{if } x > 0 \end{array} \right\}, x \right),$$

of $N_\rho(0) = \{x \in \mathbb{R} : |x| < \rho\}$ must be greater than 2, whatever $\rho > 0$ may be; this situation changes, if we replace $\mathcal{K}_{\overline{x}}$ with its conic extension (Fig. 3.4.10b). The characterization of problems for which such an equivalence holds is of interest. In particular, it is useful to know when the image, through the map $A_{\overline{x}}(x)$, of $N(\overline{x})$ is a neighbourhood $\mathcal{N}(\overline{z})$; this also is not true in general, as Example 3.4.1 shows: the image of $N_\rho(O) = \{x \in \mathbb{R}^2 : ||x|| < \rho\}$ is contained in the 2nd and 4th quadrants of the plane (u,v) (see Fig. 3.4.1). Strictly related to such questions are the characterizations, in terms of the given data X, f, and g, of interior, frontier, isolated accumulation points of the image set and of its conic extension. For instance, if $\hat{z} \in (\mathcal{H} \cap \partial\mathcal{H})\backslash\mathcal{K}_{\overline{x}}$ is an accumulation point of $\mathcal{K}_{\overline{x}}\backslash\mathcal{H}$ (see Example 3.4.13) and if there is no element of $\mathcal{H} \cap \mathcal{K}_{\overline{x}}$ having u-coordinate greater than that of $\hat{z}$, then every hyperplane, which separates $\mathcal{H}$ and $\mathcal{K}_{\overline{x}}$, must contain a face of $\mathcal{H}$, so that the related optimality conditions are "irregular" (see Chapter 5). If the conic extension $\mathcal{E}(\mathcal{K}_{\overline{x}})$ is convex, like in Example 3.4.13, then Theorem 2.2.7 provides a way of overcoming the analysis of points $\hat{z}$; their characterizations still remain interesting and pending. If $\mathcal{E}(\mathcal{K}_{\overline{x}})$ is not convex, then the question is completely alive. Several other properties deserve to be investigated, like closure and boundedness of $\mathcal{K}_{\overline{x}}$, semicontinuity, connectedness, concavity, quasiconcavity and differentiability of the superexposed part (see (3.5.20)) of $\mathcal{K}_{\overline{x}}$.

19.According to the results of Sect. 3.2, it is trivial to note that, if f and $-g$ are convex, then (3.2.7) holds for both (1.1.1) and (1.1.4). Conditions for (3.2.7) to hold (and their corresponding conditions in $\mathbb{R}^n$ or B) would be of great interest. To this end, the case where $\mathcal{Z}$ is the nonpositive level set of a suitable functional $\Phi : \mathbb{R}^{1+m} \to \mathbb{R}$ might help; Φ might depend on a finite number of real parameters and, in particular, be a quadratic form (in which case, we would apply the concept (3.2.9) in the IS). This way might be useful also to define upper bounds for (3.2.3) and thus lower bounds for (1.1.1) or (1.1.4). Of course, when the image set $\mathcal{K}_{\overline{x}}$ can be expressed as a family of manifolds which admits an envelope (like in Examples 3.4.8 and 3.4.9), then the application of condition (3.2.7) may become easy; therefore, conditions under which $\mathcal{K}_{\overline{x}}$ is a family of manifolds that admits an envelope (and, in particular, an envelope that is the frontier of a convex set) would be extremely interesting. The analysis of image coerciveness in the sense of (3.2.7) may be useful also for achieving results about duality and, in particular, duality gap (see Vol. 2); for the beginning of the analysis see [41]. All the above outlined ways have the advantage (already noted and common to the entire IS analysis) of working in a finite dimensional space — namely, the IS $\mathbb{R}^{1+m}$ — independently of the fact that the given problem be (1.1.1) or (1.1.4); this beside the possibility of achieving more general results than those which exist or can be obtained in the given space ($\mathbb{R}^n$ or B).

20.Investigation in the IS might take advantage from the introduction of a partition of the family, say $\mathcal{P}$, of problems, which are either of type (1.1.1) or of type (1.1.4). In the same class we put all members of $\mathcal{P}$ which, up to a translation with respect to the axis u, have the same image set: $\overline{P}, \hat{P} \in \mathcal{P}$, having respectively $\mathcal{K}_{\overline{x}}$ and $\mathcal{K}_{\hat{x}}$ as the image sets, belong to the same class, iff $\exists \tilde{u} \in \mathbb{R}$ s.t. $\mathcal{K}_{\overline{x}} + (O_m, \tilde{u}) = \mathcal{K}_{\hat{x}}$. In Examples 3.4.4, problems (3.4.1) and (3.4.4) have the same image set; therefore, notwithstanding the fact that their unknowns run, respectively, in a Banach and in a Euclidean space, in the IS they can be analysed with the same mathematical arguments (those of (3.4.4), of course). Such examples show that, in a same class, we can find both problems of type (1.1.1) and problems of type (1.1.4) as previously claimed. It would be interesting to define, for each class, a canonical problem to be characterized as a "simplest" problem in its class; the term simplest should receive a definition in terms of the properties of X, f, g (for instance, convexity, differentiability, and so on); can a canonical problem be always of type (1.1.1)? A related question consists in asking for conditions under which a given set of IS is the image of a problem of type (1.1.1) (and hence of type (1.1.4)). Let us come back to Examples 3.4.4: problems (3.4.1) and (3.4.5) do not have the same image set and thus they do not belong to a same class, if this is defined as above. However, they have the same conic extension (Fig. 3.4.9). Therefore, the previous concept of class can be extended by requiring that two problems have the same conic extension (Sect. 3.2) of the image set (and not necessarily the same image set) in order to be admitted to the same class (see Sect. 4.10).

21.Every transformation performed in the IS has, of course, a corresponding transformation in the given space B. The conic extension of a problem (1.1.3) induces a family of problems of the same type. More precisely, let P be a given problem, whose image set be $\mathcal{K}_{\overline{x}}$; its conic extension $\mathcal{E}(\mathcal{K}_{\overline{x}})$ induces a family, say $\{P(\xi)\}_{\xi \in \Xi}$, of problems of type (1.1.3), whose image sets are $\{\mathcal{K}_{\overline{x}}(\xi)\}_{\xi \in \Xi}$, s.t. $\exists \hat{\xi} \in \Xi$ for which

$$P(\hat{\xi}) = P; \quad \mathcal{K}_{\overline{x}}(\hat{\xi}) = \mathcal{K}_{\overline{x}}; \quad \mathcal{E}(\mathcal{K}_{\overline{x}}(\xi)) = \mathcal{E}(\mathcal{K}_{\overline{x}}), \quad \forall \xi \in \Xi,$$

while for $\xi \neq \hat{\xi}$ the equality $\mathcal{K}_{\overline{x}}(\xi) = \mathcal{K}_{\overline{x}}$ does not necessarily hold. Therefore, P can be equivalently replaced by any problem $P(\xi)$ of the family. This may be used to achieve a property which P does not enjoy. Let us give an example, by emphasizing, first of all, that the conic extension, which widens the image set (Sect. 3.2), implies a change of the given data, i.e. f and g. Let $(\hat{u}, \hat{v})$ denote a generic element of $\mathcal{K}_{\overline{x}}$, so that $\exists \hat{x} \in X$ s.t. $(\hat{u}, \hat{v}) = A_{\overline{x}}(\hat{x})$. We have:

$$\mathcal{E}(\mathcal{K}_{\overline{x}}) = \bigcup_{\hat{x} \in X} [A_{\overline{x}}(\hat{x}) - \mathrm{cl}\ \mathcal{H}] = \bigcup_{\hat{x} \in X} [(\hat{u}, \hat{v}) - \mathrm{cl}\ \mathcal{H}]. \tag{3.5.12}$$

Such a decomposition of the conic extension into the unions of cones suggests a sequential change of the data of (1.1.1) or (1.1.4). The counterimage of the set in the last square brackets of (3.5.12) is

$$X(\hat{x}) := \{x \in X : \ f(x) \geq f(\hat{x}), \ g(x) \leq g(\hat{x})\}. \tag{3.5.13}$$

Because of Proposition 3.2.7, if on $X(\hat{x})$ $f(x)$ is replaced by any function which is $\geq f(\hat{x})$, then the minimum of (1.1.1) or (1.1.4) does not change. Since in these problems we minimize, it is convenient, of course, to replace, on $X(\hat{x})$, $f(x)$ with the constant function which equals $f(\hat{x})$. This may have some advantages. For instance, in Example

3.4.11, at $\hat{x} = 1/2$ we find $X(1/2) = [1/2, +\infty[$, $f(1/2) = 2$. Therefore, $f(x)$ can be replaced by

$$f(x; \hat{x}) = \begin{cases} 4x & ,\text{if } x < \frac{1}{2}, \\ 2 & ,\text{if } x \geq \frac{1}{2}, \end{cases}$$

while g remains the same. The minimum (and now also the m.p.) of $f(x; \hat{x})$ s.t. $g(x) \geq 0$ continues to be -4 (and the m.p. equal to -1). Without changing the minimum, a transcendental function has been turned into a piece-wise linear one. Again in Example 3.4.11, at $\hat{x} = -1$ we find $f(-1) = -4$, $g(-1) = 0$, $X(-1) = \{-1\} \cup [1, +\infty[$. Then, we can replace $f(x)$ with

$$f(x; \hat{x}) = \begin{cases} 4x & ,\text{if } x < 1, \\ -4 & ,\text{if } x \geq 1. \end{cases} \tag{3.5.14}$$

Hence, the minimum of $f(x; \hat{x})$ s.t. $g(x) \geq 0$ continues to be -4, while now the set of m.p. is $\{-1\} \cup [1, +\infty[$ and contains points which are not m.p. of the given problem. $f(x; \hat{x})$ is now discontinuous. Since this might be undesirable and since any function ≥ -4 on $X(-1)$ can be chosen, we might consider, instead of (3.5.14),

$$f(x; \hat{x}) = \begin{cases} 4x, & \text{if } x < 1, \\ 2x^2 + 2, & \text{if } x \geq 1. \end{cases}$$

This way, we achieve continuity and differentiability (as it was in the given problem) and convexity; no enlargement of the set of m.p. now happens (see next comment). The use of such extensions might be useful for the methods of solutions. To this aim, conditions under which a certain property is enjoyed by the transformed (extended) problem would be very interesting. The enlargement of the set of m.p. can occur iff $f(\hat{x})$ is the minimum; therefore, if a priori we know that uniqueness holds, and if we are not yet able to obtain a m.p. of (1.1.1) or (1.1.4), but we can solve a problem of type

$$\min\ f(x; \hat{x}), \quad \text{s.t.} \quad x \in R, \tag{3.5.15}$$

and the set of m.p. of this problem is not a singleton, then we can conclude that, with the minimum of (3.5.15), we have achieved the minimum of the given problem. Other kinds of conic extensions, besides $\mathcal{E}(\mathcal{K}_{\bar{x}})$, might be conceivable. One might think of a cone which, unlike $\mathrm{cl}\,\mathcal{H}$, depends on the point of $\mathcal{K}_{\bar{x}}$ where it is applied. Indeed, Proposition 3.2.7 can continue to hold, if at some points of $\mathcal{K}_{\bar{x}}$ we extend with a cone "wider" than $\mathrm{cl}\,\mathcal{H}$. Note that (3.2.6b) is a conic extension of the graph of Φ. As well as it is useful to investigate about $\mathcal{E}(\mathcal{K}_{\bar{x}})$, it is interesting to analyse the conic extension of the homogeneization, or $\mathcal{E}(\mathcal{K}_{\bar{x}}^h)$. More generally, the topological properties of a generic conic extension (Definition 3.2.1) are of interest. For instance, conditions under which $\mathcal{E}$ turns out to be such that, $\forall z \in \mathcal{E}(\mathcal{K}_{\bar{x}})$, $\mathcal{E}(\mathcal{K}_{\bar{x}}) \cap (\{z\} + \mathrm{cl}\,\mathcal{H})$ is closed, or bounded, or a polytope, and so on (the convexity of the above intersection happens iff $\mathcal{E}(\mathcal{K}_{\bar{x}})$ is convex; this is characterized by Proposition 3.2.8). Furthermore, it is important to know conditions under which a topological property of $\mathcal{K}_{\overline{x}}$ is invariant with respect to the conic extension. For instance, to know conditions under which a boundary point of $\mathcal{K}_{\overline{x}}$ (belonging or not to $\mathcal{K}_{\overline{x}}$) is a boundary point of $\mathcal{E}(\mathcal{K}_{\overline{x}})$ too (see Examples 2.2.1 and 3.4.2 for the two alternatives), a point of the positive u-semi-axis is of accumulation for $\mathcal{K}_{\overline{x}} \setminus \mathcal{H}$ (or for $\mathcal{E}(\mathcal{K}_{\overline{x}}) \setminus \mathcal{H}$; see Examples 3.4.12 and 3.4.13), and then to express such conditions in terms of X, f and g. The previous questions exist also for the homogenization of $\mathcal{K}_{\overline{x}}$ and the related conic extension. There already exist very interesting results in the analysis of the image of a function, which are fundamental for investigating the previous topics; see for instance [1, 23, 43].

The concept of conic extension should be useful also for problems of type (1.1.5). For a first attempt in this direction see Definitions 4.8.1 and 4.8.2, and a related comment in Sect. 4.10.

22.With regard to the above mentioned possible use of the conic extensions in the numerical methods for solving (1.1.1), consider the following:

Example 3.5.1. (Continuation of Example 3.4.21). Assume that, by means of a descent method (or any other method), the local m.p. $\overline{x} = 2$ has been reached. This being an isolated m.p. at which f is locally strictly convex, it is not trivial to get away from $\overline{x}$ and go towards the global one. The partial conic extension performed in Example 3.4.21 (see Fig. 3.4.25b), which is suggested by the decomposition (3.5.12), allows us to escape from the hole. Consider the function:

$$\varphi(x; \overline{x}) := f(\overline{x}) + \frac{1}{2}[f(x) - f(\overline{x}) - |f(x) - f(\overline{x})|] =$$

$$\frac{3}{4} + \frac{1}{2}\left[x^4 - \frac{13}{3}x^3 + 5x^2 - \frac{4}{3} - \left| x^4 - \frac{13}{3}x^3 + 5x^2 - \frac{4}{3} \right| \right], \quad x \leq 2,$$

which corresponds to the counterimage of the partial extension of Example 3.4.21. Taking into account the expression in the last square brackets of (3.5.12), we have:

$$\operatorname{epi} \varphi = \operatorname{epi} f_{/]-\infty,2]} \cup \{(x, y) \in \mathbb{R}^2 : x \leq 2, \, y \geq f(\overline{x})\}.$$

Therefore, by replacing, in Example 3.4.21, $f(x)$ with

$$f(x; \overline{x}) := \begin{cases} \varphi(x; \overline{x}), & \text{if } x \leq 2, \\ f(x), & \text{if } x > 2, \end{cases}$$

we have that $\overline{x}$ is no longer an isolated m.p., and f is no longer strictly convex at $\overline{x}$, so that it is less difficult to escape from $\overline{x}$, since we can travel towards the global m.p. without being obliged to initially increase f with respect to $f(\overline{x})$, but we can go down in a non-increasing way. However, we can improve it and obtain to go in a decreasing way by means of a perturbation method. For instance, on $]-\infty, 2]$, φ can be replaced with

$$\tilde{\varphi}(x; \overline{x}; \varepsilon) := \varphi(x; \overline{x}) - \varepsilon ||x - \overline{x}||^2,$$

with $\varepsilon > 0$ and small enough. Then $\overline{x}$ is no longer a m.p. of

$$\tilde{f}(x;\overline{x};\varepsilon) := \begin{cases} \tilde{\varphi}(x;\overline{x};\varepsilon), & \text{if } x \leq 2 \\ f(x), & \text{if } x > 2. \end{cases}$$

Now we go on and decrease $\tilde{f}$ until we meet a value of $\tilde{f}$ which is also a value of f; in the example, this happens close to $x = 1/2$. Then we repeat the method. $\qquad\square$

23.Previously in this section, a comment has been made about local extrema and neighbourhoods in the given space B and in the IS. This is related to the above remark on the boundary points. As a further comment on both aspects, consider the following. $\hat{z} \in \mathcal{K}_{\overline{x}}$ will be called *exposed point* of $\mathcal{K}_{\overline{x}}$, iff $\hat{x} \in r\partial\, \mathcal{E}(\mathcal{K}_{\overline{x}})$. With regard to (1.1.1) or (1.1.4), we have the following instances of possible results.

Proposition 3.5.1. If $\hat{z} := (\hat{u}, \hat{v})$ is a local maximum point of (3.2.15), then $\hat{z}$ is an exposed point of $\mathcal{K}_{\overline{x}}$.

Proof. The assumption implies that $\exists \delta_1 > 0$ s.t.

$$\mathcal{H} \cap [\mathcal{E}(\mathcal{K}_{\overline{x}}) - (\hat{u}, O)] \cap \mathcal{N}_{\delta_1}(\hat{z} - (\hat{u}, O)) = \varnothing, \tag{3.5.16}$$

where $\mathcal{N}_r(z)$ denotes a neighbourhood (of the IS) with centre at z and radius r. Ab absurdo, suppose that $\hat{z} \in \text{ri}\, \mathcal{E}(\mathcal{K}_{\overline{x}})$, so that $\exists \delta_2 > 0$ s.t.

$$\mathcal{N}_{\delta_2}(\hat{z} - (\hat{u}, O)) \cap \text{aff}\, \mathcal{E}(\mathcal{K}_{\overline{x}}) \subseteq \mathcal{E}(\mathcal{K}_{\overline{x}}) - (\hat{u}, O). \tag{3.5.17}$$

Set $\delta = \min\{\delta_1, \delta_2\}$. Of course, (3.5.17) implies:

$$\mathcal{N}_{\delta}(\hat{z} - ((\hat{u}, O)) \cap \text{aff}\, \mathcal{E}(\mathcal{K}_{\overline{x}}) \subseteq \mathcal{E}(\mathcal{K}_{\overline{x}}) - (\hat{u}, O). \tag{3.5.18}$$

Taking into account that $\hat{z} - (\hat{u}, O) = (0, \hat{v})$ and that $\hat{v} \in D$, we draw:

$$\mathcal{N}_{\delta}(\hat{z} - (\hat{u}, O)) \cap \text{aff}\, \mathcal{E}(\mathcal{K}_{\overline{x}}) \cap \mathcal{H} \neq \varnothing. \tag{3.5.19}$$

(3.5.18) and (3.5.19) contradict (3.5.16). $\qquad\square$

Corollary 3.5.2. If $\hat{x}$ is a local m.p. of (1.1.1) or (1.1.4), then there exists a neighbourhood $N(\hat{x})$ of $\hat{x}$, whose image is not a neighbourhood of the image $\hat{z}$ of $\hat{x}$, namely

$$A_{\overline{x}}(N(\hat{x})) \neq \mathcal{N}(\hat{z}).$$

Proof. Obvious consequence of Proposition 3.5.1 and of Proposition 3.2.7. $\qquad\square$

Proposition 3.5.1 cannot be inverted, as the 1st of Examples 3.4.16 shows: $\overline{z} := A_{\overline{x}}(\overline{x}) = (0, 0, 0)$ is an exposed point of $\mathcal{K}_{\overline{x}}$, but is not maximum point of (3.2.3). However, something can be stated by strengthening the above definition. $\hat{z} \in \mathcal{K}_{\overline{x}}$ will be called *superexposed point* of $\mathcal{K}_{\overline{x}}$, iff

$$\hat{z} \notin \{z\} - \mathcal{H}, \quad \forall z \in \mathcal{K}_{\overline{x}}. \tag{3.5.20}$$

$\hat{z}$ is exposed if it is superexposed. In fact, ab absurdo, suppose that $\exists z^0 \in \mathcal{K}_{\overline{x}}$ s.t. $\hat{z} \in \text{ri}\,(\{z^0\} - \text{cl}\,\mathcal{H})$; being this equivalent to $\hat{z} \in \text{ri}\,(\{z^0\} - \mathcal{H})$, (3.5.20) is contradicted.

In the 2nd of Examples 3.4.16, $\bar{z} := A_{\bar{x}}(\bar{x}) = (0, 0, 1)$ is exposed, but not superexposed, and is not maximum point of (3.2.16) at $\xi = O$; $z^1 := (1, 0, 0)$ and $z^2 := (1, 0, 2)$ are superexposed and maximum points of (3.2.16) at $\xi = O$.

Proposition 3.5.2. If $\hat{z} := (\hat{u}, \hat{v})$ with $\hat{v} \in D$ is a superexposed point of $\mathcal{K}_{\bar{x}}$, then $\hat{z}$ is a maximum point of (3.2.16) with $\xi = \hat{v}$.

Proof. Ab absurdo, suppose that $\exists \tilde{z} := (\tilde{u}, \tilde{v}) \in \mathcal{E}(\mathcal{K}_{\bar{x}})$ with $\tilde{v} \in \{\hat{v}\} + D$ and s.t. $\tilde{u} > \hat{v}$. Therefore, $\exists z^0 \in \mathcal{K}_{\bar{x}}$ s.t. $\tilde{z} \in \{z^0\} - \mathrm{cl}\,\mathcal{H}$. Then, we have the following implications:

$$\left. \begin{array}{c} \tilde{v} \in \{\hat{v}\} + D \quad \Rightarrow \quad \hat{v} \in \{\tilde{v}\} - D \\ \hat{u} < \tilde{v} \end{array} \right\} \quad \Rightarrow \quad \left\{ \begin{array}{c} \hat{z} \in \{\tilde{z}\} - \mathcal{H} \\ \tilde{z} \in \{z^0\} - \mathrm{cl}\,\mathcal{H} \end{array} \right\} \quad \Rightarrow \quad \hat{z} \in \{z^0\} - \mathcal{H},$$

the last of which contradicts (3.5.20). $\square$

A Corollary of this proposition can be stated, as well as Corollary 3.5.2 has been obtained from Proposition 3.5.1. The deepening of these aspects in the IS and their correspondences with the given space may lead to useful results.

24.In Sect. 3.2 the image set has been defined in such a way that it intersects the halfspace $\{(u, v) \in \mathbb{R}^{1+m} : u \geq 0\}$. This does not allow us to achieve results on (finite) lower bounds for the minimum of (1.1.1) or (1.1.4), if any exist. Such a scheme has been generalized in [V29, V30], where, among many nice results, some statements deal with lower bounds; this is very important for the applications. In the same papers, several interesting questions are opened for deepening the IS Analysis.

25.As noted in Sect. 3.2, when there are bilateral constraints ($p > 0$), then $\mathrm{int}\,\mathcal{H} = \varnothing$ and (i) of Theorem 3.2.1 becomes useless; the investigation and extension of (i) to the case $p > 0$, as well as the comparison of (4i) with Theorem 2.2.7, would be interesting. Starting with Theorem 3.2.1, it would be useful to achieve equivalent propositions in terms of X, f and g.

26.Theorem 3.2.1 shows the importance of the tangent cone to the image set and to its conic extension. Of course, as noted in Sect. 2.1, the tangent cone may be a bad representative of the image set, so that, in some cases, the reachable and admissible cones (Definitions 2.1.10 and 2.1.11) may be preferable. Therefore, it is crucial to know the properties of the sets ($\mathcal{E}(\mathcal{K}_{\bar{x}})$ is shorten as $\mathcal{E}$):

$$TC(\bar{z}; \mathcal{K}_{\bar{x}}), \ TC(\bar{z}; \mathcal{E}), \ RC(\bar{z}; \mathcal{K}_{\bar{x}}), \ RC(\bar{z}; \mathcal{E}), \ AC(\bar{z}; \mathcal{K}_{\bar{x}}), \ AC(\bar{z}; \mathcal{E}).$$

The literature about cones is wide. Besides, there already exist some investigations made in the IS. For instance, the differential and derived cones introduced and exploited by Hestenes in the IS [24, 25], and the theory of tents introduced by Boltyanski (see [I6]), have showed to be very fruitful concepts; their embedding in the general IS scheme and comparison with the above cones should lead to useful results.

27. The homogenization (or linearization) of the image set and the related properties, especially the so-called Homogenization Lemma (see Proposition 3.2.6 (iii) and Theorem 3.3.4), are of fundamental importance for establishing necessary optimality conditions.

It would be useful to generalize the above propositions by weakening the assumptions, and to extend them to the case of bilateral constraints (for which a strengthening of the assumptions seems unavoidable).

28.From Theorem 3.2.3 several sufficient conditions for the existence of the minimum can be drawn. In [47] there are some instances of such conditions, which are not of Weierstrass type. The deepening of this field and the extension of the analysis to other kinds of problems, like (1.1.7)-(1.1.12) and systems of Sect. 1.3, would be very interesting. To this end, the condition expressed by Corollary 3.2.3 is too restrictive and should be weakened. The concept of cono-compactness of a set — namely, the compactness of the intersections of the set with a given cone applied to every (or some) element of the set — might be useful. However, the mere $(\mathrm{cl}\,\mathcal{H})$-compactness of $\mathcal{K}_{\overline{x}}$ (or its intersection with $\mathrm{cl}\,\mathcal{H}$) is not enough, as shown by Examples 3.2.4 and 3.2.5, where $\mathcal{K}_{\overline{x}}$ is $(\mathrm{cl}\,\mathcal{H})$-compact: indeed, $\forall z \in \mathcal{K}_{\overline{x}}$, we have $\mathcal{K}_{\overline{x}} \cap (z + \mathrm{cl}\,\mathcal{H}) = \{z\}$.

29.An important topic is, obviously, the uniqueness of solutions. In this sense, all known conditions for uniqueness can be applied to problem (3.2.3). Among the several advantages of the IS Analysis, there is not that of uniqueness. Indeed, uniqueness of maximum points of (3.2.3) does not imply, of course, that of m.p. of (1.1.1) or (1.1.4). It would be interesting to overcome such a drawback. The following remarks aim to give hints for overcoming it.

Proposition 3.5.3. A m.p. $\overline{x}$ of (1.1.1) is unique iff, $\forall \xi \in \mathbb{R}^n$, the (parametric) system (in the unknown x)

$$\langle \xi, x - \overline{x} \rangle > 0, \quad f(\overline{x}) - f(x) \geq 0, \quad g(x) \in D, \quad x \in X \qquad (3.5.21)$$

is impossible.

Proof. Only if. Ab absurdo, let $\hat{x}$ be a solution of (3.5.21). The last 2 conditions of (3.5.21) imply the feasibility of $\hat{x}$ for (1.1.1); the 3rd says that $\hat{x}$ is a m.p.; and the 1st implies $\hat{x} \neq \overline{x}$, which contradicts the assumption. **If.** Let $\hat{x} \neq \overline{x}$ be a further m.p. of (1.1.1); then we have:

$$f(\hat{x}) - f(\overline{x}) = 0, \quad g(\hat{x}) \in D, \quad \hat{x} \in X.$$

If we set $\hat{\xi} := \hat{x} - \overline{x}$, so that $\langle \hat{\xi}, \hat{x} - \overline{x} \rangle > 0$, system (3.5.21) turns out to admit the solution $\hat{x}$ at $\xi = \hat{\xi}$, and this contradicts the assumption. $\qquad\square$

Each of systems (3.5.21) is quite analogous to (3.2.1), and thus they can receive the same analysis and development as (3.2.1). Now we have a family of image sets depending on ξ; the IS is $\mathbb{R}^{2+m}$; a condition like (3.2.2) must be considered for each ξ. Application of the existence Theorem 3.2.3 to the present scheme should produce uniqueness conditions recovering known ones. For instance, in (1.1.1) set $n = m = 1$, $p = 0$, $X = \mathbb{R}$, $f(x) = (x^2 - 1)^2$, $g(x) = x + 1$. $\overline{x} = 1$ is obviously a (global) m.p. System (3.5.21) becomes:

$$\xi(x - 1) > 0, \quad -(x^2 - 1)^2 \geq 0, \quad x + 1 \geq 0, \quad x \in \mathbb{R},$$

and is possible in agreement with the fact that $\overline{x}$ is not unique. For the same problem system (3.2.1) becomes:

$$-(x^2 - 1)^2 > 0, \quad x + 1 \geq 0, \quad x \in \mathbb{R},$$

and is impossible. The image set associated to the former system is in $\mathbb{R}^3$, while that associated to the latter is in $\mathbb{R}^2$. Note that the image sets associated with (3.2.1) and with (3.5.21) are strictly related; in a certain sense (up to deletion of one coordinate), the former is a projection of the latter into a subspace. Note that system (3.5.21) can be viewed as system (3.2.1) associated to a constrained extremum problem, whose feasible region is the intersection of R with a (lower) level set of f, and whose objective function is linear. Of course, the above scheme profits of special conditions, like convexity of X, f and $-g$. Proposition 3.5.3 can be extended to (1.1.4).

Example 3.4.5 shows a problem, where f is not l.s.c. even in every neighbourhood of the global m.p., and suggests to investigate problems (1.1.3) $-$ as well as (1.1.4) and (1.1.5) $-$, in the case where the minimum exists, but f is not l.s.c. on X (even if it does on R). The $(\mathrm{cl}\,\mathcal{H})$-compactness of $\mathcal{K}_{\overline{x}}$ or $\mathcal{E}(\mathcal{K}_{\overline{x}})$ might be of help.

30. When (1.1.3) looks too much difficult, a classic approach consists in decomposing it into subproblems which look easier than it. Since the decomposition has assumed most variegated forms, a general definition of decomposition would be cumbersome and, perhaps, useless. Therefore, by considering one type of decomposition at a time, due to the fact that the decompositions have been conceived in the given space ($\mathbb{R}^n$ or B), it would be interesting to apply them to (3.2.3). This should lead one to weaken the assumptions under which the considered decomposition works. As an instance, let us consider one of the most famous decompositions, known as Bellman Recurrence Equation or Maximum Principle [II4]. Indeed, in his writings $-$ which are a mine of ideas $-$ Bellman exposed the general feature of his decomposition method (called, appropriately, principle and not theorem), and applied it to several types of (1.1.1) and, later, to problems of types (1.1.4) and (1.1.5). A rigorous formulation of Bellman Principle $-$ called Fundamental Theorem of Dynamic Programming $-$ has been given by M. Volpato in 1961 for problems of type (1.1.1) and, subsequently, extended to problems of types (1.1.4) and (1.1.5) [V35]. Let us now outline this theorem. With regard to (1.1.3), let us consider the following conditions:

(C_1) $\exists R_i \subset \mathbb{R}^i$, $i = 1, ..., n$, s.t. $\forall i = 1, ..., n$ there exists a partition of R_i, say $R_i(\xi_i)$, $\xi_i \in \Xi_i$, where Ξ_i is s.t. $\underset{\xi_i \in \Xi_i}{\cup} R_i(\xi_i) = R_i$ and s.t. $\exists \xi_n^0 \in \Xi_n$ s.t. $R_n(\xi_n^0) = R$;

(C_2) $\forall i = 2, ..., n$ there exists a homeomorphism, say h_i (the dependence on ξ_{i-1} and ξ_i is understood), which sends a part of R_{i-1} onto a part of R_i;

(C_3) $\forall i = 2, ..., n$ there exists a function $\varphi_i : R_i \to \mathbb{R}$, whose restriction to $R_i(\xi_i)$ is called φ_{ξ_i}, and a function $\Phi_i : \mathbb{R} \times \Xi_{i-1} \times \Xi_i \to \mathbb{R}$, increasing with respect to the 1st argument, s.t.

$$\varphi_{\xi_i}(h_i(R_{i-1}(\xi_{i-1}))) = \Phi_i(\varphi_{\xi_{i-1}}(R_{i-1}(\xi_{i-1})), \xi_{i-1}, \xi_i).$$

The following theorem gives the above mentioned result, whose proof is based on Theorem 1.1.3 [V35].

Theorem 3.5.2. If conditions (C_1)-(C_3) are fulfilled, R is compact together with $R_1, ..., R_n$, and $f, \varphi_{\xi_i}, i = 2, ..., n$ are l.s.c., then,

$$\min_{z^i \in R_i(\xi_i)} \varphi_{\xi_i}(x^i) = \min_{\xi_{i-1} \in \Xi_{i-1}} \Phi\big(\min_{\xi_{i-1} \in \Xi_{i-1}} \varphi_{\xi_{i-1}}(\xi_{i-1}), \xi_{i-1}, \xi_i\big),$$

$$\min_{\xi_{i-1} \in \Xi_{i-1}} \varphi_{\xi_{i-1}}(\xi_{i-1}) = \min_{x^{i-1} \in R_{i-1}} \varphi_{i-1}(x^{i-1}).$$

The strict isotonicity assumption on Φ_i is only sufficient. From the above recurrence equation, as said at the beginning of Sect. 3.2, in the Bellman approach, the unknown of each subproblem is in the corresponding IS. It would be interesting to apply such an approach in the IS associated to (1.1.3). This might be done either by decomposing the image set $\mathcal{K}_{\overline{x}}$ (or its conic extension) according to the above conditions (C_1)-(C_2), or by considering the sequence of the image sets associated to the n subproblems. The scope should be to overcome the two assumptions, existence of a homeomorphism and strict isotonicity, which are restrictive for the applications. The above decomposition scheme is strictly related to the classic theory of partial maxima and minima [V98, V99], and to the "Method of bundle" developed in the 1940 by L. Campedelli for finding multiple points of algebraic curves (see L.Campedelli, "Lezioni di Geometria (Lectures on Geometry; in Italian)", Vol. II, 2nd part, pages 61, 98-105, Published by CEDAM, Padova, 1958).

31.With regard to the function $A_{\overline{x}}(x) = (f(\overline{x}) - f(x), g(x))$ of Sect. 3.2, the condition of Definition 2.4.5 for $H = \mathrm{cl}\,\mathcal{H}$ becomes: $A_{\overline{x}}$ is $(\mathrm{cl}\,\mathcal{H})$-concavelike, iff $\forall x^1, x^2 \in X$, and $\forall \alpha \in [0, 1]$, $\exists \hat{x} \in X$, s.t.

$$\begin{aligned}
f(\hat{x}) &\leq (1 - \alpha)f(x^1) + \alpha f(x^2), \\
g_i(\hat{x}) &= (1 - \alpha)g_i(x^1) + \alpha g_i(x^2), \quad i \in \mathcal{J}^0, \\
g_i(\hat{x}) &\geq (1 - \alpha)g_i(x^1) + +\alpha g_i(x^2), \quad i \in \mathcal{J}^+,
\end{aligned} \qquad (3.5.22)$$

or, in terms of (2.5.15),

$$(\mathrm{lev}_{\leq z_0(\alpha)} f) \cap \bigcap_{i \in \mathcal{J}^0} (\mathrm{lev}_{= z_i(\alpha)} g_i) \cap \bigcap_{i \in \mathcal{J}^+} (\mathrm{lev}_{\geq z_i(\alpha)} g_i) \neq \varnothing,$$

where the $z_i(\alpha)$'s denote the right-hand sides of (3.5.22), respectively. What has been said for the general case (2.5.15) holds of course for this special − but extremely important − case, as Proposition 3.2.8 shows. What we need are conditions for the map $A_{\overline{x}}$ to be $(\mathrm{cl}\,\mathcal{H})$-concavelike or (3.5.22) to hold; the final destination of our logic chain is the convexity of the conic extension $\mathcal{E}(\mathcal{K}_{\overline{x}})$ which, as we will see in Chapter 5, guarantees the existence of multipliers. However, due to the difficulty of establishing such conditions, the following hint might be of some help. Even if it seems a nonsense, one might start from the final point, i.e. $\mathcal{E}(\mathcal{K}_{\overline{x}})$, and, in the IS, try to detect conditions on the image set $\mathcal{K}_{\overline{x}}$ for $\mathcal{E}(\mathcal{K}_{\overline{x}})$ to be convex; then, in a backward movement, to achieve conditions on the map $A_{\bar{x}}$. Only to explain this, consider the simple Examples 3.4.17. In the 1st case, the nonconvexity of $\mathcal{K}_{\bar{x}}$ is eliminated by the conic extension. In fact, taking into account the remark around (3.5.12)-(3.5.13), consider for instance the interval [-1,1], where f is nonconvex. At $\hat{x} = 1$, the set (3.5.13) becomes:

$$X(1) = \{x \in \mathbb{R} : f(x) \geq 0,\, g(x) \leq 1\} =]-\infty, 1].$$

Then f can be replaced, on $]-\infty, 1]$, by the constant function which equals 0, so that $f(x)$ can be replaced by:

$$f(x; 1) := \left\{ \begin{array}{ll} 0 & \text{, if } x \leq 1, \\ f(x) & \text{, if } x > 1, \end{array} \right.$$

which is convex. This ensures that the map $A_{\bar{x}} = (f(\bar{x}) - f(x), g(x))$ is (cl $\mathcal{H}$)-concavelike or simply concavelike, since now cl $\mathcal{H} = \mathbb{R}^2_+$. The strict relationship between the problem of finding the "counterimage" of $\mathcal{E}(\mathcal{K}_{\bar{x}})$ and that of finding condition for (3.5.22) to hold might be of help. In the 2nd of Examples 3.4.17, even if we consider the lowest value of $\tilde{f}$, the nonconvexity is not eliminated by the conic extension. In fact, at $\hat{x} = 1$, the set (3.5.13) becomes $X(1) =]-\infty, 1]$ and, on it, $\tilde{f}$ can be replaced by the constant function which equals 0; the nonconvexity of $\tilde{f}$ remains untouched. At $\hat{x} = 3$, we achieve the maximal removal of the nonconvexity of $\tilde{f}$. Thus, in $[1,3]$, $\tilde{f}$ can be replaced by:

$$\tilde{f}(x; \hat{x} = 3) := \left\{ \begin{array}{ll} \tilde{f}(x) & \text{, if } x \in [1, v^0[, \\ \frac{4}{3} & \text{, if } x \in [v^0, 3], \end{array} \right.$$

where v^0 is as in Examples 3.4.17. Unfortunately $\tilde{f}$ is still nonconvex; the map $A_{\bar{x}}$ is not, in the present case, concavelike. Example 3.4.18 shows a problem, where both f and $-g$ are not convex even in every neighbourhood of the m.p. $\bar{x} = (0,0)$. Notwithstanding this, $\mathcal{K}_{\bar{x}}$ and $\mathcal{E}(\mathcal{K}_{\bar{x}})$ turn out to be convex. According to Proposition 3.2.8(i), the map

$$A_{\bar{x}}(x) = (f(\bar{x}) - f(x),\, g(x)) = (-x_1^2 + 3x_1 x_2 + x_2^2,\, -x_1 x_2)$$

is concavelike. We note that, in this case, f is of the following type:

$$f(x) = \phi(x) + kg(x), \quad k \in \mathbb{R}_+,$$

with φ strictly convex. Thus, in the above order of ideas, it would be interesting to find classes of functions ϕ and g, s.t. $\mathcal{E}(\mathcal{K}_{\bar{x}})$ is convex, even if $\mathcal{K}_{\bar{x}}$ is not.

32. With the homogenization of Definition 3.2.2, we replace (1.1.3) with a "conic approximation", namely (3.2.10); then we consider the image set of (3.2.10), namely $\mathcal{K}_{\bar{x}}^h$. Alternatively, we might consider the image set of (1.1.3), namely $\mathcal{K}_{\bar{x}}$ and, then, perform a conic approximation (by means, for instance, of one of the cones of Sect. 2.1) of $\mathcal{K}_{\bar{x}}$. In general, these 2 operations are not permutable. For instance, in Example 3.4.1, the former operation leads to the set $\{(u, v) \in \mathbb{R}^2 : u = 0\}$, while the latter

leads to a wider set, whatever cone of Sect. 2.1 is adopted (the union of 2nd and 4th quadrants of Fig. 3.4.1, if the tangent cone is chosen). It would be interesting to find conditions which, depending on given conic approximations, ensure the permutability of the 2 operations or, at least, an inclusion between their outcomes.

33.Decomposition of the feasible region R of (1.1.3) and analysis of the implications on the image set $\mathcal{K}_{\bar{x}}$ or, vice versa, decomposition of $\mathcal{K}_{\bar{x}}$ and analysis of the implications on R may be fruitful. For instance, if $\{R(\xi)\}_{\xi\in\Xi}$ is a partition or a cover of R and $\mathcal{K}_{\bar{x}}(\xi)$ is the image set of (1.1.3) where R has been replaced by $R(\xi)$, then obviously

$$\bigcup_{\xi\in\Xi}\mathcal{K}_{\bar{x}}(\xi) = \mathcal{K}_{\bar{x}}.$$

The sets $\mathcal{K}_{\bar{x}}(\xi)$ or their conic extensions may turn out to be convex, if f and R, or even $R(\xi)$, do not. In Example 3.4.1, if X is decomposed into $X_1 := \{x \in \mathbb{R}^2 : x_2 \leq 0\}$ and $X_2 := \{x \in \mathbb{R}^2 : x_2 \geq 0\}$, then $\mathcal{K}_{\bar{x}}$ is decomposed into 2 convex sets, whose closures are the 2nd and 4th quadrants; unlike $\mathcal{K}_{\bar{x}}$, each of these sets is now linearly separable from $\mathcal{H}$. Instead of a decomposition, a composition can be considered. For instance, $R(\xi)$ and $\mathcal{K}_{\bar{x}}(\xi)$ can be s.t.

$$\bigcap_{\xi\in\Xi}R(\xi) = R \quad \text{or} \quad \bigcap_{\xi\in\Xi}\mathcal{K}_{\bar{x}}(\xi) = \mathcal{K}_{\bar{x}}.$$

This can happen, if the constraints of (1.1.3) are considered in a sequential way, by introducing m subproblems obtained from (1.1.3) where we consider the first constraint, the first two,..., all the constraints. Another form consists in considering R as intersection of m sets, each defined by one of the m constraints, as in the previously mentioned Dubovitskii-Milyutin Theory. A further form consists in a composition, as previously described in Bellman Theory.

34.In [4] the Authors consider a problem of type (1.1.7), and associate it with an integer vector, called "image" of the given problem. Then they show some interesting applications of this vector and how to compute it for several special combinatorial problems. In accordance with what has been said at the beginning of Sect. 3.2, the Authors introduce such a concept independently of the existing literature. Now, let us express their definition of image in our notation. Let $\xi \in \mathbb{R}_+$. In (3.2.3) add a constraint to obtain the problem:

$$\max(u), \quad \text{s.t.} \quad (u,v) \in \mathcal{K}_{\bar{x}}, \ v \in D, \ u \leq u_{\bar{x}} - \xi, \qquad (3.5.23)$$

which is a perturbed problem, the perturbation being in the objective function (translation of $\mathcal{K}_{\bar{x}}$ along u-axis), unlike (3.2.5) where the perturbation affects the constraints (translation along the hyperplane $u = 0$). According to (3.2.1), (3.5.23) is equivalent to consider, instead of (1.1.3), the problem:

$$\min f(x), \quad \text{s.t.} \quad x \in R, \ f(x) \geq f^{\downarrow} + \xi. \qquad (3.5.24)$$

The maximum of (3.5.23) and the minimum of (3.5.24), as functions of ξ, are strictly related to the perturbation function $u_{\overline{x}}(\xi, D)$ in (3.2.5) and $f^{\downarrow}(\xi)$ in (3.2.4), respectively. If ξ is restricted to be positive and integer, and if f is integer-valued with a finite image on the feasible region, then by solving (3.5.24) for the several values of ξ we obtain, with the corrisponding minima, an integer vector, which is precisely the image in the sense of [4]. A systematic application to (1.1.7) of the IS analysis outlined in Sect. 3.2 should lead to useful results (see also Sect. 4.10).

35.It might be useful to look at problem (1.1.1) $-$ as well as to problems (1.1.4) and (1.1.5) $-$ as the envelope of the family of its homogenizations (in particular, linearizations), which are given by (3.2.10). Of course, we must give a sense of this sentence. First of all, let us note that, in order to define (3.2.10), we have assumed the $\mathcal{C}$-differentiability of f and $-g$ precisely at $\overline{x}$, this being the point which has been used to construct the image set $\mathcal{K}_{\overline{x}}$. Such a coincidence is due to the fact that, if we only look for optimality, we need $\mathcal{C}$-differentiability only at the point which is candidate for becoming m.p., i.e. $\overline{x}$. Now, the scope is different. Indeed, we want to analyse the entire $\mathcal{K}_{\overline{x}}$, in the sense that we want to homogenize it at every point, and not only at $\overline{x}$. Therefore, the point adopted to define the image set and that where we homogenize f and g must not necessarily coincide. Hence, $\mathcal{K}_{\overline{x}}^h$ is now replaced by $\mathcal{K}_{\overline{x}}^h(\xi)$, which is defined by the following parametric equations:

$$\begin{cases} u = f(\overline{x}) - f(\xi) - \mathcal{D}_{\mathcal{C}}f(\xi; x - \xi) \\ v_i = g_i(\xi) + \mathcal{D}_{-\mathcal{C}}\, g_i(\xi; x - \xi), \quad i \in \mathcal{I}, \\ \xi \in X. \end{cases} \tag{3.5.25}$$

It would be interesting to find conditions under which $\mathcal{K}_{\overline{x}}^h$ is the envelope of the family $\{\mathcal{K}_{\overline{x}}^h(\xi)\}_{\xi \in X}$. If the dependence of $\mathcal{K}_{\overline{x}}^h(\xi)$ on ξ is smooth, then we can use the classic theory of envelopes [53]. For instance, in Example 3.4.19, f and g are derivable, so that (3.5.25) become the 2 equations which appear in (3.4.11); $\mathcal{K}_{\overline{x}}^h$ is the envelope of its linearizations. Unfortunately, to the best of our knowledge, the classic theory of envelopes (used mainly for detecting singular solutions of differential equations) has not received a substantial development. Independently of the applications to extremum problems, extensions of the theory of envelopes would be of great interest. As a hint, let us make some remarks with reference to extremum problems. Consider a generic $\xi \in X$ and a perturbation $\xi + \triangle \in X$; this consists in giving a perturbation to the set (3.5.25). If there exists a point in $\mathcal{K}_{\overline{x}}^h(\xi) \cap \mathcal{K}_{\overline{x}}^h(\xi + \triangle)$, then there must exist x_ξ and $x_{\xi+\triangle}$ which are the counterimages of such a point, so that

$$\begin{cases} f(\overline{x}) - [f(\xi) + \mathcal{D}_{\mathcal{C}}f(\xi; x_\xi - \xi)] = f(\overline{x}) - [f(\xi + \triangle) + \mathcal{D}_{\mathcal{C}}f(\xi + \triangle; x_{\xi+\triangle} - (\xi + \triangle))], \\ g_i(\xi) + \mathcal{D}_{-\mathcal{C}}\, g_i(\xi; x_\xi - \xi) = g_i(\xi + \triangle) + \mathcal{D}_{-\mathcal{C}}\, g_i(\xi + \triangle; x_{\xi+\triangle} - (\xi + \triangle)), i \in \mathcal{I}, \\ \xi, \xi + \triangle \in X. \end{cases} \tag{3.5.26}$$

Hence, we must have:

$$
\begin{cases}
\displaystyle\lim_{\|\triangle\|\downarrow 0} \frac{1}{\|\triangle\|}\{[f(\xi+\triangle)-f(\xi)]+[\mathcal{D}_{e}f(\xi+\triangle;x_{\xi+\triangle}-(\xi+\triangle))- \\
\qquad\qquad\qquad\qquad \mathcal{D}_{e}f(\xi;x_{\xi}-\xi)]\}=0, \\
\\
\displaystyle\lim_{\|\triangle\|\downarrow 0} \frac{1}{\|\triangle\|}\{[g_i(\xi+\triangle)-g_i(\xi)]+[\mathcal{D}_{-e}\,g_i(\xi+\triangle;x_{\xi+\triangle}-(\xi+\triangle))- \\
\qquad\qquad\qquad\qquad \mathcal{D}_{-e}g_i(\xi;x_{\xi}-\xi)]\}=0,\, i\in \mathfrak{I}.
\end{cases}
\tag{3.5.27}
$$

A nonsmooth example can be contructed by means of Example 3.1.2. In (1.1.1), set $g(x)=x\in X=\mathbb{R}$, and let f be that of Example 3.1.2. Then, at $\overline{x}=0$, $\mathcal{K}^h_{\overline{x}}(\xi)$ is either a line (linear) or 2 halflines which have a common apex (superlinear). This can be found through (3.5.27). However, in general, the use of (3.5.27) to eliminate the parameter ξ and then find the equations of the envelope (if any) is impracticable. The natural way would be, like in the smooth case, to have at disposal a mean value theorem which be applicable to the square brackets of (3.5.27). The above remarks are based on $\mathcal{C}$-differentiability; of course, one might consider any other kind of conic approximation of the image set $\mathcal{K}_{\overline{x}}$. Now, let us extend such remarks to other models. As an instance, consider (1.3.5) and the following array:

	ENVELOPE	APPROXIMATION
EXTREMUM PROBLEMS	P	$P(\xi)$
VARIATIONAL INEQUALITIES	?	$VI(\xi)$

In the above array, in the entry (1,1) we have located an extremum problem like (1.1.1), denoted generically by P, and in the entry (1,2) its homogenization, namely a problem of type (3.2.10), where the functions f and g have been expanded at ξ (not necessarily coincident with $\overline{x}$), which is denoted by $P(\xi)$. Note that, $\overline{x}$ being fixed, unlike the entry (1.1), the entry (1,2) is associated with a family of image sets, which differ one from the other by a transformation which is not necessarily a mere translation. The object in entry (1,1) looks as the envelope of those in the entry (1,2); this is true under smooth assumptions and it would be interesting to enlarge the thruth of such a claim. Now, let us jump to the other entries. Since (1.3.5), which is here denoted by $VI(\xi)$ (ξ acts as $\overline{x}$), is the given model (like (1.1.1)), the natural temptation would be to locate it in the entry (2,1), in correspondence of P. If we look at the image set of (1.3.5) (in the case where $I\!K$ is defined as R of (1.1.3)), we see that it behaves like that of $P(\xi)$, so that it seems more appropriate to locate $VI(\xi)$ in the entry (2,2). As consequence, we must look for the envelop of the family of linear folds $VI(\xi)$, if any, and locate in the entry (2,1).

36.Consider (1.1.1) with $p = 0$. The IS Analysis receives a substantial simplification, if $m = 1$. A classic device to reduce (1.1.1) (with $p = 0$) to the case $m = 1$ consists in replacing (1.1.1) with the equivalent problem:

$$\min \ f(x), \quad \text{s.t.} \ \ \tilde{g}(x) := \min \ \{g_1(x), ..., g_m(x)\} \geq 0, \quad x \in X. \tag{3.5.28}$$

The related image set is:

$$\tilde{\mathcal{K}}_{\overline{x}} := \{(u, v) \in I\!R^2 : u = f(\overline{x}) - f(x), \ v = \tilde{g}(x), \ x \in X\}.$$

Example 3.5.2. (continuation of Example 3.4.22). Problem (3.5.28) becomes:

$$\min x \ \ \text{s.t.} \ \ \tilde{g}(x) = \min\{x - x^3, x\} = \left\{ \begin{array}{l} x, \ \text{if } x \leq 0 \\ x - x^3, \ \text{if } x > 0 \end{array} \right\} \geq 0.$$

The related image set is (see Fig. 3.4.26b):

$$\tilde{\mathcal{K}}_{\overline{x}} = \{(u, v) \in I\!R^2 : v = -u \ \text{if } u \geq 0, \ v = -u + u^3 \ \text{if } u < 0\},$$

and its conic extension, unlike that of $\mathcal{K}_{\overline{x}}$ in Example 3.4.22, is easily recognized to be convex. The halfspace $u + v_2 \leq 0$, which contains $\mathcal{K}_{\overline{x}}$ of Example 3.4.22, and the halfsplane $u + v \leq 0$, which contains the present $\tilde{K}_{\overline{x}}$, are related by an obvious relation; the interesting question consists in finding the former, when only the latter is known.$\Box$

All the previously discussed questions about the IS Analysis are valid also here, where stronger answers should be obtained. For instance, we have seen that the conic extension may be convex, even if (1.1.1) is not. However, it may happen that the conic extension itself remains not convex (as it happens in Example 3.4.22), while that of $\tilde{\mathcal{K}}_{\overline{x}}$ be convex (see Example 3.5.1). It would be extremely interesting to give conditions under which the conic extension of $\tilde{\mathcal{K}}_{\overline{x}}$ is convex. Similar remarks can be done for the differentiability, for the exposed part of the image set, and so on. Also the numerical questions discussed in Example 3.5.1 might take advantage of (3.5.28). When there are only bilateral constraints ($p = m$ in (1.1.1)), then $\tilde{g}$ of (3.5.28) might be replaced by

$$\tilde{g}(x) := \min\{g_1(x) - g_1(x), ..., g_m(x) - g_m(x)\}.$$

37. It has been previously noted that there is a great variety of methods for decomposing (1.1.3) into subproblems which should turn out to be easier than it. Bellman's approach has been shortly discussed. Now we want to note that the IS Analysis might lead to improvements also in other decomposition methods. For instance, consider the following decomposition:

$$X = \bigcup_{i=1}^{k} X_i, \tag{3.5.29}$$

which induces that of f into its restrictions

$$f_i : \; X_i \to \mathbb{R}, \quad i = 1, ..., k, \;\; (f_i \equiv f \text{ on } X_i). \tag{3.5.30}$$

Therefore, (1.1.3) is decomposed into the subproblems:

$$f_i^{\downarrow} := \min \; f_i(x), \quad \text{s.t.} \quad x \in R_i := \{x \in X_i : \; g(x) \in D\}, \; i = 1, ..., k. \tag{3.5.31}$$

Of course, $f^{\downarrow} = \min\{f_1^{\downarrow}, ..., f_k^{\downarrow}\}$. Let us concentrate our attention to the convexity, even if several other aspects, like differentiability, deserve to be considered. If (1.1.3) is not convex, the decomposition (3.5.29) might lead to convex subproblems (3.5.31); in this case, the goal has been achieved. However, not all (3.5.31) might turn out to be convex, since, for instance, not all (3.5.30) are; in this case, some transformations suggested by IS Analysis, as conic extension, might improve the situation.

Example 3.5.3. (continuation of Example 3.4.21). Here $X = \mathbb{R}$. To obtain (3.5.29) for $k = 2$, any real might be selected. Since f is not convex, a suitable choice is a point at which f is not convex; for instance $\tilde{x} = 1$. Therefore, we have $X_1 =] - \infty, 1]$, $X_2 = [1, +\infty[$, $f_1 = f_{/X_1}$, $f_2 = f_{/X_2}$, so that (3.5.31) become:

$$\min \; f_1(x), \quad \text{s.t.} \quad x \geq 0; \quad \min \; f_2(x), \quad \text{s.t.} \quad x \geq 0. \tag{3.5.32}$$

At $\overline{x}_1 = 0$ for the former, and $\overline{x}_2 = 2$ for the latter (global m.p.), the corresponding image sets are, respectively the following curves (which appear in Figs. 3.4.25c,d):

$$\mathcal{K}_0^1 = \{(u, v) \in \mathbb{R}^2 : u = -v^4 + \frac{13}{3}v^3 - 5v^2, \; v \leq 1\},$$

$$\mathcal{K}_2^2 = \{(u, v) \in \mathbb{R}^2 : u = -v^4 + \frac{13}{3}v^3 - 5v^2 + \frac{4}{3}, \; v \geq 1\}.$$

Of course, $f_1^{\downarrow} = 0$, $f_2^{\downarrow} = \frac{4}{3}$, $f^{\downarrow} = 0$. Both subproblems (3.5.32) are not convex; the former is quasiconvex, while this does not happen to the latter. The conic extension, shown in Figs. 3.4.25c,d, does not change the status of the former, while makes convex the latter. Hence, an improvement, even if partial, has been achieved. $\qquad\square$

Another interesting case is that where R is not convex or not connected or both. Such a case happens when combinatorial problems, in special (1.1.7), are decomposed into subproblems, by alternatively fixing the variables at an integer value [II1]. Conditions under which IS transformations, as conic extension, let the subproblems have a convex extended image, even if they are not convex, would be useful and avoid a too large decomposition.

38. Examples 3.2.7-3.2.10 show that, in some problems of type (1.1.5), the theory of Lagrange multipliers (see Chapter 5) needs to be enlarged. The introduction of SM is a possible answer. Sect. 3.2 contains only some preliminary results for the development of a theory of SM.

39.In Sect. 3.2, it has been shown that the convexity of the conic extension of the image is a crucial property. Proposition 3.2.8(i) gives a characterization of its convexity. Any condition for $\mathcal{E}(\mathcal{K}_{\overline{x}})$ to be convex is of great importance. As a starting point for this investigation, let us consider the following simple sufficient condition [IV13].

Proposition 3.5.4. Consider (3.2.1) for $p = 0$. Assume that there exists $Y \subseteq X$, such that $-f_{/Y}$ and $g_{/Y}$ be concave, and that, $\forall x \in X, \exists y \in Y$, such that $f(x) \geq f(y)$ and $g(x) \leq g(y)$. Then $\mathcal{E}(\mathcal{K}_{\overline{x}})$ is convex, whatever $\overline{x}$ may be.

Proof . $\hat{\mathcal{K}}_{\overline{x}} := \{(u, v) \in \mathbb{R}^{1+m} : u = f(\overline{x}) - f(y), v = g(y), y \in Y\}$ is the image set of the "restricted" problem. $\forall x \in X, \exists y \in Y$, s.t. $u = f(\overline{x}) - f(x) \leq f(\overline{x}) - f(y)$, $v = g(x) \leq g(y)$. Hence, we have $\mathcal{E}(\mathcal{K}_{\overline{x}}) \subseteq \mathcal{E}(\hat{\mathcal{K}}_{\overline{x}})$. The thesis follows by noting that, because of Proposition 3.2.8(i) (and the subsequent remark), $\mathcal{E}(\hat{\mathcal{K}}_{\overline{x}})$ is convex. $\qquad\square$

When the assumptions of the above proposition are fulfilled , then we can treat (1.1.1) as if it were a convex problem, even if it is not. Of course, we can also replace (1.1.1) with its restriction identified by Y.

Example 3.5.4 (continuation of Examples 3.4.17). Consider again the 1st of Examples 3.4.17. It is immediate to see that, with $Y = [\alpha, +\infty[, \alpha \geq 0$, the assumptions of Proposition 3.5.4 are fulfilled. This is easily interpreted in Fig. 3.4.21a. $\qquad\square$

The above question has been discussed for (1.1.1). Obviously, it extends to all the other problems which lead to (3.2.1) or, more generally, to (1.3.16), especially to (1.1.5), (1.1.7), (1.1.22), (1.1.23).

Besides the convexity of $\mathcal{E}(\mathcal{K}_{\overline{x}})$, it would be useful to investigate that of the conic extension of the homogenization of the image set; namely, the convexity of $\mathcal{E}(\mathcal{K}_{\overline{x}}^h)$, when $\mathcal{C}$-differentiability is not assumed.

40.Proposition 3.5.4 aims to stimulate investigation on conditions for ensuring the convexity of the conic extension of the image set. It is also important to find conditions for proving (3.2.2) through linear disjunctive separation, when the conic extension $\mathcal{E}(\mathcal{K}_{\overline{x}})$ is not convex. The following proposition (see [IV13]), which has been a 1st hint for this kind of investigation, deals with (1.1.1) for $p = 0$ and refers to (3.2.1).

Proposition 3.5.5. Assume that there exist $a_i \in \mathbb{R}\backslash\{0\}$, $b_i \in \mathbb{R}, i = 1, ..., m, a_0 \in \mathbb{R}$, and $\hat{x} \in X$, such that:

$$f_{\overline{x}}(x) \leq a_0 - \sum_{i \in \mathcal{I}} a_i[g_i(x) - b_i]^2, \quad \forall x \in X, \qquad (3.5.33a)$$

$$f(\hat{x}) = a_0 - \sum_{i \in \mathcal{I}} a_i \delta_i b_i^2, \quad g_i(\hat{x}) = (1 - \delta_i)b_i, \ i \in \mathcal{I}, \qquad (3.5.33b)$$

where $\delta_i = 0$ if $b_i > 0$ and $\delta_i = 1$ if $b_i \leq 0$. Then (3.2.1) is impossible, if and only if there exist $\theta \in \mathbb{R}_+$ and $\lambda \in \mathbb{R}_+^m$, with $(\theta, \lambda) \neq O$, such that:

$$\theta f(x) + \langle \lambda, g(x) \rangle \leq 0, \quad \forall x \in X. \tag{3.5.34}$$

Proof. Let $\mathcal{P}$ denote the (elliptic) hyperparaboloid of the IS, whose equation is:

$$u = a_0 - \sum_{i \in \mathcal{I}} a_i (v_i - b_i)^2.$$

It is easy to show that the condition (3.2.2) holds, iff $\mathcal{H} \cap \text{hypo}\,\mathcal{P} = \varnothing$. **If.** (3.5.33a) implies $\mathcal{K}_{\bar{x}} \subseteq \text{hypo}\,\mathcal{P}$, so that (3.2.2) holds. **Only if.** Ab absurdo, let us suppose that $\mathcal{H} \cap \text{hypo}\,\mathcal{P} \neq \varnothing$. Then $\exists\,\tilde{u} > 0$, $\exists\,\tilde{v} \geq 0$, s.t.

$$\tilde{u} \leq a_0 - \sum_{i \in \mathcal{I}} a_i (\tilde{v}_i - b_i)^2,$$

and hence

$$a_0 - \sum_{i \in \mathcal{I}} a_i \delta_i b_i^2 > 0.$$

Account taken of (3.5.33b), this inequality implies $f_{\bar{x}}(\hat{x}) > 0, g(\hat{x}) \geq 0, \hat{x} \in X$, so that (3.2.2) is contradicted. $\qquad\square$

Note that assumption (3.5.33a) can be written as $Q(f_{\bar{x}}(x), g(x)) \leq 0$, where Q is a particular positive-semidefinite quadratic form, whose rank is m. Therefore, (3.5.33) can be interpreted as *coerciveness* (in the sense of being dominated by a strictly concave form; a sense tighter than that of (1.3.20); indeed, $\mathcal{P}$ is not necessarily strict) of the image set $\mathcal{K}_{\bar{x}}$. By exploiting a more general form Q, an interesting result has been achieved in [41].

Example 3.5.5 (continuation of Example 2.3.7). Consider (1.1.1), for $n = 1, p = 0$, $m = 1, X = \mathbb{R}, g(x) = x + 1$, and f as in Example 2.3.7. Then, at $\bar{x} = 0$, we have:

$$\mathcal{K}_{\bar{x}} = \{(u, v) \in \mathbb{R}^2 : u = -f(v - 1), v \in \mathbb{R}\},$$

and (3.5.33) are satisfied for $a_0 = 0, a_1 = b_1 = 1$, so that $\mathcal{P}$ (in the above proof) has equation $u = -(v - 1)^2$. $\qquad\square$

References

[1] Auslander A., "Closedness criteria for the image of a closed set by a linear operator". Numerical Functional Analysis and Optimization, Vol.17, No.1, 1996, pp.503-515.

[2] Bair J., "Sur la Séparation de Familles Fines d'Ensembles Convexes". Bulletin de la Societé Royal des Sciences, Liège, Vol.41, 1972, pp.281-291.

[3] Cambini A., "Non-linear separation theorems, duality and optimality conditions". In [I 12], pp.57-93.

[4] Camerini P.M., Galbiati G. and Maffioli F., "The image of weighted combinatorial problems". Annals of Operation Research, Vol.33, 1991, pp.181-197.

[5] Carathéodory C., "Calculus of Variations and Partial Differential Equations of the First Order", Chelsea Publ.Co., New York, 1982 (translation of the volume "Variationsrechnung und Partielle Differential Gleichungen Erster Ordnung"-B.G, Teubner, Berlin, 1935).

[6] Clarke F.H., "Methods of dynamic and nonsmooth optimization". Regional Conference Series in Appl. Mathematics. SIAM, Philadelphia, 1989, pp.1-90.

[7] De Figueiredo D.G., "Lectures on the Ekeland variational principle with applications and detours". Springer-Verlag, Berlin, 1989.

[8] Dem'yanov V.F. and Rubinov A.M., "Constructive Nonsmooth Analysis". Verlag Peter Lang, Frankfurt a.m., Berlin, 1995.

[9] Dem'yanov V.F. and Rubinov A.M., "An Introduction to Quasidifferential Calculus". In [I 16], pp.1-31.

[10] Dem'yanov V.F. and Vasiliev L.V., "Nondifferentiable Optimization". Optimization Software, Inc., New York, 1984.

[11] Dien P.H., Mastroeni G., Pappalardo M. and Quang P.H., "Regularity Conditions for Constrained Extremum Problems via Image Space". Jou. of Optimiz. Theory and Appls., Vol.80, No.1, 1994, pp.19-37.

[12] Dini U., "Memoria sopra le serie di Fourier" (in Italian), Annali delle Università Toscane, Vol.14, parte 2°, 1874, pp.161-176.

[13] Dini U., "Fondamenti per la teorica delle funzioni di variabili reali (Foundations of the theory of functions of real variables)" (in Italian). T.Nistri and Co.Publisher, Pisa, 1878. Reprinted on 1990 by The Italian Mathematical Society (Piazza S.Donato, 1-Bologna-Italy).

[14] Elster K.-H. and Thierfelder J., "On cone approximations and generalized directional derivatives". In [I 9], pp.133-154.

[15] Ferrero O., "Dini Sequences and Semidifferentials. I: General Results". Bollettino Unione Matematica Italiana, Vol.7, No.9-B, 1995, pp. 257-280.

[16] Ferrero O., "Dini Sequences and semidifferentials-II: Applications and numerical results". Bollettino Unione Matematica Italiana, Vol. 7, No.9-B, 1995, pp.541-552.

[17] Gauvin J. and Dubeau F., "Differential properties of the marginal function in mathematical programming". Mathem. Programming Studies, Vol.19, North-Holland Publ. Co., 1982, pp.101-119.

[18] Giannessi F., "Semidifferentiable Functions and Necessary Optimality Conditions". Jou. of Optimiz. Theory and Appls., Vol.60, No.2, 1989, pp.191-241.

[19] Giannessi F., "Image Space Approach to Optimization". In [V 32], vol.II, pp.457-464.

[20] Giannessi F., Pappalardo M. and Pellegrini L., "Necessary Optimality Conditions via Image Problem". In [I 9], pp.185-217.

[21] Giannessi F. and Rapcsák T., "Images, separation of sets and extremum problems", in "Recent Trends in Optimization Theory and Applications", Agarwal R.P. (Ed.), World Sc.Series in Applied Analysis, World Sc.Publ.Co., Singapore, 1995, pp.79-106.

[22] Giannessi F. and Uderzo A., "A Multifunction Approach to Extremum Problems having Infinite Dimensional Image. I: Composition and Selection". Proceedings of "Seminario Matematico-Fisico" of University of Modena, Via Campi 213/A-Modena-Italy, Suppl. Vol. XLVI, 1998, pp.771-785.

[23] Gwinner J., "Closed images of convex multivalued mappings in linear topological spaces with applications". Jou. of Math. Analysis and Appls., Vol.60, 1977, pp.75-86.

[24] Hestenes M.R., "Calculus of Variations and Optimal Control Theory". J. Wiley, New York, 1966.

[25] Hestenes M.R., "Optimization Theory: The finite dimensional case". J. Wiley, New York, 1975.

[26] Ioffe A.D., "Nonsmooth Analysis: Differential Calculus of Nondifferentiable Mappings". Transactions of the Amer. Mathem. Soc., Vol.266, 1981, pp.1-56.

[27] Ioffe A.D., "On the local surjection property". Nonlinear Analysis, Theory, Methods and Appls., Vol.II, No.5, 1987, pp.565-592.

[28] Jeyakumar V. and Luc D.T., "An open mapping theorem using unbounded generalized Jacobians". Appls. Math. Report AMR99/20, School of Mathem., The Univ. of New South Wales, Sydney, Australia, Oct. 1999, pp.1-19.

[29] Komlósi S., "On generalized upper quasidifferentiability". In [I 25], pp.189-200.

[30] Madani K., "Séparation Non Linéaire et Problèmes extrémaux". Thèse de Magister en Mathématiques, Universite D'Oran Es-Sénia, Departement de Mathématiques, 2000.

[31] Mastroeni G., "Stability studies in the image space". In [V 26], pp.69-76.

[32] Mastroeni G., Pappalardo M. and Yen N.D., "Image of a parametric optimization problem and continuity of the perturbation function". Jou. of Optimiz. Theory and Appls., Vol.81, No.1, 1994, pp.193-202.

[33] Oettli W. and Théra M., "Equivalents of Ekeland's principle". Bull. Australian Mathem. Soc., Vol.48, 1993, pp.385-392.

[34] Pappalardo M., "Some calculus rules for semidifferentiable functions and related topics". In [I 25], pp.281-294.

[35] Pappalardo M., "Image space approach to penalty methods". Jou. of Optimiz. Theory and Appls., Vol.64, No.1, 1990, pp.141-152.

[36] Pappalardo M., "Tangent cones and Dini derivatives". Jou. of Optimiz. Theory and Appls., Vol.70, 1991, pp.97-107.

[37] Pappalardo M., "On semidifferentiable functions". Optimization, Vol.24, 1992, pp.207-217.

[38] Pappalardo M., "Semidifferentiability and extremum problems". In [V 26], pp.77-92.

[39] Pappalardo M. and Uderzo A., "G-semidifferentiability in Euclidean Spaces". Jou. of Optimiz. Theory and Appls., Vol.101, No.1, 1999, pp.221-229.

[40] Pellegrini L., "Some Remarks on Semistationarity and Optimality Conditions". In [I 25], pp.295-302.

[41] Pellegrini L., "Coercivity and Image of constrained Extremum Problems". Jou. Optimiz. Theory and Appls., Vol. 89, No.1, 1996, pp.175-188.

[42] Penot J.-P., "Second-order generalized derivatives: relationships with convergence notions". In [I 25], pp.303-322.

[43] Pomérol J.-Ch., "Is the image of a closed convex set by a continuous linear mapping closed?". 5° Symposium on Operations Research, Heidelberg, Verlag-Hain, 1976, pp.412-419.

[44] Quang P.H., "Some notes on semidifferentiability and generalized subdifferentials". Acta Mathematica Vietnamica, Vol.18, No.1, 1993, pp.79-90.

[45] Rockafellar R.T., "The Theory of Subgradients and its Applications to Problems of Optimization. Convex and Nonconvex Functions". Heldermann Verlag, Berlin, 1981.

[46] Schaible S., "Generalized monotone maps". In [I 25], pp.392-408.

[47] Tardella F., "On the image of a constrained extremum problem and some applications to the existence of a minimum". Jou. of Optimiz. Theory and Appls., Vol.60, No.1, 1989, pp.93-104.

[48] Uderzo A., "On a generalized differentiability of operators". In [V 4], pp.205-224.

[49] Warga J., "Derivative containers, inverse function and controllability". In "Calculus of Variations and Control Theory", Russel D.L. (Ed.), Academic Press, New York, 1976, pp.13-46.

[50] Warga J., "Fat Homeomorphisms and Unbounded Derivative Containers". Jou. of Mathem. Analysis and Appls., Vol.81, 1981, pp.545-560.

[51] Yen N.D., "A mean value theorem for semidifferentiable functions". Vietnam Jou. of Mathematics, Vol.23, 1995, pp.221-228.

[52] Yen N.D., "On G-semidifferentiable functions in Euclidean spaces". Jou. of Optimiz. Theory and Appls., Vol.85, No.2, 1995, pp.377-392.

[53] Zalgaller V.A.," Theory of envelopes" (in Russian). Publishing House "Nauka", Dept.of Physics and Mathematics, Moscow, 1975.

CHAPTER 4. ALTERNATIVE AND SEPARATION

4.1. Introduction

In a general format a Theorem of the Alternative (for short, TA) claims that, between two given propositions, say S and S*, one and only one is true; namely, never neither nor both. In Mathematics, S and S* are, in general, systems of equalities or inequalities. A TA for linear algebraic systems was established as early as 1873 by Gordan [19]; then there was the celebrated Farkas Lemma in 1902 [11]; indeed, such a lemma does not appear as a TA, but an obvious reformulation shows it as a TA. Some further important TA were established in 1915 by Stiemke [41], in 1936 by Motzkin [34], in 1951 by Slater [40], in 1956 by Tucker in [I43] and by Duffin in [I43]. Subsequently, due mainly to the development of the Optimization Theory, there has been a blooming of TA; they have been extended to not necessarily algebraic systems, to systems in an infinite dimensional space, to systems in a complex space, and even to systems for point-to-set maps. TA (sometimes called transposition theorems) have been conceived as tools for proving some theorems of Linear Algebra (this is the reason why the TA by Farkas is known as lemma) or to prove the existence and uniqueness of solutions of differential and integral equations [42].

It is interesting to note that, a few years later, in a completely different field of Mathematics, some ideas mature − mainly due to Minkowski [II36] −, which lead to state so-called Separation Theorems (for short, ST). Indeed, here too, the first important result does not look like a ST: on the basis of some ideas of Helly in 1912 (see [II27, II29]), Banach in 1925 [2] and Hahn in 1927 [20], independently of each other, establish the celebrated Hahn-Banach Linear Extension Theorem; by means of an obvious reformulation it shows itself to be a ST. Here too the purpose is to have lemmas for proving other theorems − in Functional Analysis and Geometry.

Over several years TA and ST have been carried out as disjoint theories. Recently, thanks to the great development of Optimization and to the increasing use of TA and ST in the Theory of Optimization, it has been recognized that TA and ST are different "languages" for expressing the same "structural" property (this does not imply that one of them should be deleted; on the contrary, different languages let us achieve more properties) and, overall, that they are not only tools for proving theorems; indeed, they have been raised to the basis for the theory of constrained extrema.

While alternative − even in a general meaning − receives only the above definition, separation of two sets has been split into several kinds. Only a few of them will be considered here; several excellent books allow one to deepen the subject; see for instance [I34, I39, I45, I56].

In Sect. 2.2, linear separation between two sets has been considered. When the hyperplane of Definition 2.2.5 does not exist, it may happen that a non-affine manifold exists, which makes a partition of the space, such that Definition 2.2.5 holds with H^- and H^+ replaced by the closures of the two parts; this is called *nonlinear separation*. By jumping to a functional language, the two parts of the space split by the manifold (hyperplane) can be seen as level sets of a nonlinear function. This allows us to adopt a general format for both alternative and separation, and extend the results of Sect. 2.2.

4.2. Separation Functions

As we have seen in Sect. 3.2, when the images of the functions which appear in a system of type (1.3.16) have infinite dimension − as it happens to the system associated to problem (1.1.5) − , then the corresponding IS is, of course, infinite dimensional. Therefore, a TA which aims to be applied to such a system should be in an infinite dimensional space. However, in Sect. 3.2, we have adopted a different approach, which consists in postponing the infinite dimensionality to the introduction of the IS, which therefore remains finite dimensional. This is the reason why the following TA are located in $\mathbb{R}^\nu$, even if they might be stated in a more general space.

Assume we are given the positive integers ν, ℓ, the nonempty set $\mathcal{H} \subset \mathbb{R}^\nu$, and the non-empty, convex, closed, pointed cone $C \subset \mathbb{R}^\ell$ with apex at the origin and with nonempty interior; $\mathcal{H} = \mathbb{R}^\nu$ makes next (4.2.2) meaningless; $C = \mathbb{R}^\ell$ is excluded by C being pointed (see Definition 2.1.7). C identifies a partial order; as in Sect.1.1, we will set $C_o := C \backslash \{O\}$ and $\overset{o}{C} :=\mathrm{int}\ C$. Consider a function $w : \mathbb{R}^\nu \times \Pi \to \mathbb{R}^\ell$, where Π is a set of parameters to be specified case by case. For each $\pi \in \Pi$ and for each set $S \subset \mathbb{R}^\ell$, the set

$$\mathrm{lev}_S\ w(\bullet\ ;\pi) := \{z \in \mathbb{R}^\nu : w(z;\pi) \in S\} \tag{4.2.1}$$

will be called *level set of w with respect to S*.

Definition 4.2.1. The class of all the functions $w : \mathbb{R}^\nu \times \Pi \to \mathbb{R}^\ell$, such that:

$$\mathrm{lev}_C\ w(\bullet\ ;\pi) \supseteq \mathcal{H}, \quad \forall \pi \in \Pi \tag{4.2.2a}$$

and

$$\bigcap_{\pi \in \Pi} \mathrm{lev}_{C_o}\ w(\bullet\ ;\pi) \subseteq \mathcal{H}, \tag{4.2.2b}$$

will be called *class of weak separation functions* and will be denoted by $\mathcal{W}(\Pi)$.

The left-hand side of (4.2.2b) may be empty; otherwise, usually, (4.2.2b) is verified as equality. At $\ell = 1$ and $C = [0, +\infty[$, (4.2.2) become:

$$\operatorname{lev}_{\geqslant_o} w(\bullet\;;\pi) \supseteq \mathcal{H}, \quad \forall \pi \in \Pi, \tag{4.2.3a}$$

$$\bigcap_{\pi \in \Pi} \operatorname{lev}_{>_o} w(\bullet\;;\pi) \subseteq \mathcal{H}, \tag{4.2.3b}$$

since the level sets with respect to C and C_o collapse to the usual non-negative and positive level sets, respectively.

As an instance, set $\nu = 2, \ell = 1, \mathcal{H} = \{(u,v) \in \mathbb{R}^2 : u > 0, v \geq 0\}$ ($\mathcal{H}$ is as in Sect. 3.2), $C = [0, +\infty[, \Pi = \{(1,0), (0,1)\}, \pi = (\theta, \lambda), \mathcal{W}(\Pi) = \{w = \theta u + \lambda v : (\theta, \lambda) \in \Pi\} = \{u, v\}$. Then (4.2.3a) becomes:

$$\{(u,v) \in \mathbb{R}^2 : u \geq 0\} \supseteq \mathcal{H}, \quad \{(u.v) \in \mathbb{R}^2 : v \geq 0\} \supseteq \mathcal{H},$$

and (4.2.3b) becomes:

$$\{(u,v) \in \mathbb{R}^2 : u > 0\} \cap \{(u,v) \in \mathbb{R}^2 : v > 0\} \subseteq \mathcal{H},$$

and (4.2.3) are trivially verified, so that $\mathcal{W}(\Pi)$ is a class of weak separation functions. However, it is so "poor" that it is difficult to expect any application of it. Hence, $\mathcal{W}(\Pi)$ must be required to enjoy some properties, like the following one:

$$w_1, w_2 \in \mathcal{W}(\Pi), \;\; \alpha_1, \alpha_2 \in \mathbb{R}^2 \backslash \{O\} \;\; \Rightarrow \;\; \alpha_1 w_1 + \alpha_2 w_2 \in \mathcal{W}(\Pi).$$

Note also that (4.2.2a) and hence (4.2.3a) are not redundant, as simple examples show. Take for instance $\ell = 1, C = [0, +\infty[, \nu = 2,$ and $w(z;\pi) = \langle \pi, z \rangle = \pi_1 z_1 + \pi_2 z_2$ with:

$$\Pi = \{\pi \in \mathbb{R}^2 : \pi_1 < 0, \pi_2 > 0, 2\pi_1 + \pi_2 > 0\} \cup \{\pi \in \mathbb{R}^2 : \pi_1 > 0, \pi_2 < 0, \pi_1 + 2\pi_2 > 0\}.$$

The left-hand side of (4.2.3b) becomes the angle defined by $z_2 \geq \frac{1}{2}z_1$ and $z_2 \leq 2z_1$, so that (4.2.3b) is satisfied, while (4.2.3a) is not. This example is easily extended to the case $\ell > 1$. (4.2.2a) and hence (4.2.3a) become redundant, if in (4.2.2b) and hence in (4.2.3b) the inclusion must be verified as equality; this leads to the following Definition 4.2.2.

If $C \subseteq \mathbb{R}_+^\ell$ or $C \supseteq \mathbb{R}_+^\ell$, then lev_C and $\operatorname{lev}_{\overset{o}{C}}$ have the flavour of "non-negative" and "positive" level sets, respectively. In any case, when $\ell > 1$, C_o does not represent the only way of introducing a "positive" level set: any cone obtained from C by cutting off a part of its boundary might play such a role; among these cones, $\overset{o}{C}$ deserves a special attention. If $\overset{o}{C}$ is adopted, then in Definition 4.2.1 (4.2.2b) is replaced by:

$$\bigcap_{\pi \in \Pi} \operatorname{lev}_{\overset{o}{C}} w(\bullet\;;\pi) \subseteq \mathcal{H}. \tag{4.2.2b$'$}$$

At $\ell = 1$ and $C = [0, +\infty[$, both (4.2.2b) and (4.2.2b)$'$ collapse to (4.2.3b).

Besides "poorness", another drawback of $\mathcal{W}(\Pi)$ is that it may contain undesirable elements. In the above instance, in spite of the fact that $\mathcal{W}(\Pi)$ has only 2 elments, that identified by $\pi = (0,1)$ is undesiderable for the applications of Sect. 3.2: it corresponds to separate $\mathcal{H}$ and $\mathcal{K}_{\bar{x}}$ with a not disjunctive separation, since the separation line defined by $\pi = (0,1)$ has the equation $v = 0$ and intersects $\mathcal{H}$. Therefore, besides $\mathcal{W}(\Pi)$, we need the introduction of further classes of separation functions.

In some applications we will use the subclass, say $\mathcal{W}^c(\Pi)$, of $w(\bullet\,;\pi)$ which are continuous $\forall \pi \in \Pi$, and that, say $\mathcal{W}^{us}(\Pi)$, of $w(\bullet;\pi)$ which are u.s.c. with respect to C, $\forall \pi \in \Pi$ (w is said u.s.c. with respect to C iff $(\mathrm{gr}\ w) - C$ is closed). The introduction of $\mathcal{W}^{us}(\Pi)$ is motivated by the following Proposition 4.2.2. ($N_\epsilon(\hat{z})$ is neighbourhood of $\hat{z}$).

Proposition 4.2.1. If $w \in \mathcal{W}^{us}(\Pi)$, then:

$$\mathrm{cl}\ \mathrm{lev}_{C_0}\ w(\bullet\,;\pi) \subseteq \mathrm{lev}_C\ w(\bullet\,;\pi). \tag{4.2.4}$$

Proof. Let $\hat{z} \in \mathrm{cl}\ \mathrm{lev}_{C_0} w(\bullet\,;\pi)$. Then, $\forall \epsilon > 0$, $\exists z_\epsilon \in N_\epsilon(\hat{z})$, $z_\epsilon \neq \hat{z}$, such that $w(z_\epsilon;\pi) \in C_o$. This condition obviously implies $O_\ell \in w(z_\epsilon;\pi) - C$, $\forall \epsilon > 0$, so that $(\hat{z}, O_\ell)$ is a limit point of the family $\{(z_\epsilon, w(z_\epsilon;\pi) - C)\}_{\epsilon>0}$. Ab absurdo, suppose that $w(\hat{z};\pi) \notin C$. This implies $(\hat{z}, O_\ell) \notin (\hat{z}, w(\hat{z};\pi) - C)$ and contradicts the assumption of upper semicontinuity. $\square$

Proposition 4.2.2. If $w \in \mathcal{W}^{us}(\Pi)$, then the condition

$$\mathrm{cl}\ \mathrm{lev}_{C_0}\ w(\bullet\,;\pi) \supseteq \mathrm{cl}\ \mathcal{H}. \tag{4.2.5}$$

is sufficient for (4.2.2a) to hold.

Proof. (4.2.5) and (4.2.4) imply (4.2.2a). $\square$

Note that (4.2.5) is not necessary, as simple examples show where the zero level set of $w(\bullet\,;\pi)$ is of positive measure.

In the above proposition, C_o can be replaced by $\overset{o}{C}$:

Proposition 4.2.3. If $w \in \mathcal{W}^{us}(\Pi)$, then the condition

$$\mathrm{cl}\ \mathrm{lev}_{\overset{o}{C}}\ w(\bullet\,;\pi) \supseteq \mathrm{cl}\ \mathcal{H}. \tag{4.2.6}$$

is sufficient for (4.2.2a) to hold.

Proof. Since obviously $\mathrm{cl}\ \mathrm{lev}_{\overset{o}{C}}\ w(\bullet\,;\pi) \subseteq \mathrm{cl}\ \mathrm{lev}_{C_o}\ w(\bullet\,;\pi)$, from (4.2.6) and (4.2.4) we achieve (4.2.2a). $\square$

As already said, the class $\mathcal{W}(\Pi)$ is too large. Therefore, another subclass is now introduced by strengthening Definition 4.2.1.

Definition 4.2.2. The class of all the functions $w : \mathbb{R}^\nu \times \Pi \to \mathbb{R}^\ell$, such that:

$$\bigcap_{\pi \in \Pi} \mathrm{lev}_{C_o}\ w(\bullet\,;\pi) = \mathcal{H}. \tag{4.2.7}$$

is called *class of regular weak separation functions* and will be denoted by $\mathcal{W}_R(\Pi)$.

The subclasses of $\mathcal{W}_R(\Pi)$ identified by the continuity or u.s.c. of w will be denoted by $\mathcal{W}_R^c(\Pi)$ or $\mathcal{W}_R^{us}(\Pi)$, respectively.

Like before, instead of C_o we can consider $\overset{o}{C}$; in such a case (4.2.7) is replaced by:

$$\bigcap_{\pi\in\Pi}\mathrm{lev}_{\overset{o}{C}}\,w(\bullet;\pi)=\mathcal{H} \tag{4.2.7$'$}$$

The notation of the above classes does not contain C_o or $\overset{o}{C}$; this will be clarified case by case. Obviously, (4.2.7) implies that $\mathcal{H}\subseteq\mathrm{lev}_{C_o}\,w(\bullet\,;\pi)$, $\forall\pi\in\Pi$, so that (4.2.2a) is satisfied; the same happens for (4.2.7)$'$.

At $\ell=1$ and $C=[0,+\infty[$, (4.2.7) and (4.2.7)$'$ collapse to:

$$\bigcap_{\pi\in\Pi}\mathrm{lev}_{>o}\,w(\bullet;\pi)=\mathcal{H}, \tag{4.2.8}$$

since the level sets with respect to C_o and $\overset{o}{C}$ shrink to the usual positive level set.

Besides the weak separation functions, we will consider another type of separation functions:

Definition 4.2.3. The class of all the functions $s:\mathbb{R}^{\nu}\times\Pi\to\mathbb{R}^{\ell}$, such that:

$$\mathrm{lev}_{C_o}\,s(\bullet\,;\pi)\subseteq\mathcal{H},\quad\forall\pi\in\Pi \tag{4.2.9a}$$

and

$$\bigcup_{\pi\in\Pi}\mathrm{lev}_{C_o}\,s(\bullet\,;\pi)=\mathrm{ri}\,\mathcal{H}, \tag{4.2.9b}$$

is called *class of strong separation functions* and will be denoted by $\mathcal{S}(\Pi)$.

Note that, if $s\in\mathcal{S}(\Pi)$, then also $\alpha s\in\mathcal{S}(\Pi)$, $\forall\alpha\in\mathbb{R}_+\backslash\{0\}$. When $\overset{o}{C}$ is adopted instead of C_o, then (4.2.9) are replaced respectively by:

$$\mathrm{lev}_{\overset{o}{C}}\,s(\bullet\,;\pi)\subseteq\mathcal{H},\quad\forall\pi\in\Pi \tag{4.2.10a}$$

and

$$\bigcup_{\pi\in\Pi}\mathrm{lev}_{\overset{o}{C}}\,s(\bullet\,;\pi)=\mathrm{ri}\,\mathcal{H}, \tag{4.2.10b}$$

At $\ell=1$ and $C=[0,+\infty[$, (4.2.9) and (4.2.10) come to:

$$\mathrm{lev}_{>o}\,s(\bullet;\pi)\subseteq\mathcal{H}, \tag{4.2.11a}$$

$$\bigcup_{\pi\in\Pi}\mathrm{lev}_{>o}\,s(\bullet;\pi)=\mathrm{ri}\,\mathcal{H}. \tag{4.2.11b}$$

The subclasses of $\mathcal{S}(\Pi)$ identified by the continuity or l.s.c. of w will be denoted by $\mathcal{S}^c(\Pi)$ or $\mathcal{S}^{\ell c}(\Pi)$, respectively.

From Definitions 4.2.1, 4.2.2 and 4.2.3, the following inclusions are evident:

$$\mathcal{W}^c(\Pi)\subset\mathcal{W}^{us}(\Pi)\subset\mathcal{W}(\Pi),\ \ \mathcal{W}_R(\Pi)\subset\mathcal{W}(\Pi), \tag{4.2.12}$$

$$\mathcal{S}^c(\Pi)\subset\mathcal{S}^{\ell s}(\Pi)\subset\mathcal{S}(\Pi); \tag{4.2.13}$$

simple example show the non-coincidence of the above classes. It is suitable to introduce the further subclasses:

$$\mathcal{W}_R^{us}(\Pi):=\mathcal{W}^{us}(\Pi)\cap\mathcal{W}_R(\Pi),\quad\mathcal{W}_R^c(\Pi):=\mathcal{W}^c(\Pi)\cap\mathcal{W}_R(\Pi).$$

The classes of separation functions introduced in the present section have been denoted by symbols which, for the sake of simplicity, contain the same argument Π. Of course, the set of parameters is not necessarily the same for the several classes. Indeed, as we will see, we can go from one class to another by simply restricting the set Π of parameters. This is also the reason why the classes are denoted by $\mathcal{W}(\Pi), \mathcal{W}_R(\Pi), ...,$ and not merely by $\mathcal{W}, \mathcal{W}_R, ...$. Without any fear of confusion, the notation does not contain the cone; the context clarifies it. Analogous remark holds also for the classes of strong separation functions.

4.3. Special Separation Functions

In view of the applications to the problems of Chapter 1, taking into account the notation of Sects. 1.1 and 1.3, we will now consider some special classes of separation functions. To this end, we set $\nu = \ell + m$, $z = (u, v)$, $\pi = (\theta, \gamma)$ and $\Pi = \mathbb{R}_+^\ell \times \Gamma$. The case $\ell = 1$ is of interest to (1.1.1), (1.3.1) and (1.3.6); the case $\ell > 1$ is of interest to (1.1.8), (1.1.10), (1.3.8) and (1.3.9). We will consider first the case $\ell = 1$ with the notation of Sect.1.1. Now we can have either $C = \mathbb{R}_-$ or $C = \mathbb{R}_+$ (the case $C = 0$ is of few interest); being equivalent, we set $C = \mathbb{R}_+$, so that $C_o = \overset{o}{C} = \text{int}\, C = \overset{o}{\mathbb{R}_+}$, and $\mathcal{H} = \overset{o}{\mathbb{R}_+} \times D$. A wide class is given by $(\pi = (\theta, \gamma), \Pi = \mathbb{R}_+ \times \Gamma)$:

$$w(u, v; \theta, \gamma) = \theta u + \underline{w}(v; \gamma), \quad \theta \geqslant 0, \gamma \in \Gamma, \tag{4.3.1a}$$

with

$$\theta \geqslant 0, \ \gamma \in \Gamma \ \text{ s.t. } \ w(\bullet, \bullet; \theta, \gamma) \not\equiv 0, \tag{4.3.1b}$$

and where $\underline{w} : \mathbb{R}^m \times \Gamma \to \mathbb{R}$ must be such that:

$$\bigcap_{\gamma \in \Gamma} \text{lev}_{\geq 0}\, \underline{w}(\bullet; \gamma) = D, \tag{4.3.1c}$$

$$\forall \gamma \in \Gamma, \ \forall \alpha \in \mathbb{R}_+, \ \exists \gamma_\alpha \in \Gamma \ \text{ s.t. } \ \alpha \underline{w}(v; \gamma) = \underline{w}(v; \gamma_\alpha), \ \forall v. \tag{4.3.1d}$$

Note that (4.3.1c,d) imply that:

$$\forall v \not\subseteq D, \ \exists \gamma \in \Gamma \ \text{ s.t. } \ \underline{w}(v; \gamma) < 0, \tag{4.3.1e}$$

$$\exists \tilde{\gamma} \in \Gamma \ \text{ s.t. } \ \underline{w}(\bullet; \tilde{\gamma}) \equiv 0, \tag{4.3.1f}$$

$$\bigcap_{\gamma \in \Gamma} \text{lev}_{>0}\, \underline{w}(\bullet; \gamma) \subseteq D, \tag{4.3.1g}$$

where in the last intersection we stipulate to take $\underline{w}(\cdot; \gamma) \not\equiv 0$. In fact, if, ab absurdo, there exists $\hat{v} \not\subseteq D$, such that $\forall \gamma \in \Gamma$ either $\underline{w}(\hat{v}; \gamma) \geq 0$ or $\underline{w}(\hat{v}; \gamma) > 0$, then (4.3.1c) is contradicted; hence, (4.3.1c)$\Rightarrow$(4.3.1e,g). At $\alpha = 0$, (4.3.1d)$\Rightarrow$(4.3.1f).

Proposition 4.3.1. The functions (4.3.1) are weak separation functions, namely the class (4.3.1) is a subclass of $\mathcal{W}(\Pi)$ with $\Pi = \mathbb{R}_+ \times \Gamma$.

Proof. Because of (4.3.1c), $v \in D \Rightarrow \underline{w}(v; \gamma) \geq 0 \ \forall \gamma \in \Gamma$. Therefore,

$$(u, v) \in \mathcal{H} = \overset{o}{\mathbb{R}_+} \times D \ \Rightarrow \ \theta u + \underline{w}(u; \gamma) \geq 0,$$

so that (4.2.3a) is satisfied. Taking into account (4.3.1f,g), we have:

$$\bigcap_{\theta \geq 0, \gamma \in \Gamma} \mathrm{lev}_{>0} \ w(u,v;\theta,\gamma) \subseteq [\mathrm{lev}_{>0} w(u,v;1,\tilde{\gamma})] \cap \bigcap_{\gamma \in \Gamma} \mathrm{lev}_{>0} \ w(u,v;0,\gamma) =$$

$$= (\mathring{\mathbb{R}}_+ \times \mathbb{R}^m) \cap [\mathbb{R} \times \bigcap_{\gamma \in \Gamma} \mathrm{lev}_{>0} \underline{w}(\bullet;\gamma)] \subseteq (\mathring{\mathbb{R}}_+ \times \mathbb{R}^m) \cap (\mathbb{R} \times D) = \mathcal{H},$$

so that (4.2.3b) is fulfilled. $\qquad\qquad\square$

There are some particular cases of (4.3.1), which are useful for the applications. A very important case is that where $\underline{w}$ is separable with respect to the elements of v; to consider it, we must assume that Γ be a Cartesian product, namely $\Gamma = \underset{i \in \mathcal{I}}{\times} \Gamma_i$ with the Γ_i's given, so that (4.3.1a) becomes:

$$w(u,v;\theta,\gamma) = \theta u + \sum_{i \in \mathcal{I}} \underline{w}_i(v_i;\gamma_i), \quad \gamma_i \in \Gamma_i, \ i \in \mathcal{I}, \tag{4.3.2}$$

where $\gamma = (\gamma_1, ..., \gamma_m)$ and under the same conditions (4.3.1b-d) as for the class (4.3.1) (where, of course, we set $\underline{w}(v;\gamma) = \sum_{i \in \mathcal{I}} \underline{w}_i(v_i;\gamma_i)$). The above class can be furtherly particularized by setting $\lambda = (\lambda_1, ..., \lambda_m), \mu = (\mu_1, ..., \mu_m), \gamma_i = (\lambda_i, \mu_i), \ i \in \mathcal{I}, \ \Gamma_i = \mathbb{R} \times \mathbb{R}_+$ if $i \in \mathcal{I}^0$ and $\Gamma_i = \mathbb{R}_+ \times \mathbb{R}_+$ if $i \in \mathcal{I}^+$, and, $\forall i \in \mathcal{I}$,

$$\underline{w}_i = \underline{w}_i^{pe}(v_i;\lambda_i,\mu_i) := \begin{cases} \lambda_i v_i - \mu_i v_i^2, & \text{if } i \in \mathcal{I}^0, \\[2em] \begin{cases} \lambda_i v_i - \mu_i v_i^2, & \text{if } v_i \leq 0, \\ \frac{1}{2\lambda_i \mu_i}(\sqrt{1+4\lambda_i^2 \mu_i v_i} - 1), & \\ & \text{if } v_i > 0 \\ \text{with } \lambda_i, \mu_i > 0 \end{cases} \\ \text{or} \\ \lambda_i v_i \exp(-\mu_i v_i) \\ \text{or} \\ \lambda_i[1 - \exp(-\mu_i v_i)] \end{cases} \quad \text{if } i \in \mathcal{I}^+, \tag{4.3.3}$$

so that (4.3.2) becomes:

$$w(u,v;\theta,\lambda,\mu) = \theta u + \sum_{i \in \mathcal{I}} \underline{w}_i^{pe}(v_i;\lambda_i,\mu_i) \tag{4.3.4a}$$

with

$$\begin{cases} \theta \geq 0, \ \lambda \in \mathbb{R}^p \times \mathbb{R}^{m-p}, \ \mu \in \mathbb{R}_+^m, \\ (\lambda_i, i \in \mathcal{I}^0, \ \mu_i, i \in \mathcal{I}^0) \neq O \ \text{ and } \ (\lambda_i, i \in \mathcal{I}^+) \neq O \ \text{ if } \theta = 0. \end{cases} \tag{4.3.4b}$$

Several other algebraic or transcendental functions can be adopted, in place of those of (4.3.3), when a special numerical or analytical purpose is pursued.

In (4.3.3) $-$ and hence in (4.3.4) $-$ the term corresponding to the unilateral constraints (identified by $\mathcal{I}^+$) has 3 alternative forms: the 1st is of parabolic type and the others of exponential type; they have different $-$ analytic or numerical $-$ advantages; however, as next proposition shows, they all guarantee weak separation, so that, from this point of view, it is equivalent to adopt either form. On the contrary, if there is a special purpose, then one may be more convenient than the others, even if all of them are derivable. For instance, between the two exponential forms in (4.3.3), the former recovers the linear one at $\mu = 0$, and the latter is monotone and concave. The 1st form — defined by two arcs of parabolas — is, in general, computationally preferable; moreover, if in the image $\mathcal{K}_{\bar{x}}$ the element v is bounded from above, or

$$v_i^{\text{sup}} := \sup\{v_i \in \mathbb{R} : (u, v) \in \mathcal{K}_{\bar{x}}\} < +\infty, \quad i \in \mathcal{J}^+, \qquad (4.3.3)'$$

then the former arc can be adopted also for the interval $0 < v_i \leq v_i^{\text{sup}}$ under the condition $\frac{\lambda_i}{\mu_i} \geq v_i^{\text{sup}}$, and the latter arc can be disregarded; therefore, the 1st expression of (4.3.3) (which is for unilateral constraints) can be replaced by:

$$\underline{w}_i = \underline{w}_i^p(v_i; \lambda_i, \mu_i) := \lambda_i v_i - \mu_i v_i^2, \; \frac{\lambda_i}{\mu_i} \geq v_i^{\text{sup}}, \; \lambda_i, \mu_i > 0, \; i \in \mathcal{J}^+, \qquad (4.3.3)''$$

under the assumption $(4.3.3)'$; the domain is therefore contained into a hyperectangle.

The bounds defined by $(4.3.3)'$ may be very loose. Hence, in the applications, either tighter bounds are available, or one can start by disregarding $(4.3.3)'$ (and hence $\frac{\lambda_i}{\mu_i} \geq v_i^{\text{sup}}$ in $(4.3.3)''$) and, eventually, check for separability (see, e.g., the last part of Example 4.2.5).

The functions given by (4.3.4) will be called *weak parabolic-exponential separation functions*; they become simply *parabolic* in case $(4.3.3)''$; the term "weak separation" is motivated by the following:

Proposition 4.3.2. The functions (4.3.4) are weak separation functions, namely the class (4.3.4) is a subclass of $\mathcal{W}(\Pi)$ at $\Pi = \mathbb{R}_+ \times D^* \times \mathbb{R}_+^m$.

Proof. It is enough to show that (4.3.1b-d) are satisfied. (4.3.1b) is obviously implied by (4.3.4b). $v \in D$ implies $v_i = 0$ $i \in \mathcal{J}^0$ and $v_i \geq 0$ $i \in \mathcal{J}^+$ which, in their turn, imply the non-negativity of (4.3.3) whatever of the three alternative forms one may consider; this happens also in the case $(4.3.1)' - (4.3.3)''$, where D is replaced by its subset, say $\tilde{D}$, given by $v_i = 0$, $i \in \mathcal{J}^0$, $0 \leq v_i \leq v_i^{\text{sup}}$, $i \in \mathcal{J}^+$. Hence, $\underline{w}_i(\bullet; \lambda, \mu) \geq 0$, $\forall \lambda \in D^*$, $\forall \mu \in \mathbb{R}_+^m$ ($\lambda_i, \mu_i > 0$ $i \in \mathcal{J}^+$, if the 1st of the 3 alternatives is chosen), so that D (or $\tilde{D}$ in the case $(4.3.3)' - (4.3.3)''$) is contained in the left-hand side of (4.3.1c). Ab adsurdo, suppose that $\exists \hat{v} \notin D$, s.t. (in case $(4.3.3)' - (4.3.3)''$ we have not to consider elements of $D \backslash \tilde{D}$):

$$\underline{w}(\hat{v}; \lambda, \mu) \geq 0, \quad \forall \lambda \in D^*, \quad \forall \mu \in \mathbb{R}_+^m.$$

$\hat{v} \notin D$ implies the existence of either $i' \in \mathcal{J}^0$ s.t. $\hat{v}_{i'} \neq 0$ or $i'' \in \mathcal{J}^+$ s.t. $\hat{v}_{i''} < 0$ or both. By taking arbitrarily $\hat{\lambda}_{i'} \geq 0$ and $\hat{\mu}_{i'} > \max\{0, \hat{\lambda}_{i'}/\hat{v}_{i'}\}$, we have $\underline{w}_{i'}^{pe}(\hat{v}_{i'}; \hat{\lambda}_{i'}, \hat{\mu}_{i'}) < 0$; the same happens for any other $i \in \mathcal{J}^0$ s.t. $\hat{v}_i \neq 0$. For each $i \in \mathcal{J}^0$ s.t. $\hat{v}_i = 0$, of course, $\underline{w}_{i'}^{pe}(\hat{v}_i; \lambda_i, \mu_i) = 0, \forall \lambda_i, \forall \mu_i$. Whatever $\lambda_{i''}, \mu_{i''} > 0$ may be, we have $\underline{w}_{i''}(\hat{v}_{i''}, \lambda_{i''}, \mu_{i''}) < 0$ for each of the forms in (4.3.3) and in the case$(4.3.3)' - (4.3.3)''$; the same happens for any other $i \in \mathcal{J}^+$ s.t. $\hat{v}_i < 0$. For each $i \in \mathcal{J}^+$ s.t. $\hat{v} \geq 0$, in the case (4.3.3) we have:

$$
\underline{w}_i = \begin{cases}
\frac{1}{2\lambda_i\mu_i}\left(\sqrt{1+4\lambda_i^2\mu_i\hat{v}_i}-1\right) < \varepsilon, & \varepsilon > 0, \text{ if either } \hat{v}_i = 0 \text{ or } \hat{v}_i > 0 \text{ and } \lambda_i > 0, \\
& \qquad\qquad \mu_i > \max\{0, \tfrac{\lambda_i\hat{v}_i-\varepsilon}{\lambda_i\varepsilon^2}\}, \\[4pt]
\text{or} \\[4pt]
\lambda_i\hat{v}_i\exp(-\mu_i\hat{v}_i) < \varepsilon, & \varepsilon > 0, \text{ if either } \hat{v}_i = 0 \text{ or } \hat{v}_i > 0 \text{ and } \lambda_i > 0, \\
& \qquad\qquad \mu_i > \tfrac{1}{\hat{v}_i}\ln\tfrac{\lambda_i\hat{v}_i}{\varepsilon}, \\[4pt]
\text{or} \\[4pt]
\lambda_i[1 - \exp(-\mu_i\hat{v}_i)] < \varepsilon, & \varepsilon > 0, \text{ if either } \hat{v}_i = 0 \text{ or } \hat{v}_i > 0 \text{ and } \lambda_i < \varepsilon, \mu_i > 0;
\end{cases}
$$

and in the case$(4.3.3)' - (4.3.3)''$ we have (since $\lambda^2/4\varepsilon \geq (\lambda v - \varepsilon)/v^2$):

$$
\underline{w}_i = \lambda_i\hat{v}_i - \mu_i\hat{v}_i^2 < \epsilon, \quad \forall \epsilon > 0 \text{ if either } \hat{v}_i = 0 \text{ or } \hat{v}_i > 0 \text{ and } \lambda_i > 0, \mu_i > \frac{\lambda_i^2}{4\epsilon}.
$$

Therefore, we have achieved the existence of $\hat{\lambda} \in D^*$ and of $\hat{\mu} \in \mathbb{R}_+^m$, s.t. $\underline{w}(\hat{v}; \hat{\lambda}, \hat{\mu}) = \sum_{i \in \mathcal{J}} \underline{w}_i(\hat{v}_i; \hat{\lambda}_i, \hat{\mu}_i) < 0$, which contradicts the absurd assumption; hence (4.3.1c) follows. Of course, (4.3.3) is, in all the cases, s.t. either $\alpha\underline{w}_i(v_i; \lambda_i, \mu_i) = \underline{w}_i(v_i; \alpha\lambda_i, \mu_i)$ or $\alpha\underline{w}_i(v_i; \lambda_i, \mu_i) = \underline{w}_i(v_i; \alpha\lambda_i, \alpha\mu_i)$ or $\alpha\underline{w}_i(v_i; \lambda_i, \mu_i) = \underline{w}_i(v_i; \alpha\lambda_i, \mu_i/\alpha^2)$, $\forall \alpha \in \mathbb{R}_+\backslash\{0\}$, so that (4.3.1d) is satisfied. $\qquad\square$

For the sake of simplicity, with slight abuse of notation, here and in the sequel we use the functional symbols $w, \underline{w}$ even if the parameter arguments change.

At $\mu = 0$, (4.3.4) collapses to the *weak linear separation functions:*

$$
w(u, v; \theta, \lambda) = \theta u + \langle \lambda, v \rangle, \quad \theta \geq 0, \ \lambda \in D^*, \ (\theta, \lambda) \neq 0; \tag{4.3.5}
$$

the same happens, if $(4.3.3)'$ is adopted.

Note that, in the previous classes of w, if $\theta > 0$, without any loss of generality, we can set $\theta = 1$. Indeed, the replacement of w with $\frac{1}{\theta}w$ — and hence $\underline{w}$ with $\frac{1}{\theta}\underline{w}$ — does not change the stated properties.

Now, let us consider subclasses of the preceding ones, which are regular, by setting $\theta = 1$. With a slight abuse of notation with respect to (4.3.1), its regular subclass is the following one:

$$
w(u, v; \gamma) = u + \underline{w}(v; \gamma), \quad \gamma \in \Gamma, \tag{4.3.6}
$$

where $\underline{w} : \mathbb{R}^m \times \Gamma \to \mathbb{R}$ must fulfil the same (4.3.1c,d).

Proposition 4.3.3. The functions (4.3.6) are regular weak separation functions, namely the class (4.3.6) is a subclass of $\mathcal{W}_R(\Pi)$ at $\Pi = \{1\} \times \Gamma$.

Proof. We must prove (4.2.8). Since the positive level set of (4.3.6) is defined by $\underline{w}(v; \gamma) > -u$, then (u, v) belongs to the left-hand side of (4.2.8) iff $\underline{w}(v; \gamma) > 0 \ \forall \gamma \in \Gamma$ if $u \leq 0$ and $\underline{w}(v; \gamma) \geq 0$ if $u > 0$. If γ is s.t. $\underline{w}(v; \gamma) \not\equiv 0$, then, because of (4.3.1d), the former of the above inequalities obliges $\underline{w}$ to be, for each fixed v, arbitrarily great, so that no vector of type $(u \leq 0, v)$ can belong to the left-hand side of (4.2.8); the latter, because of (4.3.1c), makes the left-hand side of (4.2.8) equal to $\overset{\circ}{\mathbb{R}}_+ \times D$. Hence, taking into account (4.3.1f), we have:

$$\bigcap_{\gamma\in\Gamma}\mathrm{lev}_{>0}w(\bullet,\bullet;\gamma)=[\mathrm{lev}_{>0}w(\bullet,\bullet;\tilde{\gamma})]\cap\bigcap_{\gamma\in\Gamma\setminus\{\tilde{\gamma}\}}\mathrm{lev}_{>0}w(\bullet,\bullet;\gamma)=$$

$$(\overset{o}{\mathrm{I\!R}}_{+}\times\mathrm{I\!R}^{m})\cap(\overset{o}{\mathrm{I\!R}}_{+}\times D)=\mathcal{H},$$

where in the last intersection we stipulate to cut off all $\tilde{\gamma}$ s.t. $\underline{w}(v;\tilde{\gamma})\equiv 0$. $\qquad\square$

As a consequence of the introduction of (4.3.6), at $\theta=1$, (4.3.2) becomes:

$$w(u,v;\gamma)=u+\sum_{i\in\mathcal{J}}\underline{w}_i(v_i;\gamma_i),\quad \gamma=(\gamma_1,...,\gamma_m)\in\underset{i\in\mathcal{J}}{\times}\Gamma_i;\tag{4.3.7}$$

(4.3.4) becomes:

$$w(u,v;\lambda,\mu)=u+\sum_{i\in\mathcal{J}}\underline{w}_i^{pe}(v_i;\lambda_i,\mu_i),\quad \lambda\in D^*,\mu\in\mathrm{I\!R}_+^m,\tag{4.3.8}$$

where $\underline{w}_i^{pe}$ is given by (4.3.3); in $\underline{w}_i^{pe}$, the 1st expression for unilateral constraints can be replaced by (4.3.3)$''$ if (4.3.3)$'$ holds; (4.3.5) becomes:

$$w(u,v;\lambda)=u+\langle\lambda,v\rangle,\quad \lambda\in D^*.\tag{4.3.9}$$

Proposition 4.3.4. The classes (4.3.7), (4.3.8) and (4.3.9) are subclasses of $\mathcal{W}_R(\Pi)$ at $\Pi=\{1\}\times\Gamma,\Pi=\{1\}\times D^*\times\mathrm{I\!R}_+^m$ and $\Pi=\{1\}\times D^*$, respectively.

Proof. The claim about (4.3.7) is a consequence of Proposition 4.3.3. Taking into account that $\mathcal{W}(\{1\}\times D^*\times\mathrm{I\!R}_+^m)\subset\mathcal{W}(\mathrm{I\!R}_+\times D^*\times\mathrm{I\!R}_+^m)$, because of Propositions 4.3.2, 4.3.3, in order to prove (4.2.8) for the class (4.3.8) $-$ and thus to achieve the claim $-$ it is enough to show that $\mathcal{H}\subseteq\mathrm{lev}_{>0}w(\bullet,\bullet;\lambda,\mu)$ with w given by (4.3.8). This inclusion holds since $(u,v)\in\mathcal{H}$ implies $u>0$ and $v\in D$, and $v\in D$ implies the non-negativity of (4.3.3) and of (4.3.3)$''$. The claim about (4.3.9) is now trivial. $\qquad\square$

4.4. A General Setting for a Theorem of the Alternative

Sometimes, it is not easy or suitable to establish a TA between two propositions S and S^*. Therefore $-$ as recalled in Sect. 4.1 $-$ the definition of a TA is split. We say that *weak alternative* holds between S and S^* iff they are not both true; we say that *strong alternative* holds between S and S^* iff they are not both false. When both weak and strong alternative hold, then evidently *alternative* holds [28,14]. As announced in Sect. 4.1, we will see that there is a tight relationship between weak (or strong) alternative and weak (or strong) separation. The terminology weak and strong for the two "half pieces" of alternative is a classical one; to adopt the same terminology for separation $-$ as it has been done in Sect. 4.2 $-$ has been a Hobson's choice. In the sequel, only a special case of S will be considered, namely a mathematical system of type (1.3.16).

Theorem 4.4.1. Let the sets $\mathcal{H},X,\mathcal{W}(\Pi),\mathcal{S}(\Pi)$, the cone C with apex at the origin, the element $\xi\in\Xi$, and the function A be given.

(i) The system (1.3.16) in the unknown x, and the system:

$$w(A(x; \xi); \pi) \notin C, \quad \forall x \in X, \tag{4.4.1}$$

are not simultaneously possible, whatever the weak separation function $w \in \mathcal{W}(\Pi)$ might be.

(ii) The system (1.3.16) in the unknown x, and the system:

$$s(A(x; \xi); \pi) \notin C_0, \quad \forall x \in X, \tag{4.4.2a}$$

or the system:

$$s(A(x; \xi); \pi) \notin \overset{o}{C}, \quad \forall x \in X, \tag{4.4.2b}$$

are not simultaneously impossible, whatever the strong separation function $s \in \mathcal{S}(\Pi)$ might be.

Proof. (i) If (1.3.16) is possible, i.e. if $\exists \hat{x} \in X$ s.t. $\hat{z} := A(\hat{x}; \xi) \in \mathcal{H}$, then (4.2.2.a) implies $w(\hat{z}; \pi) \in C$, $\forall \pi \in \Pi$, so that (4.4.1) is false. **(ii)** If (1.3.16) is impossible, i.e. if $A(x; \xi) \notin \mathcal{H}$, $\forall x \in X$, then (4.2.9a) implies $s(A(x; \xi); \pi) \notin C_0$, $\forall x \in X$, and (4.2.10a) implies $s(A(x; \xi); \pi) \notin \overset{o}{C}$, $\forall x \in X$, so that (4.4.2) are true. $\qquad\square$

At $\ell = 1$ and $C = [0, +\infty[$, (4.4.1) and (4.4.2) become:

$$w(A(x; \xi); \pi) < 0, \quad \forall x \in X, \tag{4.4.3}$$

$$s(A(x; \xi); \pi) \leq 0, \quad \forall x \in X, \tag{4.4.4}$$

respectively.

Note that (4.2.2b) $-$ or (4.2.2b)$'$ $-$ have not been exploited in the proof of Theorem 4.4.1; indeed, they are not necessary to achieve alternative, while classes of separation functions which do not satisfy them are, in general, too poor for the applications as we will see in Chapter 5.

Theorem 4.4.2. Let the sets $\mathcal{H}, X, \mathcal{W}_R(\Pi)$, the cone C, the element $\xi \in \Xi$, and the function A be given. The system (1.3.16) in the unknown x, and the system:

$$\exists \pi \in \Pi \qquad \text{s.t.} \qquad w(A(x; y); \pi) \notin C_0, \quad \forall x \in X, \tag{4.4.5a}$$

or the system:

$$\exists \pi \in \Pi \qquad \text{s.t.} \qquad w(A(x; y); \pi) \notin \overset{o}{C}, \quad \forall x \in X, \tag{4.4.5b}$$

are not simultaneously possible, whatever the regular weak separation function $w \in \mathcal{W}_R(\Pi)$ might be.

Proof. If (1.3.16) is possible, i.e. if $\exists \hat{x} \in X$ s.t. $\hat{z} := A(\hat{x}; \xi) \in \mathcal{H}$, then (4.2.7) implies $w(\hat{z}; \pi) \in C_0$, $\forall \pi \in \Pi$, and (4.2.7)$'$ implies $w(\hat{z}; \pi) \in \overset{o}{C}$, $\forall \pi \in \Pi$, so that (4.4.5) are false. $\qquad\square$

Theorems 4.4.1 and 4.4.2 show the deep relationship between alternative and separation.

When both weak and strong alternative hold, then obviously alternative holds. If it is possible, under suitable assumptions on A and X, to guarantee the existence, within a family $\mathcal{W}(\Pi)$ of weak separation functions, of an element w, such that the impossibility of (1.3.16) − ξ being fixed − implies the possibility of (4.4.1) (or (4.4.5a), or (4.4.5b)) so that w acts also as strong separation function (in the sense of assuring that both systems cannot be impossible), then we have alternative besides weak one. The following Theorem 4.4.3 shows an instance of this. Same remark can be done for $\mathcal{W}_R(\Pi)$. Analogously, if it is possible, under suitable assumptions on A and X, to guarantee the existence, within a family $\mathcal{S}(\Pi)$ of strong separation functions, of an element s, such that the possibility of (4.4.2a) (or (4.4.2b)) implies the possibility of (1.3.16) − at a suitable ξ − so that s acts also as a weak separation function (in the sense of assuring that both systems cannot be possible), then we have alternative besides strong one.

Proposition 4.4.1 [8,14,17]. Let X be convex, $A : X \to \mathbb{R}^\nu$, and $H \subset \mathbb{R}^\nu$ be a convex, closed and pointed cone with apex at the origin. The set $A(X) - H$ is convex if and only if A is H-concavelike.

Proof. A is H-concavelike (see Definition 2.4.5) iff

$$\forall x^1, x^2 \in X, \quad \forall \alpha \in [0,1], \quad \exists \hat{x} \in X \text{ s.t. } A(\hat{x}) - [(1-\alpha)A(x^1) + \alpha A(x^2)] \in H,$$

or iff

$$\exists h \in H \quad \text{s.t.} \quad A(\hat{x}) - [(1-\alpha)A(x^1) + \alpha A(x^2)] = h,$$

or, by introducing $h(\alpha) := h + (1-\alpha)h^1 + \alpha h^2$, iff

$$\forall h^1, h^2 \in H, \quad (1-\alpha)[A(x^1) - h^1] + \alpha[A(x^2) - h^2] = A(\hat{x}) - h(\alpha),$$

or, account taking that $h(\alpha) \in H$, iff $A(X) - H$ is convex. $\qquad\square$

When H is an orthant or a suborthant of $\mathbb{R}^\nu$, then H-concavelike collapses to concavelike, which is a broader concept than concave. Note that $A(X) - H$ is the conic extension, in the sense of Sect. 3.2, of the image set $A(X)$.

As a special case of (1.3.16), consider system (3.2.1), where now we understand the dependence of $f_{\bar{x}}(x)$ on $\bar{x}$ to stress the fact that next Theorem 4.4.3 holds independently of $\bar{x}$; namely, it holds for any $f(x)$ and not only for $f_{\bar{x}}(x)$.

Theorem 4.4.3. Let X be convex, $f : X \to \mathbb{R}$ concave and $g : X \to \mathbb{R}^m$ concavelike. Assume that $\exists x^1, ..., x^{r+1} \in X$, such that: **(i)** the vectors

$$(g_i(x^j), \ i \in \mathcal{J}^0) \in \mathbb{R}^r, \quad j = 1, ..., r+1 \tag{4.4.6a}$$

are affinely independent and the interior (in $\mathbb{R}^r$) of their convex hull contains the origin of $\mathbb{R}^r$; **(ii)** we have:

$$\sum_{j=1}^{r+1} g_i(x^j) > 0, \quad \forall i \in \mathcal{J}^+. \tag{4.4.6b}$$

Then, the system (3.2.1) is impossible, if and only if $\exists \lambda \in D^*$, such that:

$$f(x) + \langle \lambda, g(x) \rangle \leq 0, \quad \forall x \in X. \tag{4.4.7}$$

Proof. By understanding the dependence of $f, \mathcal{K}$ and A on $\bar{x}$, with a small abuse of notation, we can say that (3.2.1) is impossible iff $\mathcal{H} \cap \mathcal{K} = \mathcal{H} \cap A(X) = \varnothing$ or (due to Proposition 3.2.7) iff $\mathcal{H} \cap [A(X) - \operatorname{cl} \mathcal{H}] = \varnothing$. Because of the Proposition 4.4.1 and of the assumptions on X, f, g, $A(x) - \operatorname{cl} \mathcal{H}$ is convex; $\mathcal{H}$ is obviously convex. **If.** Let (4.4.7) hold. Let π^- denote the halfspace whose equation is $u + \langle \lambda, v \rangle \leq 0$. (4.4.7) means $A(X) \subseteq \pi^-.v \in D \Rightarrow \langle \lambda, v \rangle \geq 0, \forall \lambda \in D^*$; $(u, v) \in \mathcal{H} = \overset{o}{\mathbb{R}}_+ \times D$ and $\lambda \in D^*$ imply $u + \langle \lambda, v \rangle > 0$ or $(u, v) \in \sim \pi^-$. Hence $\mathcal{H} \cap A(X) = \varnothing$ or (3.2.1) is impossible. **Only if.** Let (3.2.1) be impossible, so that $\mathcal{H} \cap [A(X) - \operatorname{cl} \mathcal{H}] = \varnothing$. Then, because of Theorem 2.2.4(i), there exists a hyperplane, say π, s.t. $\mathcal{H} \subseteq \pi^+$ and $A(X) - \operatorname{cl} \mathcal{H} \subseteq \pi^-$, where π^+ and π^- are the closed (opposite) halfspaces defined by π. Since the apex of $\mathcal{H}$ is the origin O of $\mathbb{R}^\nu = \mathbb{R}^{1+m}$, it is not restrictive to assume that $O \in \pi$, so that the equation of π is $\theta u + \langle \lambda, v \rangle = 0$ with $(\theta, \lambda) \neq O$. It is also not restrictive to identify π^+ with the non-negative level set (with respect to (u, v)) of $\theta u + \langle \lambda, v \rangle$, so that (being $\mathcal{H} = \overset{o}{\mathbb{R}}_+ \times D$) $\mathcal{H} \subset \pi^+$ implies $(\theta, \lambda) \in \mathcal{H}^*$ or $\theta \in \mathbb{R}_+$ and $\lambda \in D^*$. To achieve the thesis it is enough to show that $\theta = 0$ cannot occur. Ab absurdo, suppose $\theta = 0$. Set $v_i^j := g_i(x^j)$, $i \in \mathcal{I}, j = 1, ..., r + 1$. Since $A(X) \subseteq \pi^-$, we have:

$$\sum_{i \in \mathcal{I}^0} \lambda_i v_i^j + \sum_{i \in \mathcal{I}^+} \lambda_i v_i^j \leq 0, \quad j = 1, ..., r + 1. \tag{4.4.8}$$

If $\lambda_i = 0 \ \forall i \in \mathcal{I}^0$, so that $(\lambda_i, i \in \mathcal{I}^+) \neq O$, then (4.4.8) and (4.4.6b) imply:

$$0 \geq \sum_{j=1}^{r+1} \sum_{i \in \mathcal{I}^+} \lambda_i v_i^j = \sum_{i \in \mathcal{I}^+} \lambda_i \sum_{j=1}^{r+1} v_i^j > 0,$$

and a contradiction has been achieved. If $(\lambda_i, i \in \mathcal{I}^0) \neq O$, taking into account that $\lambda \in D^* \Rightarrow \lambda_i \geq 0, \forall i \in \mathcal{I}^+$, from (4.4.6b) we draw $\sum_{i \in \mathcal{I}^+} \lambda_i v_i^j \geq 0$, so that (4.4.8) implies:

$$\sum_{i \in \mathcal{I}^0} \lambda_i v_i^j \leq 0, \quad j = 1, ..., r + 1; \quad (\lambda_i, \ i \in \mathcal{I}^0) \neq O. \tag{4.4.9}$$

Because of (i), $\exists \varepsilon > 0$ s.t. the non-null vector $(\varepsilon \lambda_i, \ i \in \mathcal{I}^0)$ belongs to the interior of the convex hull of vectors $(v_i^j, \ i \in \mathcal{I}^0), j = 1, ..., r + 1$, so that $\exists \alpha_j > 0, j = 1, ..., r + 1$ with $\sum_{j=1}^{r+1} \alpha_j = 1$, such that $\varepsilon \lambda_i = \sum_{j=1}^{r+1} \alpha_j v_i^j, \ i \in \mathcal{I}^0$. Therefore, from (4.4.9) we deduce:

$$0 \geq \sum_{j=1}^{r+1} \alpha_j \sum_{i \in \mathcal{I}^0} \lambda_i v_i^j = \sum_{i \in \mathcal{I}^0} \lambda_i \sum_{j=1}^{r+1} \alpha_j v_i^j = \varepsilon \sum_{i \in \mathcal{I}^0} \lambda_i^2 > 0,$$

and a contradiction has been achieved. $\square$

Note that the "if part" of Theorem 4.4.3 does not require any assumption. This is due to the fact that the weak separation function, underlying (4.4.7), is regular, namely an element of $\mathcal{W}_R(D^*)$. Furthermore, the cone D might be any convex, closed and pointed cone with apex at the origin, and not necessarily that of Sect.1.1; in such a case, g should be assumed to be D-concavelike. If $m = 0$, then Theorem 4.4.3 becomes trivial.

Example 4.4.1. In Theorem 4.4.3 set $X = \mathbb{R}^2$, $f(x) = -2 - x_1 - 2x_2$, $p = 1$, $m = 2$, $g_1(x) = 1 - x_1^2 - x_2^2$, $g_2(x) = x_1 + x_2 + 1$. Then, system (3.2.1) becomes $(f_{\bar{x}}(x) \equiv f(x))$:

$$-2 - x_1 - 2x_2 > 0, \qquad 1 - x_1^2 - x_2^2 = 0, \qquad x_1 + x_2 + 1 \geq 0.$$

Consider the vectors $x^1 = (1/2, 0), x^2 = (0, 3/2)$. Since

$$g_1(x^1) - g_1(x^2) = 2 \neq 0, \quad 0 \in]g_1(x^2), g_1(x^1)[=] - \tfrac{5}{4}, \tfrac{3}{4}[, \quad g_2(x^1) + g_2(x^2) = 4 > 0,$$

(i) and (ii) are satisfied. Since $D = \mathbb{R} \times \mathbb{R}_+$, (4.4.7) becomes:

$$\lambda_1 x_1^2 + \lambda_1 x_2^2 + (1 - \lambda_2)x_1 + (2 - \lambda_2)x_2 + 2 - \lambda_1 - \lambda_2 \geq 0 \ , \ \forall(x_1, x_2) \in \mathbb{R}^2,$$

and should hold for some $\lambda_1 \in \mathbb{R}$ and $\lambda_2 \in \mathbb{R}_+$. The above inequality is impossible for both $\lambda_1 = 0$ and $\lambda_1 < 0$: in the former case we should have simultaneously $\lambda_2 = 1$ and $\lambda_2 = 2$; in the latter one the left-hand side is a negative definite form. For $\lambda_1 > 0$ it becomes

$$x_1^2 + x_2^2 + \frac{1 - \lambda_2}{\lambda_1}x_1 + \frac{2 - \lambda_2}{\lambda_1}x_2 + \frac{2 - \lambda_1 - \lambda_2}{\lambda_1} \geq 0, \ \forall(x_1, x_2) \in \mathbb{R}^2,$$

and is satisfied for $\lambda_1 = 1/2$, $\lambda_2 = 1$, since its left-hand side has

$$\left(x_1 = \frac{\lambda_2 - 1}{2\lambda_1} \ , \ x_2 = \frac{\lambda_2 - 2}{2\lambda_1} \right)$$

as global m.p. at which achieves the (global) minimum

$$\frac{1}{4\lambda_1^2}(-4\lambda_1^2 - 2\lambda_2^2 - 4\lambda_1\lambda_2 + 8\,\lambda_1 + 6\lambda_2 - 5);$$

such a minimum is zero for $\lambda_1 = 1/2$ and $\lambda_2 = 1$. Hence, the thesis of Theorem 4.4.1 is fulfilled. Note that (1.1.2) is now an arc of circumference of length $3\pi/2$ (contained in the 1st, 2nd and 4th quadrants), and (for $\bar{x} = 2$) $\mathcal{K}_{\bar{x}}$ of Sect. 3.2 is now a circular paraboloid given by

$$v_1 = -2u^2 - 5v_2^2 - 6uv_2 - 2u - 2v_2.$$

The plane, whose equation is $u + \frac{1}{2}v_1 + v_2 = 0$ (the coefficients of v_1 and v_2 being, respectively, the above values of λ_1 and λ_2), separates such a paraboloid from the set $\mathcal{H} = (\mathbb{R}_+/\{0\}) \times D$. $\qquad\square$

Let us now consider some particular cases of Theorem 4.4.3. The first is obtained at $p = 0$, so that assumption (i) disappears and (ii) is simplified.

Corollary 4.4.1. Let X be convex, $f : X \to \mathbb{R}$ concave, and $g : X \to \mathbb{R}^m$ concavelike. Assume that $\exists \hat{x} \in X$, such that:

$$g(\hat{x}) > 0. \tag{4.4.10}$$

Then, the system

$$f(x) > 0, \qquad g(x) \geq 0, \qquad x \in X \tag{4.4.11}$$

is impossible, if and only if $\exists \lambda \in \mathbb{R}_+^m$, such that:

$$f(x) + \langle \lambda, g(x) \rangle \leq 0, \ \ \forall x \in X. \tag{4.4.12}$$

Proof. It is enough to set $p = 0$ in Theorem 4.4.3, so that $D = \mathbb{R}_+^m$, and note that concavity implies concavitylike. $\qquad\square$

Note that the assumption (4.4.10), due to Slater [40], is a constraint qualification, as well as the more general form expressed by (4.4.6). Another particular case is obtained at $p = m$; it is a straightforward consequence of Theorem 4.4.3.

Corollary 4.4.2. Let X be convex, $f : X \to \mathbb{R}$ concave, and $g : X \to \mathbb{R}^p$ concavelike. Assume that $\exists x^1, ..., x^{p+1} \in X$, such that the vectors:

$$(g_i(x^j),\ i \in \mathcal{I}^0) \in \mathbb{R}^p, \quad j = 1, ..., p+1 \tag{4.4.13}$$

are affinely independent and that the interior (in $\mathbb{R}^p$) of their convex hull contains the origin of $\mathbb{R}^p$. Then, the system:

$$f(x) > 0, \qquad g(x) = 0, \qquad x \in X \tag{4.4.14}$$

is impossible, if and only if $\exists \lambda \in \mathbb{R}^p$, such that:

$$f(x) + \langle \lambda, g(x) \rangle \leq 0, \quad \forall x \in X. \tag{4.4.15}$$

At $m = p = 0$, Corollaries 4.4.1 and 4.4.2 become trivial. When a TA can be established, then of course the best has been achieved; however also a weak (or strong) TA can be useful.

Theorem 4.4.4. Let X, f, g and D be as in Sect. 1.1. System (3.2.1) is impossible, if $\exists \theta \in \mathbb{R}_+, \lambda \in D^*$, with $(\lambda, \theta) \neq O$, such that:

$$\theta f(x) + \langle \lambda, g(x) \rangle \leq 0, \quad \forall x \in X, \tag{4.4.16}$$

where, if $\theta = 0$ necessarily, either the inequality must be verified in strict sense, or

$$\chi(\lambda) := \{x \in X : f(x) > 0, g(x) \in D, \langle \lambda, g(x) \rangle = 0\} = \emptyset. \tag{4.4.17}$$

Proof. If (4.4.16) is verified with $\theta > 0$, then $-$ being here, as in Sect. 3.2., $C = [0, +\infty[, \mathcal{H} = \overset{o}{\mathbb{R}}_+ \times D$ and $w(u, v; \lambda) := \theta u + \langle \lambda, v \rangle \in \mathcal{W}_R(\overset{o}{\mathbb{R}}_+ \times D^*) - $ (4.4.16) can be identified with (4.4.5a) (or (4.4.5b)), and hence Theorem 4.4.2 gives the thesis. If (4.4.16) is verified necessarily at $\theta = 0$, then now $w \in \mathcal{W}(\mathbb{R}_+ \times D^*)$, since merely (4.2.2) are satisfied. Therefore, we can apply Theorem 4.4.1 (i) and (4.4.1) $-$ which now becomes (4.4.16) with strict inequality sign $-$ shows the sufficiency of (4.4.16) as strict inequality. It remains to show $-$ again when $\theta = 0$ necessarily $-$ (4.4.16) under the condition (4.4.17). It means that no element of the image $\mathcal{K}$ of (3.2.1) lies in the (separating) hyperplane $w = \theta u + \langle \lambda, v \rangle = 0$, so that separation implies disjunctive separation, or $\mathcal{K} \subseteq \text{lev}_{\leq 0}\, w(\bullet; \theta, \lambda)$ and $\chi(\lambda) = \emptyset$ imply $\mathcal{K} \subseteq \text{lev}_{<0}\, w(\bullet; \theta, \lambda)$, so that, being $\mathcal{H} \subset \text{lev}_{\geq 0}\, w(\bullet; \theta, \lambda)$, we have $\mathcal{H} \cap \mathcal{K} = \emptyset$ or the impossibility of (3.2.1). $\quad\square$

Condition (4.4.17) can be weakened, by deleting $f(x) > 0$ and $g(x) \in D$. In the preceding two theorems and two corollaries, f was single-valued; now we consider a case where f is vector-valued; e will denote a vector with entries $=1$ whose order will be clear from the context; any vector with positive entries might be used, instead of e, in next corollary.

Corollary 4.4.3. Let X be convex, $f : X \to \mathbb{R}^\ell$ and $g : X \to \mathbb{R}^m$ be concavelike, and D be as in Sect. 1.1. Then, the system:

$$f(x) > O, \qquad g(x) \in D, \qquad x \in X \tag{4.4.18}$$

is impossible, if $\exists \theta \in \overset{o}{\mathbb{R}^{\ell}_{+}}$ with $\langle e, \theta \rangle = 1$, and $\lambda \in D*$, such that:

$$\langle \theta, f(x) \rangle + \langle \lambda, g(x) \rangle \leq 0, \quad \forall x \in X. \tag{4.4.19}$$

Proof. (4.4.18) is equivalent to the system:

$$\langle e, f(x) \rangle > 0, \quad f(x) \geq O, \quad g(x) \in D, \quad x \in X. \tag{4.4.20}$$

With obvious positions, (4.4.20) is identified with (3.2.1). As a consequence, taking into account that assumptions (4.4.6) are redundant for the sufficiency of Theorem 4.4.3, the "if part" of such a theorem holds also here and shows that (4.4.20) is impossible, if $\exists \tau \in \mathbb{R}^{\ell}_{+}$ and $\exists \lambda \in D^{*}$, such that:

$$\langle e, f(x) \rangle + \langle \tau, f(x) \rangle + \langle \lambda, g(x) \rangle \leq 0, \quad \forall x \in X,$$

or, setting $\theta := e + \tau \in \overset{o}{\mathbb{R}^{\ell}_{+}}$ if (4.4.19) holds. $\square$

The above corollary cannot be inverted, as simple examples show:

Example 4.4.2. Let us set $X = \mathbb{R}_{+}, \ell = 2, m = 0, f_1(x) = \sqrt{x}, f_2(x) = -x$. If (and only if) $\theta_1 = 0, \theta_2 = 1$, (4.4.19) is verified; then (4.4.18) is impossible (as it is immediate to see). $\square$

4.5. Special Theorems of the Alternative

Most of TA, including the classic ones, can be deduced from those of Sect. 4.4 or proved by exploiting the above separation scheme. Only some TA will now be considered as an instance of such an approach. In this section, the notations of matrices and vectors are independent of those of the other sections; furthermore, for the sake of simplicity and without any fear of confusion, A and D denote both matrices and the map and the set introduced in Sect. 1.1 and used throughout the book.

Theorem 4.5.1. Let X be convex, $f : X \to \mathbb{R}^{\ell}$ and $g : X \to \mathbb{R}^{m}$ be concavelike; D as in Sect. 1.1. If the system:

$$f(x) > 0, \qquad g(x) \in D, \qquad x \in X \tag{4.5.1}$$

is impossible, then $\exists \theta \in \mathbb{R}^{\ell}_{+}$ and $\exists \lambda \in D^{*}$, with $(\theta, \lambda) \neq O$, such that:

$$\langle \theta, f(x) \rangle + \langle \lambda, g(x) \rangle \leq 0, \quad \forall x \in X. \tag{4.5.2}$$

Proof. Here $\mathcal{H} = \overset{o}{\mathbb{R}^{\ell}_{+}} \times D$ and $\nu = \ell + m$. The obvious convexity of $\mathcal{H}$ and, due to Proposition 4.4.1, that of $A(X) - \text{cl } \mathcal{H} = \mathcal{K} - \text{cl } \mathcal{H}$, imply (because of Theorem 2.2.4(i)) the existence of an hyperplane, say π, s.t. $\mathcal{H} \subseteq \pi^{+}$ and $A(X) - \text{cl } \mathcal{H} \subseteq \pi^{-}$, where π^{+} and π^{-} are the closed (opposite) halfspaces defined by π. Since the apex of

$\mathcal{H}$ is the origin O of $\mathbb{R}^\nu$, without any loss of generality we can assume that $O \in \pi$, so that the equation of π is given by $\langle \theta, u \rangle + \langle \lambda, v \rangle = 0, (\theta, \lambda) \neq O$. Now note that $\mathcal{H} \subset \pi^+ \Rightarrow \theta \geq 0$ and $\lambda \in D^*$; $A(X) - \mathrm{cl}\ \mathcal{H} \subseteq \pi^- \Rightarrow A(X) \subseteq \pi^-$ which is equivalent to (4.5.2). $\qquad\square$

Corollary 4.5.1. Let X be convex and $f : X \to \mathbb{R}^\ell$ be concavelike. The system

$$f(x) > O, \qquad x \in X \tag{4.5.3}$$

is impossible, if and only if $\exists \theta \in \mathbb{R}^\ell_+ \backslash \{O\}$, such that:

$$\langle \theta, f(x) \rangle \leq 0, \quad \forall x \in X. \tag{4.5.4}$$

Proof. Only if. It is enough to set $m = 0$ in Theorem 4.5.1. **If.** It comes trivially from Theorem 4.1.2(i) at $\nu = \ell$, $\mathcal{H} = \overset{o}{\mathbb{R}}{}^\ell_+$, $C = [0, +\infty[$, and $w(u, v; \theta) := \langle \theta, u \rangle$. However, it is easy to show directly that both (4.5.3) and (4.5.4) cannot be possible, by observing that, if $\hat{x}$ is a solution of (4.5.3), then $\theta \in \mathbb{R}^\ell_+ \backslash \{O\} \Rightarrow \langle \theta, f(\hat{x}) \rangle > 0$ which contradicts (4.5.4). $\qquad\square$

When f is concave, Corollary 4.5.1 is due to Fan-Glicksberg-Hoffman [II22]. If $X = \mathbb{R}^n$ and f is linear, so that $f(x) = Ax$, with A any matrix with real entries, then Corollary 4.5.1 collapses to the following one, due to Gordan [19].

Corollary 4.5.2. Let A be a $\ell \times n$ matrix with real entries. The system:

$$Ax > O, \qquad x \in \mathbb{R}^n \tag{4.5.5}$$

is impossible, if and only if there exists a row vector $\theta \in \mathbb{R}^\ell_+ \backslash \{O\}$, such that:

$$\theta A = O. \tag{4.5.6}$$

Proof. It is enough to note that (4.5.4) is equivalent to $\theta A x \leq 0, \forall x \in \mathbb{R}^n$, and this holds iff (4.5.6) does. $\qquad\square$

Corollary 4.5.2 can be formulated this way: *let $A_1, ..., A_\ell$ be the rows of A; the set of solutions to (4.5.5) is non-empty, if and only if*

$$O \notin \mathrm{conv}\{A_1, ..., A_\ell\}.$$

Under suitable assumptions, Theorem 4.5.1 can be "inverted". A classic instance is offered by the linear case.

Corollary 4.5.3. Let A, B, C be matrices with real entries and of dimensions $\ell \times n$, $p \times n$, $(m - p) \times n$, respectively; a, b, c be column vectors, all with real entries. The system

$$Ax > a, \qquad Bx = b, \qquad Cx \geq c, \qquad x \in \mathbb{R}^n \tag{4.5.7}$$

is impossible, if and only if there exist row vectors $\theta \in \mathbb{R}^\ell_+$, $\mu \in \mathbb{R}^p$, $\tau \in \mathbb{R}^{m-p}_+$, with $(\theta, \mu, \tau) \neq O$, such that:

$$\theta A + \mu B + \tau C = O, \quad \langle \theta, a \rangle + \langle \mu, b \rangle + \langle \tau, c \rangle \geq 0, \tag{4.5.8}$$

where the inequality in (4.5.8) must be verified in strict sense, if $\theta = 0$ necessarily, and where $\theta \neq O$ necessarily if $a = O, b = O, c = O$ (in which case the inequality is trivially verified); if $\theta \neq 0$ and $\ell = 1$, it is not restrictive to assume $\theta = 1$.

Proof. If. At $X = \mathbb{R}^n$, $\mathcal{H} = \overset{o}{\mathbb{R}^\ell_+} \times D$, and

$$w(u, v; \theta, \mu, \tau) = \langle \theta, u \rangle + \langle (\mu, \tau), v \rangle \in \mathcal{W}(\mathbb{R}^\ell_+ \times \mathbb{R}^p \times \mathbb{R}^{m-p}_+ \backslash \{O\}),$$

Theorem 4.4.1(i) gives the thesis. However, it can be obtained in an elementary way: ab adsurdo, suppose that (4.5.7) has a solution, say $\hat{x}$. If $\theta \in \mathbb{R}^\ell_+ \backslash \{O\}$, then:

$$\langle \theta, A\hat{x} - a \rangle > 0, \quad \langle \mu, B\hat{x} - b \rangle = 0, \quad \langle \tau, C\hat{x} - c \rangle \geq 0,$$

so that, taking into account (4.5.8), we find: $0 \leq \langle \theta, a \rangle + \langle \mu, b \rangle + \langle \tau, c \rangle = \langle \theta, a \rangle + \langle \mu, b \rangle + \langle \tau, c \rangle - (\theta A + \mu B + \tau C)\hat{x} = -\langle \theta, A\hat{x} - a \rangle - \langle \mu, B\hat{x} - b \rangle - \langle \tau, C\hat{x} - c \rangle < 0$. If $\theta = O$, then we have $0 < \langle \theta, a \rangle + \langle \mu, b \rangle + \langle \tau, c \rangle = -\langle \mu, B\hat{x} - b \rangle - \langle \tau, C\hat{x} - c \rangle = -\langle \tau, C\hat{x} - c \rangle \leq 0$.
Only if. Now

$$\mathcal{K} = A(\mathbb{R}^n) = \left\{ (u, v) \in \mathbb{R}^\ell \times \mathbb{R}^m : u = Ax - a, v = \begin{pmatrix} Bx - b \\ Cx - c \end{pmatrix}, x \in \mathbb{R}^n \right\}$$

is an affine manifold, which does not intersect the convex and pointed (see (2.1.8)) cone $\mathcal{H} = \overset{o}{\mathbb{R}^\ell_+} \times D$. Because of Theorem 2.2.3 and Corollary 2.2.4, there exists an hyperplane, say π, which separates $\mathcal{H}$ and $\mathcal{K}$. It is not restrictive to assume that $\mathcal{K} \subseteq \pi^-$ and $\mathcal{H} \subset \pi^+$, where π^- and π^+ are the opposite closed halfspaces defined by π, and that π^- be identified by $\langle \theta, u \rangle + \langle \lambda, v \rangle \leq 0$ with $\lambda := (\mu, \tau)$, and $(\theta, \lambda) \neq O$. $\mathcal{H} \subset \pi^+ \Rightarrow \theta \in \mathbb{R}^\ell_+$ and $\lambda \in D^*$, where D is as in Sect. 1.1. Then $\mathcal{K} \subseteq \pi^-$ becomes:

$$\langle \theta, Ax - a \rangle + \langle \mu, Bx - b \rangle + \langle \tau, Cx - c \rangle \leq 0, \quad \forall x \in \mathbb{R}^n, \tag{4.5.9}$$

and implies (4.5.8). If $\theta = O$ necessarily, because of Theorem 2.2.7 (here $S = \mathcal{K}, K = \mathcal{H}, TC(S - \mathrm{cl}\, K) = \mathcal{K} - \mathrm{cl}\, \mathcal{H}$; $\mathcal{K}$ is an affine manifold which cannot contain any face of $\mathcal{H}$), the inequality in (4.5.9) must be verified in strict sense, and hence the same must occur to the inequality in (4.5.8). If $a = O, b = O, c = O$, then $\theta = O$ would imply that the (open) face $U := \{(u, v) \in \mathcal{H} : v = O\}$ of $\mathcal{H}$ is in π and intersects $\mathcal{K}$, since $\mathcal{K}$ in this case is a linear manifold. $\qquad\square$

Note that, if $a = O, b = O, c = O$, then Corollary 4.5.3 is Motzkin TA [34,145]. Therefore, Corollary 4.5.3 is a nonhomogeneous version of Motzkin TA; most of TA for linear algebraic systems can be drawn from it; we will consider some of them as an instance of how TA can be obtained. At $a = O, B = O, b = O, C = O, c = O$, Corollary 4.5.3 collapses to Gordan TA [19]. If we set $\gamma := \langle \theta, a \rangle + \langle \mu, b \rangle + \langle \tau, c \rangle \geq 0$, then the 2nd of (4.5.8) and the related condition (written in the subsequent row) can be written as $\theta_1 + \dots + \theta_\ell + \gamma > 0$.

If the matrix A is square and $\begin{pmatrix} B \\ C \end{pmatrix} A^{-1} = O$, then $\theta = O$ in (4.5.8). In fact, in such a case, there is an one-to-one correspondence between x and u, so that u is surjective, and $v = \begin{pmatrix} b \\ c \end{pmatrix}$; therefore $\mathcal{K} = \left\{ (u, v) \in \mathbb{R}^\ell \times \mathbb{R}^m : v = \begin{pmatrix} -b \\ -c \end{pmatrix} \right\}$ is an hyperplane parallel to the subspace $\mathbb{R}^\ell$.

Corollary 4.5.4 (Slater [40]). Let A, B, C, D be matrices with real entries and of dimensions $m \times n$, $p \times n$, $q \times n$, $r \times n$, respectively. Between the systems (in the unknowns x and (t, u, v, z)):

$$Ax > O, \quad Bx = O, \quad Cx \geq O, \quad Dx \geq O, \qquad\qquad Dx \neq O, \qquad (4.5.10)$$

and

$$\begin{cases} tA + uB + vC + zD = O \\ \text{and} \\ \text{either} \quad t \geq O, t \neq O, v \geq O, z \geq O, \\ \text{or} \qquad t \geq O, v \geq O, z > O, \end{cases} \qquad (4.5.10)^*$$

one and only one has solutions.

Proof. $(4.5.10)$ is equivalent to the system (the subfix of vector e denotes its dimension):

$$\begin{pmatrix} A \\ \langle e_r, D \rangle \end{pmatrix} x > O, \quad Bx = O, \quad \begin{pmatrix} C \\ D \end{pmatrix} x \geq O, \qquad (4.5.11)$$

which is easily identified with $(4.5.7)$ at $a = O, b = O, c = O$. Because of Corollary 4.5.2, $(4.5.10)$ is impossible iff there exist row vectors $t \in \mathbb{R}^m_+$, $t' \in \mathbb{R}_+$, $u \in \mathbb{R}^p$, $v \in \mathbb{R}^q_+$, $v' \in \mathbb{R}^r_+$, with $(t, t') \neq O$, such that:

$$tA + t'\langle e_r, D \rangle + uB + vC + v'D = O.$$

This inequality becomes that in $(4.5.10)^*$ by setting $z := v' + t'e_r$. $t' = O \Rightarrow t \neq O$ and we find the 1st row of inequalities in $(3.5.10)^*$; $t' > O \Rightarrow z > O$ and the 2nd row of inequalities in $(4.5.10)^*$ follows. $\qquad\qquad\square$

Note that $(4.5.10)$ could have been reduced to the following system, in the unknowns $x \in \mathbb{R}^n$ and $y \in \mathbb{R}$, instead of $(4.5.11)$:

$$y > O, \ Bx = O, \ Ax - e_m y \geq O, \ Cx \geq O, \ Dx - e_r y \geq O, Dx \geq O.$$

If $A = O$ or $D = 0$, then obviously $(4.5.10)$ is impossible and $(4.5.10)^*$ is possible, since admits at least the null solution. This is the reason why next corollary cannot be drawn from Corollary 4.5.2.

Corollary 4.5.5 (Tucker [I43]). Let B, C, D be matrices with real entries and of orders $p \times n$, $q \times n$, $r \times n$, respectively. Between the systems (in the unknowns x and (u, v, z)):

$$Bx = O, \quad Cx \geq O, \quad Dx \geq O, \quad Dx \neq O , \qquad (4.5.12)$$

and

$$uB + vC + zD = O, \ u \in \mathbb{R}^p , \ v \in \mathbb{R}^q_+ , z \in \overset{o}{\mathbb{R}}{}^r_+ \qquad (4.5.12)^*$$

one and only one has solutions.

Proof. $(4.5.12)$ is equivalent to the system: $\langle e_r D, x \rangle > 0, \quad Bx = O, \quad \begin{pmatrix} C \\ D \end{pmatrix} x \geq O,$ which is easily identified with $(4.5.7)$ at $a = O, b = O, c = O$ (and $\ell = 1$). Because of Corollary 4.5.3, $(4.5.12)$ is impossible iff there exists row vectors $u \in \mathbb{R}^p, v \in \mathbb{R}^q_+,$

$v' \in \mathbb{R}^r_+$, such that $e_r D + uB + vC + v'D = O$. This equality becomes that in $(4.5.12)^*$ by setting $z := v' + e_r \in \overset{o}{\mathbb{R}}{}^r_+$. $\square$

$D = 0$ makes $(4.5.12)$ impossible and $(4.5.12)^*$ trivially possible. At $B = O$, $C = O$, Corollary 4.5.5 collapses to the following:

Corollary 4.5.6 (Stiemke [41]). Let D be a matrix with real entries and of order $r \times n$. Between the systems (in the unknowns x and z):

$$Dx \geq O, \qquad Dx \neq O, \tag{4.5.13}$$

and

$$zD = O, \qquad z \in \overset{o}{\mathbb{R}}{}^r_+, \tag{4.5.13}^*$$

one and only one has solutions.

Corollary 4.5.6 can be interpreted this way: let $D_1, ..., D_r$ be the rows of D; the set of non−trivial solutions to $(4.5.13)$ is non−empty, if and only if

$$O \notin \text{int conv}\{D_1, ..., D_r\}.$$

Corollary 4.5.7 (Farkas [11]). Let A be a matrix of order $m \times n$ and α be a row n-vector, both with real entries. Between the systems (in the unknowns x and v):

$$Ax \geq O, \qquad \langle \alpha, x \rangle < 0, \tag{4.5.14}$$

and

$$vA = \alpha, \quad v \in \mathbb{R}^m_+, \tag{4.5.14}^*$$

one and only one has solutions.

Proof. $(4.5.14)$ is easily identified with the homogeneous case of $(4.5.7)$. Then, the thesis follows from Corollary 4.5.3. $\square$

Indeed, the original statement of the so-called Farkas Lemma is: "if $Ax \geq O$ implies $\langle \alpha, x \rangle \geq 0$, then there exists $\tau \in \mathbb{R}^m_+$, such that $\tau A = \alpha$". This statement, which does not look as a TA, is obviously equivalent to Corollary 4.5.7. In the paper [11] there is much more than Corollary 4.5.7: Farkas considers also an infinite dimensional case; some decades ago, Corollary 4.5.7 has been extended to an infinite dimensional space; this work seems not to be aware of the results by Farkas. Corollary 4.5.7 is immediately generalized:

Corollary 4.5.8 (Farkas [11]). Let A, B be matrices with real entries and of orders $m \times n$, $p \times n$, respectively. Between the systems (in the unknowns x and (u, v)):

$$Ax \geq O, \qquad Bx > O, \tag{4.5.15}$$

and

$$uA + vB = O, \quad u \in \mathbb{R}^m_+, \quad v \in \mathbb{R}^p_+ \setminus \{O\}, \tag{4.5.15}^*$$

one and only one has solutions.

Proof. (4.5.15) is easily identified with the homogeneous case of (4.5.7). Then, the thesis follows from Corollary 4.5.3. $\square$

When $m = n$, $A = I_n$, then Corollary 4.5.8 says that either $x \geq O$, $Bx > 0$ or $vB \leq O$, $v \geq 0$, $v \neq 0$ is possible, but not both; it seems that this theorem have been established by J.von Neumann in 1944.

Corollary 4.5.9 (Gale [I45]). Let B be a matrix of order $p \times n$, and b a column p-vector, both with real entries. Between the systems (in the unknowns x and u):

$$Bx = b, \tag{4.5.16}$$

and

$$uB = O, \qquad \langle u, b \rangle = 1, \qquad u \in \mathbb{R}^p, \tag{4.5.16*}$$

one and only one has a solution.

Proof. (4.5.16) is equivalent to the system:

$$y > 0, \quad Bx - by = O, \quad x \in \mathbb{R}^n, \quad y \in \mathbb{R},$$

which is easily identified with the homogeneous case of (4.5.7). Then, because of Corollary 4.5.3 (at $a = O, C = O, c = O; \ell = 1$) (4.5.16) is impossible iff $\exists \theta \in \overset{o}{\mathbb{R}}_+$ (and hence $\theta = 1$) and $\exists \mu \in \mathbb{R}^p$, s.t.

$$(0, 1) + \mu(B, -b) = O,$$

which, setting $u = \mu$, becomes (4.5.16)*. It is interesting to note that the thesis could be deduced directly from Theorem 4.4.1(i) at $X = \mathbb{R}^p, \mathcal{H} = \{O_p\}, \Pi = \mathbb{R}^p \backslash \{O\}$, $C = [0, +\infty[, w(u; \omega) := \langle u, \omega \rangle$; now also strong alternative is guaranteed; (4.4.1) is equivalent to (4.5.16)*, and the thesis is achieved. $\square$

It is easy to show that (4.5.16)* is impossible, iff $\text{rank} B = \text{rank}(B, b)$. Therefore, the impossibility of (4.5.16)* is equivalent to the classic Rouché-Capelli condition for the existence of solutions to (4.5.16).

Corollary 4.5.10 (Gale[I45]). Let C be a matrix of order $r \times n$, and c a column r-vector, both with real entries. Between the systems (in the unknowns x and v):

$$Cx \geq c, \tag{4.5.17}$$

and

$$vC = O, \qquad \langle v, c \rangle = 1, \qquad v \in \mathbb{R}^r_+, \tag{4.5.17*}$$

one and only one has solutions.

Proof. (4.5.17) is equivalent to the system:

$$y > 0, \quad Cx - cy \geq O, \quad x \in \mathbb{R}^n, \quad y \in \mathbb{R},$$

which is easily identified with the homogeneous case of (4.5.7). Then, Corollary 4.5.3 (at $a = O, B = O, b = O, c = O; \ell = 1$) gives the existence of $\theta \in \overset{o}{\mathbb{R}}_+$ (and hence $\theta = 1$)

and $\tau \in \mathbb{R}_+^r$, s.t.

$$(0,1) + \tau(C, -c) = O,$$

which, setting $v = \tau$, becomes (3.5.17)*. Here too we might apply Theorem 4.4.1. $\square$

Corollary 4.5.11. Let A be a matrix of order $m \times n$, and a a column m-vector, both with real entries. Between the systems (in the unknowns x and u):

$$Ax > a, \tag{4.5.20}$$

and

$$\left\{ \begin{array}{llll} uA = O, & & & \\ \text{and} & & & \\ \text{either} & \langle u, a \rangle > 0\ , & u \in \mathbb{R}_+^m\ , & \\ \text{or} & \langle u, a \rangle = 0\ , & u \in \mathbb{R}_+^m \setminus \{O\}\ , & \end{array} \right. \tag{4.5.20*}$$

one and only one has solutions.

Proof. (4.5.20) is easily identified with (4.5.7) at $B = O$, $b = O$, $C = O$, $c = O$. Because of Corollary 4.5.3, (4.5.20) is impossible iff $\exists \theta \in \mathbb{R}_+^m \setminus \{O\}$, s.t. $\theta A = O$, $\langle \theta, a \rangle \geq 0$. These relations, setting $u = \theta$, lead to (3.5.20)*, when we split $\langle \theta, a \rangle \geq 0$ into either $\langle \theta, a \rangle > 0$ or $\langle \theta, a \rangle = 0$. $\square$

The above corollary is the homogeneous version of Gordan TA [19]. Next corollary is a nonhomogeneous version of Farkas TA.

Corollary 4.5.12 (Duffin [I45]). Let C be a matrix of order $r \times n$, c a column p-vector, α a row n-vector, a a scalar, all with real entries. Between the systems (in the unknowns x and v):

$$\langle \alpha, x \rangle > a, \quad Cx \geq c, \tag{4.5.21}$$

and

$$\left\{ \begin{array}{llll} \text{either} & vC = \alpha, & \langle v, c \rangle \geq a, & v \in \mathbb{R}_+^r, \\ \text{or} & vC = O, & \langle v, c \rangle > 0, & v \in \mathbb{R}_+^r, \end{array} \right. \tag{4.5.21*}$$

one and only one has solutions.

Proof. (4.5.21) is easily identified with (4.5.7). Corollary 4.5.3 gives the existence of $\theta \in \mathbb{R}_+$, $\tau \in \mathbb{R}_+^r$, with $(\theta, \tau) \neq O$, s.t.

$$\theta\alpha + \tau C = O, \quad \theta a + \langle \tau, c \rangle \geq 0,$$

which leads to (4.5.21), by splitting $\theta \geq 0$ into $\theta > 0$ (and hence, due to the homogeneity of the relations, $\theta = 1$) and $\theta = 0$ (in which case, the inequality in (4.5.8) must be verified in strict sense), and setting $v := \tau$. $\square$

Corollary 4.5.13. Let D be a matrix of order $r \times n$, and d a column r-vector, both with real entries. Between the systems (in the unknowns x and z):

$$Dx \geq d, \quad Dx \neq d, \tag{4.5.22}$$

and

$$\left\{\begin{array}{ll} zD = O \\ \text{and} \\ \text{either} & \langle z, d \rangle = 0, \qquad z \in \overset{o}{\mathbb{R}^r_+}, \\ \text{or} & \langle z, d \rangle > 0, \qquad z \in \mathbb{R}^r_+, \end{array}\right. \qquad (4.5.22)^*$$

one and only one has solutions.

Proof. (4.5.22) is equivalent to the system:

$$\langle e_r D, x \rangle > \langle e_r, d \rangle, \qquad Dx \geq d,$$

which is easily identified with (4.5.7). Then, because of Corollary 4.5.3 (4.5.22) is impossible iff $\exists \theta \in \mathbb{R}_+, \exists \tau \in \mathbb{R}^r_+$, with $(\theta, \tau) \neq O$, s.t.

$$(\theta e_r + \tau)D = O, \qquad \langle \theta e_r + \tau, d \rangle \geq 0. \qquad (4.5.23)$$

If $\theta = 0$, by setting $z := \tau$, and recalling that the inequality in (4.5.8) must be verified in strict sense, then (4.5.23) becomes the 2nd part of (4.5.22)*. If $\theta = 1$, by setting $z := \theta e_r + \tau \in \overset{o}{\mathbb{R}^r_+}$, and splitting the inequality into positive and zero cases, (4.5.23) turns out to be equivalent to the union of the systems:

$$zD = O, \ \langle z, d \rangle > 0, \ z > O; \quad \text{and} \quad zD = O, \ \langle z, d \rangle = 0, \ z > O,$$

the former of which can be disregarded, since it implies the 2nd part of (4.5.22)*; the latter is the first part of (4.5.22)*. $\qquad \square$

Corollary 4.5.13 is a nonhomogeneous version of Stiemke TA.

Sometimes we are not given a system, but a "union" of systems. The approach developed in the previous part of this chapter can be exploited in such a case too. Next theorem is an instance of this.

Theorem 4.5.2 (Mangasarian [145]). Let A, B, C, D be matrices with real entries and of orders $m \times n, p \times n, q \times n, r \times n$, respectively. Between the systems (in the unknowns x and (t, u, v, z)):

$$\left\{\begin{array}{llllll} \text{either} & Ax \geq O, & Ax \neq O, & Bx = O, & Cx \geq O, & Dx \geq O, \\ \text{or} & Ax \geq O, & & Bx = O, & Cx > O, & Dx \geq O, \end{array}\right. \qquad (4.5.24)$$

and

$$tA + uB + vC + zD = O, t \in \overset{o}{\mathbb{R}^m_+}, u \in \mathbb{R}^p, v \in \mathbb{R}^q_+ \backslash \{O\}, z \in \mathbb{R}^r_+, \qquad (4.5.24)^*$$

one and only one has solutions.

Proof. (4.5.24) is equivalent to the system:

$$\begin{cases} \text{either} \quad \langle e_m A, x\rangle > 0, \quad Bx = O, \quad \begin{pmatrix} A \\ C \\ D \end{pmatrix} x \geq O, \\[2em] \text{or} \qquad Cx > O, \qquad Bx = O, \quad \begin{pmatrix} A \\ D \end{pmatrix} x \geq O. \end{cases} \qquad (4.5.25)$$

Obviously, (4.5.25) — and hence (4.5.24) — is impossible, iff each of its "subsystems" is impossible; each of them is easily identified with (4.5.7) (at $a = O$, $b = O$, $c = O$; $\ell = 1$ in the former and $\ell = q$ in the latter). Therefore, because of Corollary 4.5.3, we have that (4.5.24) is impossible, iff $\exists \theta \in \overset{o}{\mathbb{R}}_+$, $\exists \mu \in \mathbb{R}^p$, $\exists \tau_A \in \mathbb{R}^m_+$, $\exists \tau_C \in \mathbb{R}^q_+$, $\exists \tau_D \in \mathbb{R}^r_+$, s.t.

$$\theta e_m A + \mu B + \tau_A A + \tau_C C + \tau_D D = O, \qquad (4.5.26a)$$

and $\exists \theta' \in \mathbb{R}^q_+ \backslash \{O\}$, $\exists \mu' \in \mathbb{R}^p$, $\exists \tau'_A \in \mathbb{R}^m_+$, $\exists \tau'_D \in \mathbb{R}^r_+$, s.t.

$$\theta' C + \mu' B + \tau'_A A + \tau'_D D = O. \qquad (4.5.26b)$$

Since it is not restrictive to assume $\theta = 1$, (4.5.26) are equivalent, respectively, to:

$$(e_m + \tau_A)A + \mu B + \tau_C C + \tau_D D = O, \qquad (4.5.27a)$$

$$\tau'_A A + \mu' B + \theta' C + \tau'_D D = O. \qquad (4.5.27b)$$

Summing up side by side (4.5.27), setting $t := e_m + \tau_A + \tau'_A$, $u := \mu + \mu'$, $v := \tau_C + \theta'$, $z := \tau_D + \tau'_D$, and recalling the signs of the variables of (4.5.27), we find that (4.5.27) imply the existence of $t \in \overset{o}{\mathbb{R}}^m_+$, $u \in \mathbb{R}^p$, $v \in \mathbb{R}^q_+ \backslash \{O\}$, $z \in \mathbb{R}^r_+$, s.t. (4.5.24)*holds. Vice versa, if (4.5.24)* holds, then (4.5.27) trivially holds; in fact, (t, u, v, z) is a solution of (4.5.24)*, so that $(\alpha t, \alpha u, \alpha v, \alpha z)$ is still a solution of (4.5.24)* $\forall \alpha \in \overset{o}{\mathbb{R}}_+$, we can choose α large enough in order that $(\tau_A := \alpha t - e_m \in \overset{o}{\mathbb{R}}^m_p$, $\mu := \alpha u \in \mathbb{R}^p$, $\tau_C := \alpha v$, $\tau_D := \alpha z)$ and $(\tau'_A := \alpha t \in \overset{o}{\mathbb{R}}^m_+$, $\mu' := \alpha u$, $\theta' := \alpha v$, $\tau'_D := \alpha z)$ be solutions of (4.5.27), respectively. $\qquad \square$

If $A = O$ and $C = O$, then (4.5.24) is obviously impossible, and (4.5.24)* is possible since it allows at least the null solution.

It may happen that, in the given system, only a part of the elements of the unknown vector must be non-negative or nonpositive. By means of a simple and classic device, we can easily reduce ourselves to the previous TA. As an instance, consider the following

Corollary 4.5.14. Let A_1, A_2, B_1, B_2 be matrices of orders, respectively, $r_1 \times n_1$, $r_1 \times n_2$, $r_2 \times n_1$, $r_2 \times n_2$, a be a column r_1-vector, and b a column r_2-vector, all with real entries. Between the systems (in the unknowns $\begin{pmatrix} x^1 \\ x^2 \end{pmatrix}$ and (v^1, v^2)):

$$\begin{cases} A_1 x^1 + A_2 x^2 \geq a \\ B_1 x^1 + B_2 x^2 = b \\ x^1 \geq O \end{cases}$$

and

$$\begin{cases} v^1 A_1 + v^2 B_1 & \leq \ O \\ v^1 A_2 + v^2 B_2 & = \ O \\ \langle v^1, a \rangle + \langle v^2, b \rangle & = \ 1 \\ v^1 \geq O \end{cases}$$

one and only one has solutions.

Proof. It is enough to set

$$C = \begin{pmatrix} A_1 & A_2 \\ B_1 & B_2 \\ -B_1 & -B_2 \\ I_{n_1} & O_{n_2} \end{pmatrix}, \qquad c = \begin{pmatrix} a \\ b \\ -b \\ O_{n_1} \end{pmatrix},$$

and apply Corollary 4.5.10. $\qquad\qquad\square$

4.6. A Special Separation Theorem

As seen in Sect.3.2, optimality can be reduced to show disjunction between two suitable sets; if we want to prove (3.2.2) through separation arguments, then disjunctive separation must be adopted. However, if we look for necessary conditions — by replacing, for instance, the image set $\mathcal{K}_{\bar{x}}$ with a suitable approximation —, then mere separation is enough. Now, separation will be considered between a generic set of the IS, say K, which can play several roles including that of approximation of the image set, and a special set, namely $\mathcal{H}$ [V38]. To this end,

$$z := (u, \ v_i, i \in \mathcal{I}^0, \ v_i, i \in \mathcal{I}^+) \in \mathbb{R}^{1+m}$$

will denote an $(1+m)$-vector of the IS, and $\text{proj}\, z := (v_i, i \in \mathcal{I}^0) \in \mathbb{R}^p$ its projection on the coordinate subspace associated to the equations (1.1.1b) and whose origin is O_p.

In the following statement, if $p = 0$, then both sides of (4.6.1) are "empty"; hence, we stipulate that (4.6.1)-(4.6.2) shrinks to (4.6.2). When $p > 0$ and affinely independent $z^1, ..., z^{k+1} \in K$, such that (4.6.1) is fullfilled, do not exist, then, of course, the condition (4.6.1)-(4.6.2) is meant to be satisfied. Obviously, int denotes interior with respect to $\mathbb{R}^p$ in (4.6.1) and with respect to $\mathbb{R}^{1+m}$ in (4.6.2).

Theorem 4.6.1. Let $K \subset \mathbb{R}^{1+m}$ and $k := \dim K$. If and only if for every set $\{z^1, ..., z^{k+1}\}$ of affinely independent vectors of K, such that:

$$O_p \in \text{int conv}\{\text{proj}\, z^1, ..., \text{proj}\, z^{k+1}\}, \qquad\qquad (4.6.1)$$

we have

$$(\text{ri } \mathcal{H}) \cap \text{ri conv}\{z^1, ..., z^{k+1}\} = \varnothing, \qquad\qquad (4.6.2)$$

then $\mathcal{H}$ and K are (linearly) separable.

Proof. If. The proof will be split up into parts. **(A)** $k = 0$. The thesis is trivial, since K is a singleton. **(B)** $1 \leq k \leq p - 1$. Of course, this implies $p \geq 1$ and that $p + 1$ affinely independent vectors of K cannot exist. Let $B_{\mathcal{H}}$ and B_K be bases for $\mathcal{H}$ and K, respectively; $|B_{\mathcal{H}}| = 1 + m - p$, $|B_K| = k$, $|B_{\mathcal{H}} \cup B_K| = 1 + m - p + k \leq m$. This shows that there exists a hyperplane of $\mathbb{R}^{1+m}$, which contains $\mathcal{H}$ and is parallel to K, so that separation holds. **(C)** $k \geq 1$, $k \geq p \geq 1$ and (4.6.1) does not hold, in the sense that no set considered in the assumption verifies (4.6.1). There exists at least one set of $k+1 \geq p+1$ affinely independent vectors of K; denote by $\{z^1, ..., z^{k+1}\}$ a generic one of such sets. Denote by $\mathrm{proj}\, K$ the projection of K into the coordinate subspace $\mathbb{R}^p$, whose origin is 0_p. Since for every affinely independent set $\{z^1, ..., z^{k+1}\} \subseteq K$ (4.6.1) does not hold, then $O_p \notin \mathrm{int}\,\mathrm{conv}\,\mathrm{proj}\, K$ (otherwise, because of Theorem 2.2.1, O_p could be expressed as convex combination, with positive coefficients, of $p + 1$ affinely independent vectors of $\mathbb{R}^p$ which should be projections of $p + 1$ vectors of K; since $\dim K = k$, these vectors could be augmented to form a set $\{z^1, ...z^{k+1}\}$ of affinely independent vectors of K which would satisfy (4.6.1)). Because of the well known Hahn-Banach Linear Extension Theorem (see Theorem 2.2.1 and [2]), the last relation implies the existence of a hyperplane of $\mathbb{R}^p$ through O_p, whose equation be $\sum_{i \in \mathcal{J}^0} \tilde{\lambda}_i v_i = 0$ with $(\tilde{\lambda}_1, ..., \tilde{\lambda}_p) \neq O_p$, which does not intersect $\mathrm{int}\,\mathrm{conv}\,\mathrm{proj}\, K$, so that $\mathrm{conv}\,\mathrm{proj}\, K$ belongs to one of the two halfspaces defined by it. It is not restrictive to assume that:

$$\sum_{i \in \mathcal{J}^0} \tilde{\lambda}_i v_i \leq 0 \, , \quad \forall (v_1, ..., v_p) \in \mathrm{conv}\ \mathrm{proj} K.$$

Set $\tilde{\theta} = 0$, $\tilde{\lambda}_i = 0, i \in \mathcal{J}^+$, $\tilde{\lambda} := (\tilde{\lambda}_1, ..., \tilde{\lambda}_m)$. We have :

$$\tilde{\theta} u + \langle \tilde{\lambda}, v \rangle \leq 0, \quad \forall (u, v) \in \mathrm{conv}\, K$$

since conv and proj are permutable, and obviously

$$\tilde{\theta} u + \langle \tilde{\lambda}, v \rangle = 0, \quad \forall (u, v) \in \mathcal{H}.$$

The separability of $\mathcal{H}$ and $\mathcal{K}$ follows. **(D)** $k \geq 1$, $k \geq p$ and (4.6.1) holds, in the sense that there exists a set $\{z^1, ...z^{k+1}\}$ of affinely independent vectors of K which verifies (4.6.1); this includes the case $p = 0$. We shows that (4.6.2) implies:

$$\mathrm{ri}\ \mathcal{H} \cap \mathrm{ri}\ \mathrm{conv}\ K = \emptyset. \tag{4.6.3}$$

If we deny (4.6.3), so that there is

$$\hat{z} = (\hat{u} > 0, \quad \hat{v}_i = 0, \ i \in \mathcal{J}^0, \quad \hat{v}_i > 0, \ i \in \mathcal{J}^+) \in \mathrm{ri}\ \mathrm{conv}\, K,$$

then, because of Theorem 2.1.1, $\hat{z}$ can be expressed as convex combination, with positive coefficients, of $k + 1$ affinely independent vectors of K, say $\{w^1, ..., w^{k+1}\}$. If this set verifies (4.6.1), then (4.6.2) is contradicted. Otherwise, such a set verifies

$$O_p \notin \mathrm{int}\ \mathrm{conv}\{\mathrm{proj}\, w^1, ...,\mathrm{proj}\, w^{k+1}\}.$$

Because of Theorem 2.2.1, $\exists \tilde{\lambda} \in \mathbb{R}^p$, s.t. $\tilde{\lambda} \neq O_p$ and

$$\sum_{i \in \mathfrak{I}^0} \tilde{\lambda}_i (\operatorname{proj} w^j)_i \leq 0 , \qquad \forall j = 1, ..., k+1.$$

We set $\tilde{\theta} = 0$, $\tilde{\lambda}_i = 0$, $i \in \mathfrak{I}^+$, $\tilde{\lambda} = (\tilde{\lambda}_1, ..., \tilde{\lambda}_m)$ and denote by H the hyperplane of $\mathbb{R}^{1+m}$ defined by $\{x \in \mathbb{R}^{1+m} : \langle (\tilde{\theta}, \tilde{\lambda}), x \rangle = 0\}$. We have:

$$\langle (\tilde{\theta}, \tilde{\lambda}), w^j \rangle \leq 0, \qquad \forall j = 1, ..., k+1; \quad \langle (\tilde{\theta}, \tilde{\lambda}), \tilde{z} \rangle = 0.$$

Since $\tilde{z}$ can be expressed as convex combination of $\{w^1, ..., w^{k+1}\}$, with positive coefficients, then:

$$\langle (\tilde{\theta}, \tilde{\lambda}), w^j \rangle = 0 , \qquad \forall j = 1, ..., k+1;$$

thus the set K is contained in H, and in particular we have:

$$\sum_{i \in \mathfrak{I}^0} \tilde{\lambda}_i (\operatorname{proj} z^j)_i = 0 , \qquad \forall j = 1, ..., k+1;$$

therefore

$$\operatorname{int} \operatorname{conv}\{\operatorname{proj} z^1, ..., \operatorname{proj} z^{k+1}\} = \varnothing,$$

which contradicts the assumption. Because of Theorem 2.2.4(ii), (4.6.3) implies (even proper) separation between $\mathcal{H}$ and K. **Only if.** By assumption, $\exists \theta \in \mathbb{R}_+$ and $\exists \lambda \in D^*$, $(\theta, \lambda) \neq O$, s.t.

$$\theta u + \langle \lambda, v \rangle \leq 0 , \quad \forall (u, v) \in K. \tag{4.6.4}$$

Denote by H^- the halfspace (of $\mathbb{R}^{1+m}$) identified by (4.6.4). Ab absurdo, suppose that there exist affinely independent vectors $z^1, ..., z^{k+1} \in K$, s.t. (4.6.1) holds (if no set of $k+1$ affinely independent vectors of K exists, s.t. (4.6.1) is satisfied, then the thesis is trivial), so that $k \geq p$, while (4.6.2) is violated, so that there exist scalars $\alpha_j > 0$, $j \in J := \{1, ..., k+1\}$ with $\sum_{j=1}^{k+1} \alpha_j = 1$, and $J^0 \subseteq J$ with $|J^0| = p+1$, s. t.:

$$\operatorname{proj} z^j \neq O_p, j \in J^0; \quad \operatorname{proj} \left(\sum_{j \in J} \alpha_j z^j\right) = \sum_{j \in J} \alpha_j \operatorname{proj} z^j = O_p ; \tag{4.6.5}$$

$$\sum_{j \in J} \alpha_j z^j \in \operatorname{ri} \mathcal{H} = (\mathbb{R}_+ \backslash \{0\}) \times O_p \times \operatorname{int} \mathbb{R}_+^{m-p}. \tag{4.6.6}$$

Set $z^j = (u^j, v_i^j, i \in \mathfrak{I}), j \in J, \lambda = (\lambda_i, i \in \mathfrak{I})$. Since $z^j \in K \subseteq H^-$, we have:

$$\theta u^j + \sum_{i \in \mathfrak{I}^0} \lambda_i v_i^j + \sum_{i \in \mathfrak{I}^+} \lambda_i v_i^j \leq 0 , \quad j \in J.$$

Multipling by α_j both sides of the above inequality and summing up side by side with respect to j, we obtain the inequality:

$$\theta \sum_{j \in J} \alpha_j u^j + \sum_{i \in \mathfrak{I}^0} \lambda_i \sum_{j \in J} \alpha_j v_i^j + \sum_{i \in \mathfrak{I}^+} \lambda_i \sum_{j \in J} \alpha_j v_i^j \leq 0 ,$$

which, because of (4.6.6), implies $\theta = \lambda_i = 0$, $i \in \mathfrak{I}^+$, and, because of $(\theta, \lambda) \neq O$, $(\lambda_i, i \in \mathfrak{I}^0) \neq O$. Therefore, (4.6.4) becomes:

$$\sum_{i \in \mathfrak{I}^0} \lambda_i v_i \geq 0 \quad . \tag{4.6.7}$$

Because of (4.6.7), the projections of $z^1, ..., z^{k+1}$ into the coordinate subspace $\mathbb{R}^{p'}$ considered as points of $\mathbb{R}^{1+m}$, namely the points :

$$y^j := (u = 0, \text{proj } z^j, \ v_i = 0, i \in \mathfrak{I}^+) = (u = 0, \ v_i = v_i^j, i \in \mathfrak{I}^0, \ v_i = 0, i \in \mathfrak{I}^+), j \in J,$$

belong to H^- and, obviously, $\text{conv}\{y^1, ..., y^{k+1}\} \subseteq H^-$. From (4.6.5) we deduce that $O_{1+m} \in \text{ri conv}\{y^1, ..., y^{k+1}\}$. Therefore, if $\beta > 0$ is small enough (for instance, less than $\min\{\|(v_1^j, ..., v_p^j)\|, j \in J\}$), then:

$$(u = 0, \ v_i = \beta\lambda_i, i \in \mathfrak{I}^0, \quad v_i = 0, i \in \mathfrak{I}^+) \in \text{ri conv}\{y^1, ..., y^{k+1}\} \subseteq H^-,$$

so that we achieve the inequality $\beta \sum_{i \in \mathfrak{I}^0} \lambda_i^2 \leq 0$, which contradicts $(\lambda_i, i \in \mathfrak{I}^0) \neq O$. $\quad\square$

Theorem 4.6.1 can be formulated as a Helly-type theorem (see Theorem 2.1.3 and subsequent remarks). Let $\mathcal{F}$ be the family of all nonempty subsets of K, and $\mathcal{F}_r$ be the family of subsets of K of $r + 1$ affinely independent elements of K; of course, $\mathcal{F}_r \subseteq \mathcal{F}, r = 1, 2, ...$Consider the property "an element of $\mathcal{F}$ and the set $\mathcal{H}$ are separable". Theorem 4.6.1 says that, if the property is enjoyed by the elements of $\mathcal{F}_{1+\dim K}$, then it is enjoyed by all the elements of $\mathcal{F}$.

Example 4.6.1 In Theorem 4.6.1, set $m = 2$, $p = 1$, so that $\mathfrak{I}^0 = \{1\}, \mathfrak{I}^+ = \{2\}$, $z = (u, v_1, v_2)$, $O_p = \{0\}$, $D = \{0\} \times \mathbb{R}_+$, and set:

$$\mathcal{H} = \left\{ (u, v_1, v_2) \in \mathbb{R}^3 : u > 0, v \in D \right\},$$

$$K = \left\{ (u, v_1, v_2) \in \mathbb{R}^3 : u + v_2 = 1 \right\}.$$

Obviously $\mathcal{H}$ and the plane K have nonempty intersection, so that the thesis of Theorem 4.6.1 does not hold. We have $k = \dim K = 2$. If we consider the vectors $z^1 = (1, 1, -1)$, $z^2 = (2, -1, -2)$, $z^3 = (1, -2, -3)$, which are affinely independent, then we find proj $z^1 = 1$, proj $z^2 = -1$, proj $z^3 = -2$, so that (4.6.1) becomes $0 \in]-2, 1[$ and ri $\text{conv}\{z^1, z^2, z^3\}$ is a triangle which lies in the open dihedron defined by $u > 0$, $v_1 < 0$; therefore, (4.6.1)and (4.6.2) are satisfied. This does not mean that the assumption of Theorem 4.6.1 be satisfied. In fact, if we choose the vectors $z^1 = (1, 1, 1)$, $z^2 = (2, -1, -2)$, $z^3 = (1, -1, -3)$, which again are affinely independent, then we find proj $z^1 = 1$, proj $z^2 = -1$, proj $z^3 = -1$, so that (4.6.1) becomes $0 \in]-1, 1[$ and is fulfilled; ri $\text{conv}\{z^1, z^2, z^3\}$ is a triangle which intersects $\mathcal{H}$, so that the thesis does not hold, and (4.6.2) is violated. $\quad\square$

4.7. Theorems of the Alternative for Multifunctions

In the previous sections, we have considered systems of functions. This does not cover all the applications. Indeed, as shown in Sect. 3.2 with regard to (1.1.5), we may be faced with point-to-set maps.

As in the previous sections and in Sect. 1.1, ν denotes a positive integer, $\mathcal{H} \subset \mathbb{R}^\nu$ is assumed to be non-empty; $\mathcal{H}$ is any set, even if the special case considered in Sect. 3.2 is very important. Furthermore, we consider the non-empty sets $X \subseteq B$, $Y \subseteq \mathbb{R}^\nu$ and the point-to-set map $A : X \rightrightarrows Y$, so that $A(x) \subseteq Y$, $\forall x \in X$; unlike Sects. 1.3 and 4.4, for the sake of simplicity, the dependence of A on the parameter ξ is here understood. We want to study conditions under which the generalized system

$$\exists x \in X \quad , \quad \text{s.t.} \quad A(x) \subseteq \mathcal{H}, \qquad\qquad , \qquad\qquad (4.7.1)$$

has or has not solutions. When A is the map $A_{\overline{x}}$ of Sect. 3.2 and $\mathcal{H}$ that of the same section, then the impossibility of (4.7.1) is equivalent to (3.2.27). If, in particular, A is single-valued, then the impossibility of (4.7.1) is equivalent to (3.2.2); as noted in Sect. 3.2, in the general case the impossibility of (4.7.1) does not imply

$$\mathcal{H} \cap \mathcal{K} = \varnothing, \qquad\qquad (4.7.2)$$

where now $\mathcal{K} := \bigcup_{x \in X} A(x)$; $\mathcal{K}$ shrinks to $\mathcal{K}_{\overline{x}}$ of Sect.3.2, when A becomes $A_{\overline{x}}$ (see also Definition 4.8.1). However, in the vein of the selection approach of Sect. 3.2, it is possible to reduce ourselves to study (4.7.1) in terms of disjunction of two sets and hence to exploit separation arguments.

Consider the function $\Phi : 2^Y \times \Omega \to \mathbb{R}^\nu$, where Ω must be regarded as a set of parameters, such that, $\forall x \in X$, we have:

$$A(x) \subseteq \mathcal{H} \;\Leftrightarrow\; \Phi(A(x); \omega) \in \mathcal{H} \;, \;\; \forall \omega \in \Omega. \qquad (4.7.3)$$

A special case of such a function is given by (3.2.28); indeed, (3.2.29) specifies (4.7.3). When Φ can be chosen in such a way that $\Phi \in A(x)$, then Φ is precisely what is called a **selection function.** If $\mathcal{H}$ is convex, then $A(x)$ can be replaced by $\mathrm{conv}A(x)$, since obviously $A(x) \subseteq \mathcal{H}$ iff $\mathrm{conv}A(x) \subseteq \mathcal{H}$, so that Φ can be defined by replacing (4.7.3) with:

$$A(x) \subseteq \mathcal{H} \;\Leftrightarrow\; \Phi(\mathrm{conv}\, A(x); \omega) \in \mathcal{H} \;, \;\; \forall \omega \in \Omega; \qquad (4.7.3)'$$

hence, without any loss of generality, we might assume that $\Phi \in A(x)$. However, as it happens in the case of problems of type (1.1.5) (see also Examples 3.2.7−3.2.10), such an assumption may not be practically meaningful. If $\mathcal{H}$ is not convex , as it happens in some problems of type (1.1.7)− (1.1.11), then $A(x)$ cannot be replaced by $\mathrm{conv}A(x)$, since obviously $A(x) \subseteq \mathcal{H}$ does not imply $\mathrm{conv}\, A(x) \subseteq \mathcal{H}$. Example 4.7.2 will show a case, where Φ is a **generalized selection function** (GSF; see (3.2.28) and the sequel), but not a selection function; in the applications, as those to problems of type (1.1.5) (see Sect. 3.2), Φ is not in general a selection function, and to make it so would require a heavy assumption. Condition (4.7.3) can be equivalent written as:

$$A(x) \not\subseteq \mathcal{H} \quad \Leftrightarrow \quad \exists \omega \in \Omega, \ \text{s.t.} \ \Phi(A(x); \omega) \notin \mathcal{H} . \tag{4.7.3}''$$

A function fulfilling (4.7.3) or (4.7.3)$''$ will be referred to as a GSF and ω as a **selection multiplier** (SM).

Of course, in (4.7.3)$''$ ω will depend in general on x, and hence x will appear as argument of ω, while this is not necessary when (4.7.3) is exploited. In the particular case where ω does not depend on x, then Φ can be characterized by the following condition:

$$A(x) \not\subseteq \mathcal{H} , \ \forall x \in X \quad \Leftrightarrow \quad \exists \overline{\omega} \in \Omega, \ \text{s.t.} \ \Phi(A(x); \overline{\omega}) \notin \mathcal{H}, \ \forall x \in X. \tag{4.7.3}'''$$

As in the previous sections, w and s denote a weak and a strong separation function, respectively. A TA can now be stated for point-to-set maps [15,16]. The notation is as in Sects. 4.1-4.4; the dependence of w and s on the parameter π is understood, since it does not play any role.

Theorem 4.7.1 Let the sets $\mathcal{H}, X, \mathcal{W}(\Pi), \mathcal{S}(\Pi)$, the cone C with apex at the origin, and the point-to-set map A be given. (**i**) The system (4.7.1) and the system

$$\forall x \in X , \ \exists \omega \in \Omega, \ \text{s.t.} \ w(\Phi(A(x); \omega)) \notin C \tag{4.7.4}$$

are not simultaneously possible, whatever the weak separation function $w \in \mathcal{W}(\Pi)$ and the generalized selection function Φ may be. (**ii**) The system (4.7.1) and the system

$$\forall x \in X , \ \exists \omega \in \Omega, \ \text{s.t.} \ s(\Phi(A(x); \omega)) \notin C_0, \tag{4.7.5a}$$

or the system

$$\forall x \in X , \ \exists \omega \in \Omega, \ \text{s.t.} \ s(\Phi(A(x); \omega)) \notin \overset{o}{C}, \tag{4.7.5b}$$

are not simultaneously impossible, whatever the strong separation function $s \in \mathcal{S}(\Pi)$ and the generalized selection function Φ may be.

Proof. (**i**) If (4.7.1) is possible, i.e. if $\exists \hat{x} \in X$ s.t. $A(\hat{x}) \subseteq \mathcal{H}$, then (4.7.3) implies:

$$\Phi(A(\hat{x}); \omega) \in \mathcal{H} , \ \forall \omega \in \Omega;$$

then, because of (4.2.2a), we have:

$$w(\Phi(A(\hat{x}); \omega)) \in C, \ \forall \omega \in \Omega,$$

so that (4.6.4) is impossible. (**ii**) If (4.7.1) is impossible, i.e. if

$$A(x) \not\subseteq \mathcal{H} , \ \forall x \in X,$$

then (4.7.3)$''$ implies that:

$$\forall x \in X , \ \exists \omega \in \Omega \quad \text{s.t.} \quad \Phi(A(x); \omega) \notin \mathcal{H},$$

hence, because of (4.2.9a) and (4.2.10a), we have, respectively, that, $\forall x \in X , \ \exists \omega \in \Omega$, s.t.:

$$s(\Phi(A(x); \omega)) \notin C_0$$

and

$$s(\Phi(A(x);\omega)) \notin \overset{o}{C},$$

so that (4.7.5) are possible.   □

Example 4.7.1. Let us set $\nu = \ell = 1, Y = \mathbb{R},\ C = \mathbb{R}_+,\ \mathcal{H} = \mathbb{R}_+ \setminus \{0\} = C_0 = \overset{o}{C},$

$$X = \{x(t) \in C^0([0,1]) : 0 \leq x(t) \leq 2\ ,\quad \forall t \in [0,1]\},$$

$$A(x) = \{y \in \mathbb{R} : y = 1 - x(t)\,;\, t \in [0,1]\}.$$

The system (4.7.1) is evidently possible and its set of solutions is:

$$\{x(t) \in X\ :\ x(t) < 1\,,\ \forall t \in [0,1]\}.$$

In the present case, the choice of Φ (independent of ω) is natural:

$$\Phi(A(x);\omega)\ =\ \min_{t\in[0,1]} A(x(t))\ =\ \min_{t\in[0,1]}[1 - x(t)],\quad x(t) \in X.$$

Taking into account the fact that $A(x)$ is a closed and convex subset of $\mathbb{R}$ it is easy to show that $\Phi(A(x);\omega)$, which here does not depend on ω, fulfils (4.7.3) and that here Φ is precisely a selection function. Since $\mathcal{H} \subset C$, a family of weak separation functions, namely $\mathcal{W}(\Pi)$, is given by functions $w : \mathbb{R} \times \Pi \to \mathbb{R}$ which are positive if the 1st argument is positive, and whose positive level sets have $\mathcal{H}$ as intersection. Let us set, for instance, $\Pi = \mathbb{R}_+$ and

$$w(z;\pi) = z + \pi.$$

Thus, the systems (4.7.1) and (4.7.4) become, respectively,

$$\exists x(t) \in X \quad \text{s.t.}\quad 1 - x(t) > 0,\ \forall t \in [0,1],$$

and

$$\min_{t\in[0,1]}[1 - x(t)]\ \leq\ -\pi,\ \forall x(t) \in X,$$

and, according to (i) of Theorem 4.7.1, cannot be both possible: the former is possible and the latter impossible. Analogously,

$$s(z\,;\pi) = z - \pi,\ \pi \in \Pi = \mathbb{R}_+\,,$$

is a strong separation function; since (4.7.1) is possible, we are in agreement with (ii) of Theorem 4.7.1, whatever $\pi \in \Pi$ may be; now the conditions (4.7.5) coincide and become:

$$\min_{t\in[0,1]}[1 - x(t)] \leq \pi\ ,\ \forall x(t) \in X,$$

which is possible or impossible, according to $\pi \geq 1$ or $\pi < 1$, respectively.   □

Example 4.7.2. Let us set $\nu = 2,\ \ell = 1,\ Y = \mathbb{R}^2,\ \mathcal{H} = \mathbb{R}_+^2,\ C = \mathbb{R}_+,\ C_0 = \overset{o}{C} =$

$\mathbb{R}_+ \setminus \{0\}$,

$$X = \{x(t) \in C^0([0,4]) : 0 \le x(t) \le 1, \ \forall t \in [0,4], x(0) = x(4) = 0\},$$

$$A(x) = \{(z_1, z_2) \in \mathbb{R}^2 : z_i = f_i(t, x(t)), i = 1, 2, \text{ for some } t \in [0,4]\},$$

with

$$f_1(t, x(t)) = t^2 - 3t + 2 + x(t),$$

$$f_2(t, x(t)) = t^2 - 5t + 6 + x(t).$$

Simple calculations show us that (4.7.1) is possible and that any one of its solutions represents a continuous curve joining two points and overcoming two obstacles. The set $A(x)$ is not convex, and there is no particular evidence (as in Example 4.7.1) to suggest a minimum (or maximum) criterion for choosing Φ. However, a different way is provided by exploiting the structure of (4.7.1); this means that f_i must be non-negative $\forall t \in [0,4]$. Introduce the class Ω_i of continuous functions $\omega_i : [0,4] \to \mathbb{R}_+, i = 1, 2$, and set:

$$\omega = \omega(t) = (\omega_1(t), \omega_2(t)), \qquad \Omega = \Omega_1 \times \Omega_2.$$

When we consider a condition like (4.7.3)″ or the possibility of a system like (4.7.4) or (4.7.5), and we require the existence of a selection multiplier ω_i, then obviously such a multiplier depends in general on x, so that we adopt the notation $\omega_i(t, x(t))$ (a general dependence of type $\omega_i(t, x) : \mathbb{R} \times X \to \mathbb{R}_+$ is here unnecessary), instead of $\omega_i(t)$. Now, it is easy to show that:

$$f_i(t, x(t)) \ge 0, \qquad \forall t \in [0,4],$$

iff

$$\int_0^4 \omega_i(t) f_i(t, x(t)) \, dt \ge 0, \qquad \forall \omega_i \in \Omega_i, \ i = 1, 2.$$

It follows that:

$$\Phi(A(x); \omega) := \left(\int_0^4 \omega_1(t) f_1(t, x(t)) \, dt, \int_0^4 \omega_2(t) f_2(t, x(t)) \, dt \right), \qquad \omega \in \Omega,$$

fulfills (4.7.3). Since $C = \mathbb{R}_+$, then every $w : \mathbb{R}^2 \to \mathbb{R}$ of the type $w = \lambda_1 z_1 + \lambda_2 z_2$, with $\lambda_1, \lambda_2 \ge 0$, is evidently a weak separation function. The system (4.7.4) becomes: $\exists \lambda_i$ and $\exists \omega_i(t, x(t)), i = 1, 2$, such that:

$$\int_0^4 \sum_{i=1}^2 \lambda_i \omega_i(t, x(t)) f_i(t, x(t)) \, dt < 0, \qquad \forall x \in C^0([0,4]).$$

One easily sees the nonexistence, $\forall x \in X$, of multipliers λ_i and selection multipliers $\omega_i(t, x(t))$, such that the last inequality be satisfied, in agreement with (i) of Theorem 4.7.1. Note that the above Φ is a generalized selection function, but not a selection function. In fact, if we set $x(t) = -\frac{t^2}{4} + t$, then we have:

$$f_1 = \frac{3}{4}t^2 - 2t + 2, \qquad f_2 = \frac{3}{4}t^2 - 4t + 6,$$

so that

$$f_i \geq \frac{2}{3}, \qquad \forall t \in [0,4], \quad i = 1,2.$$

If each ω_i is chosen constant, for instance $\omega_i = \frac{1}{16}$, $i = 1,2$, then we obtain:

$$\int_0^4 \omega_i f_i(t, x(t)) dt = \frac{1}{2}, \quad i = 1,2.$$

Therefore,

$$\Phi(A(x); \omega) = (\tfrac{1}{2}, \tfrac{1}{2}) \notin A(x), \quad \text{since} \quad A(x) \subseteq (\tfrac{2}{3}, \tfrac{2}{3}) + \mathbb{R}_+^2. \qquad \square$$

When A is a single-valued function, then Φ trivially collapses to $\Phi(A(x); \omega) = A(x)$, so that (4.7.4)-(4.7.5) become (4.4.1)-(4.4.2), respectively; that is, Theorem 4.7.1 collapses to Theorem 4.4.1.

The system (4.7.1) does not contain all the kinds of systems which one can meet in the field of multifunctions. Instead of system (4.7.1), one may be interested in studying a more general form, namely:

$$\exists x \in X \text{ s.t. } A_1(x) \nsubseteq \mathcal{H}_1, \ A_2(x) \subseteq \mathcal{H}_2, \tag{4.7.1$'$}$$

where , for each $i = 1,2$, the multifunction $A_i : X \rightrightarrows Y_i$ and the set $\mathcal{H}_i \subset Y_i \subseteq \mathbb{R}^{\nu_i}$ are given. Let $\Phi_i : 2^{Y_i} \times \Omega_i \to \mathbb{R}^{\nu_i}$ be a function which fulfills (4.7.3). Then, instead of Theorem 4.7.1, we have the following theorem, which will be stated for $\ell = 1$, the extension to the case $\ell > 1$ being now straightforward; on the other hand, we consider a more general form for the system in alternative, by letting the images of the separation functions belong to generic sets Z_1, Z_2 instead of to halflines of the real line.

Theorem 4.7.2. Let the sets $X, \mathcal{H}_1, \mathcal{H}_2, Z_1, Z_2$, and the multifunctions A_1, A_2 be given. (**i**) The system (4.7.1)$'$ and the system

$$\forall x \in X, \quad \text{either} \quad s(\Phi_1(A_1(x); \omega^1)) \notin Z_1, \qquad \forall \omega^1 \in \Omega_1,$$
$$\text{or} \quad \exists \bar{\omega}^2 \in \Omega_2 \qquad \text{s.t.} \qquad w(\Phi_2(A_2(x); \bar{\omega}^2)) \in Z_2, \tag{4.7.6}$$

are not simultaneously possible, whatever the weak separation function w, the strong separation function s, and the generalized selection functions Φ_1, Φ_2 may be. (**ii**) The system (4.7.1)$'$ and the system

$$\forall x \in X, \quad \text{either} \quad w(\Phi_1(A_1(x); \omega^1)) \notin Z_1, \qquad \forall \omega^1 \in \Omega_1 \tag{4.7.7}$$

$$\text{or} \quad \exists \bar{\omega}^2 \in \Omega_2 \text{ s.t. } s(\Phi_2(A_2(x); \bar{\omega}^2)) \in Z_2,$$

are not simultaneously impossible, whatever the weak separation function w, the strong separation function s, and the generalized selection functions Φ_1, Φ_2 may be.

Proof. (**i**) If (4.7.1)$'$ is possible, i.e., if $\exists \bar{x} \in X$ s.t. $A_1(\bar{x}) \nsubseteq \mathcal{H}_1$ and $A_2(\bar{x}) \subseteq \mathcal{H}_2$, then

$$(4.7.3)' \ \Rightarrow \ \exists \bar{\omega}^1 \in \Omega_1, \quad \text{so that} \quad \Phi_1(A_1(\bar{x}); \bar{\omega}^1) \notin \mathcal{H}_1,$$

and

$$(4.7.3) \ \Rightarrow \ \Phi_2(A_2(\bar{x}); \omega^2) \in \mathcal{H}_2, \ \forall \omega^2 \in \Omega_2.$$

Thus, (4.2.3a) and (4.2.2a) imply, respectively,

$$s(\Phi_1(A_1(\bar{x}); \bar{\omega}^1)) \in Z_1$$

and

$$w(\Phi_2(A_2(\bar{x}); \omega^2)) \notin Z_2, \qquad \forall \omega^2 \in \Omega_2.$$

Hence, (4.7.6) is impossible . **(ii)** If system (4.7.1.)$'$ is impossible, then,$\forall x \in X$, either $A_1(x) \subseteq \mathcal{H}_1$ or $A_2(x) \nsubseteq \mathcal{H}_2$. In the former case,

$$(4.7.3) \;\Rightarrow\; \Phi_1(A_1(x); \omega^1) \in \mathcal{H}_1, \;\; \forall \omega^1 \in \Omega_1;$$

in the latter case, (4.7.3)$''$ implies that $\exists \bar{\omega} \in \Omega_2$, so that:

$$\Phi_2(A_2(x); \bar{\omega}^2) \notin \mathcal{H}_2.$$

Hence, $\forall x \in X$, (4.2.2a) and (4.2.3a) imply, respectively,

$$w(\Phi_1(A_1(x); \omega^1)) \notin Z_1, \;\; \forall \omega^1 \in \Omega_1 \,,$$

and

$$s(\Phi_2(A_2(x); \bar{\omega}^2)) \in Z_2.$$

Thus, (4.7.7) is possible. $\square$

Note that, by introducing the set $\mathcal{K}_1 := \bigcup_{x \in X} A_1(x)$, the feasibility of (4.7.1)$'$ implies the feasibility of the system:

$$\mathcal{K}_1 \cap (\sim \mathcal{H}_1) \neq \varnothing, \qquad \exists x \in X \;\; \text{s.t.} \;\; A_2(x) \subseteq \mathcal{H}_2,$$

whose first part can be treated in terms of ordinary separation of two sets, as in Sect. 3.2. The above relation may be useful when $\mathcal{K}_1$ is convex.

A system for single-valued functions has been generalized to multifunctions in several ways. In the literature, two types of systems are considered, which generalize the well-known Farkas Lemma (see Corollaries 4.5.7 and 4.5.8). The first is:

$$x \in X \;\; \text{and} \;\; A_1(x) \cap S_1 \neq \varnothing \;\Rightarrow\; A_2(x) \cap S_2 \neq \varnothing, \tag{4.7.8}$$

where A_1 and A_2 are again given multifunctions and S_1, S_2 are given sets. If we set $\mathcal{H}_i = \sim S_i$, $i = 1, 2$, then (4.7.8) is true iff (4.7.1)$'$ is impossible. The second is:

$$x \in X \;\; \text{and} \;\; f_1(x) \in S_1 \;\Rightarrow\; A_2(x) \subseteq S_2, \tag{4.7.9}$$

where f_1 is a given single-valued function. If we set:

$$A(x) := \{f_1(x)\} \times A_2(x), \quad \mathcal{H} := S_1 \times (\sim S_2),$$

then (4.7.9) can be reduced to a system which will be studied in Section 4.9; namely, (4.7.9) is true iff (4.9.1) is impossible.

For each of the two systems (4.7.8) and (4.7.9), it has been shown that, under suitable assumptions, there exists a map which sends the second function onto the former. Hence, the purpose has been a generalization of the Farkas Lemma. Reducing (4.7.8) and (4.7.9) respectively to (4.7.1)$'$ and (4.9.1) is a different aim, in the sense that we generalize the TA which underlies the Farkas Lemma; these generalizations, namely Theorems 4.7.2 and 4.9.1, should enable one to get, under suitable assumptions, some generalizations of the above type.

4.8. Cone Multifunctions

In order to establish TA for other types of systems or systems more general than (4.7.1), some previously introduced concepts need to be extended. To this end, with the notations of Sect. 4.7, let us introduce some concepts. First of all, let us consider the following subclass of the class of cones (2.1.14):

$$\mathcal{H} \text{ is a convex cone with apex at the origin } O \notin \mathcal{H} , \qquad (4.8.1a)$$

$$\mathcal{H} + \mathrm{cl}\mathcal{H} = \mathcal{H} . \qquad (4.8.1b)$$

Note that the condition (4.8.1), which is fulfilled in most of applications to extremum problems, characterizes the cones $\mathcal{H}$, which fulfil (4.8.1a) and are such that every face of cl $\mathcal{H}$ either does not intersect $\mathcal{H}$ or its relative interior is contained in $\mathcal{H} \cup \{O\}$.

In the case where A is a single-valued function, it has been shown in Sect. 3.2 that the set $\mathcal{K}_{\bar{x}}$, namely the image of X through A, can be extended without modifying the intersection between it and $\mathcal{H}$. Because of the importance of such an extension in the applications to extremum problems and related fields, we will extend this concept to the present case of multifunctions. Unlike Sect. 3.2, here the dependence of the conic extension on the cone is made explicit, since it plays a role. With a slight abuse of notation, without any fear of sonfusion, in the next definition we use the symbol $\mathcal{K}(\omega)$ which disagrees with $\mathcal{K}_{\bar{x}}(X)$ used in Sect. 3.2 a few rows before (3.2.26).

Definition 4.8.1. The sets:

$$\mathcal{K} := \bigcup_{x \in X} A(x), \qquad \mathcal{K}(\omega) := \bigcup_{x \in X} \{\Phi(A(x); \omega)\}$$

are called, respectively, the *image* and the *selected image* of the set X through the multifunction A ($\mathcal{K}$ is the same as in Sect. 4.7). The dependence of $\mathcal{K}$ on $\bar{x}$ (like in Sect. 3.2) is omitted, since $\mathcal{K}$ may be the image of a generalized system, like (1.3.16), and not necessarily of an extremum problem, and $\bar{x}$ does not play any role. Note that ω is a function of x; for the sake of simplicity, such a dependence is understood.

When A is single-valued, then the above concepts shrink to those of Sect. 3.2; namely, we have:

$$\mathcal{K}(\omega) = \mathcal{K}, \quad \text{since} \quad \Phi(A(x); \omega) = A(x).$$

Definition 4.8.2 Let $C \subseteq \mathbb{R}^\nu$ be a convex cone with apex at the origin. The sets

$$\mathcal{E}(\mathcal{K}; C) := \mathcal{K} - C, \qquad \mathcal{E}(\mathcal{K}(\omega); C) = \mathcal{K}(\omega) - C$$

are called *conic extension of the image* and *of the selected image with respect to the cone* C, respectively.

Note that these extensions can be decomposed in this way:

$$\mathcal{E}(\mathcal{K}; C) = \bigcup_{x \in X} [A(x) - C], \qquad \mathcal{E}(\mathcal{K}(\omega); C) = \bigcup_{x \in X} \{\Phi(A(x); \omega) - C\}.$$

Definition 4.8.2 extends Definition 3.2.1.

Proposition 4.8.1.   Assume that $\mathcal{H}$ be a convex cone which fulfils (4.8.1) and let $\omega : X \to \Omega$. The system (4.7.1) is impossible if and only if there exists $\overline{\omega}(x) \in \Omega$, such that:

$$\mathcal{H} \cap \mathcal{E}(\mathcal{K}(\overline{\omega}); \operatorname{cl} \mathcal{H}) = \varnothing.$$

Proof.   If. Since evidently $\mathcal{K}(\overline{\omega}) \subseteq \mathcal{E}(\mathcal{K}(\overline{\omega}); \operatorname{cl} \mathcal{H})$, we have that:

$$\mathcal{H} \cap \mathcal{E}(\mathcal{K}(\overline{\omega}); \operatorname{cl} \mathcal{H}) = \varnothing \;\Rightarrow\; \mathcal{H} \cap \mathcal{K}(\overline{\omega}) = \varnothing$$

$$\Rightarrow \;\; \Phi(A(x); \overline{\omega}) \notin \mathcal{H}, \quad \forall x \in X;$$

hence, because of (4.7.3)″, (4.7.1) is impossible. **Only if.** The impossibility of (4.7.1) means that $A(x) \not\subseteq \mathcal{H}, \forall x \in X$, so that, because of (4.7.3)″,

$$\forall x \in X, \;\; \exists \overline{\omega} \in \Omega, \;\; \text{s.t.} \;\; \Phi(A(x); \overline{\omega}) \notin \mathcal{H};$$

it follows that $\mathcal{H} \cap \mathcal{K}(\overline{\omega}) = \varnothing$, and hence:

$$\mathcal{H} \cap \mathcal{E}(\mathcal{K}(\overline{\omega}); \operatorname{cl} \mathcal{H}) = \varnothing. \qquad\qquad \square$$

Definition 2.2.4 is immediately estended to the case of a point-to-set map; the present cone C corresponds to the cone H of Definition 2.4.4.

Definition 4.8.3. Let X be convex and C be a convex cone with apex at the origin. A is called a *C-multifunction* iff

$$(1 - \alpha)A(x^1) + \alpha A(x^2) \subseteq A(x(\alpha)) + C, \qquad \forall \alpha \in [0, 1], \; \forall x^1, x^2 \in X, \qquad (4.8.2)$$

where $x(\alpha) := (1 - \alpha)x^1 + \alpha x^2$. It is called a *C-convex multifunction*, when (4.8.2) holds with $C \supseteq \mathbb{R}^\nu_+$, and a *convex multifunction*, when (4.8.2) holds with $C = \mathbb{R}^\nu_+$. A is called a *C-concave* (or *concave*) *multifunction* iff $- A$ is C-convex (or convex). When (4.8.2) holds with $C = \{O\}$, A is called *preaffine*; a preaffine multifunction is called *affine* iff X is the entire space and (4.8.2) holds as equality $\forall \alpha \in \mathbb{R}$; an affine multifunction A such that $O \in A(O)$ is called *linear*. Other existing and related concepts, which are useful in the present context, are the following. A is called (positively) *homogeneous*, iff X is the entire space and

$$O \in A(O), \quad A(\alpha x) = \alpha A(x), \qquad \forall \alpha > 0, \forall x \in X. \tag{4.8.2}'$$

A homogeneous multifunction A (which extends (2.3.15b)), having all sets $A(x)$ convex, and such that:

$$A(x + y) \subseteq \mathrm{cl}\,[A(x) + A(y)], \qquad \forall x, y \in X, \tag{4.8.2}''$$

is called a *fan*.

By invoking the definitions, it is easy to show that A is C-concave iff it is $(-C)$-convex; that is, if A is both C-convex and C-concave, we have:

$$(1 - \alpha)A(x^1) + \alpha A(x^2) \subseteq [A(x(\alpha)) + C] \cap [A(x(\alpha)) - C],$$

$$\forall \alpha \in \mathbb{R}, \ \forall x^1, x^2 \in X;$$

that is, if A is affine, $G(x) := A(x) - A(O)$ is linear, namely:

$$O \in G(O), \qquad G(\alpha x) = \alpha G(x), \qquad \forall \alpha \in \mathbb{R}, \ \forall x.$$

As a consequence, it turns out that a linear multifunction is homogeneous.

If A is a single-valued function, we recover the notion of cone function of Definition 2.4.4; if in addition $\nu = 1$, then the concepts of convex, concave, and affine multifunctions collapse to those of ordinary convex, concave, and affine functions.

Between the 2 equivalent ways that Definition 2.3.1 shows for characterizing a single-valued affine function, the latter is that which characterizes the same concept for multifunctions, while the former is no longer valid when we go from functions to multifunctions.

The image $\mathcal{K}$, and the graph

$$\Gamma := \{(x, z) : x \in X, z \in A(x)\}$$

of an affine multifunction, as well as the epigraph

$$\mathrm{epi}\, A := \{(x, z) : x \in X; z \in A(x) + C\}$$

of a C-multifunction, are convex, as is easy to check. In fact, if $z^1, z^2 \in \mathcal{K}$, then $\exists x^1, x^2 \in X$ such that $z^i \in A(x^i)$, $i = 1, 2$. Let us set:

$$z(\alpha) := (1 - \alpha)z^1 + \alpha z^2 \quad \text{and} \quad x(\alpha) := (1 - \alpha)x^1 + \alpha x^2.$$

Thus, since $x(\alpha) \in X$, from (4.8.2) at $C = \{O\}$ we have:

$$z(\alpha) \in (1 - \alpha)A(x^1) + \alpha A(x^2) \subseteq A(x(\alpha)) \subseteq \mathcal{K}, \qquad \forall \alpha \in [0, 1].$$

This shows the convexity of the image of an affine multifunction. Now, let $(x^1, z^1), (x^2, z^2)$ belong to Γ. Then, $\forall \alpha \in [0, 1]$, from (4.8.2) we have:

$$z(\alpha) \in (1 - \alpha)A(x^1) + \alpha A(x^2) \subseteq A(x(\alpha)) + C.$$

Since $x(\alpha) \in X$, it follows that $(x(\alpha), z) \in$ epi $A, \forall \alpha \in [0, 1]$. This shows the convexity of epi A and, since epi $A = \Gamma$ at $C = \{O\}$, also the convexity of the graph Γ of an affine multifunction.

Proposition 4.8.2. If X is convex, C a convex cone with apex at the origin, and A a multifunction, then $\mathcal{E}(\mathcal{K}; -C)$ is convex.

Proof. Consider any $z^1, z^2 \in \mathcal{E}(\mathcal{K}; -C)$ and any $\alpha \in [0, 1]$. For $i = 1, 2$,

$$z^i \in \mathcal{E}(\mathcal{K}; -C) \quad \Rightarrow \quad \exists\, x^i \in X \text{ s.t. } z^i \in A(x^i) + C.$$

Let us set:

$$z(\alpha) := (1 - \alpha)z^1 + \alpha z^2, \quad x(\alpha) := (1 - \alpha)x^1 + \alpha x^2,$$

$$a(\alpha) := (1 - \alpha)A(x^1) + \alpha A(x^2).$$

Note that:

$$z^i \in A(x^i) + C \quad \Rightarrow \quad \begin{cases} \exists\, \bar{z}^i \in A(x^i) \\[2mm] \exists\, \hat{z}^i \in C \end{cases} \quad \text{s.t. } z^i = \bar{z}^i + \hat{z}^i.$$

It follows that:

$$z(\alpha) = [(1 - \alpha)\bar{z}^1 + \alpha\bar{z}^2] + [(1 - \alpha)\hat{z}^1 + \alpha\hat{z}^2];$$

and hence, since

$$(1 - \alpha)\bar{z}^1 + \alpha\bar{z}^2 \in a(\alpha) \quad \text{and} \quad (1 - \alpha)\hat{z}^1 + \alpha\hat{z}^2 \in C,$$

we have $z(\alpha) \in a(\alpha) + C$. Then, $\exists\, k' \in a(\alpha)$ and $\exists\, k'' \in C$, s.t. $z(\alpha) = k' + k''$ or $z(\alpha) - k' = k'' \in C$. From (4.8.2), we have $a(\alpha) \subseteq A(x(\alpha)) + C$, so that $\exists\, k \in A(x(\alpha))$ s.t. $k' - k \in C$. Since C is a convex cone, it follows that:

$$(z(\alpha) - k') + (k' - k) \in C, \quad \text{or} \quad z(\alpha) - k \in C;$$

this means that $z(\alpha) \in A(x(\alpha)) + C$; and hence, since $x(\alpha) \in X$, $z(\alpha) \in \mathcal{E}(\mathcal{K}; -C)$. $\square$

Proposition 4.8.3. Let X be a convex set, C a convex cone with apex at the origin, and $\omega : X \to \Omega$. If $\Phi(A(x); \omega)$ is a C-function, then $\mathcal{E}(\mathcal{K}(\omega); -C)$ is convex.

Proof. If the cardinality of $\mathcal{E}(\mathcal{K}(\omega); -C)$ is ≤ 1, then the thesis is trivial. Consider any $z^1, z^2 \in \mathcal{E}(\mathcal{K}(\omega); -C)$, so that $\exists\, x^1, x^2 \in X$ s.t.

$$z^i \in \{\Phi(A(x^i); \omega(x^i))\} - C \;, \quad i = 1, 2.$$

Because of Definition 2.4.4 with $H = -C$ and the assumption on Φ, we have:

$$(1 - \alpha)\Phi(A(x^1); \omega(x^1)) + \alpha\Phi(A(x^2); \omega(x^2)) - \Phi(A(x(\alpha)); \omega(x(\alpha))) \in -C,$$

where $x(\alpha) := (1 - \alpha)x^1 + \alpha x^2 \in X, \forall \alpha \in [0, 1]$. Set $z(\alpha) := (1 - \alpha)z^1 + \alpha z^2$. Because of the convexity of C, from the above relations we have:

$$z(\alpha) \in \{\Phi(A(x(\alpha)); \omega(x(\alpha)))\} - C \subseteq \mathcal{K}(\omega(x(\alpha))) - C, \ \forall \alpha \in [0,1],$$

and, according to Definitions 4.8.1 and 4.8.2, the thesis follows. $\qquad\square$

4.9. Systems of Intersection Type

A system for point-to-set maps may occur in several forms; not all of them are of inclusion type, like (4.7.1), requiring the introduction of a GSF. An instance is offered by a kind of system (considered in the literature), which differs from (4.7.1), since the inclusion is replaced by noninclusion; such a system is a special case of the following one:

$$\exists x \in X, \quad \text{s.t.} \quad A(x) \cap \mathcal{H} \neq \varnothing, \tag{4.9.1}$$

whose feasibility — which is implied by that of (4.7.1) — can be equally well expressed as impossibility of system (4.7.2). The case where A is single-valued was reduced (in Sect. 3.2; see (3.2.2)) to state disjunction or conjunction of 2 sets. Then, it is natural, in the present case (4.9.1), to expect to be able to use the same tools which have been adopted for (3.2.2). As a matter of fact, it is possible to prove a theorem concerning (4.9.1), which is quite analogous to Theorem 4.4.1 proved for (1.3.16) and generalized by Theorem 4.7.1. To this end, let us set:

$$w(A(x)) := \bigcup_{z \in A(x)} w(z), \qquad s(A(x)) := \bigcup_{z \in A(x)} s(z).$$

Theorem 4.9.1. Let the sets $\mathcal{H}, X, \mathcal{W}(\Pi), \mathcal{S}(\Pi)$, the cone C with apex at the origin, and the point-to-set map A be given. **(i)** the system (4.9.1) and the system

$$w(A(x)) \subseteq {\sim} C, \qquad \forall x \in X, \tag{4.9.2}$$

are not simultaneously possible, whatever the weak separation function $w \in \mathcal{W}(\Pi)$ may be. **(ii)** The system (4.9.1) and either the system

$$s(A(x)) \subseteq {\sim} C_0, \quad \forall x \in X, \tag{4.9.3a}$$

or the system

$$s(A(x)) \subseteq {\sim} \overset{o}{C}, \quad \forall x \in X, \tag{4.9.3b}$$

are not simultaneously impossible, whatever the strong separation function $s \in \mathcal{S}(\Pi)$ may be.

Proof. **(i)** If (4.9.1) is possible, i.e. if $\exists \bar{x} \in X$ s.t.

$$A(\bar{x}) \cap \mathcal{H} \neq \varnothing,$$

so that $\exists \bar{z} \in A(\bar{x}) \cap \mathcal{H}$, then (4.2.2a) implies $w(\bar{z}) \in C$, and hence (4.9.2) is impossible.
(**ii**) If (4.9.1) is impossible, i.e. if

$$A(x) \subseteq \sim \mathcal{H}, \qquad \forall x \in X,$$

then, from (4.2.9a) we draw that

$$x \in X \quad \Rightarrow \quad s(z) \notin C_0, \quad \forall z \in A(x),$$

and from (4.2.10a) we draw that

$$x \in X \quad \Rightarrow \quad s(z) \notin \overset{o}{C}, \quad \forall z \in A(x).$$

Therefore, (4.9.3) are possible. $\qquad\square$

At $\ell = 1$, (4.9.1) is equivalent to (4.7.1), and Theorem 4.9.1 is equivalent to Theorem 4.7.1, since we can set $\Phi(A(x); \omega) = A(x)$.

It is possible to prove a TA for (4.9.1) by following an approach like that adopted for (4.7.1) or (4.7.1)$'$; but in this way we would find results unnecessarily more complicated.

We will now prove, for the system (4.9.1), a property like Proposition 4.8.1. Indeed, in spite of the fact that (4.9.1) is not equivalent to (4.7.1), we might deduce next Proposition 4.9.1 from Proposition 4.8.1, since the conic extension $\mathcal{E}(\mathcal{K}(\bar{\omega}); \mathrm{cl}\,\mathcal{H})$ of Proposition 4.8.1 can be replaced, from a formal point of view, by the conic extension $\mathcal{E}(\mathcal{K}; \mathrm{cl}\,\mathcal{H})$ associated with (4.9.1). However, the proof of Proposition 4.9.1 will be given independently of Proposition 4.8.1, as it is short.

Proposition 4.9.1. Let (4.8.1) hold. The system (4.9.1) is impossible, if and only if

$$\mathcal{H} \cap \mathcal{E}(\mathcal{K}; \mathrm{cl}\,\mathcal{H}) = \varnothing. \tag{4.9.4}$$

Proof. If. Since evidently $\mathcal{K} \subseteq \mathcal{E}(\mathcal{K}; \mathrm{cl}\,\mathcal{H})$, the assumption implies $\mathcal{H} \cap \mathcal{K} = \varnothing$, and hence (4.9.1) is impossible. **Only if.** Ab absurdo, suppose that (4.9.4) be false. Then, $\exists\, x' \in X$ s.t.

$$\mathcal{H} \cap [A(x') - \mathrm{cl}\,\mathcal{H}] \neq \varnothing \,,$$

and thus $\exists\, a \in A(x')$ and $\exists\, z \in \mathcal{H} \cap [\{a\} - \mathrm{cl}\,\mathcal{H}]$. Note that

$$z \in \{a\} - \mathrm{cl}\,\mathcal{H} \quad \Rightarrow \quad \exists\, z' \in \mathrm{cl}\,\mathcal{H} \text{ s.t. } z = a - z'.$$

Hence, because of (4.8.1), $a = z + z' \in \mathcal{H}$. It follows that $A(x') \cap \mathcal{H} \neq \varnothing$, which contradicts the impossibility of (4.9.1). $\qquad\square$

Now, let us consider a special case of the system (4.9.1). To this end, let ν_1 and ν_2 be positive integers s.t. $\nu_1 + \nu_2 = \nu$, and let $K \subseteq \mathbb{R}^{\nu_1}$ be a convex cone with apex at the origin and such that $\mathrm{int}\, K \neq \varnothing$. Let $Y_i \subseteq \mathbb{R}^{\nu_i}, i = 1, 2$. For each $i = 1, 2$, consider the point-to-set map $A_i : X \rightrightarrows Y_i$, so that $A_i(x) \subseteq Y_i, \forall x \in X$; let y be a given element of Y_1. Consider the following system:

$$\exists\, x \in X \quad \text{s.t.} \quad y \in A_1(x) + \mathrm{int}\, K, \quad O \in A_2(x), \tag{4.9.5}$$

where O denotes (without any fear of confusion with other $O's$) a ν_2-dimensional null vector. The first part of (4.9.5) can be written also as

$$[\{y\} - A_1(x)] \cap \operatorname{int} K \neq \varnothing. \tag{4.9.6}$$

In fact, $y \in A_1(x) + \operatorname{int} K$ implies that $\exists\, y' \in A_1(x)$ and $\exists\, y'' \in \operatorname{int} K$ s.t. $y = y' + y''$ or $y'' = y - y'$. Since $y'' \in \{y\} - A_1(x)$, (4.9.6) follows. Vice versa, $y'' \in [\{y\} - A_1(x)] \cap \operatorname{int} K$ implies that $\exists\, y' \in A_1(x)$ s.t. $y'' = y - y'$, so that $y = y' + y'' \in A_1(x) + \operatorname{int} C$. Now, taking into account the equivalence between (4.9.5) and (4.9.6), by setting

$$A(x) = [\{y\} - A_1(x)] \times A_2(x), \quad \mathcal{H} = (\operatorname{int} K) \times \{O\}, \tag{4.9.7}$$

the system (4.9.5) becomes a particular case of (4.9.1). Note that the special case of $\mathcal{H}$, defined by (4.9.7), fulfils (4.8.1).

Theorem 4.9.1, which consists in a weak alternative statement and in a strong alternative one, can be specialized to the particular case (4.9.5); moreover, if we take a convexity assumption, we get a TA for (4.9.5). To this end, denote by $\mathcal{H}^*$ the positive polar of $\mathcal{H}$ (see (2.2.12b)), and observe that:

$$\mathcal{H}^* = \{(u, v) \in \mathbb{R}^{\nu_1} \times \mathbb{R}^{\nu_2} : \langle u, u' \rangle \geq 0, \ \forall u' \in K\} = K^* \times \mathbb{R}^{\nu_2}, \tag{4.9.8}$$

where K^* is the positive polar of K considered obviously as a subset of $\mathbb{R}^{\nu_1}$ (and not of $\mathbb{R}^{\nu}$). The cone $\mathcal{H}$ is contained in a halfspace, iff the gradient of the frontier of such a halfspace belongs to $\mathcal{H}^*$; (4.9.8) means that this gradient can be decomposed. Thus, we are led to state the following:

Proposition 4.9.2. At $\ell = 1$ and $C =]0, +\infty[$, the function $w : \mathbb{R}^{\nu_1} \times \mathbb{R}^{\nu_2} \to \mathbb{R}$, with

$$w(u, v) := \langle \theta, u \rangle + \langle \lambda, v \rangle \ , \quad \theta \in K^* \setminus \{O\} \ , \quad \lambda \in \mathbb{R}^{\nu_2} \ , \tag{4.9.9}$$

is a weak separation function for the system (4.9.1) in case (4.9.7), and hence also for (4.9.4).

Proof. According to (4.2.3a), we must show that $\mathcal{H} = (\operatorname{int} K) \times \{O\}$ is contained in the positive level set of the function (4.9.9), $\forall \theta \in K^* \setminus \{O\}$, $\forall \lambda \in \mathbb{R}^{\nu_2}$. Let $(\bar{u}, \bar{v}) \in \mathcal{H}$, i.e. $\bar{u} \in \operatorname{int} K, \bar{v} = O$. Then, $\forall \theta \in C^* \setminus \{O\}$, we have $\langle \theta, \bar{u} \rangle > 0$, and hence $w(\bar{u}, \bar{v}) > 0$. $\quad\square$

Now, we are in the position to obtain, as a consequence of Theorem 4.9.1 and of Propositions 4.9.1 and 4.9.2, a TA for the system (4.9.5).

Theorem 4.9.2. Let X be convex , K be a convex cone with apex at the origin and such that $\operatorname{int} C \neq \varnothing$; let A_1 be a point-to-set map, and A_2 an affine point-to-set map.
(i) If the system (4.9.5) is impossible, then there exist multipliers $\bar{\theta} \in K^*$ and $\bar{\lambda} \in \mathbb{R}^{\nu_2}$, with $(\bar{\theta}, \bar{\lambda}) \neq O$, such that:

$$\bigcup_{\substack{u \in A_1(x) \\ v \in A_2(x)}} \{\langle \bar{\theta}, y - u \rangle + \langle \bar{\lambda}, v \rangle\} \subseteq [0, +\infty[, \ \forall x \in X. \tag{4.9.10}$$

(ii) If there exist multipliers $\bar{\theta} \in K^* \setminus \{O\}$ and $\bar{\lambda} \in \mathbb{R}^{\nu_2}$, so that (4.9.10) holds, then (4.9.5) is impossible.

Proof. (i) Let (4.9.5), i.e. (4.9.1) in case (4.9.7), be impossible, so that $\mathcal{H} \cap \mathcal{K} = \varnothing$. Hence, because of Proposition 4.9.1,

$$\mathcal{H} \cap \mathcal{E}(\mathcal{K}; \mathrm{cl}\,\mathcal{H}) = \varnothing. \tag{4.9.11}$$

In the present case (4.9.7), $\mathcal{H}$ is convex and A is a $(-\mathrm{cl}\,\mathcal{H})$-multifunction; hence, because of Proposition 4.8.2, $\mathcal{E}(\mathcal{K}; \mathrm{cl}\,\mathcal{H})$ is convex, and thus its closure is the intersection of its supporting halfspaces (Theorem 2.2.2); these are of type:

$$\langle \theta, u \rangle + \langle \lambda, v \rangle \leq \alpha \in \mathbb{R}, \quad \text{with } \theta \in K^*, \ \lambda \in \mathbb{R}^{\nu_2}, \ \pi := (\theta, \lambda) \neq O. \tag{4.9.12}$$

In fact, $\mathcal{E}(\mathcal{K}; \mathrm{cl}\,\mathcal{H})$ contains a translation of the cone $-\mathrm{cl}\,\mathcal{H}$ and hence the gradient of the left-hand side of (4.9.12) must belong to $\mathcal{H}^*$, which can be decomposed according to (4.9.8). It is impossible that each halfspace (4.9.12) have $\alpha > 0$; in fact, if this happens, then, $\forall \pi \in \mathcal{H}^*$, we have:

$$\sup_{z \in \mathcal{E}(\mathcal{K};\mathrm{cl}\,\mathcal{H})} \langle \pi, z \rangle > 0,$$

so that $O \in \mathrm{int}\,\mathcal{E}(\mathcal{K}; \mathrm{cl}\,\mathcal{H})$, which contradicts (4.9.11). It follows that, among the halfspaces (4.9.12), at least one has $\alpha \leq 0$. In other words, $\exists \overline{\theta} \in K^*$ and $\exists \overline{\lambda} \in \mathbb{R}^{\nu_2}$, with $(\overline{\theta}, \overline{\lambda}) \neq O$, s.t. $\mathcal{E}(\mathcal{K}; \mathrm{cl}\,\mathcal{H})$ is contained in the halfspace defined by $\langle \overline{\theta}, u \rangle + \langle \overline{\lambda}, v \rangle \leq \alpha$, with $\alpha \leq 0$, and hence in the one, say S, defined by $\langle \overline{\theta}, u \rangle + \langle \overline{\lambda}, v \rangle \leq 0$, namely, $\mathcal{E}(\mathcal{K}; \mathrm{cl}\,\mathcal{H}) \subseteq S$. Thus,

$$\mathcal{K} - \mathrm{cl}\,\mathcal{H} \subseteq S \quad \text{or} \quad \bigcup_{x \in X} A(x) - \mathrm{cl}\,\mathcal{H} \subseteq S;$$

this inclusion is contradicted if, ab absurdo, we assume that $\exists \overline{z} \in \left[\bigcup_{x \in X} A(x)\right] \setminus S$ (since, taking $\hat{z} \in \mathrm{cl}\,\mathcal{H}$ with $\|\hat{z}\|$ small enough, we have $\overline{z} - \hat{z} \notin S$), and hence we have:

$$\bigcup_{x \in X} A(x) \subseteq S,$$

so that:

$$\langle (\overline{\theta}, \overline{\lambda}), z \rangle \leq 0, \qquad \forall z \in A(x), \ \ \forall x \in X;$$

now, account being taken of (4.9.7), we achieve (4.9.10). **(ii)** The system (4.9.5) and the condition (4.9.10) are particular cases (defined by (4.9.7)) of (4.9.1) and (4.9.2), respectively. Thus the thesis follows from Theorem 9.4.1, observing that the function w, used to define the left-hand side of (9.4.10) and equipped with $C =]0, +\infty[$, is (according to Proposition 4.9.2) a weak separation function. $\qquad\qquad\square$

The following simple example shows that in (ii) of Theorem 4.9.2, we cannot admit the multiplier $\overline{\theta} = O$, as we did in (i).

Example 4.9.1. Let us set $\nu_1 = \nu_2 = 1$, $X = [0, 1]$, $y = 0$, $K = [0, +\infty[$, $A_1(x) = [-2, x-1]$, $A_2(x) = [x/2, x+1]$. The proof that A_1 is an $(\mathbb{R}_+)$-multifunction and that A_2 is an affine multifunction is trivial, since each $A_i(x)$ is an interval of $\mathbb{R}$, whose extrema depend on x linearly (indeed , also A_1 is an affine multifunction). Then, the system (4.9.5) becomes: to find an $x \in \mathbb{R}$, such that:

$$0 \in [-2, x-1] +]0, +\infty[, \quad 0 \in \left[\frac{1}{2}, x+1\right], \quad x \in [0, 1], \tag{4.9.13}$$

and is possible; $x = 0$ is obviously the solution. The inclusions (4.9.10) become:

$$\bigcup_{\substack{u\in[-2,x-1] \\ v\in[x/2,x+1]}} \{-\overline{\theta}u + \overline{\lambda}v\} \subseteq \,]-\infty, 0], \qquad \forall x \in [0, 1], \tag{4.9.14}$$

and are equivalent to

$$\sup(-\overline{\theta}u + \overline{\lambda}v) \leq 0, \qquad \text{s.t.} \; -u + 2v \geq 1, \; -2 \leq u \leq 0, \; 0 \leq v \leq 2;$$

hence, taking into account that $\overline{\theta} \in K^*$ now means $\overline{\theta} \geq 0$, that $\overline{\lambda}$ is not sign-constrained, and that $(\overline{\theta}, \overline{\lambda}) \neq O$, it follows that (4.9.14) is satisfied by $\overline{\theta} = 0$ and $\overline{\lambda} < 0$. Thus, if we admit the value $\overline{\theta} = 0$, the possibility of (4.9.10) does not exclude that of (4.9.5). $\square$

4.10. Comments

1. The class of separation functions (4.3.2) (which often and improperly are called "additive" or "separable"; a less improper term might be "sum-decomposable") is extremely important for the applications. Therefore, it is crucial to deepen their role in the IS Analysis. For instance, it would be important to split the class of problems, which enjoy the continuity (or differentiability, or C^k property), into two classes, according to which, respectively, $\mathcal{H}$ and $\mathcal{K}_{\overline{x}}$ of Sect. 3.2 can or cannot be separated disjunctively by a function of type (4.3.2) when (3.2.2) holds. The same question exists also when $\mathcal{K}_{\overline{x}}$ is replaced by $\mathcal{K}_{\overline{x}}^h$.

2. In the paper [II46], there is an excellent treatment of Lagrangian theory for constrained extremum problems. The class of strictly increasing positively homogeneous functions is introduced and exploited. Connections between this theory and that outlined in Sects. 4.2-4.3, especially conditions (4.3.1c,d), should lead to interesting results.

3. By exploiting some generalizations of convexity, as geodetic convexity [I56], it would be useful to extend Theorem 4.6.1 to nonconvex cases. In (4.6.1)-(4.6.2), the convex hull might be replaced by the "geodesic convex hull", obtained by replacing, in Definition 2.1.2, the linear combination with the geodesic combination. In addition to this, the extension might be achieved by using separation functions, which are suprema of linear (or convex) functions (see (2.3.20)). Alternatively, one might search for conditions under which a partition (or a cover) of K (of Theorem 4.6.1) into $K_1, ..., K_r$ exists, s.t. for each K_i Theorem 4.6.1 holds. When this happens, the application r times of Theorem 4.6.1 produces r separation hyperplanes; they enable one to define a nonlinear separation function, which is a (piece-wise linear) sup-function (see (2.3.20)).

4. In Theorem 4.6.1, when K is convex either (4.6.1) does not occur or (4.6.2) is trivially verified (if $p = 0$, the entire assumption shrinks to $(\text{int } \mathcal{H}) \cap \text{ri } K = \varnothing$). Therefore, for $K_1 = K$ and $K_2 = \mathcal{H}$, Theorem 2.2.4(i) follows from Theorem 4.6.1. Indeed, the fact that $\mathcal{H}$ has a special form does not play any substantial role in the proof of

Theorem 4.6.1; hence, it would be useful to free such a proof from the peculiarity of $\mathcal{H}$ and then to recover known separation theorems. In this order of ideas, the cases $p > 0$ and $p = 0$ of Theorem 4.6.1 should be replaced by int $\mathcal{H} = \varnothing$ and int $\mathcal{H} \neq \varnothing$, respectively.

5. In Vol.2, it will be shown that exterior and interior penalty theories correspond to weak and strong separation functions, respectively. For this reason, it would be interesting to develop the latter as well as the former.

6. With regard to the so called Farkas Lemma (Corollary 4.5.7), note that its 1st extension to infinite dimensional spaces has been done by Farkas himself. More precisely, at page 17 of [11] (titled "Infinitesimale Systeme"), Farkas considers systems of integrals of differential forms. The existing literature on generalizations of Corollary 4.5.7 to infinite dimensional spaces seems not to be aware of this fact: the achieved results (even in locally convex spaces [7], or in topological vector spaces [21]), though extremely interesting, do not recover, as particular case, that considered by Farkas. Due, perhaps, to the fact that it is in German, the paper by Farkas, notwithstanding the fact it is diffusely quoted (but always with the same wrong final page), seems to have remained untouched since the 2nd world war when TA have started to grow. This is a drawback. As a consequence, there exists a disproportion between the existing mathematical results and their exploitation by, for instance, Engineering. In 1977 (see [37], pages 242-243), Prager made an interesting application of a theorem of Farkas (from Sect. IX of [11]) to a problem of Structural Engineering. As far as I know, this has remained an "isolated point". A systematic investigation of TA for systems of type considered by Farkas would be extremely useful. With reference to Gordan Theorem (see Corollary 4.5.3 for $a = O, B = O, b = O, C = O, c = O$), in [46] there is a good result in the above desirable direction.

7. As noted in Sect. 3.5 with reference to the paper [III4], the IS Analysis might be useful also in the field of Combinatorial and Discrete Optimization. To this end, the development of TA and ST is of fundamental importance. An early hint can be found in [13]. Consider (1.1.1) in the case, where $X = \overline{X} \cap \mathbb{Z}^n$ with $\overline{X}$ compact, and f, g are integer valued, or

$$x \in \mathbb{Z}^n \;\Rightarrow\; f(x) \in \mathbb{Z}, \;\; g(x) \in \mathbb{Z}^m.$$

In this case, it is not restrictive to replace $\mathcal{H}$ of Sect. 3.2 with $\mathcal{H} = [1, +\infty[\times D$, so that strict separation can be achieved between $\mathcal{H}$ and $\mathcal{K}_{\overline{x}}$, when they are disjoint. In this case the theory of Sects. 4.2-4.6 can be sharpened.

8. In order to investigate the image of a combinatorial problem (see a comment around (3.5.23)-(3.5.24)), the following theorem might be useful. Consider a pseudomanifold, namely a simplicial complex, s.t. it is non-branching (each $(n-1)$-dimensional simplex is a face of two n-dimensional simplices), strongly connected (any two n-dimensional simplices can be joined by a chain of n-dimensional simplices in which each pair of neighbouring simplices have a common $(n-1)$-dimensional face), and it has dimensional homogeneity (each simplex is a face of some n-dimensional simplex. The Jordan-Brower Theorem (see [1], pages 54 and 94) says: "every $(n-1)$-dimensional closed

pseudo-manifold in $\mathbb{R}^n$ is orientable, separates $\mathbb{R}^n$ into precisely two domains and is the common boundary of these two domains".

9. Theorems 4.7.1 and 4.7.2 are general statements, which become effective when the functions w, s, Φ receive some specifications. In other words, such theorems must be considered as sources for deriving TA for given classes of systems of type (4.7.1); to each class, we have to assign a family of weak or strong separation functions as well as a family of GSF, if we want to give the above theorems a form which is suitable for the applications. This has been already done in Sect. 3.2 for an important class of problems of type (1.1.5).

10. The thesis of Proposition 4.8.2, which is crucial for the applications to extremum problems, can be ensured under weaker assumptions, as has been done for the single-valued case [III47]. Proposition 4.8.2 states the convexity of $\mathcal{E}(\mathcal{K}; -C)$; it is important to have conditions under which the convexity of the conic extension of the selected image $\mathcal{E}(\mathcal{K}(\omega); -C)$ is also ensured. Proposition 4.8.3 is a starting point for such developments.

11. In the applications to problems of type (1.1.5) (see Sect.3.2), A is precisely only one multifunction. However, it is conceivable to extend Definition 4.8.2 to embody the case where A is a vector of multifunctions, say $A_i; \forall x \in X$, the elements (sets) A_i of $A(x)$ can be viewed as coordinates of a set $A(x)$, to be defined, of $\mathbb{R}^\nu$, as well as in the case where A is a vector of single-valued functions and these are the coordinates of a point of $\mathbb{R}^\nu$. One possible way consists in defining $A(x)$ as the Cartesian product of the sets $A_i(x)$. When, in Sect. 3.2 in correspondence of system (3.2.26) $-$ associated to (1.1.5) $-$, we have introduced the multifunction $A_{\overline{x}}$, we might have considered it as a vector of $1+m$ multifunctions, namely $f(x)$ (which, indeed, is a single-valued one) and

$$A_i(x) := \{v_i \in \mathbb{R} : v_i = \psi_i(t, x(t), x'(t)), t \in [0, 1]\}, \ i \in \mathcal{I}.$$

Then, we might have set $A(x) = \{f(x)\} \times A_1(x) \times ... \times A_m(x)$. The set $A(x)$ defined in this way does not necessarily coincide with that defined in Sect. 3.2; however, since in the case of (1.1.5) the closure of $\mathcal{H}$ is a suborthant of $\mathbb{R}_+^{1+m}$, then (4.7.1) is possible in one case, iff it is in the other one.

12. In Proposition 4.8.3, the main assumption has been made on the GSF Φ. This has been done to have a preliminary statement. The analysis needs to be deepened to achieve statements which be based on assumptions made on the point-to-set A, on the function ω and on the GSF Φ, and not on their composition, as done in Proposition 4.8.3. Due to the importance of the subject, besides necessary and/or sufficient conditions for the convexity of the conic extensions in the general case, it would be useful to consider classes of problems $-$ like (1.1.5) and related GSF as (3.2.28) $-$ and for them to deepen the general analysis.

13. Both in the case of functions and in that of multifunctions, the TA of this chapter (apart from those of Sect. 4.5.4) have been achieved by following an approach different from the existing literature. Therefore, the comparison between the present theorems and those of the literature should lead to new results. As an instance, in the

case of multifunctions, it would be interesting to consider the results of the paper by J.Borwein: "Multivalued convexity and Optimization: a unified approach to Inequality and Equality Constraints" in Mathematical Programming, Vol.13, 1977, pp. 183-199. We can observe that, when the images Y and Z of the maps A and H, which are contained in Theorem 2.1 of such a paper, have finite dimension, then such a theorem becomes a corollary of Theorem 4.9.2 (i), which is a strong alternative statement and does not require the lower semicontinuity and opennes assumptions made in the above paper. It is conceivable to extend Theorem 4.9.2 to the infinite-dimensional case, and the result should hold under mild assumptions. Theorem 2.1 of Borwein might be generalized by adopting the present approach and by starting with Theorem 4.9.1. Of course, in trying to state a proposition like (ii) of Theorem 4.9.2 in the context of the above paper, we cannnot hope to admit the multiplier $\bar{\theta} = 0$ even if the assumptions of Theorem 2.1 of Borwein are made. In fact, Example 4.9.1 fulfils such assumptions: A_1 is l.s.c. in every $x \in [0,1]$, in the sense of Borwein, since, $\forall\,]a,b[\subseteq \mathbb{R}$ and such that $[-2, x-1]\cap]a,b[\neq \varnothing$ (this happens iff $a < x-1, b > -2$ and obviously $a < b$), we have:

$$[-2, r-1]\cap]a,b[\,\neq \varnothing, \quad \forall r \in]x - \varepsilon, x + \varepsilon[\cap[0,1],$$

with ε positive and small enough. A_2 is open in every $x \in [0,1]$, still in the sense of Borwein, since $\bigcup_r]r/2, r + 1[\neq \varnothing$, with $r \in]x - \varepsilon, x + \varepsilon[$, and with ε positive and small enough. Thus the assumptions of Theorem 2.1 of Borwein are fulfilled; the system (I) of such a theorem is (4.9.13); the system (II) (account being taken of the fact that λ is sign-free, so that A_2 can be replaced by $-A_2$) becomes (4.9.14); hence (I) and (II) are both possible.

14. The thesis of Theorem 4.9.2 can be achieved under weaker assumptions. In fact, the crucial part of its proof is the convexity of $\mathcal{E}(\mathcal{K}; \mathrm{cl}\,\mathcal{H})$. As is shown in Sect. 3.2 (see also [III47,14]), we can weaken the assumption that A be a cone-multifunction. Moreover, A_2 can be a cone-multifunction, and not necessarily an affine multifunction; in this case, $\mathcal{H}$ has not necessarily empty interior with respect to $\mathbb{R}^\nu$.

15. By considering a suitable subset of $\mathcal{S}^c(\Pi)$, defined by conditions corresponding to (4.11.1)-(4.11.4), it is possible to state a theorem of asymptotic alternative for the strong case.

16. When (3.2.2) or (3.2.39) are proved by showing that the related sets belong to opposite level sets of a functional w (i.e., by using a ST or a TA), then the properties enjoyed by w are crucial for subsequent developments, including the numerical and computational aspects. A classification of problems or systems according to the "simplest" separation function w necessary to show disjunction between the above sets − one of which conically extended, so that (3.2.14)is considered, instead of (3.2.2) − would be extremely useful. For instance, the set of linear w can be associated − up to regularity − with convexlike problems or systems (see Definition 2.4.5, Propositions 3.2.8 and Corollary 4.4.1). Given a constrained extremum problem, the "simplest" separation function necessary to show optimality would define the *"analytical complexity"* of it. Such a classification is strictly connected with the investigation of the (numerical)

behaviour of penalty functions (see Sect. 3.5 and Vol. 2).

17. In the previous sections, the separation of sets has been considered with respect to two sets only. In the literature, the concept of separation among several sets has been introduced, and some applications have been done; see [III2, I34, II18]. In Sect. 2.5, it has been shown that, by means of a suitable transformation, the separation of several sets can be reduced to that of precisely two sets. Since the separation of two sets has beeen deeply investigated and several results are available, it would be useful to express, through the above mentioned transformation, the properties of the previous sections in terms of separation of several sets.

18. In Chapter 2, the connections among the concepts of linear support, linear separation, cone polarity and conjugacy for functions have been outlined; detailed relationships among them are a classical part of the literature. Their extension to the general nonlinear case, which, for the separation, has been introduced in Sect. 4.2, would be interesting. In this case, the polarity to be exploited should be that of Definition 2.2.4.

19. In Sect. 3.2, the conic extension of the image set of problems (1.1.1) and (1.1.4) has shown to be useful. For instance, it has been instrumental for establishing Theorem 3.2.3. Definition 4.8.2 gives the conic extension for (1.1.5) through the selection. It would be interesting to establish an existence theorem for (1.1.5), like Theorem 3.2.3, by using Definition 4.8.2, and to recover known existence conditions. With respect to this investigation, it might be useful to extend the image set of (1.1.5), before performing a selection on it. For instance, the image set $\mathcal{K}$ of Definition 4.8.1 might be replaced equivalently by (D is as in Sect. 1.1):

$$\mathcal{K} - D = \bigcup_{x \in X} [A(x) = D],$$

and $\mathcal{K}(\omega)$ by:

$$\bigcup_{x \in X} \{\Phi(A(x) - D; \omega)\}.$$

This might lead to "regularize" the problem, as for (1.1.1.) and (1.1.4). The regularization might deal with the elimination of some nonconvexities, nondifferentiabilities, discontinuities; or it might lead to numerical advantages, as discussed in Sect. 3.5.

20. The theory of constrained extremum problems in Complex Spaces, like (1.1.21)-(1.1.23), deserves to be investigated. Until now, it has not received much attention. Some research has beeen done on TA for systems in Complex Spaces; see [V31] and references therein. The development of TA and ST for such systems is fundamental and, perhaps, preliminary for achieving the above theory. The TA of Sect. 4.5 − as well as the preceding ones − have been obtained through an approach different from known ones: by means of the one-to-one correspondence, defined in Sect.1.1 for problems (1.1.21)-(1.1.23), systems on a complex space have been reduced to a real space. By exploiting the same correspondence, it would be interesting to extend the IS Analysis of Sect. 3.2 to systems and problems in complex spaces. To this end, the development of TA and ST in a complex space is of fundamental importance; the approach of this

chapter might be of help. In other words, as well as in Sect. 4.5 most of existing TA in a real space have been drawn from Theorem 4.5.1, it should be possible to deduce, from Theorem 4.7.1 the few existing TA in a complex space (see [II5], Chapter I, Sect. 4 and next Theorem 4.10.1) and to state new ones like those obtained in Sects. 4.7-4.9. The following theorem represents only a starting point of a research which should be carried on intensively.

Consider the set $S := \{(\zeta, \zeta^*) : \zeta \in \mathbb{C}^n\}$, and note that it is a linear manifold on $\mathbb{R}$, but not on $\mathbb{C}$, since:

$$\left. \begin{array}{l} \zeta, \eta \in S \\ \alpha, \beta \in \mathbb{C} \end{array} \right\} \quad \not\Rightarrow \quad \alpha(\zeta, \zeta^*) + \beta(\eta, \eta^*) = (\alpha\zeta + \beta\eta, \alpha\zeta^* + \beta\eta^*) \in S;$$

the implication holds, if $\alpha, \beta \in \mathbb{R}$. S is convex, since:

$$\zeta \in S, \ \alpha \in \mathbb{R}_+ \quad \Rightarrow \quad \alpha(\zeta, \zeta^*) \in S,$$

$$\left. \begin{array}{l} (\zeta, \zeta^*) \in S \\ \eta, \eta^* \in S \\ \alpha \in [0, 1] \end{array} \right\} \Rightarrow (1 - \alpha)(\zeta, \zeta^*) + \alpha(\eta, \eta^*) = ((1 - \alpha)\zeta + \alpha\eta, ((1 - \alpha)\zeta + \alpha\eta)^*) \in S).$$

Moreover, the (complex positive) polar of S (see (2.2.12c)) is given by: $S^* = \{(\zeta, -\zeta^*) : \zeta \in \mathbb{C}^n\}$. Taking into account Definition 2.4.5, we can state the following TA, where D is a closed and convex cone of $\mathbb{C}^m$.

Theorem 4.10.1. Let $X \subseteq \mathbb{C}^n$ be convex, $f : X \times X \to \mathbb{C}$ be such that Re $f(\zeta, \zeta^*)$ is concave, and $g : X \times X \to \mathbb{C}^m$ be D-concavelike on S.

(i) If the system

$$\text{Re} f(\zeta, \zeta^*) > 0, \quad g(\zeta, \zeta^*) \in D, \quad \zeta \in X \tag{4.10.1}$$

is impossible, then $\exists \theta \in \mathbb{R}_+$ and $\exists \lambda \in D^*$, with $(\theta, \lambda) \neq O$, such that:

$$\theta \text{Re} f(\zeta, \zeta^*) + \text{Re}\langle \lambda, g(\zeta, \zeta^*) \rangle \leq 0, \ \forall \zeta \in X. \tag{4.10.2}$$

(ii) System (4.10.1) is impossible, if $\exists \theta \in \mathbb{R}_+$ and $\exists \lambda \in D^*$, with $(\theta, \lambda) \neq O$, such that (4.10.2) holds, where, if $\theta = O$ necessarily, either the inequality must be verified in strict sense, or

$$\chi(\lambda) := \{\zeta \in X : \text{Re} f(\zeta, \zeta^*) > 0, g(\zeta, \zeta^*) \in D, \text{Re}\langle \lambda, g(\zeta, \zeta^*) \rangle = 0\} = \varnothing. \tag{4.10.3}$$

Proof. Taking into account of the one-to-one correspondence, defined in Sect. 1.1 for problems (1.1.21)-(1.1.23), systems (4.10.1) and (4.10.2) can be put in the formats (3.2.1) and (4.4.16). Therefore, we can apply Theorems 4.5.1 and 4.4.4 to achieve (i) and (ii), respectively. $\square$

The above theorem shrinks to Theorems 4.5.1 and 4.4.4, when the imaginary parts of its data are zero. In problems (1.1.21) and (1.1.23), as well as in systems (4.10.1) and (4.10.2), f and g depend on both ζ and ζ^* to have a sufficiently general format. In fact, if f and g are independent of ζ^*, then the convexity/concavity of the real parts with respect to $\mathbb{R}$ cannot happen.

Obviously, $\overline{\zeta} \in R$ is a complex global m.p. of (1.1.21), iff the system

$$\operatorname{Re} f(\overline{\zeta}, \overline{\zeta}^*) - \operatorname{Re} f(\zeta, \zeta^*) > 0, \quad g(\zeta, \zeta^*) \in D, \ \zeta \in X \qquad (4.10.4)$$

is impossible. $\operatorname{Re} f(\overline{\zeta}, \overline{\zeta}^*)$ being a constant, system (4.10.4) is easily recognized to be of type (4.10.1). Hence, Theorem 4.10.1 can be applied to (1.1.21).

21. Another field, which has received only a few attention and, on the contrary, deserves to be deepened, is that of *asymptotic alternative*. As it will be seen in Vol. 2, it is fundamental for giving the penalty methods a general and theoretical background. Here we will give only a hint for further investigation.

When a weak separation function $w \in \mathcal{W}(\Pi)$ is adopted, in the general case (i) of Theorem 4.4.1 does not enable one to claim anything about (1.3.16), if (4.4.1) is impossible. However, in some cases a claim is still possible, if the concept of weak alternative is enlarged. This will be shown in the particular case, where in Definition 4.2.1 we have $\ell = 1, C =\,]0, +\infty[\,= C_0 = \overset{o}{C}$. Assume that a sequence, say $\{\pi^r\}_1^\infty$, can be drawn out from Π, s.t.

$$w(\bullet; \pi^r) \text{ is continuous}, \quad r = 1, 2, ..., \qquad (4.10.5)$$

$$\operatorname{lev}_{>0} w(\bullet; \pi^r) \supset \operatorname{lev}_{>0} w(\bullet; \pi^{r+1}), \quad r = 1, 2, ..., \qquad (4.10.6)$$

$$\bigcap_{r=1}^{\infty} \operatorname{lev}_{>0} w(\bullet; \pi^r) = \mathcal{H}, \qquad (4.10.7)$$

$$\forall z \in \mathcal{H}, \exists k(z) > 0, \text{ s.t. } w(z; \pi^r) \geq k(z), \quad \forall r = 1, 2, ... \qquad (4.10.8)$$

(4.10.5) means that $\{w(\bullet; \pi^r)\}_1^\infty \subseteq \mathcal{W}^c(\Pi)$. (4.10.6) says that the positive level sets are nested. (4.10.7) means that the intersection of such a nested sequence is precisely $\mathcal{H}$, so that $w(\bullet; \pi^r)$ is "less and less weak" as r grows, and tends to be a sort of indicator function of $\mathcal{H}$, in the sense that:

$$\lim_{r \to +\infty} w(z; \pi^r) = \left\{ \begin{array}{ll} +\infty, & \text{if} \quad w(z; \pi^r) \leq 0, \\ 0, & \text{otherwise}, \end{array} \right\} = \delta(z; \mathcal{H}),$$

where δ is defined by (2.3.16). (4.10.8) is a boundedness assumption, which is fulfilled by most of separation functions. An instance of separation function, which satisfy (4.10.5)-(4.10.8), is offered by a suitable extraction of a denumerable set from $\mathcal{W}_R(\Pi)$, in particular the functions (4.3.6); for the sake of simplicity and without any fear of confusion, we continue to adopt the stipulation of Sect. 4.2, according to which Π denotes the domains of all of $\mathcal{W}, \mathcal{W}^c, \mathcal{W}_R$, even if they are not equal. Note that conditions (4.10.5)-(4.10.8) do not imply that $\mathcal{H}$ be closed.

Theorem 4.10.2. Let $w \in \mathcal{W}(\Pi)$ and fulfil (4.10.5)-(4.10.8). Then, for a given (fixed) ξ, system (1.3.16) is impossible, if and only if

$$\inf_{r \in \mathbb{N}} \sup_{x \in X} w(A(x; \xi); \pi^r) \leq 0. \qquad (4.10.9)$$

Proof. If (1.3.16) is possible, i.e. if $\exists \hat{x} \in X$, s.t. $\hat{z} := A(\hat{x}; \xi) \in \mathcal{H}$, then (4.10.8) implies:

$$w(A(\hat{x}; \xi); \pi^r) = w(\hat{z}; \pi^r) \geq k(\hat{z}) > 0, \quad \forall r \in \mathbb{N}. \qquad (4.10.10)$$

Ab absurdo, suppose that (4.10.9) holds. Then, $\forall \varepsilon > 0, \exists r_\varepsilon \in \mathbb{N}$ s.t.

$$\sup_{x \in X} w(A(x;\xi);\pi^{r_\varepsilon}) < \varepsilon. \tag{4.10.11}$$

Because of (4.10.10) and (4.10.11) for $\varepsilon = k(\hat{z})$, we meet the contradiction:

$$k(\hat{z}) \leq w(\hat{z};\pi^{r_\varepsilon}) \leq \sup_{x \in X} w(A(x;\xi);\pi^{r_\varepsilon}) < \varepsilon = k(\hat{z}).$$

Hence, the impossibility of (4.10.9) follows. To show that (4.10.9) holds when (1.3.16) is impossible, i.e. when $z := A(x;\xi) \notin \mathcal{H}, \forall x \in X$, it is enough to prove that, $\forall \varepsilon > 0$, $\exists r_\varepsilon \in \mathbb{N}$ s.t.

$$w(A(x;\xi);\pi^{r_\varepsilon} < \varepsilon, \quad \forall x \in X. \tag{4.10.12}$$

Because of (4.10.5)-(4.10.7), given $\delta > 0, \exists r_\delta \in \mathbb{N}$ s.t.

$$z \in [\ \mathrm{lev}_{>0}\, w(\bullet;\pi^{r_\delta})] \setminus \mathcal{H} \quad \Rightarrow \quad \exists z'(z) \text{ s.t.} \begin{cases} w(z'(z);\pi^{r_\delta}) = 0, \\ \|z - z'(z)\| < \delta. \end{cases}$$

Hence, because of (4.10.5), given any $\varepsilon > 0$, by making a suitable choice of δ, we obtain:

$$w(z;\pi^{r_\delta}) = w(z;\pi^{r_\delta}) - w(z'(z);\pi^{r_\delta}) < \varepsilon, \ \forall z \in [\mathrm{lev}_{>0}\, w(\bullet;\pi^{r_\delta})] \setminus \mathcal{H},$$

and thus (4.10.12) follows. $\square$

Theorem 4.10.2 can be interpreted in terms of a sequence of weak separation functions, whose level sets approximate more and more $\mathcal{H}$, so that the weak alternative is obtained asymptotically. More precisely, because of (4.10.5)-(4.10.8), alternative is achieved. In the particular case where $\exists r^0 \in \mathbb{N}$ s.t. $w(z;\pi^{r^0})$ guarantees weak alternative, then at $r = r^0$ the sufficiency of Theorem 4.10.2 becomes (i) of Theorem 4.4.1 for $\ell = 1$. If the class of w is enlarged by deleting (4.10.7), then the necessity is lost in Theorem 4.10.2.

In other words, trying to separate two sets — at least one of them not convex — by means of a nonlinear manifold to be found within a given class $\{w(\pi)\}$, it may happen that none of the elements of it separates the sets. However, it might be possible to extract, from the class, a sequence converging to an element which allows us to separate.

22. As it will be seen in Vol. 2, the class (4.3.1) is important for duality and penalty theories. The main feature of (4.3.1) is to express w as a sum of a function of u and a function of v. Unlike the latter, the former is linear; this is important for the above reason. However, for other applications, it should be important that the former were not necessarily linear. In other words, it would be important to investigate the class of weak separation functions (4.3.1), where (4.3.1a) is replaced by

$$w(u,v;\delta,\gamma) := \theta(u;\delta) + \underline{w}(v;\gamma), \quad \delta \in \Delta, \gamma \in \Gamma, \tag{4.10.13}$$

where Δ is a suitable set of parameters, $\theta : \mathbb{R} \times \Delta \to \mathbb{R}$, and Γ as in (4.3.1). The function θ might be important in simplifying the optimality conditions (Chapter 5). As an instance, consider the following elementary example.

Example 4.10.1. In (1.1.1) set $m = n = 1, p = 0, X = \mathbb{R}, f(x) = \tan^{-1}x, g(x) = x - 1$. On X, f is not convex and hence $f_{\overline{x}}$ of (3.2.1) is not concave. Therefore, to prove (3.2.2) we cannot expect to use (4.3.1a). Thus, we try to use a nonlinear θ by setting $\theta(u; \delta) = 1 - e^{-\delta u}, \delta > 0$. Hence, for $\overline{x} = 1$, we have $f_{\overline{x}}(x) = \pi/4 - \tan^{-1}x$ and thus $\theta(f_{\overline{x}}(x); \delta) = 1 - e^{-\delta(\frac{\pi}{4} - \tan^{-1}x)}$; this function is concave for $\delta > 1$. Therefore, for $\delta > 1$, (4.10.13) allows us to go on as if f were convex. We have chosen $\overline{x}$ as the m.p.; however, note that the above result is independent of the value of $f(\overline{x})$. $\qquad\square$

The function θ, adopted in the above example, enjoys the following property: $\forall \delta \in \Delta, \theta(u; \delta) \gtreqless 0$ according to, respectively, $u \gtreqless 0$. Therefore, $f_{\overline{x}}(x) > 0$ iff $\theta(f_{\overline{x}}(x); \delta) > 0$, and then the application of such a θ does not change the solutions of (3.2.1). Hence, it would extremely useful to apply, to the scheme (4.10.13), the results existing in the literature on convexification (here concavification). Since a θ which fulfils such a property can be obviously replaced by $k\theta$ with $k > 0$, then the application of $\theta(u; \delta)$ can be split into 2 stages: in the former we try to concavify $f_{\overline{x}}(x)$; in the latter we use (4.3.1a). A quite similar 2-stage view can be considered also for $\underline{w}$. The above questions extend to all problems and systems introduced in Chapter 1 besides (1.1.1).

23. The development of TA or ST in a discrete space is of extreme importance in order to have a theory for Discrete Optimization — as well as for the Combinatorial one —, which be not a pulling of that for so-called Continuous Optimization. As an instance, let us consider the following system:

$$f(x) \notin \mathbb{Z}, \quad g(x) \in \mathbb{Z}^m, \quad x \in \mathbb{R}^n, \tag{4.10.14}$$

where $f : \mathbb{R}^n \to \mathbb{R}, g : \mathbb{R}^n \to \mathbb{R}^m$. (4.10.14) extends (3.2.1) to the discrete case. The set $\mathcal{H}$ of Sect. 3.2 is now replaced by $\mathcal{H} := \{(u, v) \in \mathbb{R}^{1+m} : u \notin \mathbb{Z}^\ell, v \in \mathbb{Z}^m\}$. Consider the function:

$$w(u, v; \lambda) := u + \langle \lambda, \lfloor v \rfloor \rangle + \delta(v)(\lceil u \rceil - u), \lambda \in \mathbb{Z}^m, \tag{4.10.15}$$

where $\delta(v) = 0$ if $v \in \mathbb{Z}^m$ and $\delta(v) = 1$ if $v \notin \mathbb{Z}^m$. It is easy to show the inclusion:

$$\text{lev}_{\mathbb{R}\backslash\mathbb{Z}}\, w(\bullet, \bullet; \lambda) = \mathcal{H}, \quad \forall \lambda \in \mathbb{Z}^m. \tag{4.10.16}$$

Indeed, from the fact that $(u, v) \in \mathcal{H}$ we draw $u \notin \mathbb{Z}, \delta(v) = 0, \lfloor v \rfloor = v$, and hence $w(u, v; \lambda) = u + \langle \lambda, v \rangle \notin \mathbb{Z}, \forall \lambda \in \mathbb{Z}^m$. Then, the inclusion $\supseteq$ holds for the sides of (4.10.16). Now, let $u \in \mathbb{R}, v \in \mathbb{R}^m$ and $\lambda \in \mathbb{Z}^m$ be s.t. $(u, v) \notin \mathcal{H}$. Therefore, $u \in \mathbb{Z}, v \notin \mathbb{Z}^m, \delta(v) = 1, w(u, v; \lambda) = u + \langle \lambda, \lfloor v \rfloor \rangle \in \mathbb{Z}$, so that: $(u, v) \notin \text{lev}_{\mathbb{R}\backslash\mathbb{Z}}\, w$. Hence, the equality in (4.10.16) follows. According to Sect.4.2, (4.10.15) is both weak and strong separation function and then alternative is guaranteed. A particular form of function δ is:

$$\delta(v) = \left\lceil \frac{\sigma}{1 + \sigma} \right\rceil, \quad \text{with } \sigma := \sum_{i=1}^{m}(v_i - \lfloor v_i \rfloor).$$

Just to have an idea of this completely open area of study, consider the following particular case: set $f(x) = \langle c^o, x \rangle, g_i(x) = \langle c^i, x \rangle, i = 1, ..., m$, where c^i is an n-vector with real entries. In this particular case, (4.10.14), i.e.

$$\langle c^o, x \rangle \notin \mathbb{Z}, \quad \langle c^i, x \rangle \in \mathbb{Z}, \quad i = 1, ..., m, \quad x \in \mathbb{R}^n,$$

is in alternative with the following system: $\exists \lambda \in \mathbb{Z}^m$ s.t.

$$\left\langle c^o + \sum_{i=1}^{m} \lambda_i c^i,\, x \right\rangle \in \mathbb{Z}, \quad \forall x \text{ s.t. } g(x) \in \mathbb{Z}^m, \qquad (4.10.17a)$$

$$\lceil \langle c^o, x \rangle \rceil + \sum_{i=1}^{m} \lambda_i \lfloor \langle c^i, x \rangle \rfloor \in \mathbb{Z}, \quad \forall x \text{ s.t. } g(x) \notin \mathbb{Z}^m. \qquad (4.10.17b)$$

Because of (4.10.16), (4.10.14) is in alternative with the system: $\exists \lambda \in \mathbb{Z}^m$ such that, $w \in \mathbb{Z}, \forall x \in \mathbb{R}^n$; this system becomes : $\exists \lambda \in \mathbb{Z}^m$ s.t.

$$w(\langle c^o, x \rangle, \langle c^1, x \rangle, ..., \langle c^m, x \rangle\,; \lambda) \in \mathbb{Z}, \quad \forall x \in \mathbb{R}^n,$$

and, taking into account (4.10.15), is equivalent to (4.10.17).

When (4.10.17b) is redundant (this can be achieved, if all the entries belong to $\mathbb{Q}$), then (4.10.17) collapses to:

$$c^o + \sum_{i=1}^{m} \lambda_i c^i, = O. \qquad (4.10.18)$$

In fact, in the contrary case, if $\bar{x}$ is s.t. $g(\bar{x}) \in \mathbb{Z}^m$, by a well known property of diophantine systems, the condition

$$\left\langle c^o + \sum_{i=1}^{m} \lambda_i c^i,\, \bar{x} \right\rangle \in \mathbb{Z}$$

is contradicted. The result (4.10.18) is known as *integer Farkas* TA, whose early version is due to Kronecker (1899). At last, note that, if we replace w with $u + \langle \lambda, v \rangle$, then alternative is no longer ensured; however, weak alternative is still guaranteed. In fact, if system (4.10.14) is possible, that is if $\exists\, \bar{x} \in \mathbb{R}^n$ such that $f(\bar{x}) \notin \mathbb{Z}$ and $g(\bar{x}) \in \mathbb{Z}^m$, then $f(\bar{x}) + \langle \lambda, g(\bar{x}) \rangle \notin \mathbb{Z}, \forall \lambda \in \mathbb{Z}^m$.

24. As it will be seen in Vol. 2, the class of separation functions of type (4.3.7) is of fundamental importance in the theory of duality for constrained extremum problems. One of the reason is that each constraint can be associated with an addend of (4.3.7), which can play the role of multiplier (Chapter 5). If (1.1.1) belongs to a certain class (for instance, that of $f, -g$ convex and differentiable), can we expect that (3.2.2) be proved by means of a separation function of type (4.3.7) belonging to the same class?

References

[1] Aleksandrow P.S., "Combinatorial Topology".Graylock Press, Albany, N.Y., 1960.

[2] Banach S., "Sur les lignes rectifiables et les surfaces dont l'aire est finie". Fundam. Math., Vol.7, 1925, pp.225-237.

[3] Bazaraa M.S., "A Theorem of the Alternative with Applications to Convex Programming: Optimality, Duality, and Stability". Jou.of Mathem. Analysis and Appls., Vol.41, 1973, pp.701-715.

[4] Bigi G. and Pappalardo M., "Regularity Conditions for the Linear Separation of Sets". Jou.Optimiz.Th.Appls., Vol.99, No.2,1998, pp.533-540.

[5] Castellani G. and Giannessi F., "Decomposition of Mathematical Programs by means of Theorems of the Alternative for Linear and Nonlinear Systems". Proceedings of the 9th Inter. Symposium on Mathematical Programming, Hungarian Academy of Sciences, Budapest, 1979, pp.423-439.

[6] Craven B.D., Gwinner G. and Jeyakumar V., "Nonconvex Theorems of the Alternative and Minimization". Optimization, Vol.18, No.2, 1987, pp.151-163.

[7] Craven B.D. and Koliha J.J., "Generalizations of Farkas Theorem". SIAM Jou. Mathem. Analysis, Vol.8, No.6, 1977, pp.983-997.

[8] Craven B.D. and Mond B., "Transposition theorems for cone-convex functions". SIAM Jou.Appl.Mathem., Vol.24, No.4, 1973, pp.603-612.

[9] Dax A. and Sreedharan V.P., "Theorems of the Alternative and Duality". Jou. Optimiz. Th. Appls., Vol.94, No.3, 1997, pp.561-590.

[10] Dinh The Luc, "Theorems of the Alternative and their applications in Multiobjective Optimization". Acta Math.Hungarica, Vol.45, No.3-4, 1985, pp.311-320.

[11] Farkas J., "Über die Theorie der einfachen Ungleichungen". Jou. Reine Angew. Mathem., Vol. 124, 1902, pp.1-27.

[12] Ferrero O., "Theorems of the Alternative for Set-Valued Functions in Infinite-Dimensional Spaces". Optimization, Vol.20, No.2, 1989, pp.167-175.

[13] Giannessi F., "Theorems of the Alternative, Quadratic programs, and Complementarity Problems". In [I 13], pp.151-186.

[14] Giannessi F., "Theorems of the Alternative and Optimality Conditions". Tech. Report No.83, Dept. of Mathem., Univ.of Pisa, Sect. of Optimization, 1982, pp.1-30. Published with the same title in Jou. Optimiz. Th. Appls., Vol.42, No. 3, 1984, pp.331-365.

[15] Giannessi F., "Theorems of the Alternative for multifunctions with applications to Optimization. Necessary conditions". Tech. Paper No.131, Optimiz. Series, Dept. of Mathem., Univ. of Pisa, Pisa, Italy, 1986, pp.1-127.

[16] Giannessi F., "Theorems of the Alternative for Multifunctions with Applications to Optimization: General Results". Jou.Optimiz.Th.Appls., Vol. 55, No.2, 1987, pp.233256.

[17] Giannessi F., "Theorems of the Alternative and Optimization". In [V 26], Vol.V, pp.437-444.

[18] Golikov A.I. and Evtushenko Yu.G., "Theorems of the Alternative and their Applications in Numerical Methods". Computational Mathematics and Mathematical Physics, Vol.43, No.3, 2003, pp.338-358.

[19] Gordan P., "Über die Auflösungen linearer Gleichungen mit reelen Coefficienten". Mathematishe Annalen, Vol.6, 1873, pp.23-28.

[20] Hahn H., "Über lineare Gleichungen in lineare Räumen". Jou.Mathem., Vol.157, 1927, pp.214-229.

[21] Heinecke G. and Oettli W., "A Nonlinear Theorem of the Alternative without Regularity Assumption". Jou.of Mathem.Analysis and Appls., Vol.146, No.2, 1990, pp.580-590.

[22] Illés T, and Kassay G., "Farkas Type Theorems for Generalized Convexities". Report No.94-23 of Tech.Univ.Delft, Fac.of Tech.Mathem.and Informatics, 1994, pp.1-12.

[23] Jeyakumar V., "Convexlike Alternative Theorems and Mathematical Programming". Optimization, Vol.16, No.5, 1895, pp.643-652.

[24] Jeyakumar V., "A generalization of a minimax theorem of Fan via a theorem of the alternative". Jou. of Optimiz. Theory and Appls., Vol.48, No.3, 1986, pp.525-533.

[25] Jeyakumar V., "A General Farkas Lemma and Characterization of Optimality for a Nonsmooth Program involving Convex Processes". Jou.Optimiz.Th.Appls., Vol.55, No.3, 1987, pp.449-461.

[26] Lehmann R. and Oettli W., "The Theorem of the Alternative: Key-Theorem, and the Vector Maximum Problem". Mathematical Programming, Vol.8, 1975, pp.332-344.

[27] Li Z., "A Theorem of the Alternative and its Applications to the Optimization of Set-Valued Maps". Jou.Optimiz.Th.Appls., Vol.100, No.2, 1999, pp.365-375.

[28] MacLinden L., "Duality Theorems and Theorems of the Alternative". Proc.of Annals of Mathem.Soc., Vol.53, 1975, pp.172-175.

[29] Mangasarian O.L., "A stable Theorem of the Alternative: an extension of the Gordan Theorem". Linear Algebra and its Appls., Vol.41, 1981, pp.209-223.

[30] Martinez-Legaz J.E. and Seeger A., "Yuan's Alternative Theorem and the Maximization of the Minimum Eigenvalue Function". Jou.Optimiz.Th.Appls., Vol.82, No.1, 1994, pp.159-167.

[31] Mastroeni G. and Pappalardo M., "Separation and regularity in the Image Space". In "New Trends in Mathematical Programming", Series in Applied Optimization, Vol.13, Kluwer, Dordrecht, 1998, pp.181-190.

[32] Mastroeni G. and Pellegrini L., "Linear separation for G-semidifferentiable problems". Proceedings of the Conference "Convessitá e Calcolo Parallelo (Convexity and Parallel Computation)". G.Giorgi and F.Rossi Eds., Publisher Univ.of Verona, Via dell'Artigliere,19-Verona-Italy, 1997, pp.187-203.

[33] Mazzoleni P., "Some generalizations of the Theorem of the Alternative for functions and multifunctions". Proc.of the Dept.of Applied Mathem. of University of Venice, Vol.XIX, 1982, 39-51.

[34] Motzkin T.S., "Beiträge zur Theirie der Linearen Ungleichungen". Inaugural Diss. Basel, Jerusalem, 1936.

[35] Nehse von R., "Some General Separation Theorems". Mathem.Nachr., Vol.($, 1978, pp.319-327.

[36] Oettli W., "A new version of th Hahn-Banach Theorem". Proc.of the Int.Congress on Mathematical Programming, April 1981 (Rio de Janeiro), North-Holland, 1984, pp.289-295.

[37] Prager W., "Optimal arrangement of the beams of a rectangular grillage". In "Problemi attuali di Meccanica teorica e applicata (Present problems of theorical and applied Mechanics)", Proceedings of the Int.Confenence in Memory of M.Panetti, Published by Academy of Sciences of Turin, Turin, 1977, pp.239-249.

[38] Simons S., "Variational Inequalities via the Hahn-Banach Theorem". Archiv der Mathematik, Vol.31, Fasc.5, 1978, pp.482-490.

[39] Simons S., "Minimax and Variational Inequalities. Are they of Fixed-Point or Hahn-Banach type?". In "Game Theory and Mathematical Economics", O.Moeschin and D.Pallaschke Eds., North-Holland, 1981, pp.379-387.

[40] Slater M.L., "A Note on Motzkin's Transposition Theorem". Econometrica, Vol.19, 1951, pp.185-186.

[41] Stiemke E., "Über positive Lösungen homogener linearer Gleichungen". Mathematische Annalen, Vol.76, 1915, pp.340-342.

[42] Tricomi F.G., "Integral Equations". Interscience, 1957.

[43] Yang X.M., "Alternative Theorems and Optimality Conditions with weakened convexity". OPSEARCH, Vol.29, No.2, 1992, pp.125-135.

[44] Yang X.M., Yang X.Q. and Chen G.-Y., "Theorems of the Alternative and optimization with Set-Valued Maps". Jou.Optimiz.Th.Appls., Vol.107, No.3, 2000, pp.627-640.

[45] Zalinescu C., "A generalization of the Farkas Lemma and Applications to Convex Programming". Jou.Mathem.Analysis Appls., Vol.66, No.3, 1978, pp.651-678.

[46] Zalmai G.J., "A transposition Theorem with Applications to Constrained Optimal Control Problems". Optimization, Vol.20,No 3, Academic-Verlag, Berlin, 1989, pp.265-279.

CHAPTER 5. OPTIMALITY CONDITIONS.
PRELIMINARY RESULTS

5.1. Introduction

The study of optimality conditions is a very old one. In [I59] and in [I22] we find two
of the first attempts to set up a theory of maxima and minima: we are around 1640
and Torricelli had given an elegant method for solving a minimization problem posed
by Fermat. Sufficient conditions and necessary ones are the main optimality conditions.
The former aim to say whether or not a given point $\bar{x}$ is a local or global m.p. of (1.1.1)
or (1.1.4)-(1.1.6). The latter search for a (as small as possible) subset of the feasible
region R of (1.1.1), which contains the set of local or global m.p. of (1.1.1); when
this subset enjoys suitable properties, then its elements are called stationary or critical
points of (1.1.1). A sufficient condition can be used to see whether or not a stationary
point is a m.p. .

The most classic and famous optimality conditions are the necessary ones established
by Euler and Lagrange in the first half of 18th century [I6, I34, I39, II18, II45, III24,
III25, 1, 57, 105]. The so-called Euler equation turned out to be a corner stone for
building the Calculus of Variations. The method of multipliers, developed by Lagrange,
represented a revolution for finding constrained extrema, and till now is a fundamental
approach. At that time, before Lagrange ideas, the way of writing a necessary condition
for a smooth problem with bilateral constraints ($p = m$) in a Euclidean space consisted
in trying to obtain, from (1.1.1b), m variables as functions of the remaining $n - m$, to
eliminate them from $f(x)$, and then to equate to zero the gradient of the restriction
of f. This method has at least two drawbacks: in general, the above elimination is
impossible; even if possible, it implies to find **all** the solutions of (1.1.1b), namely to solve
it **analytically**. The Lagrange approach had two enormous advantages: to postpone
the resolution of (1.1.1b) to the writing of a necessary condition and, consequently, to
reduce one to solve (1.1.1b) **numerically**, in the sense that, in general, it is necessary
to find only a few solutions of it, often only one. The Lagrange Method of Multipliers
(for short, LMM) has been and is one of the most popular mathematical tools. It has
been used in many fields of Science, has received several generalizations, has suggested

many important new theories — like, for instance, Penalization Theory of Courant (see Vol. 2) —, has been treated in a huge number of books and papers.

Lagrange conceived his method for studying the equilibrium of systems; in proposing his method, he wrote "Méthode trés-simple de trouver lés equations nécessaires pour l'équilibre d'un systéme quelconque de corps regardés comme des points, ou comme des masses finies, et tirés par des puissances données" (see J.-L. Lagrange, "Mécanique Analytique". Éditions Jacques Gabay, Sceaux et Paris, 1989, p.44). Due to the strict relationship between the equilibrium of a system and the extremum of a functional (like energy), the application of his method to the constrained extremum problems was a straightforward consequence.

Recently, the introduction of IS Analysis has shown that the IS is the natural environment for introducing the LMM. Unlike the classic way, here we introduce the LMM through the IS in a more general form. This way, it turns out that separation or alternative arguments (which came much later than Lagrange) can be considered as a "root" of the Theory of Extrema of Lagrange type. Since separation or alternative can be split into two aspects (see Sect. 4.2), then also the theory of constrained extrema can be split into two aspects. This is outlined in Fig.5.1.1 where, in the same column, we find some theories which — as we will see — are substantially equivalent and differ

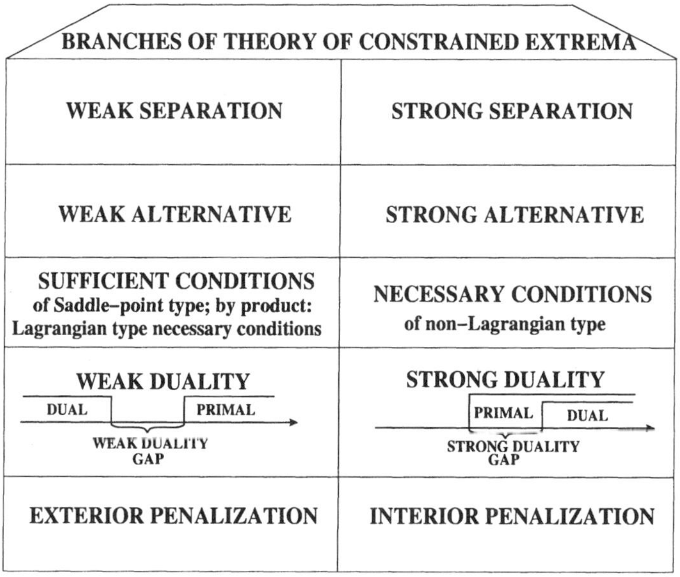

BRANCHES OF THEORY OF CONSTRAINED EXTREMA	
WEAK SEPARATION	**STRONG SEPARATION**
WEAK ALTERNATIVE	**STRONG ALTERNATIVE**
SUFFICIENT CONDITIONS of Saddle–point type; by product: Lagrangian type necessary conditions	**NECESSARY CONDITIONS** of non–Lagrangian type
WEAK DUALITY DUAL — PRIMAL WEAK DUALITY GAP	**STRONG DUALITY** PRIMAL — DUAL STRONG DUALITY GAP
EXTERIOR PENALIZATION	**INTERIOR PENALIZATION**

Fig. 5.1.1

in the language only. This has been shown in Sect. 4.2 as concerns separation and alternative. It will be shown that saddle-point type sufficient conditions express the same substantial fact as weak separation; therefore they must be located in the left column of Fig. 5.1.1. Hence, because of a "sort of symmetry", we might expect to find necessary conditions in corrispondence of strong separation; this in fact happens, but these conditions are not of classic Lagrangian type. Unexpectedly, Lagrangian-type necessary conditions do not appear in the box of necessary conditions on the right column of Fig. 5.1.1; while, they appear as a by-product of weak separation: indeed, the classic Lagrangian necessary conditions are obtained from the Lagrangian function which — to within an obvious transformation — is nothing more than a weak separation function. The above symmetry continues to hold in going to duality and penalization. With regard to the latter, the existing theory is already split into two distinct parts — namely, exterior and interior penalization — , which perfectly correspond to weak and strong separation. As concerns the former, the situation is entangled. In fact, the existing theory has been derived from the classic Lagrangian function and therefore must be located in the left column of Fig. 5.1.1; as a consequence, the related classic terminology of weak or strong duality, to mean that the duality gap is, respectively, non-negative or zero, becomes in contrast with that of weak and strong alternative and separation. Hence, one of the two terminologies must be changed. Even if the change of a well established tradition is always troublesome, since the embedding of duality in the scheme of Fig. 5.1.1 is fundamental for having a general and uniform theory and, hence, for further developments, we propose to call weak duality the classic duality, as coming from weak alternative or separation; with regard to its duality gap, there is no need of an attribute and hence we can simply use the natural terms: weak positive or non-negative or zero duality gap according to respectively the gap is positive, or weak non-negative, or weak zero. All this with respect to the duality in the left column of Fig. 5.1.1. By simmetry reasons, we may expect to find a duality corresponding to strong alternative or separation. This indeed occurs, so that the right column of Fig. 5.1.1 shows strong duality; its gap will be called strong positive, or strong non-negative, or strong zero duality gap.

5.2. Weak Separation and Sufficient Condition

Let us consider problems (1.1.1) and (1.1.4). It will be shown that weak separation leads in a straightforward way to a sufficient condition. To this aim, consider the class $W_R(\Pi)$ of regular weak separation functions (see Definition 4.2.2.). Since (1.1.1) and (1.1.4) are scalar problems —, i.e., f is scalar — then (4.2.7) shrinks to (4.2.8) ($\ell = 1, C = [0, +\infty[$). A generic element of $W_R(\Pi)$ is denoted by $w(\bullet; \pi)$ with $\pi \in \Pi$; hence W_R is described by letting π run in Π. Hence, any subclass of $W_R(\Pi)$ can be identified by a subset of Π. Set $\bar{u} := f_x(\bar{x}) = 0$, $\bar{v} := g(\bar{x})$ (Sect. 3.2).

Proposition 5.2.1. Let the class $W_R(\overline{\Pi}) \subseteq W_R(\Pi)$ and $\bar{x} \in R$ be a given. If $\exists \bar{\pi} \in \overline{\Pi}$, such that:

$$w(u, v; \pi) \leq 0, \qquad \forall (u, v) \in \mathcal{K}_{\bar{x}}, \tag{5.2.1}$$

then $(\bar{u}, \bar{v})$ is a global maximum point of (3.2.3) and, hence, $\bar{x}$ is a global minimum point of (1.1.1) or (1.1.4) or (1.1.6) with $\mathcal{B} = \mathbb{R}$.

Proof. $(5.2.1) \Rightarrow (3.2.2)$ – and hence the thesis –. In fact, since $w(\bullet; \overline{\pi})$ fulfils (4.2.8), we have that $(\hat{u}, \hat{v}) \in \mathcal{H} \Rightarrow w(\hat{u}, \hat{v}; \pi) > 0$, $\forall \pi \in \overline{\Pi}$; this would be contradicted by the absurd assumption $(\hat{u}, \hat{v}) \in \mathcal{K}_{\overline{x}}$ which, because of (5.2.1), implies $w(\hat{u}, \hat{v}; \overline{\pi}) \leq 0$. $\square$

Note that, without any further assumption, we cannot replace, in the above proposition, either $\overline{x} \in R$ with $\overline{x} \in X$ (as Examples 3.2.1 and 3.2.2 show), or $\mathcal{W}_R(\Pi)$) with $\mathcal{W}(\Pi)$ since "$(5.2.1) \Rightarrow (3.2.2)$" does not hold for $\mathcal{W}(\Pi)$. However, *if the assumption is merely $\overline{x} \in X$, then $f(\overline{x})$ is the mimimum of* (1.1.1) *or* (1.1.4) *or* (1.1.6) *with $\mathcal{B} = \mathbb{R}$, even if $\overline{x}$ is not necessarily a m.p.*. The above proposition shows a first general connection between the 1st (or the 2nd) and the 3rd boxes of the left column of Fig. 5.1.1. More specific propositions will be derived here and in the sequel. All these propositions, by themselves, are useful for investigations in the IS; from them it is easy to draw useful statements for the applications in the given space.

Consider the class (4.3.1) with $\theta = 1$, namely the functions $w(u, v; 1, \gamma)$ which satisfy (4.3.1b-e). With a slight abuse of notation, they will be denoted by $w(u, v; \gamma)$:

$$w(u, v; \gamma) = u + \underline{w}(v; \gamma), \quad \gamma \in \Gamma, \tag{5.2.2}$$

subject to (4.3.1b-e).

Proposition 5.2.2. Let a class of regular weak separation functions (5.2.2) and $\overline{x} \in R$ be given. If $\exists \overline{\gamma} \in \Gamma$, such that:

$$u + \underline{w}(v; \overline{\gamma}) \leq 0, \quad \forall (u, v) \in \mathcal{K}_{\overline{x}}, \tag{5.2.3}$$

then $(\overline{u}, \overline{v})$ is a global maximum point of (3.2.3) and, hence, $\overline{x}$ is a global minimum point of (1.1.1) or (1.1.4) or (1.1.6) with $\mathcal{B} = \mathbb{R}$.

Proof. It is enough to note that, because of Proposition 4.3.3, the class (4.2.2) is a subclass of $\mathcal{W}_R(\Pi)$, and then to apply Proposition 5.2.1. $\square$

As well as the class (4.3.1) has been particularized to (4.3.2), it is suitable to consider the following particular case of (5.2.2) (again with a slight abuse of notation with respect to (4.3.2)); namely, the class of functions:

$$w(u, v; \gamma) = u + \sum_{i \in \mathcal{I}} \underline{w}_i(v_i; \gamma_i), \quad \gamma_i \in \Gamma_i, \, i \in \mathcal{I}, \tag{5.2.4}$$

subject to (4.3.1b-e) with $\Gamma = \underset{i \in \mathcal{I}}{\times} \Gamma_i$.

Proposition 5.2.3. Let a class of separable regular weak separation functions (5.2.4) and $\overline{x} \in R$ be given. If $\exists \overline{\gamma}_i \in \Gamma_i$, $i \in \mathcal{I}$, such that:

$$u + \sum_{i \in \mathcal{I}} \underline{w}_i(v_i; \overline{\gamma}_i) \leq 0, \quad \forall (u, v) \in \mathcal{K}_{\overline{x}} \tag{5.2.5}$$

then $(\overline{u}, \overline{v})$ is a global maximum point of (3.2.3) and, hence, $\overline{x}$ is a global minimum point of (1.1.1) or (1.1.4) or (1.1.6) with $\mathcal{B} = \mathbb{R}$.

Proof. It is enough to observe that the class (5.2.4) is contained in the class (5.2.2), and then to apply Proposition 5.2.2. $\qquad\square$

The class (5.2.4) is very important for the applications. A subclass of it, which covers most of the applications, will now be set up. To this aim, consider the vectors $\lambda = (\lambda_1, ..., \lambda_m)$, $\mu = (\mu_1, ..., \mu_m)$, and set $\gamma_i = (\lambda_i, \mu_i), i \in \mathfrak{I}, \Gamma_i = \mathbb{R} \times \mathbb{R}_+$ if $i \in \mathfrak{I}^0$ and $\Gamma_i = \mathbb{R}_+ \times \mathbb{R}_+$ if $i \in \mathfrak{I}^+$. Therefore $\underline{w}_i$ of (5.2.4) is particularized as in (4.3.3), and (again with a slight abuse of notation with respect to (4.3.4a)) we consider the functions:

$$w(u.v; \lambda, \mu) = u + \sum_{i \in \mathfrak{I}} \underline{w}_i^{pe}(v_i; \lambda_i, \mu_i), \qquad \lambda \in D^*, \mu \in \mathbb{R}_+^*, \tag{5.2.6}$$

$\underline{w}_i^{pe}$ being given by (4.3.3) where the 1st term can be replaced by (4.3.3)$''$ if (4.3.3)$'$ holds.

Proposition 5.2.4. Let $\overline{x} \in R$ be given. If $\exists \overline{\lambda} \in D^*$ and $\overline{\mu} \in \mathbb{R}_+^m$, such that:

$$u + \sum_{i \in \mathfrak{I}} \underline{w}_i^{pe}(v_i; \lambda_i, \mu_i) \leq 0, \qquad \forall(u, v) \in \mathcal{K}_{\overline{x}}, \tag{5.2.7}$$

$\underline{w}_i^{pe}$ being given by (4.3.3), where the 1st term can be replaced by (4.3.3)$''$ if (4.3.3)$'$ holds, then $(\overline{u}, \overline{v})$ is a global maximum point of (3.2.3) and, hence, $\overline{x}$ is a global minimum point of (1.1.1) or (1.1.4) or (1.1.6) with $\mathcal{B} = \mathbb{R}$.

Proof. It is enough to observe that, because of Proposition 4.3.4, the class (5.2.6) is a subclass of $\mathcal{W}_R(\Omega)$, and then to apply Proposition 5.2.1. $\qquad\square$

At $\mu = 0$, the class (5.2.6) collapses to that of linear ones or to the class (4.3.5) with $\theta = 1$, and Proposition 5.2.4 to:

Proposition 5.2.5. Let $\overline{x} \in R$. If $\exists \overline{\lambda} \in D^*$, such that:

$$u + \langle \overline{\lambda}, v \rangle \leq 0, \qquad \forall(u, v) \in \mathcal{K}_{\overline{x}}, \tag{5.2.8}$$

then $(\overline{u}, \overline{v})$ is a global maximum point of (3.2.3) and, hence, $\overline{x}$ is a global minimum point of (1.1.1) or (1.1.4) or (1.1.6) with $\mathcal{B} = \mathbb{R}$.

The preceding propositions are stated in terms of the IS and, as above said, aim to support IS Analysis more than to give tools for the applications. These are the general lines and purposes of such an analysis: to achieve results in the IS, where often things are more "regular" and "general" than in the given space; once this has been accomplished, then of course the results must be "translated" in terms of given space. The first trivial translation of the previous propositions is immediately done by replacing the image variable with its meaning in terms of x (see Sect. 3.2): $(u, v) = (f(\overline{x}) - f(x), g(x))$. Therefore, the preceding five propositions are trivially equivalent, respectively, to the following ones.

Proposition 5.2.6. Let the class $\mathcal{W}_R(\overline{\Pi}) \subseteq \mathcal{W}_R(\Pi)$ and $\overline{x} \in \mathbb{R}$ be given. If $\exists \overline{\pi} \in \overline{\Pi}$, such that:

$$w(f(\overline{x}) - f(x), g(x); \overline{\pi}) \leq 0, \quad \forall x \in X, \tag{5.2.9}$$

then $\overline{x}$ is a global m.p. of (1.1.1) or (1.1.4) or (1.1.6) with $\mathcal{B} = \mathbb{R}$.

Let us consider the functions $\mathcal{L}^w : B \times \Gamma \to \mathbb{R}$, $L^w : B \times \Gamma_1 \times ... \times \Gamma_m \to \mathbb{R}$, $\mathcal{L} : B \times D^* \times \mathbb{R}_+^m \to \mathbb{R}$ and $L : B \times D^* \to \mathbb{R}$, given by:

$$\mathcal{L}^w(x; \gamma) := f(x) - \underline{w}(g(x); \gamma); \tag{5.2.10a}$$

$$L^w(x; \gamma) := f(x) - \sum_{i \in \mathcal{I}} \underline{w}_i(g_i(x); \gamma_i); \tag{5.2.10b}$$

$$\mathcal{L}(x; \lambda, \mu) := f(x) - \sum_{i \in \mathcal{I}} \underline{w}_i^{pe}(g_i(x); \lambda_i, \mu_i); \tag{5.2.10c}$$

$$L(x; \lambda) := f(x) - \langle \lambda, g(x) \rangle. \tag{5.2.10d}$$

$\underline{w}_i^{pe}$ being given by (4.3.3) where the 1st term can be replaced by (4.3.3)'' if (4.3.3)' holds. In (5.2.10c), $\underline{w}_i^{pe}$ contains some alternative forms; a short comment about them is just after (4.3.4). At $\mu = 0$, one of the alternative forms of (5.2.10c) recovers the linear one; namely, $\mathcal{L}(x; \lambda, 0) = L(x; \lambda)$. To avoid cumbersome symbols, with a slight abuse of notation but wthout any fear of confusion, the classic Lagrangian function (5.2.10d) is denoted with the same symbol as the extended one, which appears in (3.3.8); the present $L(x; \lambda)$ corresponds to $L(x; 1, \lambda)$ of Sect. 3.3.

Proposition 5.2.7. Let a class of regular weak separation functions (5.2.2), and let **(i)** $\overline{x} \in R$ be given. If **(ii)** $\exists \overline{\gamma} \in \Gamma$, such that:

$$\mathcal{L}^w(x; \overline{\gamma}) \geq f(\overline{x}), \quad \forall x \in X, \tag{5.2.11}$$

then $\overline{x}$ is a global minimum point of (1.1.1) or (1.1.4) or (1.1.6) with $\mathcal{B} = \mathbb{R}$.

Proposition 5.2.8. Let a class of separable regular weak separation functions (5.2.4) and $\overline{x} \in R$ be given. If $\exists \overline{\gamma}_i \in \Gamma_i$, $i \in \mathcal{I}$, such that:

$$L^w(x; \overline{\gamma}_1, ..., \overline{\gamma}_m) \geq f(\overline{x}), \quad \forall x \in X, \tag{5.2.12}$$

then $\overline{x}$ is a global minimum point of (1.1.1) or (1.1.4) or (1.1.6) with $\mathcal{B} = \mathbb{R}$. .

Proposition 5.2.9. Let $\overline{x} \in R$ be given. If $\exists \overline{\lambda} \in D^*$ and $\exists \overline{\mu} \in \mathbb{R}_+^m$, such that:

$$\mathcal{L}(x; \overline{\lambda}, \overline{\mu}) \geq f(\overline{x}), \quad \forall x \in X, \tag{5.2.13}$$

then $\overline{x}$ is a global minimum point of (1.1.1) or (1.1.4) or (1.1.6) with $\mathcal{B} = \mathbb{R}$.

Proposition 5.2.10. Let $\overline{x} \in R$. If $\exists \overline{\lambda} \in D^*$, such that:

$$L(x; \overline{\lambda}) \geq f(\overline{x}), \quad \forall x \in X, \tag{5.2.14}$$

then $\overline{x}$ is a global minimum point of (1.1.1) or (1.1.4) or (1.1.6) with $\mathcal{B} = \mathbb{R}$.

If in the previous 4 propositions the assumption "$\overline{x} \in R$" is replaced by the weaker one "$x \in X$", then $\overline{x}$ is not necessarily a m.p. of the given problem; however $f(\overline{x})$ is the minimum (see the remark after Proposition 5.2.1.)

The previous propositions deal with global m.p.. Of course, the corresponding propositions for a local m.p. are obtained by intersecting the domain with a neighbourhood of the point; a warning is suitable: as neighbourhood of a point $(\overline{u}, \overline{v})$ of the IS (image of $\overline{x} \in B$) we can consider either an open set of IS containing it or the image of an open set of B containing $\overline{x}$; they may be different.

Function (5.2.10d) is the classic *Lagrangian function*. (5.2.10c) is a generalization of it and can be called *parabolic-exponential Lagrangian function*. (5.2.10a) is a further generalization of the Lagrangian function, which is closer to separation arguments than to Lagrange ideas; however, it is useful both for proving general statements and to understand the mathematical structure underlying Lagrange theory. Indeed, it will now be used to prove an equivalence statement.

Proposition 5.2.11. Let a class of regular weak separation functions (5.2.2) be given.(i)-(ii) of Proposition 5.2.7 hold if and only if **(i)** $\overline{x} \in X$, and **(ii)** $\exists \overline{\gamma} \in \Gamma$ such that:

$$\mathcal{L}^w(\overline{x}; \gamma) \leq \mathcal{L}^w(\overline{x}; \overline{\gamma}) \leq \mathcal{L}^w(x; \overline{\gamma}), \quad \forall x \in X, \quad \forall \gamma \in \Gamma. \tag{5.2.15}$$

Proof. Since the class (5.2.2) is the class (4.3.1) at $\theta = 1$, we have:

$$g(\overline{x}) \in D \quad \Leftrightarrow \quad \underline{w}(g(\overline{x}); \gamma) \geq 0, \quad \forall \gamma \in \Gamma. \tag{5.2.16}$$

If. First of all, we show that the 1st of inequalities (5.2.15), which is equivalent to

$$\underline{w}(g(\overline{x}); \gamma) \geq \underline{w}(g(\overline{x}); \overline{\gamma}), \quad \forall \gamma \in \Gamma, \tag{5.2.17}$$

implies $g(\overline{x}) \in D$. In fact, if, ab absurdo, we suppose that $g(\overline{x}) \notin D$, then (4.3.1e) $\Rightarrow$ $\exists \tilde{\gamma} \in \Gamma$ s.t. $\underline{w}(g(\overline{x}); \tilde{\gamma}) < 0$; consequently, since (4.3.1d) implies that, $\forall \alpha > 0, \exists \tilde{\gamma}_\alpha \in \Gamma$ s.t. $\alpha \underline{w}(g(\overline{x}); \tilde{\gamma}) = \underline{w}(g(\overline{x}); \tilde{\gamma}_\alpha)$, then we draw that $\underline{w}(g(\overline{x}); \tilde{\gamma}_\alpha) \to -\infty$ as $\alpha \to +\infty$ which contradicts (5.2.17). Now, we want to prove that the inequality of (5.2.16) (which is true since we have achieved $g(\overline{x}) \in D$) holds as equality at $\gamma = \overline{\gamma}$. Ab absurdo, suppose that $\underline{w}(g(\overline{x}); \overline{\gamma}) > 0$. Then, because of (4.3.1d), we have that, $\forall \alpha > 0, \exists \gamma_\alpha \in \Gamma$ s.t. $\underline{w}(g(\overline{x}); \gamma_\alpha) = \alpha \underline{w}(g(\overline{x}); \overline{\gamma})$. Therefore, if $\alpha < 1$, we draw:

$$\underline{w}(g(\overline{x}); \gamma_\alpha) < \underline{w}(g(\overline{x}); \overline{\gamma}),$$

which contradicts (5.2.17). Hence, we have:

$$\underline{w}(g(\overline{x}); \overline{\gamma}) = 0. \tag{5.2.18}$$

Because of this equality, the 2nd of (5.2.15) becomes (5.2.11). **Only if.** (i) of Proposition 5.2.7 implies $g(\overline{x}) \in D$, which, because of (5.2.16) for $\gamma = \overline{\gamma}$, implies $\underline{w}(g(\overline{x}); \overline{\gamma}) \geq 0$. At $x = \overline{x}$, (5.2.11)$\Rightarrow \underline{w}(g(\overline{x}); \overline{\gamma}) \leq 0$. It follows (5.2.18). Hence, (5.2.11) is equivalent to the 2nd of (5.2.15). Being $g(\overline{x}) \in D$, the inequality of (5.2.16) and (5.2.18) imply the 1st of (5.2.15). $\qquad \square$

Definition 5.2.1. Let Y and Z be any sets of two Banach spaces, and $F : Y \times Z \to \mathbb{R}$ any function. $(\overline{y}, \overline{z}) \in Y \times Z$ is called *saddle-point* of F on $Y \times Z$, iff

$$F(\overline{y}, z) \leq F(\overline{y}, \overline{z}) \leq F(y, \overline{z}), \quad \forall y \in Y, \quad \forall z \in Z; \tag{5.2.19}$$

$F(\overline{y}, \overline{z})$ is called *saddle-value* of F on $Y \times Z$.

Example 5.2.1. Set $Y = \mathbb{R}$, $Z = \mathbb{R}_+$, $F(y, z) = y^2 - (z - 2)y + 1$. The 1st of (5.2.19) becomes $\overline{y}(z - \overline{z}) \geq 0$ $\forall z \geq 0$; it implies $\overline{y} \geq 0$ and, consequently, $\overline{z} \geq 0$ if $\overline{y} = 0$ and $\overline{z} = 0$ if $\overline{y} > 0$. The 2nd of (5.2.19) becomes $y^2 - (\overline{z} - 2)y - \overline{y}^2 + \overline{y}\,\overline{z} - 2\overline{y} \geq 0$ $\forall y \in \mathbb{R}$, and is true iff $\Delta := [2\overline{y} - (\overline{z} - 2)]^2 = 0$ or $\overline{y} = (\overline{z} - 2)/2$. This and $\overline{y} \geq 0$ imply $\overline{z} \geq 2$. $\overline{z} > 2$ would imply $\overline{y} > 0$ and hence violate the 1st of (5.2.19). It follows that $(\overline{y} = 0, \overline{z} = 2)$ is the unique saddle-point of F on $\mathbb{R} \times \mathbb{R}_+$. $\qquad\square$

Example 5.2.2. (continuation of Example 1.2.7). As matrix A, set:

$$A = \begin{pmatrix} 1 & 2 & 8 & 7 \\ 3 & 0 & 4 & 2 \\ 2 & 4 & 6 & 3 \end{pmatrix}.$$

It is easily seen that $a_{\overline{i}\,\overline{j}} = a_{23} = 4$ fulfils the last condition of Example 1.2.7, so that $(\overline{y} = (0, 1, 0), \overline{z} = (0, 0, 1, 0))$ is a saddle-point of $F(y, z) = \langle y, Az \rangle$ on $Y \times Z$, these sets being those of Example 1.2.7. a_{23} is the saddle-value of matrix A. If a matrix admits a saddle value and if A is the payoff matrix of a game, then such a game has "pure strategies" namely, a saddle-point of F is a pair of vectors, each having one element equal to 1 and the others equal to 0. $\qquad\square$

The fact that the inequalities in (5.2.19) must be verified $\forall y$ and $\forall z$ leads one to introduce suitable extrema, namely those which appear in next (5.2.20).

Proposition 5.2.12. We have:

$$\sup_{z \in Z} \inf_{y \in Y} F(y, z) \;\leq\; \inf_{y \in Y} \sup_{z \in Z} F(y, z), \tag{5.2.20}$$

whatever the sets Y, Z and the function $F : Y \times Z \to \mathbb{R}$ may be.

Proof. From the obvious inequality $\inf_{y \in Y} F(y, \overline{z}) \leq F(\overline{y}, \overline{z})$ – which holds $\forall \overline{y} \in Y$ and $\forall \overline{z} \in Z$ – we draw $\sup_{z \in Z} \inf_{y \in Y} F(y, z) \leq \sup_{z \in Z} F(\overline{y}, z)$, $\forall \overline{y} \in Y$, and consequently we achieve (5.2.20). $\qquad\square$

At first glance, it might seem possible to state the equivalence between (5.2.20) and (5.2.19) and hence to reduce a saddle-point to the search for the extrema of (5.2.20). Unfortunately, in the general case, this is fallacious. First of all, note that in (5.2.20) the equality may not occur, as shown by next example.

Example 5.2.3. Set $Y = \mathbb{R}$, $Z = \mathbb{R}_+$, $F(y, z) = (y - 1)^3 - yz + z$. We easily find:

$$\sup_{z \in \mathbb{R}_+} \inf_{y \in \mathbb{R}} F(y, z) = -\infty < \inf_{y \in \mathbb{R}} \sup_{z \in \mathbb{R}_+} f(y, z) = 1. \qquad\square$$

Proposition 5.2.13. If $(\overline{y}, \overline{z})$ is a saddle-point of $F : Y \times Z \to \mathbb{R}$ on $Y \times Z$, then we have:

$$\sup_{z \in Z} \inf_{y \in Y} F(y, z) = F(\overline{y}, \overline{z}) = \inf_{y \in Y} \sup_{z \in Z} F(y, z). \tag{5.2.21}$$

Proof. From (5.2.19) we draw $\sup_{z \in Z} F(\overline{y}, z) \leq F(\overline{y}, \overline{z}) \leq \inf_{y \in Y} F(y, \overline{z})$, and thus:

$$\inf_{y \in Y} \sup_{z \in Z} F(y, z) \leq F(\overline{y}, \overline{z}) \leq \sup_{z \in Z} \inf_{y \in Y} F(y, z).$$

This double inequality and (5.2.20) imply (5.2.21). $\qquad\square$

Without further assumptions, unfortunately, Proposition 5.2.13 cannot be inverted, showing that the relationship between the saddle-point and the extrema in (5.2.20) is one-way-only. In other words, even if the extrema in (5.2.20) are equal, F may have not a saddle-point, as shown by next example.

Example 5.2.4. Set $Y = \mathbb{R}$, $Z = \mathbb{R}_+$, $F(y, z) = e^{-y} - yz$. We find $\sup_{z \in \mathbb{R}_+} \inf_{y \in \mathbb{R}} F(y, z) = \inf_{y \in \mathbb{R}} \sup_{z \in \mathbb{R}_+} F(y, z) = 0$. Notwithstanding this, F has no saddle-points. In fact, the 2nd of (5.2.19) becomes $e^{-y} \geq \overline{z}y + e^{-\overline{y}} - \overline{y}\,\overline{z}$ and, being $\overline{z} \geq 0$, is not true $\forall y \in \mathbb{R}$. $\qquad\square$

The preceding results are summarized up in the flow-chart of Fig. 5.2.1, where $\boxed{\alpha} \to \boxed{\beta}$ means $\alpha \leq \beta$.

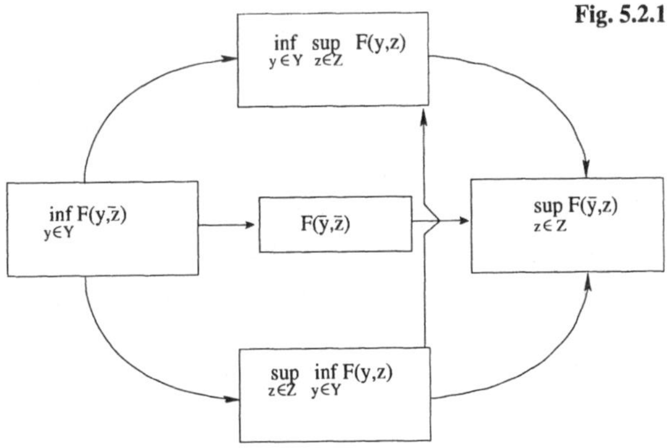

Fig. 5.2.1

The concept of saddle-point expressed by Definition 5.2.1 is not the most general. However, it is sufficient for the present applications to extremum problems. Let us note that condition (5.2.15) is a particular case of (5.2.19). Therefore, Proposition 5.2.11 can be immediately written in terms of a sufficient condition; this is done with Theorem 5.2.1. In the same vein, Theorems 5.2.2, 5.2.3 and 5.2.4 are saddle-point versions of Propositions 5.2.3, 5.2.4 and 5.2.5. All next four theorms are trivial consequences of the corresponding previous propositions.

Theorem 5.2.1. Let a class of regular weak separation functions (5.2.2) and $\overline{x} \in X$ be given. If $\exists \overline{\gamma} \in \Gamma$, such that $(\overline{x}, \overline{\gamma})$ is a saddle-point of $\mathcal{L}^{\omega}$ on $X \times \Gamma$ — or (5.2.15) is fulfilled — , then $\overline{x}$ is a global minimum point of (1.1.1) or (1.1.4) or (1.1.6) with $\mathcal{B} = \mathbb{R}$.

Theorem 5.2.2. Let a class of separable regular weak separation functions (5.2.4) and $\overline{x} \in X$ be given. If $\exists \overline{\gamma}_i \in \Gamma_i, i \in \mathcal{I}$, such that $(\overline{x}, \overline{\gamma}_1, ..., \overline{\gamma}_m)$ is a saddle- point of L^{ω} on

$X \times \Gamma_1 \times ...\Gamma_m$, then $\overline{x}$ is a global minimum point of (1.1.1) or (1.1.4) or (1.1.6) with $\mathcal{B} = \mathbb{R}$.

Theorem 5.2.3. Let $\overline{x} \in X$ be given. If $\exists \overline{\lambda} \in D^*$ and $\exists \overline{\mu} \in \mathbb{R}^m_+$, such that $(\overline{x}, \overline{\lambda}, \overline{\mu})$ is a saddle-point of $\mathcal{L}$ on $X \times D^* \times \mathbb{R}^m_+$, or

$$\mathcal{L}(\overline{x}; \lambda, \mu) \leq \mathcal{L}(\overline{x}; \overline{\lambda}, \overline{\mu}) \leq \mathcal{L}(x; \overline{\lambda}, \overline{\mu}), \quad \forall x \in X, \ \forall \lambda \in D^*, \ \forall \mu \in \mathbb{R}^m_+, \qquad (5.2.22)$$

then $\overline{x}$ is a global minimum point of (1.1.1) or (1.1.4) or (1.1.6) with $\mathcal{B} = \mathbb{R}$.

Theorem 5.2.4. Let $\overline{x} \in X$ be given. If $\exists \overline{\lambda} \in D^*$ such that $(\overline{x}, \overline{\lambda})$ is a saddle-point of L on $X \times D^*$, or

$$L(\overline{x}; \lambda) \leq L(\overline{x}; \overline{\lambda}) \leq L(x; \overline{\lambda}), \quad \forall x \in X, \ \forall \overline{\lambda} \in D^*, \qquad (5.2.23)$$

then $\overline{x}$ is a global minimum point of (1.1.1) or (1.1.4) or (1.1.6) with $\mathcal{B} = \mathbb{R}$.

In Theorems 5.2.1-5.2.4, X can be of course replaced by R itself. In fact, if in (1.1.1) — as well as in (1.1.4) or (1.1.6) with $\mathcal{B} = \mathbb{R}$ — the domain X is replaced by R, then the problem is not modified. Indeed, the scope of the above theorems is precisely that of freeing ourselves from the constraints. However, in some special cases (see the last part of Example 5.2.5) it may be useful to go back to R; this happens when we must choose the lesser of two evils, one of them being "not to free ourselves from the constraints".

Theorem 5.2.4 is a classic result [I45, I12,I143, I145]; indeed, the classic results have come from the linear Lagrangian function. Theorem 5.2.3 can be seen as a natural extension of the classic one to the case of parabolic-exponential Lagrangian function. Theorems 5.2.2 and 5.2.1 are further extensions, which are closer to separation than to Lagrangian ideas. With these four theorems we have achieved explicit relations between the 3rd box and the 1st one in the left-hand side of Fig. 5.1.1.

In order to prove Proposition 5.2.11, (and hence Theorem 5.2.1) we have achieved (5.2.18). This equation is of fundamental importance, and not merely a step within a proof. (5.2.18) is implied by weak separation or by saddle-point of $\mathcal{L}^\omega$. In the cases of Theorems 5.2.2, 5.2.3 and 5.2.4, it becomes, respectively,

$$\sum_{i \in \mathcal{I}} \underline{w}_i(g_i(\overline{x}); \overline{\gamma}_i) = 0, \qquad (5.2.24)$$

$$\sum_{i \in \mathcal{I}} \underline{w}_i^{pe}(g_i(\overline{x}); \overline{\lambda}_i, \overline{\mu}_i) = 0, \qquad (5.2.25)$$

$$\langle \overline{\lambda}, g(\overline{x}) \rangle = 0. \qquad (5.2.26)$$

Since $\overline{x} \in R$, the terms of the left-hand sides of the above equalities, which correspond to $\mathcal{I}^0$ — or to (1.1.1b) or (1.1.4b) — are of course identically zero. Relation (5.2.26) is classic: it says that existence of a saddle-point of classic Lagrangian function — namely L — implies *orthogonality* between the constraining function and the vector λ (called multiplier); we stress that such an orthogonality holds in a very general situation: X may be even a finite set. (5.2.25) and (5.2.24) generalize such an orthogonality relationship. In next section we will meet (5.2.26) — and its generalizations — as a fundamental part of a necessary condition; indeed, in this sense, it has been one of the first achievements in the study of optimality conditions, but under differentiability assumption on (1.1.1) or (1.1.4), and named *complementarity* relationship. Later it has been recognized to be the mathematical model for some equilibrium problems (see Sects. 1.4, 1.5, and Vol.2).

Theorems 5.2.1-5.2.4 have an undoubted theoretical importance; for instance, the classic Theorem 5.2.4 has stimulated the birth of some branches of Mathematics, like Game Theory, and Minimax Theory [32, Vol.III, pp.272-289]; the last has strongly influenced Mathematical Statistics. With regard to their exploitation to test whether or not a given point is local or global m.p. or even to find a m.p., it is easy to note that, in general, the numerical calculations may be heavy. However, as we will see later, such theorems can be used as sources for deriving other sufficient conditions, which are more meaningful from a numerical point view. Now, let us consider some examples.

Example 5.2.5 (continuation of Example 3.4.3). In the family of minimization problems of Example 3.4.3, consider that corresponding to $\xi = \frac{5}{2}$. We will discuss a possible use of Theorems 5.2.4 and 5.2.3. In exploiting them, one can start with any $\overline{x} \in X$; however, when possible, it is obviously convenient to begin with a feasible point, namely with $\overline{x} \in R$. This is now trivially done, due to the simplicity of the example. Then, take for instance the feasible point $\overline{x} = (\frac{3}{4}, 1); \overline{x}$ is a vertex of R, which is now a triangle. The choice of a vertex is not at random: since f is strictly concave and R is a polytope, then the minimum occurs necessarily (at least) at a vertex (see Theorem 2.3.8(ii)). The Lagrangian function (5.2.10d) becomes now:

$$L(x; \lambda) = x_1 + 2x_2 + Q(x_1, x_2) - \lambda \left(2x_1 + x_2 - \frac{5}{2} \right),$$

where $Q(x_1, x_2) := 3x_1(1 - x_1) + 3x_2(1 - x_2)$. Let us try to apply Theorem 5.2.4. Since $g(\overline{x}) = 0$, the former of inequalities (5.2.23) is identically verified, and the latter is equivalent to find $\overline{\lambda} \geq 0$, s.t.

$$Q(x_1, x_2) \geq (2\overline{\lambda} - 1)x_1 + (\overline{\lambda} - 2)x_2 + \frac{53}{16} - \frac{5}{2}\overline{\lambda}, \ \forall x \in X.$$

Since $Q(x_1, x_2) \geq 0$ on (the square) X and $= 0$ on its vertices and only there, then (5.2.27) is satisfied iff $\overline{\lambda}$ is ≥ 0 and s.t. the maximum of its right-hand side on X is ≤ 0. Such a maximum, say $q(\overline{\lambda})$, is easily found to be:

$$q(\overline{\lambda}) = \begin{cases} \frac{53}{16} - \frac{5}{2}\overline{\lambda}, & \text{if } 0 \leq \overline{\lambda} \leq \frac{1}{2}, \quad (max.\,point : x_1 = x_2 = 0), \\ \frac{37}{16} - \frac{1}{2}\overline{\lambda}, & \text{if } \frac{1}{2} < \overline{\lambda} \leq 2, \quad (max.\,point : x_1 = 1, x_2 = 0), \\ \frac{5}{16} + \frac{1}{2}\overline{\lambda}, & \text{if } 2 < \overline{\lambda}, \quad (max.\,point : x_1 = x_2 = 1). \end{cases}$$

Since the minimum of $q(\overline{\lambda})$ is positive, that is $\frac{21}{16}$, then (5.2.23) is not satisfied, and thus Theorem 5.2.4 cannot be applied. This, of course, does not imply that $\overline{x}$ be not global m.p. of the present problem. Indeed, by evaluating f at the (two) adjacent vertices, we easily see that $\overline{x}$ is not even local m.p.; we could have made this observation at the beginning and avoided to test (5.2.23); we did so for a better understanding of a saddle-point condition. Now, we replace the above $\overline{x} = (\frac{3}{4}, 1)$ with the adjacent vertex $(1, \frac{1}{2})$; and then we consider the new point $\overline{x} = (1, \frac{1}{2})$. By replacing the previous steps, in place of the above $q(\overline{\lambda})$, we find now:

$$q(\overline{\lambda}) = \begin{cases} \frac{11}{4} - \frac{5}{2}\overline{\lambda}, & \text{if } 0 \leq \overline{\lambda} \leq \frac{1}{2}, \\ \frac{7}{4} - \frac{1}{2}\overline{\lambda}, & \text{if } \frac{1}{2} < \overline{\lambda} \leq 2, \\ -\frac{1}{4} + \frac{1}{2}\overline{\lambda}, & \text{if } 2 < \overline{\lambda}. \end{cases}$$

Again, the minimum of this $q(\overline{\lambda})$ is positive, namely $\frac{3}{4}$. Hence, here too Theorem 5.2.4 cannot be applied. Now, we try to apply Theorem 5.2.3. In the present case, $(4.3.3)'$ is verified:

$$v^{\text{sup}} := \max\left\{ v \in \mathbb{R} : 2x_1 - \frac{5}{2} \leq v \leq 2x_1 - \frac{3}{2}, \ x_1 \in [0, 1] \right\} = \frac{1}{2} < +\infty.$$

Therefore, among the forms offered by (5.2.10c), we can choose the pure parabolic one; namely, we can adopt $(4.3.3)''$. Since there is no bilateral constraint, (5.2.10c) is now:

$$\mathcal{L}(x_1, x_2; \lambda, \mu) = f(x) - \lambda g(x) + \mu g(x)^2 =$$

$$= x_1 + 2x_2 + Q(x_1, x_2) - \lambda\left(2x_1 + x_2 - \frac{5}{2}\right) + \mu\left(2x_1 + x_2 - \frac{5}{2}\right)^2,$$

with $\lambda, \mu > 0$ and $\frac{\lambda}{\mu} \geq \frac{1}{2}$. Since $g(\overline{x}) = 0$, the former of (5.2.22) is identically verified, and the latter is equivalent to find $\overline{\lambda}, \overline{\mu}$ with $0 < \overline{\mu} \leq 2\overline{\lambda}$, s.t.

$$\mathcal{L}(x_1, x_2; \overline{\lambda}, \overline{\mu}) - \frac{11}{4} \geq 0, \quad \forall(x_1, x_2) \in X. \tag{5.2.28}$$

Whatever $\overline{\mu} > 0$ may be, the Hessian of $\mathcal{L}$ has one positive eigenvalue and a negative one, so that $\mathcal{L}$ is an indefinite form; therefore, problem (5.2.28) is not the most desiderable one and, in the present particular case, is even worse than the given problem (minimization of an indefinite form is, in general, worse than that of a strictly concave one). However, in order to understand the role that separation functions may play, it is useful to solve (5.2.28). To this end, find:

$$-\frac{5}{2} = \min_{x \in X}\left(2x_1 + x_2 - \frac{5}{2}\right), \qquad \frac{1}{2} = \max_{x \in X}\left(2x_1 + x_2 - \frac{5}{2}\right).$$

For $-\frac{5}{2} \leq v \leq \frac{1}{2}$, the global m.p. of problem:

$$\min(x_1 + 2x_2), \qquad \text{s.t. } 2x_1 + x_2 - \frac{5}{2} = v, \qquad (x_1, x_2) \in X,$$

is trivially found to be $(x_1 = 1, x_2 = v + \frac{1}{2})$. Then, (5.2.28) is equivalent to find $\overline{\lambda}, \overline{\mu}$ with $0 < \overline{\mu} \leq 2\overline{\lambda}$, s.t.

$$(\overline{\mu} - 3)v^2 + (2 - \overline{\lambda})v + \frac{11}{4} \geq 0, \qquad \forall v \in \left[-\frac{5}{2}, \frac{1}{2}\right];$$

$\overline{\lambda} = 2$ and $\overline{\mu} = 3$ trivially satisfy it. Hence $\overline{x} = (\overline{x}_1 = 1, \overline{x}_2 = \frac{1}{2})$ is a global m.p. of the given problem at $\xi = \frac{5}{2}$. The effect of the parabolic separation function allows one to (strictly) convexify $\mathcal{L}$ along the lines orthogonal to the constraint (which have (-1, 2) as gradient); this corresponds to the positive eigenvalue of the Hessian of $\mathcal{L}$. While, on the lines parallel to the constraint, identified by $2x_1 + x_2 - \frac{5}{2} = v$, $\mathcal{L}$ is strictly concave, since the parabolic term of the weak separation function is constant on them, so that the "structure" of $\mathcal{L}$ is the same as that of f; this corresponds to the negative eigenvalue of the Hessian of $\mathcal{L}$. These remarks suggest the following device. Since the presence of one constraint allows us to obtain the positivity of one eigenvalue of the Hessian of $\mathcal{L}$, while the eigenvalues are 2 (as the dimension of X), then we lack a 2nd constraint to hope to achieve the positivity of the 2nd eigenvalue too. Then, we adopt the device, which consists in adding a redundant (i.e., satisfied by all the elements of R) constraint to the given problem; this is replaced equivalently by the problem:

$$\min f(x), \text{ s.t. } g_1(x) = 2x_1 + x_2 - \tfrac{5}{2} \geq 0, \ g_2(x) = x_2 - \tfrac{1}{2} \geq 0, \ x \in X. \qquad (5.2.29)$$

With the same parabolic form of (5.2.10c) as above, the generalized Lagrangian function becomes now ($x = (x_1, x_2)$, $\lambda = (\lambda_1, \lambda_2), \mu = (\mu_1, \mu_2)$):

$$\mathcal{L}(x; \lambda, \mu) = f(x) - \lambda_1 g_1(x) + \mu_1 g_1(x)^2 - \lambda_2 g_2(x) + \mu_2 g_2(x)^2 = x_1 + 2x_2 + Q(x_1, x_2) -$$
$$-\lambda_1 \left(2x_1 + x_2 - \tfrac{5}{2}\right) + \mu_1 \left(2x_1 + x_2 - \tfrac{5}{2}\right)^2 - \lambda_2 \left(x_2 - \tfrac{1}{2}\right)^2 + \mu_2 \left(x_2 - \tfrac{1}{2}\right)^2,$$

with $\lambda_i, \mu_i > 0, i = 1, 2$; since the bounds (4.3.3)' may be loose, we disregard them; of course, we must keep in mind this lack. Since $g_1(\overline{x}) = g_2(\overline{x}) = 0$, the former of (5.2.22) is identically verified. Before considering the latter of (5.2.22), let us note that, in (5.2.22), X can be of course replaced by R itself; in fact, if in (1.1.1) $-$ as well as in (1.1.4) $-$ the domain X is replaced by R, then the problem is not modified. Indeed, the scope of Theorems 5.2.1-5.2.4 is precisely that of freeing ourselves from the constraints; however, in some special cases, like the present one, it may be useful to go back to R (i.e., to replace X with R in (5.2.22)); this means that we choose the lesser of two evils (these being "not to free ourselves from the constraints" and "to come up against a nonconvex problem"). Then, the latter of (5.2.22), with R in place of X, is now equivalent to find $\overline{\lambda}_i, \overline{\mu}_i > 0, i = 1, 2$, such that:

$$\mathcal{L}(x_1, x_2; \overline{\lambda}_1, \overline{\lambda}_2, \overline{\mu}_1, \overline{\mu}_2) - \tfrac{11}{4} \geq 0, \quad \forall (x_1, x_2) \in R. \qquad (5.2.30)$$

Due to the equivalence, in the present case, between (5.2.30) and a weak separation, if (5.2.3) is satisfied, then the lack of (4.3.3)' is overcome. Let us now consider the characteristic trinomial of the Hessain of $\mathcal{L}$ and its discriminant, which are, respectively,

$$t^2 = 2(5\overline{\mu}_1 + \overline{\mu}_2 - 6)t + 4(4\overline{\mu}_1\overline{\mu}_2 - 15\overline{\mu}_1 = 3\overline{\mu}_2 + 9), t \in \mathbb{R}; 4[(5\overline{\mu}_1 - \overline{\mu}_2)^2 + 4\overline{\mu}_1\overline{\mu}_2].$$

Therefore, the eigenvalues of the Hessian of $\mathcal{L}$ are both positive, iff

$$\overline{\mu}_1, \overline{\mu}_2 > 0, \quad 5\overline{\mu}_1 + \overline{\mu}_2 - 6 > 0, \quad 4\overline{\mu}_1\overline{\mu}_2 = 15\overline{\mu}_1 - 3\overline{\mu}_2 + 9 > 0.$$

To fulfil these inequalities, we choose $\overline{\mu}_1 = 1, \overline{\mu}_2 = 7$, so that $\mathcal{L}(x_1, x_2; \overline{\lambda}_1, \overline{\lambda}_2, 1, 7)$

turns out to be strictly convex, whatever $\overline{\lambda}_1, \overline{\lambda}_2$ may be. Due to this property, to find the global m.p. of $\mathcal{L}$ over R (in order to see whether or not (5.2.30) is true) is equivalent to find a stationary point of $\mathcal{L}$ over R. To this end, we consider the gradient of $\mathcal{L}(x_1, x_2; \overline{\lambda}_1, \overline{\lambda}_2, 1, 7)$:

$$\mathcal{L}'_x(x_1, x_2; \overline{\lambda}_1, \overline{\lambda}_2, 1, 7) = (2x_1 + 4x_2 - 2\overline{\lambda}_1 = 6, 4x_1 + 10x_2 - \overline{\lambda}_1 - \overline{\lambda}_2 - 7),$$

and with elementary calculations, we find that, with $\overline{\lambda}_1 = \overline{\lambda}_2 = 1, \mathcal{L}$ is stationary just at $\overline{x} = (1, \frac{1}{2})$ and $\mathcal{L}(1, \frac{1}{2}; 1, 1, 1, 7) - \frac{11}{4} = 0$. Hence, we conclude that $\overline{x}$ is a global m.p. of the given problem. $\qquad\qquad\square$

Due to the simplicity of the given problem, obviously, we could have solved it by direct inspection, avoiding the above more complicated method. However, the present purpose was to do some practice with the IS Analysis, and to give some hints for defining a strictly convex minimization able to say whether or not a point $\overline{x}$ is a global m.p. of a strictly concave function over a convex set.

5.3. Weak Separation and Necessary Conditions

According to the approach outlined in Sect.5.1, a necessary condition of Lagrangian type arises as an auxiliary step toward the fulfilment of a sufficient condition of saddle-point type (see entry 3.1 of Fig.5.1.1). Naturally, this view is possible, at the present time, after the IS Analysis has been introduced. Historically, necessary optimality conditions — as well as other topics — have been introduced autonomously under the push of specific needs. Before Lagrange ideas, it was very difficult to write a necessary condition for $\overline{x} \in R$ to be a minimum point for (1.1.6); it required, in the classic case of bilateral constraints ($p = m$) and at $B = \mathbb{R}^n$ (recall that here, as well as elsewhere, B and $\mathcal{B}$ denote Banach spaces), to express in an explicit form some variables as functions of the remaining ones, unless the problem laid itself open to some devices. Even if possible, expliciting some variables means finding *all the solutions* of the constraining system: a very heavy task. Following Lagrange, the treatment of the constraining equations is **postponed** to the writing of a necessary condition: the objective function f is replaced by a linear combination of it and the constraining functions (namely, $L(x; \lambda)$; see (5.2.10d)), and a necessary condition is written for this new function (now called Lagrangian function) which has been freed of the constraints; the above drawback is overcome. However, the resolution of the constraining system is not avoided, but only delayed, so that the advantage might seem limited. While, the consequences of this very simple idea (which has been originated by the equilibrium of mechanical forces; see [12]) are enormous and endless. In fact, after the above transformation one meets the resolution of the constraining system — indeed, an enlarged one: also the coefficients (multipliers) of the linear combination must be found —; however, now one has to find *a few* (often, *only one*) solutions of the constraining system. In other words, as a consequence of the Lagrange idea the resolution of the constrained system is lowered from **analytical** level (to find all the solutions) to **numerical** one (to find only some, often only one). The above advantage, even if huge, is only the first of a sequence.

In the sequel, the Lagrange ideas will be presented by exploiting the scheme outlined in Sect.5.1 and not according to the historical development; the present status of the theory must be considered as an evolution of Lagrange pioneering ideas. Before going on, we want to stress the main feature of Lagrange approach. With reference to the case $\mathcal{B} = \mathbb{R}$, it consists in associating, to a point x, an m-vector λ, such that the pair (x, λ) be stationary for L. A question is crucial: given a m.p. $\overline{x}$ of (1.1.1) or (1.1.4), does an m-vector $\overline{\lambda}$ always exist, such that $(\overline{x}, \overline{\lambda})$ is stationary for L? Unfortunately, the answer is no, and the problem of the *existence of Lagrange multipliers* arises. A classic way of achieving their existence, in the (Fréchet) differentiable case, consists in exploiting the Dini Implicit Function Theorem (when $B = \mathbb{R}^n$) or Ljusternik Theorem (when $B \neq \mathbb{R}^n$) [III13,III24,III25,III27]. There exist many attempts to overcome the use of these theorems, since they are powerful tools, but they are also based on strong assumptions. In the sequel, the contribution of the IS Analysis to this topic will be discussed.

From a general point of view, a necessary condition can be identified with any superset of the set of m.p. of (1.1.1). We think of a Lagrangian condition in the case of a linear combination, with coefficients θ, λ, of the (generalized) derivatives of f,g (indeed, when $\theta = 0$, we should say F.John condition [I1, I45], as we will see); when the combination is not linear or we have any composition of f,g (like when we start with the separation functions (4.3.1)−(4.3.4), then we speak of Lagrangian type conditions. A common feature of Lagrangian type conditions is the existence of a separation (or alternative) scheme where they are coming from.

Before going into details, let us outline the **"logical chain"** which will lead us to achieve a necessary condition of Lagrangian type. In this section, we will consider (1.1.6) in the case where its image is finite dimensional ($\mathcal{B} = \mathbb{R}$), so that its typical special cases are (1.1.1) and (1.1.4). The extension, which exploits the results of Sects. 3.2 and 4.7, will be carried out in Vol. 2.

The 1st **"link"** of the chain is: "$\overline{x}$ is a m.p. of (1.1.6) implies that $\mathcal{H}$ and the homogenization of $\mathcal{K}_{\overline{x}}$, namely $\mathcal{K}_{\overline{x}}^h$ are separable". In the general case, this sentence is false; this does not change even if differentiability of f and g is assumed, as Example 5.3.1 (at $\alpha, \beta > 0$) shows. Therefore, it is necessary to restrict the class of problems by making assumptions on X, f and g. The reason of this drawback lies mainly in the presence of bilateral constraints, namely $g_i(x) = 0$, $i \in \mathcal{J}^0$; in fact, as we will see, the achievement of a necessary condition is facilitated by $\mathcal{H}$ being a convex body; if in (1.1.6) $p > 0$, then int $\mathcal{H} = \varnothing$. In the classic differentiable case with only bilateral constraints ($p = m$), such a difficulty is overcome by means of assumptions which guarantee the existence of the implicit function defined by the constraining system (Dini and Ljusternik Theorems); the analysis in the IS will allow us to understand the role of such assumptions. $\mathcal{K}_{\bar{x}}^h$ is a conic approximation, at $\bar{z}$ (see Sect. 3.2), of $\mathcal{K}_{\bar{x}}$; of course, the other kinds of conic approximations of $\mathcal{K}_{\bar{x}}$ can be adopted in order to achieve a necessary condition.

The 2nd **"link"** of the chain is: "separability between $\mathcal{H}$ and $\mathcal{K}_{\bar{x}}^h$ implies that the gradient (θ, λ) of a separating hyperplane (which obviously belongs to $-(\mathcal{K}_{\bar{x}}^h - \bar{z})^*$, since it supports $\mathcal{K}_{\bar{x}}^h$) belongs also to $(\mathbb{R}_+ \times D^*) \setminus \{O\}$ (where D is as in Sect. 1.1)". This step does not offer many difficulties and can be established under very general assumptions; it "turns" separability into the existence of multipliers with suitable "sign".

The 3rd **"link"**, which is parallel to the above one, is: "separability between $\mathcal{H}$ and $\mathcal{K}_{\bar{x}}^h$ implies that the subvector λ of the gradient (normal vector) (θ, λ) of a separating hyperplane is orthogonal to the vector of the constraining functions evaluated at $\bar{x}$". This property, known as *complementarity relationship*, is the same as (5.2.26), even if now it comes out from a different context and, unlike (5.2.26), holds under less general assumptions.

The 4th **"link"** is: "the gradient (θ, λ) of a separating hyperplane, having the properties achieved with the two previous steps, makes the generalized Lagrangian function lower semistationary". This step, which does not offer any difficulty, makes use of Theorem 3.3.2

The 5th **"link"** is: "the lower semistationarity of the generalized Lagrangian function implies the non-negativity of its directional $\mathcal{C}$-derivative along all the directions admitted by X". This step — as well as the above one — transfers the analysis from the IS to the given space B.

The 6th and last **"link"** consists in recovering, from the non-negativity of the directional $\mathcal{C}$-derivative of the generalized Lagrangian function, the classic results.

Example 5.3.1. In (1.1.1) set $X = \mathbb{R}^2$, $p = 1$, $m = 2$ (so that $\mathcal{J}^0 = \{1\}$, $\mathcal{J}^+ = \{2\}$, $\mathcal{J} = \{1, 2\}$, $x = (x_1, x_2)$), $f(x) = -x_1$ and

$$g_1(x) = \begin{cases} x_2, & \text{if } x_1 \leq 0, \\ x_2 - x_1^2, & \text{if } x_1 > 0, x_2 \leq 0, \\ x_2 + x_1^2, & \text{if } x_1 > 0, x_2 > 0, \end{cases} \qquad g_2(x) = \alpha + \beta x_1,$$

where $\alpha, \beta \in \mathbb{R}_+$ are parameters. Therefore, (1.1.1) is now a family of problems described by α and β. With a slight abuse of the notation for introducing the dependence on the parameters, at the global m.p. $\bar{x} = O = (0,0)$ the image $\mathcal{K}_0(\alpha, \beta)$ of (1.1.1) (Sect.3.2) is now:

$$\left\{ (u, v_1, v_2) \in \mathbb{R}^3 : v_1 = \begin{cases} x_2, & \text{if } u \leq 0 \\ x_2 - u^2, & \text{if } u > 0, x_2 \leq 0 \\ x_2 + u^2, & \text{if } u > 0, x_2 > 0 \end{cases}, v_2 = \alpha + \beta u, \ x_2 \in \mathbb{R} \right\}.$$

For the sake of simplicity, $\bar{x}$ has been chosen as the global m.p. (trivial to be found by direct inspection), since we want to show a behaviour of the image set; up to a translation, such a behaviour is independent of $\bar{x}$; (see Sect. 3.2). In the closed halfspace of $\mathbb{R}^3$ (i.e., the IS), defined by $u \leq 0$, $\mathcal{K}_0(\alpha, \beta)$ is a closed halfplane (i.e., a set of lines parallel to v_1-axis) and through the point $(u \leq 0, v_1 = 0, v_2 = \alpha + \beta u)$; in the open halfspace of $\mathbb{R}^3$, defined by $u > 0$, $\mathcal{K}_0(\alpha, \beta)$ is a (2-dimensional) parabolic closed "hypograph" or open "epigraph" according to $x_2 \leq 0$ or $x_2 > 0$; $\mathcal{K}_0(\alpha, \beta)$ belongs to the plane defined by $-\beta u + v_2 = 0$. Fig. 5.3.1 shows $\mathcal{K}_0(1, 0)$ denoted simply as $\mathcal{K}_0$, and the above plane, which is denoted as $\mathcal{K}_0^h$. Indeed, f and g are differentiable, so that the linearization of $\mathcal{K}_0(\alpha, \beta)$ at the origin is the plane (see Definition 3.2.2):

$$\mathcal{K}_0^h(\alpha, \beta) = \{(u, v_1, v_2) \in \mathbb{R}^3 : v_1 = x_2, \, v_2 = \alpha + \beta u, \, x_2 \in \mathbb{R}\},$$

which becomes that of Fig. 5.3.1 at $\alpha = 1, \beta = 0$. This plane (as well as $\mathcal{K}_0^h(\alpha, \beta)$ with $\alpha > 0$) cannot be separated from $\mathcal{H} = \{(u, v_1, v_2) \in \mathbb{R}^3 : u > 0, v_1 = 0, v_2 \geq 0\}$; indeed, its projection on the subspace of v_1-axis contains the origin of v_1-axis in its interior, so that condition (4.6.2) (with $K = \mathcal{K}_0^h(\alpha, \beta), m = 2, k = 1$) is not fulfilled. Therefore, notwithstanding the fact that the problem be differentiable, the 1st link of the preceding chain does not hold. If $\alpha = \beta = 0$, then the separability occurs and condition (4.6.2) is satisfied. The example shows also that such a separability does not depend only on f, but also on the level at which g is constrained. $\qquad\square$

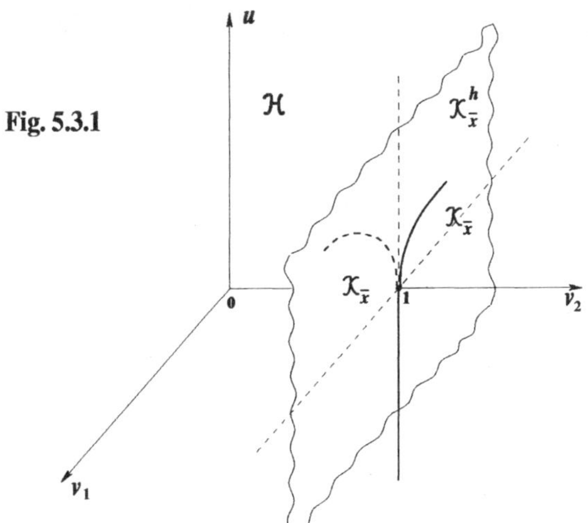

Fig. 5.3.1

Example 5.3.2. In (1.1.1) set $X = \mathbb{R}^2$, $p = m = 1$ (so that $\mathcal{I}^0 = \mathcal{I} = \{1\}, \mathcal{I}^+ = \varnothing$); f and $g \equiv g_1$ are as in Example 5.3.1. At the global m.p. $\bar{x} = O$ the image of (1.1.4) (Sect.3.2) is now (set $v = v_1$):

$$\mathcal{K}_0 = \left\{ (u, v) \in \mathbb{R}^2 : v = \left\{ \begin{array}{ll} x_2, & \text{if } u \leq 0 \\ x_2 - u^2, & \text{if } u > 0, \, x_2 \leq 0 \\ x_2 + u^2, & \text{if } u > 0, \, x_2 > 0 \end{array} \right\}, \, x_2 \in \mathbb{R} \right\},$$

and is illustrated by Fig.5.3.2, which shows that, irrespective of the spaces where the present $\mathcal{K}_0$ and that of Example 5.3.1 (at $\beta = 0$) are embedded, they are equal. The same happens to their linearizations; in fact, we have now:

$$\mathcal{K}_0^h = \left\{ (u,v) \in \mathbb{R}^2 : u = -x_1 \ , \ v = x_2 \ , \ x_i \in \mathbb{R}, \ i = 1,2 \right\} = \mathbb{R}^2.$$

However, unlike Example 5.3.1, now $\mathcal{K}_0^h$ fills the entire IS. Now,

$$\mathcal{H} = \{ (u,v) \in \mathbb{R}^2 : u > 0, \ v = 0 \},$$

so that (3.2.2) is satisfied in agreement with the minimality of $\overline{x}$; notwithstanding the differentiability of f and g, the linearization $\mathcal{K}_0^h$ behaves very badly. $\square$

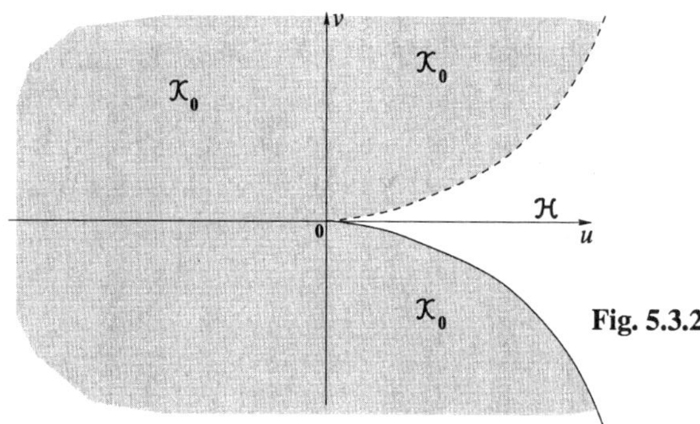

Fig. 5.3.2

Example 5.3.3. In (1.1.4) set $t_0 = 0, t_1 = 1$ (so that $T = [0,1]$), $X = \{x \in C^2([0,1]) : x_i(0) = x_i(1) = 0, \ i = 1,2\}, p = 1, m = 2$ (so that $\mathcal{J}^0 = \{1\}, \mathcal{J}^+ = \{2\}, \mathcal{J} = \{1,2\}$), $\psi_0 = -x_1(t)$, and

$$\psi_1 \begin{cases} x_2(t), & \text{if } x_1 \in X_1^-, \\ x_2(t) - x_1(t)^2, & \text{if } x_1 \in X_1^+, \quad x_2 \in X_2^-, \\ x_2(t) + x_1(t)^2, & \text{if } x_1 \in X_1^+, \quad x_2 \in X_2^+, \end{cases} \qquad \psi_2 \equiv 1, t \in T, \quad t \in T,$$

where

$$X_1^- := \left\{ x_1 : \textstyle\int_T x_1(t)\mathrm{d}t \le 0 \right\} \quad , \quad X_1^+ := \left\{ x_1 : \textstyle\int_T x_1(t)\mathrm{d}t > 0 \right\},$$
$$X_2^- := \left\{ x_2 : \textstyle\int_T x_2(t)\mathrm{d}t \le 0 \right\} \quad , \quad X_2^+ := \left\{ x_2 : \textstyle\int_T x_2(t)\mathrm{d}t > 0 \right\}.$$

Like in Example 5.3.1, by direct inspection, we easily find the global m.p., and we keep it as $\overline{x}$; i.e. $\overline{x} = (\overline{x}_1(t) \equiv 0, \overline{x}_2(t) \equiv 0)$. At $\overline{x} \equiv O$, the image of (1.1.4) (see Sect. 3.2) is now:

$$\mathcal{K}_0 = \Bigg\{ (u, v_1, v_2) \in \mathbb{R}^3 : u = -\int_T x_1(t)\mathrm{d}t,$$
$$v_1 = \int_T \left(x_2(t) + \begin{cases} 0, & \text{if } x_1 \in X_1^- \\ -x_2(t)^2, & \text{if } x_1 \in X_1^+, x_2 \in X_2^- \\ +x_2(t)^2, & \text{if } x_1 \in X_1^+, x_2 \in X_2^+ \end{cases} \right) \mathrm{d}t, \ v_2 = 1, \ x \in X \Bigg\},$$

and its linearization at the origin is;

$$\mathcal{K}_0^h = \left\{ (u, v_1, v_2) \in \mathbb{R}^3; u = -\int_T x_1(t)\mathrm{dt},\ v_1 = \int_T x_2(t)\mathrm{dt},\ v_2 = 1, x \in X \right\}.$$

In fact, $g_1(x)$ — as well as, obviously, f and g_2 — is (Fréchet) differentiable; its derivative is $\int_T x_2(t)\mathrm{dt}$, and its (infinitesimal of higher order) remainder is identically zero or $\mp \int_T x_2(t)^2\mathrm{dt}$. Since $\int_T x_2(t)^2\mathrm{dt}$ is strictly convex, it follows that (like in Example 5.3.1) $\mathcal{K}_0$ is the union of a halfspace (corresponding to $x_1 \in X_1^-$) and two 2-dimensional convex sets, one closed and the other open, and $\mathcal{K}_0^h$ is a plane of $\mathbb{R}^3$ (the IS); therefore, Fig. 5.3.1 can act for the present case too. Note that, according to the usual classification (see Fig. 1.1.1), the present problem and those of Example 5.3.1 are infinite and finite dimensional, respectively; while, all of them have finite dimensional image (which is almost the same). $\qquad\square$

Example 5.3.4. In (1.1.4), set $t_0 = 1$, $X = \{x \in C^2([0,1]); x_i(0) = x_i(1) = 0,\ i = 1, 2\}$, $p = m = 1$ (so that $\mathcal{I}^0 = \mathcal{I} = \{1\}, \mathcal{I}^+ = \varnothing$); ψ_0 and ψ_1 are as in Example 5.3.3. At the global m.p. (easily found by direct inspection) $\overline{x} \equiv 0$ the image of (1.1.5) (see Sect. 3.2; we set $v = v_1$; $X_i^{\pm}, i = 1, 2$ are as in Example 5.3.3) is now:

$$\mathcal{K}_0 = \left\{ (u, v) \in \mathbb{R}^2 : u = -\int_T x_1(t)dt, v = \int_T (x_2(t) + y(t))\,\mathrm{dt},\ x \in X \right\},$$

$$\text{where} \qquad y(t) := \left\{ \begin{array}{ll} 0 & \text{if } x_1 \in X_1^- \\ -x_2(t)^2, & \text{if } x_1 \in X_1^+,\ x_2 \in X_2^- \\ x_2(t)^2, & \text{if } x_1 \in X_1^+,\ x_2 \in X_2^+ \end{array} \right\},$$

and is illustrated by a figure quite similar to Fig.5.3.2. Now we have $\mathcal{H} = \{(u, v) \in \mathbb{R}^2 : u > 0, v = 0\}$. Therefore, notwithstanding the fact that Example 5.3.2 and the present one have been posed, respectively, in a Euclidean and Banach space, they have the same (finite dimensional) IS and quite similar image. $\qquad\square$

Example 5.3.5. In (1.1.1) set $X = \mathbb{R}^2, p = 0, m = 2$ (so that $\mathcal{I}^0 = \varnothing, \mathcal{I}^+ = \mathcal{I} = \{1, 2\}$), $x = (x_1, x_2)$, $f(x) = x_1 + x_2$, $g_1(x) = x_1^3 - x_2$, $g_2(x) = x_2$. At the global m.p. $\overline{x} = (0, 0)$ — easily found by direct inspection — the image of (1.1.1)(Sect.3.2) is now:

$$\mathcal{K}_{(0,0)} = \left\{ (u, v_1, v_2) \in \mathbb{R}^3 : u = -x_1 - x_2, v_1 = x_1^3 - x_2,\ v_2 = x_2,\ (x_1, x_2) \in \mathbb{R}^2 \right\} =$$

$$= \left\{ (u, v_1, v_2) \in \mathbb{R}^3 : u = -v_2 - x_1,\ v_1 = -v_2 + x_1^3,\ x_1 \in \mathbb{R} \right\}.$$

Since $\mathcal{H} = \left\{ (u, v_1, v_2) \in \mathbb{R}^3 : u > 0, v_1 \geq 0, v_2 \geq 0 \right\}$, (3.2.2) is satisfied. In the plane $v_2 = 0, \mathcal{K}_{(0,0)}$ is given by $v_1 = -u^3$; this shows that the positive u-semi-axis belongs to the tangent cone (Definition 2.1.9) to $\mathcal{K}_{(0,0)}$ at the origin. Therefore, we cannot expect to separate $\mathcal{K}_{(0,0)}^h$ (the linearization of $\mathcal{K}_{(0,0)}$ at the origin) and $\mathcal{H}$ by means of a plane which does not contain the positive u-semi-axis. Indeed, we have:

$$\mathcal{K}_{(0,0)}^h = \left\{ (u, v_1, v_2) \in \mathbb{R}^3 : u = -x_1 - x_2, v_1 = -x_2,\ v_2 = x_2,\ (x_1, x_2) \in \mathbb{R}^2 \right\} =$$

$$= \left\{ (u, v_1, v_2) \in \mathbb{R}^3 : \ v_1 + v_2 = 0 \right\},$$

and hence the positive u-semi-axis is contained in $\mathcal{H}^h_{(0,0)}$. $\square$

Example 5.3.6. In (1.1.1) set $X = \mathbb{R}^3, p = 0, m = 2$ (so that $\mathfrak{I}^0 = \varnothing, \mathfrak{I}^+ = \mathfrak{I} = \{1,2\}$), $x = (x_1, x_2, x_3)$, $f(x) = x_1 + x_1^2 + x_2 + x_2^2 + x_3^2$, $g_1(x) = -x_1^2 - x_2^2 + 2x_1, g_2(x) = -x_1^2 - x_2^2 - 2x_1$. At the global $m.p.\ \overline{x} = (0,0,0)$ − easily found by direct inspection − the image of (1.1.1) (Sect 3.2) is now:

$$\mathcal{K}_{(0,0,0)} = \{ (u, v_1, v_2) \in \mathbb{R}^3 : u = -x_1 - x_1^2 - x_2^2 - x_3^2, \ v_1 = x_1^2 + x_2^2 - 2x_1,$$

$$v_2 = x_1^2 + x_2^2 - 2x_1, \ (x_1, x_2, x_3) \in \mathbb{R}^3 \} =$$

$$= \{ (u, v_1, v_2) \in \mathbb{R}^3 : u = -v_1^2 - (\sin t - 3\cos t)v_1 - x_3^2 - 2,$$

$$v_2 = v_1^2 + 4(\cos t)v_1 + 3, \ t \in [0, 2\pi], \ 8(v_1 + v_2) \geq (v_1 - v_2)^2 \},$$

where we have set $x_1 = 1 + v_1 \cos t, x_2 = v_1 \sin t$. Now we have:

$$\mathcal{H} = \left\{ (u, v_1, v_2) \in \mathbb{R}^3 : u > 0, v_1 \geq 0, v_2 \geq 0 \right\}$$

and (3.2.2) is satisfied. The linearization of $\mathcal{K}_{(0,0,0)}$ at the origin is:

$$\mathcal{K}^h_{(0,0,0)} = \{ (u, v_1, v_2) \in \mathbb{R}^3 : u = -x_1 - x_2, \ v_1 = -2x_1, \ v_2 = 2x_1, \ (x_1, x_2) \in \mathbb{R}^2 \} =$$

$$= \left\{ (u, v_1, v_2) \in \mathbb{R}^3 : \ v_1 + v_2 = 0 \right\}$$

and contains the positive u-semi-axis. $\square$

Example 5.3.7. In (1.1.1) set $X = \mathbb{R}^3, p = m = 2$ (so that $\mathfrak{I}^+ = \varnothing, \mathfrak{I} = \{1,2\}$); f, g_1, g_2 are as in Example 5.3.6. It is easily seen that also the global $m.p.$ is the same $\overline{x} = (0,0,0)$. Therefore, the image set $\mathcal{K}_{(0,0,0)}$ and its linearization $\mathcal{K}^h_{(0,0,0)}$ at

$$\overline{z} = (f_{\overline{x}}(\overline{x}), g_1(\overline{x}), g_2(\overline{x})) = (0,0,0)$$

(see Sect.3.2) are the same as in Example 5.3.6. While, now we have:

$$\mathcal{H} = \left\{ (u, v_1, v_2) \in \mathbb{R}^3 : u > 0, \ v_1 = v_2 = 0 \right\}.$$

(3.2.2) is satisfied here too. $\square$

Unlike Examples 5.3.1 (at $\alpha > 0$ or $\alpha = 0, \beta > 0$) and 5.3.2-5.3.4, in Examples 5.3.1 (at $\alpha = \beta = 0$) and 5.3.5-5.3.7, $\mathcal{H}$ and $\mathcal{K}^h_{\overline{x}}$ are separable, so that the 1st link of the preceding chain holds; however, the fact that the separating hyperplane contains the u-semi-axis is, in the same cases, a drawback which will appear evident later.

Now let us consider the proposition which allows us to carry out the 6th link. To this end, consider the following (recall that we are in case $\mathcal{B} = \mathbb{R}$) :

Hypothesis H. Set $k := \dim \mathcal{K}^h_{\overline{x}}$. For every set $\{z^1, ..., z^{k+1}\}$ of affinely independent vectors of $\mathcal{K}_{\overline{x}}^h$ such that (if $p = 0$, we stipulate that condition (5.3.1)-(5.3.2) shrinks to (5.3.2)):

$$O_p \in \text{int conv}\{\text{proj } z^1, ..., \text{proj } z^{k+1}\}, \qquad (5.3.1)$$

we have:

$$(\text{ri } \mathcal{H}) \cap \text{ri conv}\{z^1, ..., z^{k+1}\} = \varnothing, \qquad (5.3.2)$$

where, for each $z = (u, v_i, i \in \mathcal{I}^0, v_i, i \in \mathcal{I}^+) \in \mathbb{R}^{1+m}$, $\text{proj } z$ denotes its projection upon the coordinate subspace of the $v_i, i \in \mathcal{I}^0$; namely, $\text{proj } z := (v_i, i \in \mathcal{I}^0) \in \mathbb{R}^p$. $\quad\square$

Example 5.3.1 (at $\alpha > 0$ or $\alpha = 0$, $\beta > 0$) and 5.3.2-5.3.4 shows that Hypothesis H may be not verified even if $\overline{x}$ is a global *m.p.*; vice versa, Hypothesis H may be satisfied without $\overline{x}$ being a *m.p.* of (1.1.6), as simple examples show; for instance, in Example 5.3.5 change $\overline{x} = (0, 0)$ into $\overline{x} = (1, 0)$, which is no longer *m.p.*; and note that $\mathcal{K}_{(1,0)}$ is obtained from $\mathcal{K}_{(0,0)}$ through a trivial translation ($u = -x_1 - x_2$ is replaced by $u = 1 - x_1 - x_2$) which does not alter the linearization; namely, $\mathcal{K}_{(1,0)}^h = \mathcal{K}_{(0,0)}^h$. Hence, Hypothesis H (which is easily seen to be fulfilled in Example 5.3.5) continues to be satisfied even if $\overline{x} = (1, 0)$ is not a *m.p.*. However, the fact that Theorem 4.6.1 (which will now be exploited with $K = \mathcal{K}_{\overline{x}}^h$) expresses a necessary and sufficient condition means that, in the statement of the 1st link, hypothesis and thesis are close. Therefore, we must think of Hypothesis H as a source for deriving more applicable conditions and as a tool for theoretical investigations. In order to use it as a source, let us note that Hypothesis H is verified if the following ones are, as it is easy to check. Recall that, in this section, we are dealing with (1.1.6) at $\mathcal{B} = \mathbb{R}$, and thus, we are referred to problems of type (1.1.1) and (1.1.4).

Hypothesis H$_1$. The homogenization $\mathcal{K}_{\overline{x}}^h$ of the image set is such that:

$$\dim \mathcal{K}_{\overline{x}}^h \leq p. \qquad (5.3.3\text{a})$$

If f and g are differentiable, then (5.3.3a) is equivalent to:

$$\text{rank } (f'(\overline{x}), g'(\overline{x})) \leq p, \qquad (5.3.3\text{b})$$

where f' and g' denote Fréchet or ordinary derivatives.

H$_1 \Rightarrow$ H. Proof. $\mathcal{H}^0 := \{(u, v) \in \text{cl } \mathcal{H} : u = 0\}$ is the projection of $\mathcal{H}$ into the subspace $u = 0$. If (5.3.3a) holds, then $\dim \mathcal{H}^0 + \dim \mathcal{K}_{\overline{x}}^h \leq (m - p) + p = m$, so that there exists a hyperplane of $\mathbb{R}^{1+m}$, which contains both $\mathcal{H}^0$ and $\mathcal{K}_{\overline{x}}^h$. It follows that condition (5.3.1)-(5.3.2) is satisfied. $\quad\square$

Hypothesis H$_1$ may require a great (local) dependence (redundancy) of the constraints; this happens if $m \gg p$. (5.3.3) improve the case (B) of the proof of Theorem 4.6.1. Indeed, unlike the general case of that theorem, now $\mathcal{K}_{\overline{x}}^h$ has an element (that is $\overline{x}$) in common with $\mathcal{H}^0$.

Hypothesis H$_2$. The homogenization $\mathcal{K}_{\overline{x}}^h$ of the image set is such that:

$$\dim \text{ proj} \mathcal{K}_{\overline{x}}^h < p < \dim \mathcal{K}_{\overline{x}}^h. \qquad (5.3.4\text{a})$$

If f and g are differentiable, then (5.3.4a) is equivalent to:

$$\operatorname{rank}(g_i'(\overline{x}),\ i \in \mathcal{J}^0) < p < \operatorname{rank}(f'(\overline{x}), g'(\overline{x})), \tag{5.3.4b}$$

where f' and g' denote Fréchet or ordinary derivatives.

$H_2 \Rightarrow$ H. Proof. $\dim \operatorname{proj} \mathcal{K}_{\overline{x}}^h < p \Rightarrow \nexists z^1, ..., z^{k+1}$ which satisfy (5.3.1), so that condition (5.3.1), so that condition (5.3.1)$-$(5.3.2) is fulfilled. $\qquad\square$

The 2nd of (5.3.4) is useless for the proof; it simply makes H_1 and H_2 disjoint.

Hypothesis H_3. There are only unilateral constraints ($p = 0, m \geq 1$). Either **(i)** (5.3.2) holds, or **(ii)** f and $-g$ are $\mathcal{C}-$differentiable and $\overline{x}$ is a global minimum point. If f and g are differentiable, then (5.3.2) can be verified only if

$$\operatorname{rank} g'(\overline{x}) \leq m, \tag{5.3.5}$$

where g' denotes Fréchet or ordinary derivative.

$H_3 \Rightarrow$ H. Proof. **(i)** Condition (5.3.1)$-$(5.3.2) is trivially fullfilled, since (5.3.1) is "identically" true. **(ii)** because of the Homogenization Lemma (Proposition 3.2.6(iii)) and of Proposition 3.2.7, $\mathcal{H}$ and $\mathcal{E}(\mathcal{K}_{\overline{x}}^h)$ are convex and disjoint. Therefore, because of Theorem 2.2.4 (i), they are separable. Easy examples (e.g.,with $\operatorname{rank}(f'(\overline{x}), g'(\overline{x})) = 1 + m$) show that (5.3.5) is not sufficient. $\qquad\square$

Hypothesis H_4. There are only bilateral constraints ($p = m \geq 1$), and

$$U := \{(u,v) \in \mathbb{R} \times \mathbb{R}^m : u > 0, v = 0\} = \mathcal{H} \not\subseteq \operatorname{int} \operatorname{conv} \mathcal{K}_{\overline{x}}^h. \tag{5.3.6a}$$

This condition is equivalent to:

$$\sum_{j=1}^{m+1} \alpha_j(-\mathcal{D}_\mathfrak{e}f(\overline{x}; d^j), ..., \mathcal{D}_\mathfrak{e}g_m(\overline{x}; d^j)) \neq (1, 0, ..., 0), \in \mathbb{R}^{1+m}, \tag{5.3.6b}$$

whatever the positive reals $\alpha_1, ... \alpha_m$ and the affinely independent vectors

$$(-\mathcal{D}_\mathfrak{e}f(\overline{x}; d^j),\ \mathcal{D}_\mathfrak{e}g_i(\overline{x}; d^j),\ i \in \mathcal{J}),\ j = 1, ..., m+1$$

may be; if no such a set of affinely independent vectors exists (like when $\dim \mathcal{K}_{\overline{x}}^h \leq m$), then obviously the condition is satisfied. If f and g are differentiable, then the above condition is equivalent to $\dim \mathcal{K}_{\overline{x}}^h \leq m$, or

$$\operatorname{rank}(f'(\overline{x}), g'(\overline{x})) \leq m, \tag{5.3.6c}$$

where f' denote Fréchet or ordinary derivatives.

$H_4 \Rightarrow$ H. Proof. Since $g(\overline{x}) = O$, $O_{1+m} \in \mathcal{K}_{\overline{x}}^h$. If $\operatorname{int} \operatorname{conv} \mathcal{K}_{\overline{x}}^h = \varnothing$, then (5.3.6a) is trivially satisfied; otherwise, (5.3.6a) means that no elements of $\operatorname{ri} \mathcal{H}$ (which equals $\mathcal{H}$) can belong to $\operatorname{int} \operatorname{conv} \mathcal{K}_{\overline{x}}^h$, so that (5.3.2) is fullfilled. (5.3.6b) is equivalent to (5.3.6a), since $\mathcal{K}_{\overline{x}}^h$ is a cone, so that it is enough to consider only one point of $\operatorname{ri} \mathcal{H} = \mathcal{H}$ (e.g., $(1,0,...,0)$), and, to identify an element of $\operatorname{int} \operatorname{conv} \mathcal{K}_{\overline{x}}^h$, to consider only $m+1$ vectors of $\mathcal{K}_{\overline{x}}^h$ because of Carathéodory Theorem for cones (see Corollary 2.1.1). If f and g are differentiable, then $\mathcal{K}^h_{\overline{x}}$ is a linear manifold. The projection $\mathcal{H}^0$ of $\mathcal{H}$ into the subspace $u = 0$ is now O_{1+m}. If the inequality (5.3.6c) holds, then dim $\mathcal{H}^0 + \dim \mathcal{K}^h_{\overline{x}} \leq 0 + m = m$, so that there exists a hyperplane of $\mathbb{R}^{1+m}$ which contains both $\mathcal{H}^0$ and $\mathcal{K}^h_{\overline{x}}$; therefore, condition $(5.3.1) - (5.3.2)$ is satisfied. $\qquad\square$

Hypothesis H_5. $k := \dim \mathcal{K}^h_{\overline{x}}$. For every set $\{z^1, ..., z^{k+1}\}$ of affinely independent vectors of $\mathcal{K}^h_{\overline{x}}$, we have:

$$O_p \notin \text{int conv} \{\text{proj} \, z^1, ..., \text{proj} \, z^{k+1}\}. \tag{5.3.7a}$$

Set $k^0 := \dim \text{proj} \, \mathcal{K}^h_{\overline{x}}$. The above condition is equivalent to: either $k^0 \leq p - 1$, or

$$\sum_{j=1}^{p+1} \alpha_j (\mathcal{D}_{\mathcal{C}} g_1(\overline{x}; d^j), ..., \mathcal{D}_{\mathcal{C}} g_p(\overline{x}; d^j)) \neq O_p, \tag{5.3.7b}$$

whatever the positive reals $\alpha_1, ..., \alpha_p$ and the affinely independent vectors

$$(\mathcal{D}_{\mathcal{C}} g_i(\overline{x}; d^j), \, i \in \mathcal{I}), \quad j = 1, ..., p+1, \quad d^j \in X - \overline{x},$$

may be. If f and g are differentiable, then the above condition can be verified only if

$$\text{rank} \, (g'_1(\overline{x}), ..., g'_p(\overline{x})) \leq p - 1, \tag{5.3.7c}$$

where g'_i denotes Fréchet or ordinary derivative.

$H_5 \Rightarrow H$. Proof. (5.3.7a) means that (5.3.1) never happens, so that $(5.3.1) - (5.3.2)$ is obviously satisfied. If $k^0 \leq p$, then we are in case (5.3.3a); if (5.3.7b) holds, then (5.3.1) is satisfied; in both cases, condition $(5.3.1) - (5.3.2)$ is fulfilled. If f and g are differentiable, then (5.3.7c) is equivalent to (5.3.3b). $\qquad\square$

Hypothesis H_6 Let $A(x) = (f_{\overline{x}}(x), g(x))$ be (cl $\mathcal{H}$)-concavelike on X and (5.3.2) hold.

$H_6 \Rightarrow H$. Proof. The (cl $\mathcal{H}$)-concavelikeness implies the convexity of the conic extension $\mathcal{E}(\mathcal{K}_{\overline{x}})$ of the image set (see Definition 2.4.5 and Proposition 3.2.8(i)). Under this, separability between $\mathcal{H}$ and $\mathcal{K}^h_{\overline{x}}$ has been proved by Theorem 2.2.4(i). $\qquad\square$

Because of Theorem 4.6.1, the 1st link holds under Hypotesis H. The following two propositions realize links 2 and 3. Recall that $\overline{z}$ is the image of $\overline{x}$ (Sect. 3.2); * and $\perp$ denote polar and orthogonal complement, respectively.

Proposition 5.3.1 (sign of multipliers). In (1.1.6) for $\mathcal{B} = \mathbb{R}$ let X be convex, f and $-g$ be $\mathcal{C}$-differentiable, and Hypothesis H hold at least at $\overline{x}$. Then:

$$(\mathcal{K}^h_{\overline{x}} - \overline{z})^* \cap (\mathbb{R}_- \times \mathbb{R}^p \times \mathbb{R}^{m-p}_-) \backslash \{O\} \neq \varnothing. \tag{5.3.8}$$

If f and g are (Fréchet) differentiable, then (5.3.8) becomes:

$$(\mathcal{K}^h_{\overline{x}} - \overline{z})^\perp \cap (\mathbb{R}_- \times \mathbb{R}^p \times \mathbb{R}^{m-p}_-) \backslash \{O\} \neq \varnothing \tag{5.3.8$'$}$$

Proof. Thanks to Hypothesis H, $\exists \, \omega \in \mathbb{R}^{1+m} \backslash \{O\}$, s.t.

$$\mathcal{H} \subset \{z \in \mathbb{R}^{1+m} : \langle \omega, z \rangle \leq 0\},$$

$$\mathcal{K}^h_{\overline{x}} - \overline{z} \subseteq \{z \in \mathbb{R}^{1+m} : \langle \omega, z \rangle \geq 0\}.$$

The former of the above inclusions implies $\omega \in -\mathcal{H}^* = \mathbb{R}_- \times \mathbb{R}^p \times \mathbb{R}^{m-p}_-$; the latter implies $\omega \in (\mathcal{K}^h_{\overline{x}} - \overline{z})^*$, so that (5.3.8) follows. (5.3.8)$'$ is now obvious. $\square$

If X is replaced by $X \cap N(\overline{x})$, $N(\overline{x})$ being a convex neighbourhood of $\overline{x}$, then, obviously, in the above proposition — as well as in the following ones — the links deal with a necessary condition for a local m.p., instead of a global one. If there are only unilateral constraints ($p = 0$) or only bilateral ones ($p = m$), the above proposition receives obvious simplifications.

Proposition 5.3.2 (complementarity). In (1.1.6) with $\mathcal{B} = \mathbb{R}$, let X be convex, f and $-g$ be $\mathcal{C}$-differentiable and Hypothesis H hold at least at $\overline{x}$. If $\overline{x}$ is a global minimum point of (1.1.6), then:

$$\exists(-\overline{\theta}, -\overline{\lambda}) \in (\mathcal{K}^h_{\overline{x}} - \overline{z})^* \cap (\mathbb{R}_- \times \mathbb{R}^p \times \mathbb{R}^{m-p}_-)\backslash\{O\}, \tag{5.3.9a}$$

s.t.

$$\overline{\lambda}_i g_i(\overline{x}) = 0, \quad i \in \mathcal{I}^+. \tag{5.3.9b}$$

Proof. Ab absurdo, suppose that (5.3.9) be false. Then,

$$\forall \omega := (-\theta, -\lambda) \in (\mathcal{K}^h_{\overline{x}} - \overline{z})^* \cap (\mathbb{R}_- \times \mathbb{R}^p \times \mathbb{R}^{m-p}_-)\backslash\{O\} \neq \varnothing$$

(where the inequality is a consequence of Proposition 5.3.1, which can be applied because of the $\mathcal{C}$-differentiability and Hypothesis H), we have that $\lambda_i g_i(\overline{x}) > 0$ $i \in \mathcal{I}^+$ and hence $\langle \omega, -\overline{z} \rangle > 0$, since $\overline{z} = (0, g(\overline{x})), g_i(\overline{x}) = 0$ $i \in \mathcal{I}^0$, and $\sum_{i \in \mathcal{I}^+} \lambda_i g_i(\overline{x}) > 0$. The last but one inequality, the convexity of $\mathcal{E}(\mathcal{K}^h_{\overline{x}} - \overline{z})$ (implied by the $\mathcal{C}$-differentiability of f and $-g$ at $\overline{x}$; see Propositions 3.2.6(ii) and 3.2.8(iii)), the inclusion $\mathcal{E}^*(\mathcal{K}^h_{\overline{x}} - \overline{z}) \subseteq (\mathcal{K}^h_{\overline{x}} - \overline{z})^*$, and $\omega \in \mathcal{E}^*(\mathcal{K}^h_{\overline{x}} - \overline{z})$ (see the proof of Proposition 5.3.1) imply (due to the closure of the polar and to Theorem 2.2.6(4i)(i)):

$$-\overline{z} \in \text{int}[\mathcal{E}^*(\mathcal{K}^h_{\overline{x}} - \overline{z}) \cap (\mathcal{D})\backslash\{O\}]^* = \text{int}[\mathcal{E}^*(\mathcal{K}^h_{\overline{x}} - \overline{z}) \cap (\mathcal{D})]^* =$$

$$= \text{int cl conv}[\mathcal{E}(\mathcal{K}^h_{\overline{x}} - \overline{z}) \cup (-\mathcal{H})] = \text{int } \mathcal{E}(\mathcal{K}^h_{\overline{x}} - \overline{z}).$$

where $\mathcal{D} := \mathbb{R}_- \times \mathbb{R}^p \times \mathbb{R}^{m-p}_-$. Denote by $N_\epsilon(z)$ an open sphere of $\mathbb{R}^{1+m}$ with centre at z and suitably small radius ε. From the last relationships we deduce that $N_\epsilon(-\overline{z}) \subset \mathcal{E}(\mathcal{K}^h_{\overline{x}} - \overline{z})$, and therefore $N_\epsilon(O_{1+m}) \subset \mathcal{E}(\mathcal{K}^h_{\overline{x}})$; this inclusion implies $\mathcal{E}(\mathcal{K}^h_{\overline{x}}) \cap \text{ri } \mathcal{H} = \varnothing$, which contradicts the minimality of $\overline{x}$, because of the equivalence between (3.2.2) and (3.2.14) (see Proposition 3.2.7). $\square$

The 4th and 5th links are realized by Theorem 3.3.2 and the next proposition, respectively.

Proposition 5.3.3 (sign of directional derivative). Let us consider (1.1.6) with $\mathcal{B} = \mathbb{R}$.

(i) Let X be convex, f and $-g$ be $\mathcal{C}$-differentiable at $\overline{x} \in X$, and suppose that their $\mathcal{C}$-derivatives $\mathcal{D}_{\mathcal{C}}, -\mathcal{D}g_i, \in \mathcal{J}$ be bounded from above by a finite constant in a neighbourhood of $\overline{x}$ if $X \subseteq B \neq \mathbb{R}^n$. If $\overline{x}$ is a lower semistationary point of $L(\bullet; \theta, \lambda)$ with $\theta \geq 0$, $\lambda \in D^*$, $(\theta, \lambda) \neq O$, then:

$$\inf_{d \in S} \mathcal{D}_{\mathcal{C}} L(\overline{x}; d; \theta, \lambda) \geq 0, \tag{5.3.10}$$

where $S := \{d \in B : ||d|| = 1, \text{ and } \exists \alpha > 0 \text{ s.t. } \alpha d \in X - \overline{x}\}$, and

$$\mathcal{D}_{\mathcal{C}} L(\overline{x}; d; \theta, \lambda) = \theta \mathcal{D}_{\mathcal{C}} f(\overline{x}; d) - \sum_{i \in \mathcal{J}} \lambda_i \mathcal{D}_{\mathcal{C}} g_i(\overline{x}; d)$$

is the $\mathcal{C}$-derivative of the Lagrangian function L at $\overline{x}$. **(ii)** If $\overline{x} \in \operatorname{int} X$, f and g are differentiable, then (5.3.10) becomes:

$$\min_{d \in S} V_L(\overline{x}; d; \theta, \lambda) = 0, \tag{5.3.11}$$

where V_L is a (continuous) linear functional (1st variation of L). If, in addition, $X = B = \mathbb{R}^n$, then, instead of (5.3.11), we find:

$$\min_{d \in S} \langle L'_x(\overline{x}; \theta, \lambda), d \rangle = 0 \tag{5.3.12}$$

where L'_x denotes gradient of L at $(\overline{x}; \theta, \lambda)$ with respect to x.

Proof. (i) Set $x = \overline{x} + \alpha d \in X$ with $\alpha \in \mathbb{R}_+$. (3.3.1), with $Y = X$ and f replaced by L, holds iff

$$\sup_{\substack{r > 0 \\ \alpha \in]0, r]}} \inf_{d \in S} \left[\mathcal{D}_{\mathcal{C}} L(\overline{x}; d; \theta, \lambda) + \frac{1}{||\alpha d||} \varepsilon_L(\overline{x}; \alpha d; \theta, \lambda) \right] \geq 0 \tag{5.3.13}$$

where $\varepsilon_L := \theta \varepsilon_f - \sum_{i \in \mathcal{J}} \varepsilon_{g_i}$ is the remainder of the expansion of L. Ab absurdo, suppose that

$$l := \inf_{d \in S} \mathcal{D}_{\mathcal{C}} L(\overline{x}; d; \theta, \lambda) < 0.$$

Since $\varepsilon_L/||\alpha d|| \to 0$ as $\alpha \downarrow 0$, we find that the infimum in (5.3.13) is less than a negative constant, so that its supremum is still negative; this fact, which holds a fortiori if $l = -\infty$, contradicts (5.3.13) and proves (5.3.10). **(ii)** Now $\mathcal{D}_{\mathcal{C}} L$ is continuous linear and S is compact so that its infimum becomes minimum; call it δ and let $\hat{d} \in X - \overline{x}$ be a global m.p. or $\mathcal{D}_{\mathcal{C}} L(\overline{x}; d; \theta, \lambda) = \delta$. Since $\hat{d} \in X - \overline{x} \Rightarrow -\hat{d} \in X - \overline{x}$, we find:

$$\delta = \min_{d \in S} \mathcal{D}_{\mathcal{C}} L(\overline{x}; d; \theta, \lambda) \leq \mathcal{D}_{\mathcal{C}} L(\overline{x}; -\hat{d}; \theta, \lambda) = -\mathcal{D}_{\mathcal{C}} L(\overline{x}; \hat{d}; \theta, \lambda) = -\delta,$$

and draw $\delta \leq 0$. This and (i) imply $\delta = 0$, which is (5.3.11) (to within the change of symbol from $\mathcal{D}_{\mathcal{C}} L$ to V_L to recall that $\mathcal{D}_{\mathcal{C}} L$ is now the variation of L), and (5.3.12) (observing that, when $X = B = \mathbb{R}^n$, $\mathcal{D}_{\mathcal{C}} L$ shrinks to a scalar product). $\square$

Theorem 5.3.1 (necessary condition). Consider (1.1.6) with $\mathcal{B} = \mathbb{R}$. Let X be convex, f and $-g_i$, $i \in \mathcal{J}$ be $\mathcal{C}$-differentiable at least at $\overline{x} \in X$ and suppose that their

$\mathcal{C}$-derivatives $\mathcal{D}_e f, -\mathcal{D}_e g_i, i \in \mathcal{I}$ be bounded from above by a finite constant in a neighbourhood of $\bar{x}$ when $X \subseteq B \neq \mathbb{R}^n$. Assume that

$$\text{ri dom}\, \mathcal{D}_e f \cap \text{ri dom}\, \mathcal{D}_e(-g_1) \cap ... \cap \text{ri dom}\, \mathcal{D}_e(-g_m) \neq \varnothing, \tag{5.3.14}$$

and that Hypothesis H be fullfilled. If $\bar{x}$ is a minimum point of (1.1.6) with $\mathcal{B} = \mathbb{R}$, then there exists multipliers $\bar{\theta} \in \mathbb{R}$ and $\bar{\lambda} \in \mathbb{R}^m$, such that:

$$\inf_{d \in S} \mathcal{D}_e L(\bar{x}; d; \bar{\theta}, \bar{\lambda}) \geq 0, \tag{5.3.15a}$$

$$g(\bar{x}) \in D, \ \bar{\theta} \in \mathbb{R}_+, \ \bar{\lambda} \in D^*, \ (\bar{\theta}, \bar{\lambda}) \neq O, \tag{5.3.15b}$$

$$\langle \bar{\lambda}, g(\bar{x}) \rangle = 0, \tag{5.3.15c}$$

where

$$\mathcal{D}_e L(\bar{x}; d; \bar{\theta}, \bar{\lambda}) = \theta \mathcal{D}_e f(\bar{x}; d) - \sum_{i \in \mathcal{I}} \lambda_i \mathcal{D}_e g_i(\bar{x}; d)$$

is the $\mathcal{C}$-derivative of the Lagrangian function at $\bar{x}$ and $S := \{d \in B : \|d\| = 1, \text{ and } \exists \alpha \in \mathbb{R}_+ \text{ s.t. } \alpha d \in X - \bar{x}\}$. (5.3.15a) is equivalent to:

$$O \in \bar{\theta} \underline{\partial}_e f(\bar{x}) + \sum_{i \in \mathcal{I}} \bar{\lambda}_i \underline{\partial}_e(-g_i(\bar{x})), \tag{5.3.15a$'$}$$

which becomes:

$$O \in \bar{\theta} \partial f(\bar{x}) + \sum_{i \in \mathcal{I}} \bar{\lambda}_i \partial(-g_i(\bar{x})), \tag{5.3.15a$''$}$$

if, in particular, f and $-g_i$ are convex. When f and g are differentiable at $\bar{x} \in \text{int} X$, then, in case (1.1.4), (5.3.15a) collapses to $V_L = 0$ along $x = \bar{x}$, V_L being the 1st variation of L, and becomes:

$$\Psi'_x(t, \bar{x}, \bar{x}'; \bar{\theta}, \bar{\lambda}) - \frac{d}{dt} \Psi'_{x'}(t, \bar{x}, \bar{x}'; \bar{\theta}, \bar{\lambda}) = 0, \tag{5.3.15a$'''$}$$

where $\Psi(t, x, x'; \theta, \lambda) := \theta \psi_0(t, x, x') - \sum_{i \in \mathcal{I}} \lambda_i \psi_i(t, x, x')$ is the integrand of the Lagrangian function; while, in case (1.1.1), (5.3.15a) collapses to:

$$L'_x(\bar{x}; \bar{\theta}, \bar{\lambda}) = 0, \tag{5.3.15a$''''$}$$

where L'_x is the gradient of L with respect to x.

Proof. The assumption on $\mathcal{C}$-differentiability and Hypothesis H, besides that on the minimality of $\bar{x}$ (which gives obviously the 1st of (5.3.15b)), allow us to apply Proposition 5.3.2, Theorem 3.3.2 and Proposition 5.3.3(i). The 1st gives the existence of the pair $(\bar{\theta}, \bar{\lambda})$ such that (5.3.9) hold, so that (5.3.15b,c) follow and, because of the 1st part of (5.3.9a), Theorem 3.3.2 can be applied to achieve the semistationarity of $\bar{x}$ for L. Now Proposition 5.3.3 gives (5.3.15a). The $\mathcal{C}$-derivative $\mathcal{D}_e L$ of the Lagrangian function is sublinear (see Sect.2.3), so that its $\mathcal{C}$-subdifferential (see Definition 3.1.2) $\underline{\partial}_e L(\bullet; \bar{\theta}, \bar{\lambda})$ at $\bar{x}$ is $\neq \varnothing$; (5.3.15a) implies the relation $O \in \underline{\partial}_e L(\bar{x}; \bar{\theta}, \bar{\lambda})$, which is equivalent to (5.3.15a)$'$ because of assumption (5.3.14). When f and $-g$ are convex, then the $\mathcal{C}$-subdifferential shrinks to the classic subdifferential, so that (5.3.15a)$'$ becomes

(5.3.15a)″. Now let us prove (5.3.15a)‴ and (5.3.15a)⁗. f and g are differentiable at $\overline{x} \in \text{int } X$, so that $d \in S \Rightarrow -d \in S$. From Proposition 5.3.3(ii) we have that (5.3.15a) collapses to (5.3.11) in the case (1.1.4) and to (5.3.12) in the case (1.1.1). In the latter case, from (5.3.12) we immediately draw (5.3.15)⁗. In the former case, from (5.3.11) we have:

$$V_L(\overline{x}; d; \overline{\theta}, \overline{\lambda}) = 0, \quad \forall d \in S. \tag{5.3.16}$$

Since L is differentiable, with expansion

$$L(\overline{x} + d; \overline{\theta}, \overline{\lambda}) = L(\overline{x}; \overline{\theta}, \overline{\lambda}) + V_L(\overline{x}; d; \overline{\theta}, \overline{\lambda}) + \varepsilon_L(\overline{x}; d; \overline{\theta}, \overline{\lambda}),$$

where V_L is linear with respect to d and ε_L is infinitesimal of higher order than d, then we have:

$$\frac{d}{d\alpha} L(\overline{x} + \alpha d; \overline{\theta}, \overline{\lambda})_{/\alpha=0} = \lim_{\alpha \downarrow 0} \frac{1}{\alpha}[L(\overline{x} + \alpha d; \overline{\theta}, \overline{\lambda}) - L(\overline{x}; \overline{\theta}, \overline{\lambda})] =$$

$$= \lim_{\alpha \downarrow 0} \frac{1}{\alpha}[V_L(\overline{x}; \alpha d; \overline{\theta}, \overline{\lambda}) + \varepsilon_L(\overline{x}; d; \overline{\theta}, \overline{\lambda})] = V_L(\overline{x}; d; \overline{\theta}, \overline{\lambda}).$$

This shows the classic fact that the variation in the sense of the continuous linear term of the expansion of L equals its directional derivative. Therefore, following Euler's approach, the variation of the functional L is reduced to an ordinary derivative, so that (5.3.16) can be replaced equivalently with

$$\frac{d}{d\alpha} L(\overline{x} + \alpha d; \overline{\theta}, \overline{\lambda})_{/\alpha=0} = 0, \quad \forall d \in S. \tag{5.3.16$'$}$$

Since d is arbitrary, we consider any $\hat{x} \in X \backslash \{\overline{x}\}$, and set $d = \hat{x} - \overline{x} := \Delta\overline{x}$, so that αd is the variation associated to the pencil of functions $x(t; \alpha) := \overline{x}(t) + \alpha[\hat{x}(t) - \overline{x}(t)]$, with $t \in T, \alpha \in [0, 1]$; of course $\Delta\overline{x}' := \hat{x}'(t) - \overline{x}'(t) = (\Delta\overline{x})' = d'$. By means of these positions, the Lagrangian function $L(\overline{x} + \alpha d; \overline{\theta}, \overline{\lambda})$ of (5.3.16)$'$, considered as a function of only α, call it $\phi(\alpha)$, becomes:

$$\phi(\alpha) := \int_T \left[\overline{\theta}\psi_0(t, x(t; \alpha), x'(t; \alpha)) - \sum_{i \in \mathcal{J}} \overline{\lambda}_i \psi_i(t, x(t; \alpha), x'(t; \alpha)) \right] dt =$$

$$= \int_T \Psi(t, x(t; \alpha), x'(t; \alpha); \overline{\theta}, \overline{\lambda}) dt.$$

Since

$$\frac{\partial}{\partial\alpha} x(t; \alpha) = \Delta\overline{x} \quad \text{and} \quad \frac{\partial}{\partial\alpha} x'(t; \alpha) = \Delta\overline{x}',$$

we find (Ψ'_x and $\Psi'_{x'}$ denote always partial derivatives of Ψ with respect to, respectively, the 2nd and 3rd argument, irrespective of wheter they depend on α or not):

$$\phi'(\alpha) = \int_T \left[\Psi'_x \cdot \frac{\partial}{\partial\alpha} x(t; \alpha) + \Psi'_{x'} \cdot \frac{\partial}{\partial\alpha} x'(t; \alpha) \right] dt =$$

$$= \int_T (\Psi'_x \cdot \Delta \overline{x} + \Psi'_{x'} \cdot \Delta \overline{x}) dt,$$

where

$$\Psi'_x = \frac{\partial}{\partial x} \Psi(t, x(t; \alpha), x'(t; \alpha); \overline{\theta}, \overline{\lambda}), \quad \Psi'_{x'} = \frac{\partial}{\partial x'} \Psi(t, x(t; \alpha), x'(t; \alpha); \overline{\theta}, \overline{\lambda}).$$

Therefore (5.3.16)$'$ becomes $\phi'(0) = 0$ or

$$\int_T \left[\Psi'_x(t, \overline{x}, \overline{x}'; \overline{\theta}, \overline{\lambda}) \, \Delta \overline{x} + \Psi'_{x'}(t, \overline{x}, \overline{x}'; \overline{\theta}, \overline{\lambda}) \, \Delta \overline{x}' \right] dt = 0,$$

which, integrating by parts the 2nd term after having recalled that $\Delta \overline{x}' = (\Delta \overline{x})'$, becomes:

$$\int_T \left(\Psi'_x - \frac{d}{dt} \Psi'_{x'} \right) \Delta \overline{x} \, dt = 0, \tag{5.3.16$''$}$$

since all the feasible solutions of problem (1.1.4) satisfy the end points conditions, so that $\Delta \overline{x}_{/t=t_0} = \Delta \overline{x}_{/t=t_1} = 0$ and,consequently, the term $[\Psi'_x \cdot \Delta \overline{x}]_{t_0}^{t_1}$ of the integration by parts vanishes. The differentiability of $\Delta \overline{x}$, its vanishing at the end points, and the inequality $|\Delta \overline{x}| < \varepsilon$ with ε positive and arbitrary, allow us to apply the so-called Fundamental Lemma of Calculus of Variations [III5, III24, 61]. According to it, $\Delta \overline{x}$ being arbitrary, (5.3.16)$'' \Rightarrow$(5.3.15a)$'''$. $\square$

5.4. Some Applications

Let us now consider some examples which clarify the previous developments.

Examples 5.4.1. (continuation of Examples 1.2.2). Consider problem (1.2.2a). Since $V > 0$, no minimum point is lost, if we intersect its feasible region with the simplex $\{x \in \mathbb{R}^n_+ : f(x) \le M\}$, where M is large enough; for instance, $M > V$ (in fact,within this simplex, f takes values less than M). Such an intersection is compact; then, f being continuous, Theorem 1.1.1 gives the existence of the global minimum of (1.2.2a). Now let us apply Theorem 5.3.1 to (1.2.2a), where we temporarily disregard the sign-constraints on x. f and g are differentiable; then having 1 bilateral constraint only, the necessary condition (5.3.15) holds and shrinks to the system of (5.3.15a)$''''$ and the equality constraint, which, being

$$L(x; \theta, \lambda) = \theta \sum_{j=1}^n x_j - \lambda \left(\prod_{j=1}^n x_j - V \right), \tag{5.4.1}$$

becomes:

$$\begin{cases} \theta - \lambda \prod_{\substack{j \neq i}}^{j-1,...,n} x_j = 0, \ i = 1,...,n; \ (\theta, \lambda) \in \mathbb{R}^2 \backslash \{O\}, \\ \prod_{j=1}^{n} x_j = V. \end{cases} \tag{5.4.2}$$

Since $\theta = 0$ makes (5.4.2) impossible, it is not restrictive to set $\theta = 1$. Then, the unique solution of (5.4.2) is easily found to be:

$$\overline{x}(V) = (\overline{x}_j(V) = V^{1/n}, j = 1,...,n), \quad \overline{\lambda}(V) = V^{-(n-1)/n}. \tag{5.4.3}$$

Since the minimum exists and (5.3.15) is necessary, and since the sign-constraints are satisfied by $\overline{x}(V)$, this is the (unique) global m.p. of (1.2.2a), and the minimum is $f(\overline{x}(V)) = nV^{1/n}$. Note that $\overline{\lambda}(V) = f'(\overline{x}(V))$; hence $\overline{\lambda}$ is the (instantaneous) velocity according to which the minimum (as a function of V) changes; $\overline{\lambda}$ is then the gradient of the perturbation function of Sect. 3.2. The above method of solution is known as *Lagrange Method of Multipliers* (LMM). $\qquad\qquad\qquad\qquad\square$

An obvious consequence of the above solution (5.4.3) is the following classic *isoperimetric inequality*. Let $\ell^{(n)}$ denote the sum of the edgelengths of any hyperrectangle which has volume equal to V. We have $\ell^{(n)} \geq nV^{1/n}$, since the minimum of $\ell^{(n)}$ is achieved when the hyperrectangle is a n-cube. For $n = 2$, denoting by ℓ the (length of the) perimeter of any quadrilateral and by V (the measure of) its area, taking into account that $\ell = 2\ell^{(2)}$, the isoperimetric inequality becomes $\ell \geq 16V$. If, in the vein of Example 1.2.2, we consider the elementary problem of a m-gon of $\mathbb{R}^2$, having given area and minimum perimeter, we find that, for any m-gon of area V and perimeter ℓ, the isoperimetric inequality $\ell^2 \geq 4mV \cdot \mathrm{tn}(\pi/m)$ holds, since the equality holds for the regular m-gon which implies the minimum perimeter. As $n \to +\infty$, we find $\ell^2 \geq 4\pi V$, which will be met agail below. The achieved results allow us to prove a well known inequality. Let $a_1,...,a_n$ be any positive reals, so that

$$G := \left(\prod_{j=1}^{n} a_j \right)^{1/n}$$

is their geometric mean. The n-tuple $\hat{x} = (a_j/G, j = 1,...,n)$ fulfils the equality constraint of (1.1.2a) for $V = 1$, so that

$$f(\hat{x}) \geq [\min_{x_1 \bullet ... \bullet x_n = 1} f(x)] = n.$$

It follows that:

$$\frac{1}{n}\sum_{j=1}^{n} a_j \geq G, \tag{5.4.4}$$

where the equality holds, iff $a_1 = a_2 = ... = a_n$. If in (1.2.2a) the objective function f is replaced by $\varphi(x) = \sum_{j=1}^{n} \frac{1}{x_j}$, and we perform the same reasoning, instead of (5.4.3), as above we find:

$$\overline{x}(V) = (\overline{x}_j(V) = V^{1/n}, j = 1, ..., n), \quad \overline{\lambda}(V) = V^{-(n+1)/n}. \tag{5.4.5}$$

The n-tuple $\hat{x}$ fulfils the equality constraint of (1.1.2a) for $V = 1$, so that

$$\varphi(\hat{x}) \geq [\min_{x_1 \bullet ... \bullet x_n = 1} \varphi(x)] = n.$$

It follows that:

$$G \geq \frac{n}{\sum\limits_{j=1}^{n} \frac{1}{a_j}}, \tag{5.4.6}$$

where the equality holds, iff $a_1 = a_2 = ...a_n$. Since the left-hand side of (5.4.4) and the right-hand side of (5.4.6) are ,respectively, the arithmetic and harmonic means of $a_1, ..., a_n$, from (5.4.4) and (5.4.6) we have:

Proposition 5.4.1. Given the positive numbers $a_1, ..., a_n$, their arithmetic mean is not less than their geometric mean, which is not less than their harmonic mean:

$$\frac{1}{n}\sum_{j=1}^{n}a_j \geq \left(\prod_{j=1}^{n}a_j\right)^{1/n} \geq \frac{n}{\sum\limits_{j=1}^{n}\frac{1}{a_j}}, \tag{5.4.7}$$

where the equalities hold if and only if $a_1 = a_2 = ... = a_n$.

The inequalities (5.4.7) have the flavour of isoperimetric inequalities. For the sake of simplicity, consider now (1.2.2) in the particular case $n = 2$; the extension to the general case is straightforward . According to Sect 3.2 (see just after (3.2.2)), the image set associated with (1.2.2a) is given by ($\overline{x}$ stands for $\overline{x}(V)$ in (5.4.3)):

$$\mathcal{K}_{\overline{x}} = \left\{(u,v) \in \mathbb{R}^2 : u = 2\sqrt{V} - x_1 - x_2, v = x_1x_2 - V, x_1x_2 \in \mathbb{R}_+\right\} =$$

$$= \left\{(u,v) \in \mathbb{R}^2 : v = -x_1u - x_1^2 + 2\sqrt{V}x_1 - V, x_1 \geq 0\right\},$$

and is a family of lines, which admit the following arc of parabola

$$v(u) = \tfrac{1}{4}u^2 - \sqrt{V}u, \quad u \leq 2\sqrt{V}, \tag{5.4.8}$$

as envelope (by disregarding $x_1 \geq 0$, we find the classic envelope, which then must be restricted to satisfy $x_1 \geq 0$). Since $\overline{x}$ is the m.p., (3.2.2) is obviously fulfilled (here $\mathcal{H} = \{(u,v) \in \mathbb{R}^2 : u > 0, v = 0\}$). Again for $n = 2$, consider now (1.2.2b). With the same $\overline{x}$, and for $\ell = f^{\downarrow}(V) = 2\sqrt{V}$, its image set turns out to be:

$$\mathcal{K}_{\overline{x}} = \left\{(u,v) \in \mathbb{R}^2 : u = x_1x_2 - V, v = 2\sqrt{V} - x_1 - x_2, x_1, x_2 \in \mathbb{R}_+\right\} =$$

$$= \left\{(u,v) \in \mathbb{R}^2 : u = -x_1v - x_1^2 + 2\sqrt{V}x_1 - V, 0 \leq x_1 \leq 2\sqrt{V} - v\right\},$$

and is a family of lines, which admit the following arc of parabola

$$u(v) = \tfrac{1}{4}v^2 - \sqrt{V}v, \quad v \leq 2\sqrt{V}, \tag{5.4.9}$$

as envelope. Since $\bar{x}$ is the maximum point, (3.2.2) is fulfilled ($\mathcal{H}$ is as above; in the definition of $\mathcal{K}_{\bar{x}}, u = f(x) = f(\bar{x})$ since now $\bar{x}$ is a maximum point; consequently we have set $v = -g(x)$ in order to let the function in the left-hand side of (3.2.5) represent the perturbation function of (1.2.2b)). If from (5.4.8) we explicit u in terms of v, then we obtain the perturbation function of (1.2.2a) (the parameter being v itself):

$$u_{\bar{x}}(v, 0) = 2\sqrt{V} - 2\sqrt{V + v}, \quad v \geq -V. \tag{5.4.10}$$

This is indeed (3.2.5) for $D = O$ and $\xi = v$. If we compose (5.4.9) (which, in its turn, is the perturbation function (3.2.5) for (1.2.2b)) with (5.4.10), we obtain the relation

$$u(v) \circ u_{\bar{x}}(v, O) = v, \tag{5.4.11}$$

which characterizes (5.4.9) as inverse of (5.4.8).

Due to the simplicity of (1.2.2a), the existence of its minimum has been proved by elementary arguments. Now, let us follow the IS approach and apply Theorem 3.2.3. Consider the image set of (1.2.2a) corresponding to any $\hat{x}$ (previously, to simplify the computation, we have considered the m.p., noticing, once more, that this choice is not necessary):

$$\mathcal{K}_{\bar{x}} = \left\{ (u, v) \in \mathbb{R}^n : u = f(\hat{x}) - \sum_{j=1}^{n} x_j, \, v = \prod_{j=1}^{n} x_j - V, \, x \in \mathbb{R}_{+}^n \right\}. \tag{5.4.12}$$

Due to the obvious *isoperimetric inequality* $V < \ell^2/4$, the set $U_{\hat{x}}$ turns out to be bounded from above: $\ell > 2\sqrt{V}$ implies $u < f(\hat{x}) - 2\sqrt{V}$. Now, let us show that $U_{\hat{x}}$ is closed. Ab absurdo, suppose that its supremum, say $\hat{z} = (\hat{u}, O)$, does not belong to $U_{\hat{x}}$. Then, $\exists \{(u_r, O)\}_1^\infty \subset U_{\hat{x}}$ s.t.

$$u_r < \hat{u}, \quad \lim_{r \to +\infty} u_r = \hat{u}. \tag{5.4.13}$$

Consequently, $\exists \{x^r\}_1^\infty \subset \mathbb{R}_{+}^n$ s.t.

$$\prod_{j=1}^{n} x_j^r = V, \, \forall r; \quad u_r = f(\hat{x}) - \sum_{j=1}^{n} x_j^r. \tag{5.4.14}$$

Since boundedness can be assumed for the feasible region of (1.2.2a), it is not restrictive to suppose that $\exists \tilde{x} := \lim_{r \to +\infty} x^r$. Then, $\tilde{x}$ being feasible because of the 1st of (5.4.14), we have:

$$\lim_{r \to +\infty} [f(\hat{x}) - \sum_{j=1}^{n} x_j^r] = f(\hat{x}) - \sum_{j=1}^{n} \tilde{x}_j \in U_{\hat{x}}.$$

From the 2nd relation of both (5.4.13) and (5.4.14), we draw:

$$\lim_{r \to +\infty} [f(\hat{x}) - \sum_{j=1}^{n} x_j^r] = \hat{u} \notin U_{\hat{x}},$$

and a contradiction has been reached. Hence, $U_{\hat{x}}$ being closed, in Theorem 3.2.3 we can set $S = U_{\hat{x}}$. Since (by Definition 3.2.1) obviously $U_{\hat{x}} \subseteq \mathcal{E}(\mathcal{K}_{\hat{x}})$, (3.2.23b) is satisfied, and the existence of the minimum follows. Note that we have exploited (5.4.12), which is the parametric form of the image set as it appears in its definition (Sect. 3.2, just after (3.2.2)); in other words, the above proof of existence has not required to explicit

v in terms of u (or vice versa), as done previously; such operations are obviously impracticable in the general case. Now, let us consider (1.2.3a), and let $\hat{x}(t)$ satisfy its constraints. Instead of (5.4.12), the image set is here:

$$\mathcal{K}_{\hat{x}} = \left\{ (u,v) \in \mathbb{R}^2 : u = f(\hat{x}) - \int_T \sqrt{1 + x'^2}\, dt, v = \int_T x\, dt - V, \right.$$

$$\left. x(t_0) = x^0, x(t_1) = x^1 \right\}. \tag{5.4.15}$$

Notwithstanding the difference between (5.4.15) and (5.4.12), also the present $U_{\hat{x}}$ is bounded from above, since the isoperimetric inequality $V < \ell^2/4$ holds for (1.2.3a) too. The proof of closedness, as well as the application of Theorem 3.2.3, are as above, in as much as we choose the norm in the given space; for instance $\| \bullet \|_\infty$ as for (1.1.4). Now let us apply Theorem 5.3.1. (5.3.15a)'''' becomes:

$$\theta\sqrt{1 + x'^2} - \lambda x - x' \frac{\theta x'}{\sqrt{1 + x'^2}} = k_1, \tag{5.4.16}$$

where k_1 is a constant. Since

$$\theta = 0 \;\Rightarrow\; \lambda \neq O \;\Rightarrow\; x(t) = \text{constant} \;\Rightarrow\; x^0 = x^1 \;\Rightarrow\; V = x^0(t_1 - t_0),$$

apart from such a very special case, it is not restrictive to assume $\theta \neq 0$ and hence $\theta = 1$. Then (5.4.16) becomes:

$$\lambda x + k_1 = \frac{1}{\sqrt{1 + x'^2}}. \tag{5.4.17}$$

$\lambda = 0$ implies that the derivative of the arclength with respect to t, namely $\sqrt{1 + x'^2}$, is constant, so that the curve is a line, and again we are in a very special case. Therefore, it is not restrictive to assume $\lambda \neq 0$. Since the image of $\tan \tau, \tau \in]-\frac{\pi}{2}, +\frac{\pi}{2}[$ is the entire $\mathbb{R}$, it is not restrictive to set $x'(t) = \tan \tau$. This is the device which allows us to overcome the fact that the unknown of (1.2.3a) cannot be any curve joining the 2 given points, but only those which intersects each vertical line 1 time only. Then (5.4.17) is equivalent to:

$$\lambda x + k_1 = \cos \tau, \qquad \frac{dx}{dt} = \tan \tau, \qquad \tau \in] = \frac{\pi}{2}, \frac{\pi}{2}[$$

from which we draw:

$$dx = -\frac{1}{\lambda} \sin \tau d\tau,$$

$$dt = \frac{dx}{\tan \tau} = -\frac{\sin \tau}{\lambda \tan \tau} d\tau = -\frac{1}{\lambda} \cos \tau.$$

It follows:

$$t = k_2 + \frac{1}{\lambda} \sin \tau, \quad x = \frac{k_1}{\lambda} + \frac{1}{\lambda} \cos \tau, \quad \tau \in]-\frac{\pi}{2}, \frac{\pi}{2}[,$$

where k_2 is a constant. Then we draw the parametric equation

$$\left(x - \frac{k_1}{\lambda} \right)^2 + (t - k_2)^2 = \frac{1}{\lambda^2}, \tag{5.4.18}$$

which represents a family of circumferences; k_1, k_2 and λ can be used to find the element of the family which fulfils the boundary conditions and encircles the prescribed area. Since the minimum exists and (5.3.15) is necessary, then the above unique (stationary) point is the (unique) global m.p. of (1.2.3a). Now let us discuss λ. Due to the previous device, instead of (1.2.3a), we have solved the more general problem settled just before it. Therefore, the Lagrange multiplier λ — being the derivative of the minimum (here the length of the circumference) with respect to the level of the constraint (here the area) — is the curvature of the circumference (5.4.18); this is an agreement with the elementary formula $\ell_c = 2\pi\sqrt{V_c/\pi}$, which gives the length of a circumference in terms of its area: if r is the radius, $d\ell_c/dV_c = \sqrt{\pi/V_c} = 1/r = $ curvature. This appear more evident in the following. By means of the same device as above, let us now solve (1.2.3b). From $(5.3.15a)''''$ — where, of course, $\phi(x)$ is replaced by $-\phi(x)$ — , instead of (5.4.18), we find:

$$(x - c_1)^2 - (t - c_2)^2 = \mu^2, \tag{5.4.19}$$

where c_1 and c_2 are constants and η is the Lagrange multiplier, which turns out to be the radius of the circumference (the reciprocal of its curvature and then of λ; this in agreement with the fact that (1.2.3b) is reciprocal of (1.2.3a); see Sect. 5.5). If, instead of considering the entire circumference, we impose the boundary conditions to (5.4.19), then the solution of (1.2.3b) is an arc of circumference of length ℓ and joining the points $(t_0, x^0), (t_1, x^1)$. The remarks made about the connections with Sect. 3.2 can be repeated here in a quite similar way.

The above result is a special case of an isoperimetric property in $\mathbb{R}^3$: a smooth manifold, whose area is minimum, has zero average curvature. As a consequence, we have that, among all convex and compact bodies of $\mathbb{R}^3$ with constant volume, the sphere is that whose boundary has minimum area. This can be extended to $\mathbb{R}^n$.

Again from the above result, we draw a classic *isoperimetric inequality*: $\ell^2 \geq 4\pi V$, where ℓ is the length of the boundary of a compact and convex body of $\mathbb{R}^2$ and V (the measure of) its area; this is a consequence of the fact that ℓ achieves its minimum when the body is a circle; this has been obtained as a limit in the 1st part of the present example. If (1.2.2b) is extended to manifolds of $\mathbb{R}^3$, so that the sphere is the solution, then the isoperimetric inequality $s^3 \geq 36\pi V^2$ is found, where V is the volume of the compact and convex body of $\mathbb{R}^3$, whose boundary is such a manifold and s the measure of the boundary. $\qquad\square$

5.5. Reciprocal Problems

As already noted, the isoperimetric problems (1.2.2a) and (1.2.3a) are modern formulations of problems, which were well known to ancient Greeks: they knew also the reciprocal problems (1.2.2b) and (1.2.3b) and the fact that, under a suitable relation between V and ℓ, problems (1.2.2) — as well as (1.2.3) — have the same solution.

Given a constrained extremum problem having finite dimensional image — namely, (1.1.1) or (1.1.4) —, it is now-a-days classic to consider the *reciprocal problem*, which is defined this way: one of the constraining functions of the given problem becomes the function to be extremized; the objective function of the given problem, constrained (as equality or inequality) to a suitable constant, becomes a costraint; the remaining constraints (of the given problem) are in common. The so-called *Principle of Reciprocity* consists in showing that the two problems have the same solution(s), under suitable assumptions, which are classically based on the twice differentiability of the Lagrangian function and on its positive definiteness on the tangential set [III25].

Now a Principle of Reciprocity will be stated under very general assumptions, which do not require any differentiability [34]. To this end, we need to specialize the notation of Sect. 1.1, and make some changes in the symbols. Consider the following problems:

$$\mathcal{P}^{\downarrow}(\xi): \quad f^{\downarrow}(\xi) := \min f(x), \quad \text{s.t.} \quad \phi(x) = \xi, \; x \in S, \tag{5.5.1}$$

and

$$\mathcal{P}^{\uparrow}(\xi): \quad f^{\uparrow}(\xi) := \max f(x), \quad \text{s.t.} \quad \phi(x) = \xi, \; x \in S, \tag{5.5.2}$$

where $\xi \in \Xi \subseteq \mathbb{R}$, $f : X \to \mathbb{R}$, $\phi : X \to \mathbb{R}$, $S \subseteq X$, X is any set, and Ξ is the image of the restriction of ϕ to S.

By means of trivial positions, the parametrized forms of (1.1.1) and (1.1.4) — namely (3.2.4) — can be put in the format (5.5.1): the set S collects all the constraints but one equation; indeed, the purpose is here the exchange betweeen the objective function and one constraining equation, while all the remaining constraints do not play any role and are swallowed by the set S. We emphasize that, in (5.5.1)-(5.5.2), x is an *abstract variable* in the sense that may be an element of any space, a differential operator, as well as any discrete variable; consequently, f and ϕ are correspondences between *a set X of any kind* and $\mathbb{R}$.

Now, let us perform, in (5.5.1)-(5.5.2), the exchange between the objective function and the constraining equation; this leads to the following reciprocal problems:

$$\mathcal{Q}^{\downarrow}(\eta): \quad \phi^{\downarrow}(\eta) := \min \phi(x), \quad \text{s.t.} \quad f(x) = \eta, \; x \in S \tag{5.5.3}$$

and

$$\mathcal{Q}^{\uparrow}(\eta): \quad \phi^{\uparrow}(\eta) := \max \phi(x), \quad \text{s.t.} \quad f(x) = \eta, \; x \in S, \tag{5.5.4}$$

where $\eta \in \Upsilon \subseteq \mathbb{R}$, Υ being the image of the restriction of f to S. The sets:

$$R_{\phi}(\xi) := \{x \in S : \phi(x) = \xi\} \quad \text{and} \quad R_{f}(\eta) := \{x \in S : f(x) = \eta\} \tag{5.5.5}$$

are the feasible regions of problems (5.5.1)-(5.5.2) and (5.5.3)-(5.5.4), respectively. The functions $f^{\downarrow}, f^{\uparrow}, \phi^{\downarrow}, \phi^{\uparrow}$ are perturbation functions; see Sect. 3.2.

Proposition 5.5.1. If $R_{\phi}(\xi) \neq \varnothing$ and $\hat{x} \in R_{\phi}(\xi)$, then

$$R_{f}(f(\hat{x})) \neq \varnothing, \tag{5.5.6}$$

$$R_{\phi}(\xi) \cap R_{f}(f(\hat{x})) \neq \varnothing \tag{5.5.7}$$

and the restriction of ϕ on $R_\phi(\xi) \cap R_f(f(\hat{x}))$ is constant and equal to ξ.

Proof. The very definition of the feasible regions gives:

$$R_\phi(\xi) \cap R_f(f(\hat{x})) = \{x \in S : \phi(x) = \xi, \ f(x) = f(\hat{x})\}. \tag{5.5.8}$$

Since the domain of f contains S, $f(\hat{x})$ exists, therefore, the equality $f(x) = f(\hat{x})$ holds at least for $x = \hat{x}$, so that $\hat{x}$ belongs to the set (5.5.8); hence (5.5.6)-(5.5.7) follow. The last part of the claim is now trivial. $\qquad\square$

Among (5.5.2) and its reciprocal problems (5.5.3) and (5.5.4) a relationship will now be established (recall Definitons 1.1.1 and 2.3.2).

Theorem 5.5.1. (i) If $\overline{x}$ is a maximum point of $\mathcal{P}^\uparrow(\xi)$ (or minimum point of $\mathcal{P}^\downarrow(\xi)$) and $f^\uparrow(\xi)$ (or of $f^\downarrow(\xi)$) is increasing on Ξ, then $\overline{x}$ is a minimum point of $\mathcal{Q}^\downarrow(f^\uparrow(\xi))$ (or a maximum point of $\mathcal{Q}^\uparrow(f^\downarrow(\xi))$). Furthermore, the equality

$$\phi^\downarrow(f^\uparrow(\xi)) = \xi \quad (\text{or } \phi^\uparrow(f^\downarrow(\xi)) = \xi) \tag{5.5.9}$$

holds, which characterizes $\phi^\downarrow(\eta)$ (or of $\phi^\uparrow(\eta)$) as inverse function of $f^\uparrow(\xi)$ (or of $f^\downarrow(\xi)$). (ii) If $\overline{x}$ is a maximum point of $\mathcal{P}^\uparrow(\xi)$ (or a minimum point of $\mathcal{P}^\downarrow(\xi)$) and $f^\uparrow(\xi)$ (or $f^\downarrow(\xi)$) is decreasing on Ξ, then $\overline{x}$ is a maximum point of $\mathcal{Q}^\uparrow(f^\uparrow(\xi))$ (or minimum point of $\mathcal{Q}^\downarrow(f^\downarrow(\xi))$). Furthermore, the equality

$$\phi^\uparrow(f^\uparrow(\xi)) = \xi \quad (\text{or } \phi^\downarrow(f^\downarrow(\xi)) = \xi) \tag{5.5.10}$$

holds, which characterizes $\phi^\uparrow(\eta)$ (or of $\phi^\downarrow(\eta)$) as inverse function of $f^\uparrow(\xi)$ (or of $f^\downarrow(\xi)$). (iii) If $\overline{\xi} \in \text{int } \Xi$ is a strict maximum point of the perturbation function $f^\uparrow(\xi)$ and $\overline{x}(\overline{\xi})$ is a maximum point of $\mathcal{P}^\uparrow(\overline{\xi})$, then $\overline{x}(\overline{\xi})$ is both a minimum point of $\mathcal{Q}^\downarrow(f^\uparrow(\overline{\xi}))$ and a maximum point of $\mathcal{Q}^\uparrow(f^\uparrow(\overline{\xi}))$. If $\overline{\xi} \in \text{int } \Xi$ is a strict minimum point of $f^\uparrow(\xi)$, and

$$f^\uparrow(\overline{\xi}) \in \left(\bigcup_{\xi \in \Xi \setminus \{\overline{\xi}\}} R_\phi(\xi) \right), \tag{5.5.11}$$

then $\overline{x}(\overline{\xi})$ is not extremum point of either $\mathcal{Q}^\downarrow(f^\uparrow(\overline{\xi}))$ or $\mathcal{Q}^\uparrow(f^\uparrow(\overline{\xi}))$.

Proof. (i) Because of Proposition 5.5.1, the assumption that $\overline{x}$ be a maximum point of (5.5.2) implies the non-emptiness of the set $R_\phi(\xi) \cap R_f(f^\uparrow(\xi))$ on which the objective function $\Phi(x)$ of (5.5.3) is constant and equal to ξ. Then, to prove the 1st part of the claim it is enough to show that:

$$x \in R_f(f^\uparrow(\xi)) \setminus R_\phi(\xi) \ \Rightarrow \ \phi(x) > \xi. \tag{5.5.12}$$

Being $\phi(x) \neq \xi$ on the set in the left-hand side of (5.5.12), suppose, ab absurdo, that this set has an element, say $\hat{x}$, s.t. $\hat{\xi} := \phi(\hat{x}) < \xi$. We have:

$$\hat{x} \in R_f(f^\uparrow(\xi)) \ \Rightarrow \ f(\hat{x}) = f^\uparrow(\xi); \quad \hat{x} \in R_\phi(\hat{\xi}) \ \Rightarrow \ f(\hat{x}) \leq f^\uparrow(\hat{\xi}).$$

These relations contradict the assumption on $f^\uparrow(\xi)$. With regard to the last part of the claim, observe that, $\overline{x}$ being both a maximum point of $\mathcal{P}^\uparrow(\xi)$ and a minimum point of

$\mathcal{Q}^{\downarrow}(f^{\uparrow}(\xi))$ implies, respectively,

$$\phi(\overline{x}) = \xi \quad \text{and} \quad \phi(\overline{x}) = \phi^{\downarrow}(f^{\uparrow}(\xi)),$$

so that (5.5.9) follows. The alternative statement within brackets is proved in a quite analogous way. **(ii)** Because of Proposition 5.5.1, to prove the 1st part of the claim, it is enough to show that the left-hand side of (5.5.12) implies $\phi(x) < \xi$; this can be proved with a quite analogous reasoning as above. With regard to the last part of the claim, observe that, $\overline{x}$ being a maximum point of both $\mathcal{P}^{\uparrow}(\xi)$ and $\mathcal{Q}^{\uparrow}(f^{\uparrow}(\xi))$ implies, respectively,

$$\phi(\overline{x}) = \xi \quad \text{and} \quad \phi(\overline{x}) = \phi^{\uparrow}(f^{\uparrow}(\xi)),$$

so that (5.5.10) follows. The alternative statement within brackets is proved in a quite analogous way. **(iii)** Because of Proposition 5.5.1, the fact that $\overline{x}(\overline{\xi})$ be a maximum point of $\mathcal{P}^{\uparrow}(\overline{\xi})$ implies the non-emptiness of the set $R_{\phi}(\overline{\xi}) \cap R_f(f^{\uparrow}(\overline{\xi}))$ on which the objective function $\phi(x)$ is constant and equal to $\overline{\xi}$. Then, to prove the 1st part of the claim, it is enough to show that:

$$R_f(f^{\uparrow}(\overline{\xi})) \backslash R_{\phi}(\overline{\xi}) = \varnothing. \tag{5.5.13}$$

Ab absurdo, suppose that (5.5.13) be false, so that there exists an element, say $\hat{x}$, of the set in the left-hand side of (5.6.13). Therefore, we have:

$$\hat{x} \in S, \quad f(\hat{x}) = f^{\uparrow}(\overline{\xi}), \quad \hat{\xi} := \phi(\hat{x}) \neq \hat{\xi}. \tag{5.5.14}$$

The 1st and last of (5.5.14) imply $\hat{x} \in R_{\phi}(\hat{\xi})$, so that $f(\hat{x}) \leq f^{\uparrow}(\hat{\xi})$; this inequality and the 2nd of (5.5.14) imply:

$$f^{\uparrow}(\overline{\xi}) \leq f^{\uparrow}(\hat{\xi}),$$

and the assumption that $\overline{\xi}$ be a strict maximum of $f^{\uparrow}(\xi)$ is contradicted. With regard to the 2nd part of the claim, (5.5.11) and the assumption that $\overline{\xi} \in \text{int}\,\Xi$ imply the existence of $\xi_1, \xi_2 \in \Xi$ with $\xi_1 < \overline{\xi} < \xi_2$, and hence of $x^i \in R_{\phi}(\xi_i), i = 1, 2$, s.t.

$$f(x^1) = f(x^2) = f^{\uparrow}(\overline{\xi}).$$

It follows that $x^i \in R_f(f^{\uparrow}(\overline{\xi})), i = 1, 2$, and that

$$\phi(x^1) = \xi_1 < \overline{\xi} = \phi(\overline{x}(\overline{\xi})), \tag{5.5.15}$$

$$\phi(x^2) = \xi_2 > \overline{\xi} = \phi(\overline{x}(\overline{\xi})), \tag{5.5.16}$$

where the equalities are implied by the assumption that $\overline{x}(\overline{\xi})$ be a maximum point of $\mathcal{P}^{\uparrow}(\overline{\xi})$. From (5.5.15)-(5.5.16) we draw that the restriction of $\phi(x)$ on $R_f(f^{\uparrow}(\overline{\xi}))$ has not either a maximum or a minimum at $\overline{x}(\overline{\xi})$. $\qquad\square$

By means of the obvious remark that (5.5.1) has the same extremum points and opposit extremum of the problem

$$\max[-f(x)], \quad \text{s.t.} \quad x \in R_{\phi}(\xi),$$

a relationship among (5.5.1) and its reciprocal problems (5.5.3)-(5.5.4) is obtained as a straightforward consequence of Theorem 5.5.1.

Theorem 5.5.2. **(i)** If $\bar{x}$ is a minimum point of $\mathcal{P}^{\downarrow}(\xi)$ and $f^{\downarrow}(\xi)$ is increasing on Ξ, then $\bar{x}$ is a maximum point of $\mathcal{Q}^{\uparrow}(f^{\downarrow}(\xi))$. Furthermore, the equality

$$\phi^{\uparrow}(f^{\downarrow}(\xi)) = \xi \qquad (5.5.17)$$

holds, which characterizes $\phi^{\uparrow}(\eta)$ as inverse function of $f^{\downarrow}(\xi)$. **(ii)** If $\bar{x}$ is a minimum point of $\mathcal{P}^{\downarrow}(\xi)$ and $f^{\downarrow}(\xi)$ is decreasing on Ξ, then $\bar{x}$ is a minimum point of $\mathcal{Q}^{\downarrow}(f^{\downarrow}(\xi))$. Furthermore, the equality

$$\phi^{\downarrow}(f^{\downarrow}(\xi)) = \xi \qquad (5.5.18)$$

holds, which characterizes $\Phi^{\downarrow}(\eta)$ as inverse function of $f^{\downarrow}(\xi)$. **(iii)** If $\bar{\xi} \in \text{int } \Xi$ is a strict minimum point of $f^{\downarrow}(\xi)$ and $\bar{x}(\bar{\xi})$ is a minimum point of $\mathcal{P}^{\downarrow}(\bar{\xi})$, then $\bar{x}(\bar{\xi})$ is both a minimum point of $\mathcal{Q}^{\downarrow}(f^{\downarrow}(\bar{\xi}))$ and a maximum point of $\mathcal{Q}^{\uparrow}(f^{\downarrow}(\bar{\xi}))$. If $\bar{\xi} \in \text{int } \Xi$ is a strict maximum point of $f^{\downarrow}(\xi)$, and (5.5.11) holds, then $\bar{x}(\bar{\xi})$ is not extremum point of either $\mathcal{Q}^{\downarrow}(f^{\downarrow}(\xi))$ or $\mathcal{Q}^{\uparrow}(f^{\downarrow}(\xi))$. $\qquad\qquad\qquad\square$

Until now, we have considered the exchange between the objective function and an equality constraint. Let us now consider the case where the objective function is exchanged with an inequality constraint. To this end, consider the problem:

$$\mathcal{P}^{\downarrow}_{\geq}(\alpha): \qquad f^{\downarrow}_{\geq}(\alpha) := \min f(x), \quad \text{s.t.} \quad \phi(x) \geq \alpha, \, x \in S, \qquad (5.5.19)$$

which replaces (5.5.1); for the sake of brevity, we do not consider the inequality case corresponding to (5.5.2); it requires only formal changes with respect to the following development.

By means of trivial positions, the parametrized forms of (1.1.1) and (1.1.4) − namely, (3.2.4) − can be put in the format (5.6.19).

While the reciprocal problems of (5.5.1) are two − namely (5.5.3) and 5.5.4 − now, having the possibility of chosing an inequality or the opposite one, (5.5.19) is associated with four reciprocal problems, according to which we choose min or max and, in the constraint, $\geq$ or $\leq$. For the sake of simplicity, we will consider only the following reciprocal problem:

$$\mathcal{Q}^{\uparrow}_{\leq}(\beta): \qquad \phi^{\uparrow}_{\geq}(\beta) := \max \phi(x), \quad \text{s.t.} \quad f(x) \leq \beta, \, x \in S. \qquad (5.5.20)$$

The extension of the results to the other reciprocal problems is a straightforward matter.

Problem (5.5.19) enjoys the following obvious equality:

$$f^{\downarrow}_{\geq}(\alpha) = \min f^{\downarrow}(\xi), \quad \text{s.t.} \quad \xi \geq \alpha, \quad \xi \in \Xi, \qquad (5.5.21)$$

which connects a problem with unilateral constraints with one having bilateral constraints. As an immediate consequence, we have:

Proposition 5.5.2. If $f^{\downarrow}(\xi)$ is increasing on Ξ and $\alpha \in \Xi$, then $\mathcal{P}^{\downarrow}_{\geq}(\alpha)$ and $\mathcal{P}^{\downarrow}(\alpha)$ are equivalent, in the sense that they have the same minimum points and the same minimum.

From this proposition, its obvious analogous for $Q^\uparrow_\leq(\beta)$ and $Q^\uparrow(\beta)$, and Theorem 5.5.2, we easily draw:

Theorem 5.5.3. If $\overline{x}$ is a minimum point of problem $\mathcal{P}^\downarrow_\geq(\alpha)$, and $f^\downarrow(\xi)$ is increasing on Ξ, then $\overline{x}$ is a maximum point of problem $Q^\uparrow_\leq(f^\downarrow(\alpha))$. Furthermore, the equality

$$\phi^\uparrow_\leq(f^\downarrow_\geq(\alpha)) = \alpha$$

holds, which characterizes $\phi^\uparrow_\leq(\beta)$ as inverse function of $f^\downarrow_\geq(\alpha)$.

Example 5.5.1. (continuation of Examples 1.2.2 and 5.4.1). Consider again problem (1.2.2a). In Example 5.4.1, we have found its unique minimum point through the LMM. Therefore, besides the desired n-tuple $\overline{x}(V)$, we have found the Lagrange multiplier $\overline{\lambda}(V)$. The derivative of the perturbation function of (1.2.2a), that is the function

$$f^\downarrow(V) = f(\overline{x}(V)) = nV^{1/n},$$

is precisely the value of the Lagrange multiplier $\overline{\lambda}(V)$, whose positivity implies that $f^\downarrow$ is increasing. Then (i) of Theorem 5.5.1 can be applied; it follows that $\overline{x}(V)$ is also global maximum point of (1.2.2b) with $\ell = f^\downarrow(V)$. The perturbation functions of the reciprocal problems $\mathcal{P}^\downarrow(V)$ and $Q^\uparrow(2\sqrt{V})$ – namely, $f^\downarrow(V)$ and $\phi^\uparrow(2\sqrt{V})$ – are shown by (5.4.8) and (5.4.9), respectively. The relation (5.5.9) is here given by (5.4.11). $\square$

The properties of reciprocal problems have found several applications and many have to come. Let us consider an instance, by introducing the following problem:

$$f^\downarrow(\xi) := \min f(x), \quad \text{s.t.} \quad g_s(x) = \xi, \quad x \in R_s, \tag{5.5.22}$$

where

$$R_s := \{x \in X : g_i(x) = 0, \ i \in \mathfrak{I}^0 \backslash \{s\}, \ g_i(x) \geq 0, \ i \in \mathfrak{I}^+\},$$

$$\xi \in \Xi := \operatorname{Im} g_s \subseteq \mathbb{R}.$$

Problem (5.5.22) is a parametrization (like (3.2.4)) for (1.1.1)) with respect to the sth constraint (assuming, therefore, $p \geq 1$). (5.5.22) points out, in (1.1.1), one constraining equation, while the others are swallowed by R_s. (5.5.22) is immediately identified with (5.5.1). Now consider the following relaxation of (5.5.22):

$$f^\downarrow_s := \min f(x), \quad \text{s.t.} \quad x \in R_s, \tag{5.5.23}$$

and denote by $x^{(s)}$ a global m.p. of (5.5.23). We will suppose that suitable conditions (as lower semicontinuity of f and compactness of the feasible region) be satisfied, so that the equality

$$\min_{\xi \in \Xi} f^\downarrow(\xi) = f^\downarrow_s \tag{5.5.24}$$

holds. Consider the problems:

$$g^\downarrow_s(\xi) := \min g_s(x), \quad \text{s.t.} \quad f(x) = f^\downarrow(\xi), \quad x \in R_s, \tag{5.5.25}$$

$$g^\uparrow_s(\xi) := \max g_s(x), \quad \text{s.t.} \quad f(x) = f^\downarrow(\xi), \quad x \in R_s, \tag{5.5.26}$$

which are reciprocal of (5.5.22) and, therefore, are easily identified with (5.5.3) and (5.5.4), respectively. Set $\xi^{(s)} := g_s(x^{(s)}) \in \Xi$.

Theorem 5.5.4. Let f be strictly convex, $-g_i, i \in \mathcal{J}^+$ be convex and $g_i, i \in \mathcal{J}^0 \backslash \{s\}$ be affine. Let $\overline{x}(\xi)$ be a global minimum point of (5.5.22). We have: **(i)** if g_s is convex, then $\overline{x}(\xi)$ is a global minimum point of (5.5.25) for $\xi \in \Xi \cap\,] - \infty, \xi^{(s)}[$; **(ii)** if g_s is concave, then $\overline{x}(\xi)$ is a global maximum point of (5.5.26) for $\xi \in \Xi \cap\,]\xi^{(s)}, +\infty[$.

Proof. By varying ξ, the feasible regions of (5.5.22) form a partition of R_s. Due to the strict convexity of f, the minima of f on the parts different from that corresponding to $\xi = \xi^{(s)}$ are greater than that achieved on the region identified by $\xi^{(s)}$. Then $f^{\downarrow}(\xi)$ turns out to be decreasing on $]-\infty, \xi^{(s)}[$ and increasing on $]\xi^{(s)}, +\infty[$. Now **(i)** and **(ii)** become consequences, respectively, of (i) and (ii) of Theorem 5.5.1. $\square$

Whether g_s be convex or concave, problem (5.5.22) is, in general, not convex, while either (5.5.25) or (5.5.26) is convex, according to g_s is convex or concave, respectively. This allows us to replace the analysis of a not convex problem with that of a convex one. Theorem 5.5.4 is a first result in this direction.

5.6. Connections between Discrete and Continuous Problems

Even though some discrete and combinatorial optimization problems have been studied since ancient times, the increase of their importance and their development has been very fast in the last few decades thanks to the possibility of practically solving them with modern computers and because of several appplications that they have found in many fields [I2, I36]. Such a recent and great interest in this branch of Mathematics has made up for a loss. Indeed, the development of Infinitesimal Analysis and its applications to Mechanics and Engineering braked the growth of disciplines related to combinatorial problems; only Number Theory has received a costant, though moderate, attention. However, even now, the remedy to the hystorical loss is not satisfactory: the field of discrete and combinatorial optimization problems − a rich mine for all mathematical activities − and that of so-called continuous problems have only a few interactions. This is a drawback for both fields and hence for the entire Mathematics.

In this section we give a connection between the two fields which, notwithstanding its simplicity and limitation, aims to be an invitation to break the "wall" which separates them. It should be clear that our point of view is that perfect "simmetry" between the two field is desirable: when a real problem rises, one should equally consider a discrete or a continuous formulation and choose the most suitable, and, in any case, to be able to jump from one to the other. Indeed, in some cases the distinction between discrete and continuous problems is artificial and counterfeit: a problem (1.1.3), with $X = \mathbb{R}^n$

and g differentiable, which has many local (but not global) minima perhaps might have had an advantage from a discrete formulation; there are problems which have been formulated as (1.1.7), while $x \in \mathbb{B}^n$ can be replaced by $x \in \text{conv}\,\mathbb{B}^n$.

The connection we want to show between the two fields is based on the penalty approach (see Vol.2), according to which a general result is achieved on the equivalence between the problem of minimizing a function on a set and that of minimizing a penalized function on a larger set [2, 41, 43]. The symbols of this section must be considered independent of those of the other sections, if they overlap; this certainly happens for G (see Sect. 3.1), z (see Sect. 3.2), μ (see Sect. 5.3), A (see Sects.1.1, 3.2, 4.4, 4.5, 4.7), C (see Sects. 1.1, 1.3, 1.4), and Φ (see Sect. 3.2, formula (3.2.28)).

Let $f : \mathbb{R}^n \to \mathbb{R}$, $G \subseteq \mathbb{R}^n$, $Z \subset \mathbb{R}^n$, and consider the following problem:

$$\min f(x), \quad \text{s.t.} \quad x \in G \cap Z. \tag{5.6.1}$$

Assume we are given a set $X \subset \mathbb{R}^n$ s.t. $Z \subseteq X$. Replacement of $G \cap Z$ with $G \cap X$ is known as relaxation of (5.6.1); it leads to a lower bound for the minimum of problem (5.6.1), which in general does not equal the minimum. Equality between them may be forced by means of a suitable penalization of the objective function of (5.6.1). To this end, let us introduce a function $\varphi : \mathbb{R}^n \to \mathbb{R}$, and consider the family $\{\mathcal{P}(\mu)\}_{\mu \in \mathbb{R}}$ of problems, where $\mathcal{P}(\mu)$ is defined as follows:

$$\min[f(x) + \mu\varphi(x)], \quad \text{s.t.} \quad x \in G \cap X; \tag{5.6.2}$$

it shows, with respect to (5.6.1), a relaxation of the feasible region and a penalization of the objective function.

It is immediate to note that (1.1.7) (or (1.1.1) with $X = \mathbb{Z}^n$) is a special case of (5.6.1) with $G = \{x \in \mathbb{R}^n : g(x) \in D\}$, $Z = \mathbb{B}^n$ (or with G as before and $Z = \mathbb{Z}^n$). (5.6.1) is of course of type (1.1.3), where $R = G \cap Z$; f and X are here as in Sect. 1.1.

We want to state conditions under which (5.6.1) and (5.6.2) are equivalent in the sense that they have the same minimum (or infimum: $+\infty$, if $G \cap Z = \varnothing$) and the same set of minimum points. If no requirement is made on φ and if f is bounded, then the answer is trivial: it is sufficient to choose:

$$\varphi(x) = \begin{cases} 0, & \text{if } x \in Z, \\ 1, & \text{if } x \in X \backslash Z, \end{cases}$$

and

$$\mu > \sup_{x \in X} f(x) - \inf_{x \in X} f(x)$$

to guarantee the equivalence between (5.6.1) and (5.6.2); this condition is an obvious deduction from the necessary and sufficient one. In fact, with the above choice of φ, and with suitable stipulations for the case of non-finiteness of the infima, the equivalence holds iff

$$\mu > \inf_{x \in G \cap Z} f(x) + \mu - \inf_{x \in G \cap (X \backslash Z)} f(x),$$

as it is easy to show: the infimum (or minimum, if it exists) in $\mathcal{P}(\mu)$ equals

$$\inf\left\{\inf_{x\in G\cap(X\setminus Z)}[f(x)+\mu],\quad \inf_{x\in G\cap Z}f(x)\right\}=$$

$$=\inf\left\{\mu+\inf_{x\in G\cap(X\setminus Z)}f(x),\quad \inf_{x\in G\cap Z}f(x)\right\};$$

hence $\mathcal{P}(\mu)$ is equivalent to $\mathcal{P}$, iff the above inequality holds.

Of course, the function φ above is discontinuous; thus the equivalence is of no much interest. The following theorem [41] gives a condition under which the above equivalence is achieved whithin the class of continuous functions φ. Fig. 5.6.1 illustrates the sets, which appear in Theorem 5.6.1; Fig. 5.6.2 illustrates (for $n=2$) the same sets, but in

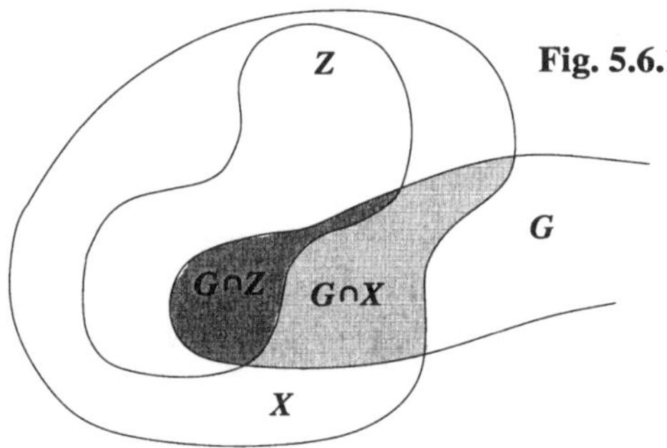

the particular case where $Z = \mathbb{B}^n$ and $X = \operatorname{conv} Z$; namely, X is the unit hypercube and Z the set of its vertices.

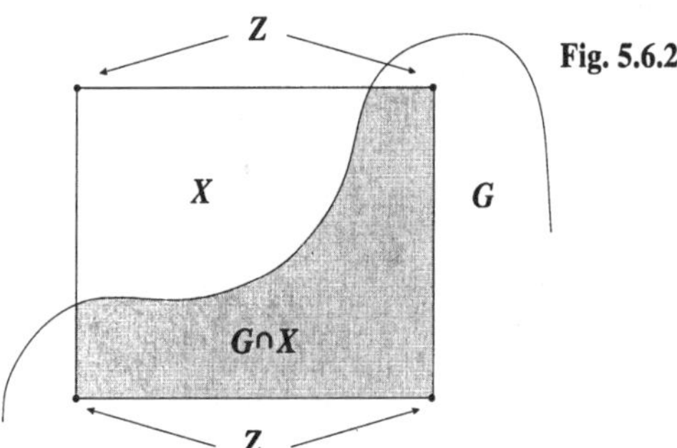

Theorem 5.6.1. Let $G \subseteq \mathbb{R}^m$ be closed, $Z \subseteq X \subset \mathbb{R}^n$, Z and X compact, and let the following hypotheses hold:

(H_1) $f : \mathbb{R}^n \to \mathbb{R}$ is bounded on X, and there exists an open set $\Omega \supset Z$ and real numbers $\alpha, L > 0$, such that $\forall x, y \in \Omega$, f fulfils the following Hölder condition:

$$|f(x) - f(y)| \ \leq \ L\|x - y\|^\alpha.$$

(H_2) It is possible to find $\varphi : \mathbb{R}^n \to \mathbb{R}$, such that:

(i) φ is continuous on X;

(ii) $\varphi(x) = 0$, $\forall x \in Z$, and $\varphi(x) > 0$, $\forall x \in X\backslash Z$;

(iii) $\forall z \in Z$, there exists a neighbourhood $N(z)$ of z, and a real $\varepsilon(z) > 0$, such that:

$$\varphi(x) \geq \varepsilon(z)\|x - z\|^{\alpha}, \quad \forall x \in N(z) \cap (X\backslash Z).$$

Then, a real μ_0 exists, such that $\forall \mu > \mu_0$ problems (5.6.1) and (5.6.2) are equivalent.

Proof. To prove the thesis, we will show that $\exists \mu_0 \in \mathbb{R}$, s.t. $\forall \mu > \mu_0$ the minimum of $f(x) + \mu\varphi(x)$ on $G \cap X$ is achieved necessarily at a point $\overline{z} \in G \cap Z$. Since $\varphi(\overline{z}) = 0$, $\forall \overline{z} \in G \cap Z$, we can conclude that the solution sets of problems (5.6.1) and (5.6.2) are the same whenever $\mu > \mu_0$. Let us introduce the sets $X_G := G \cap X$ and $Z_G := G \cap Z$. It will be shown that the function

$$F_{\overline{z}}(x) := \frac{f(\overline{z}) - f(x)}{\varphi(x)}, \quad x \in \Omega \cap (X_G\backslash Z_G), \ \overline{z} \in Z_G,$$

is bounded in some neighbourhood of $\overline{z}$. To see this, consider $\overline{N}(\overline{z}) := \Omega \cap N(\overline{z})$. $\forall x \in \overline{N}(\overline{x}) \cap (X_G\backslash Z_G)$, we have:

$$\varphi(x) \geq \varepsilon(\overline{z})\|x - \overline{z}\|^{\alpha}, \quad \varepsilon(\overline{z}) > 0,$$

$$|f(x) - f(\overline{z})| \ \leq \ L\|x - \overline{z}\|^{\alpha},$$

so that

$$|F_{\overline{z}}(x)| \ < \ \frac{L}{\varepsilon(\overline{z})} \ < +\infty.$$

The family $\{\overline{N}(z)\}_{z \in Z_G}$ is obviously a cover of Z_G. Since Z is compact and Z_G is a closed subset of Z, so that Z_G is compact, there exists a finite subfamily, say $\{\overline{N}(z^i)\}_{i=1}^{k}$, which is a cover of Z_G. Consider the set:

$$S := \left[\bigcup_{i=1}^{k} \overline{N}\left(z^i\right)\right] \bigcap X_G \supseteq Z_G.$$

It is clear that $\mu > \beta := \max\{L/\varepsilon(z^i), i = 1, ..., k\}$ implies:

$$f(x) + \mu\varphi(x) \ > \ f(z^i), \quad \forall x \in S\backslash Z_G, \ i = 1, ..., k. \tag{5.6.3}$$

On the other side, the set

$$X_0 := X_G\backslash S = X_G \bigcap \left[\mathbb{R}^n \ \backslash \ \bigcup_{i=1}^{k} \overline{N}\left(z^i\right)\right]$$

is compact, and we have:

$$X_G = X_0 \bigcup (S\backslash Z_G) \bigcup Z_G, \quad X_0 \bigcap Z_G = \varnothing,$$

$$X_0 \bigcap (S\backslash Z_G) = \varnothing, \quad (S\backslash Z_G) \bigcap Z_G = \varnothing.$$

Besides, f is bounded on X_G and thus:

$$M_f := \inf_{x \in X_G} f(x) > -\infty;$$

φ is continuous and positive on X_0, and thus:

$$M_\varphi := \inf_{x \in X_0} \varphi(x) = \min_{x \in X_0} \varphi(x) > 0.$$

Since f is bounded on X_G, we have:

$$\gamma_0 := \sup_{x \in X_G} \frac{f(x) - M_f}{M_\varphi} < +\infty$$

and, of course, $\gamma_0 \geq 0$. If $\mu > \gamma_0$, then, by the definition of γ_0, $\forall z \in Z_G$ and $\forall x \in X_0$, we have:

$$f(x) + \mu\varphi(x) > M_f + \gamma_0 M_\varphi \geq f(z). \tag{5.6.4}$$

Inequalities (5.6.3) and (5.6.4) hold, if

$$\mu > \mu_0 := \max\{\beta, \gamma_0\}.$$

Therefore, $f(x) + \mu\varphi(x)$ cannot have its minimum either at a point in X_0, for this would not agree with (5.6.4), or at one in $S \backslash Z_G$, for this would not agree with (5.6.3). $\quad\square$

In some applications of Theorem 5.6.1, where $f(x) + \mu\varphi(x)$ cannot be convex, it is useful to be able to choose φ strictly concave. This happens, for instance, for (1.1.7), as it will be shown later. An extensive treatment of both the theory and methods of concave minimization problems can be found in [102,103,138]. If f is not concave, then (5.6.2) may require the minimization of an indefinite form, which may be undesiderable. The following theorem states a condition under which the objective function of (5.6.2) is strictly concave; this will be done for the case:

$$X = X_Q := \{x \in \mathbb{R}^n : O \leq x \leq e\}, \ \varphi(x) = \langle x, e - x \rangle, \tag{5.6.5}$$

where $e := (1, ..., 1)^T$. Notwithstanding the fact that X_Q be a special case of X, note that the unit hypercube X_Q is the usual relaxation of the constraint $x \in \mathbb{B}^n$ of (1.1.7) and hence it is of great importance.

Theorem 5.6.2. If $f \in C^2(X_Q)$ and $Z \subset X_Q$, then $\exists \mu_1 \in \mathbb{R}$, such that, $\forall \mu > \mu_1$, problems (5.6.1) and (5.6.2) with $X = X_Q$ are equivalent and (5.6.2) has a strictly concave objective function.

Proof. Let $H(x)$ and $\hat{H}(x)$ be the Hessian matrices of f and of $f + \mu\langle x, e - x \rangle$, respectively. We have:

$$\hat{H}(x) = H(x) - 2\mu I_n,$$

where I_n denotes the identity matrix of order n. H is continuous. Because of a well known theorem on linear and continuous operators in a normed space, the continuity of the map $H : \mathbb{R}^n \to \mathbb{R}^{n \times n}$ implies the continuity of the function $\gamma : \mathbb{R}^n \to \mathbb{C}$, where $\gamma(x)$ denotes any eigenvalue of $H(x)$. Let $\gamma_1(x), ..., \gamma_n(x)$ be the eigenvalues of $H(x)$; these are bounded since X is compact. Thus

$$\tilde{\gamma} := \max_{i=1,\dots,n} \sup_{x \in X} |\gamma_i(x)| < +\infty.$$

$\forall x \in X, \nu$ is a (real) eigenvalue of the (symmetric) matrix $\hat{H}(x)$ iff

$$\det[\hat{H}(x) - \nu I_n] = \det[H(x) - (2\mu + \nu)I_n] = 0.$$

Therefore, ν is an eigenvalue of $\hat{H}(x)$ iff $\gamma := \nu + 2\mu$ is an eigenvalue of $H(x)$. If μ_0 is defined as in the proof of Theorem 5.6.1, and

$$\mu \geq \mu_1 := \max\{\tfrac{1}{2}\tilde{\gamma}, \mu_0\},$$

then (5.6.1) and (5.6.2) are equivalent and, furthermore,

$$\nu = \gamma - 2\mu \leq \gamma - \tilde{\gamma} < 0.$$

Hence, $\hat{H}(x)$ is negative definite. $\qquad\qquad\square$

The closedness of X_Q and the assumption $f \in C^2(X_Q)$ in Theorem 5.6.2 cannot be weakened, as the following example shows.

Example 5.6.1. Set $n = 1$, $X_Q = [0, 1]$, $\varphi(x) = x(1 - x)$, and

$$f(x) = \begin{cases} -x^3 \sin \frac{1}{x}, & \text{if } x \neq 0, \\ 0, & \text{if } x = 0, \end{cases}$$

so that

$$f'(x) = \begin{cases} -3x^2 \sin \frac{1}{x} + x \cos \frac{1}{x}, & \text{if } x \neq 0, \\ 0, & \text{if } x = 0, \end{cases}$$

$$f''(x) = \begin{cases} (\frac{1}{x} - 6x) \sin \frac{1}{x} + 4 \cos \frac{1}{x}, & \text{if } x \neq 0, \\ \text{undefined}, & \text{if } x = 0, \end{cases}$$

$$f''(x) + \mu\varphi''(x) = \begin{cases} (\frac{1}{x} - 6x) \sin \frac{1}{x} + 4 \cos \frac{1}{x} - 2\mu, & \text{if } x \neq 0, \\ \text{undefined}, & \text{if } x = 0, \end{cases}$$

$\forall \mu \in \mathbb{R}_+\backslash\{0\}$, we can find $x \in [0, 1]$ (close enough to zero), s.t. $f''(x) + \mu\varphi''(x) > 0$. Indeed, it is sufficient to choose

$$x = \frac{2}{(8\lceil\mu\rceil + 13)\pi},$$

so that the above inequality becomes:

$$\pi^2(8\lceil\mu\rceil + 13)^2 - 4\pi(8\lceil\mu\rceil + 13)\mu - 24 > 0,$$

and is verified for $\mu \geq 1$. Hence, $f(x) + \mu\varphi(x)$ is not concave in $[0, 1]$ even if $f \in C^1([0, 1])$ and f has continuous second derivative in $]0,1]$. $\qquad\square$

Now, let us consider a special case of (5.6.1), which embraces most combinatorial extremum problems, and set:

$$G = \{x \in \mathbb{R}^n : g_i(x) \geq 0, i \in \mathfrak{I}^+\} \quad \text{and} \quad Z = \mathbb{B}^n, \tag{5.6.5}*$$

so that G is the feasible region R of Sect. 1.1 for $p = 0$ and $X = \mathbb{R}^n$. Thus (5.6.2) becomes (1.1.7) for $p = 0$, namely:

$$\min f(x), \quad \text{s.t.} \quad g(x) \geq 0, \ x \in \mathbb{B}^n. \tag{5.6.6}$$

The case where, in (5.6.6.), $x \in \mathbb{B}^n$ is replaced by $x \in \mathbb{Z}^n$ can be reduced to (5.6.6) by means of well known devices, like binary expansion. The usual relaxation of $Z = \mathbb{B}^n$ and the penalization when G is defined by (5.6.5) are, respectively, the hypercube X_Q and $\varphi(x) = \langle x, e - x \rangle$ previously considered; with this choice, (5.6.2) becomes:

$$\min[f(x) + \mu\langle x, e - x \rangle], \quad \text{s.t.} \quad g(x) \geq O, \ O \leq x \leq e. \tag{5.6.7}$$

Theorem 5.6.1 becomes here:

Theorem 5.6.3. Let f verify assumption (H_1) Theorem 5.6.1 with $\alpha = 1$, namely let f be bounded on X_Q and Lipschitz continuous on an open set $\Omega \supset Z = \mathbb{B}^n$. Then, there exists $\mu_0 \in \mathbb{R}$, such that, for every $\mu \geq \mu_0$, problems (5.6.6) and (5.6.7) are equivalent.

Proof. We only need to prove that $\varphi(x) = \langle x, e - x \rangle$ satisfies assumption (H_2) of Theorem 5.6.1. Note that (i) and (ii) are trivially true. We will now prove that (iii) holds with

$$N(z) = \{x \in \mathbb{R}^n : \|x - z\| \leq \overline{\rho} < 1\}$$

and $\varepsilon(z) = 1 - \overline{\rho}$. To see this, consider $\rho \in \mathbb{R}$ and $t = (t_1, ..., t_n)$ satisfying:

$$\rho := \|x - z\| \leq \overline{\rho}, \qquad t := \frac{1}{\rho}(x - z).$$

Then

$$\varphi(x) = \sum_{j=1}^{n} (\rho t_j + z_j)(1 - z_j - \rho t_j). \tag{5.6.8}$$

Since $z + \rho t = x \in X_Q$, then $t_j > 0$ implies $z_j = 0$ and $t_j < 0$ implies $z_j = 1$; therefore, from (5.6.8) we draw:

$$\varphi(x) = \sum_{t_j > 0} \rho t_j (1 - \rho t_j) + \sum_{t_j < 0} (1 + \rho t_j)(-\rho t_j) =$$

$$= \sum_{j=1}^{n} \rho|t_j|(1 - \rho|t_j|) = \rho\sum_{j=1}^{n} |t_j| - \rho^2\sum_{j=1}^{n} |t_j|^2. \tag{5.6.9}$$

Since

$$\rho \leq \overline{\rho} < 1, \quad \sum_{j=1}^{n} |t_j|^2 = \|t\|^2 = 1, \quad \sum_{j=1}^{n} |t_j| \geq \|t\| = 1,$$

from (5.6.9) we obtain:

$$\varphi(x) \geq \rho(1 - \overline{\rho}) = \varepsilon(z)\|x - z\|. \qquad \square$$

When f is linear (or quadratic) and g affine, Theorem 5.6.3 states an *equivalence between the linear (or quadratic) combinatorial minimization* (5.6.6) *and the minimization, over the reals,* (5.6.7) which, because of Theorem 5.6.2, is a strictly concave quadratic problem, if μ is large enough. An analogous remark can be made in the more general case $f \in C^2(X_Q)$. This condition is not redundant, as it may be shown by Example 5.6.1.

When the equivalence between (5.6.6) and (5.6.7) holds, properties and methods valid for one of the two problems can be transferred to the other one. As an instance of this, consider the case:

$$f(x) = \langle c, x \rangle + \tfrac{1}{2}\langle x, Cx \rangle, \quad g(x) = Ax - b,$$

where $b \in \mathbb{R}^m$, $c \in \mathbb{R}^n$, $A \in \mathbb{R}^{m \times n}$ and $C \in \mathbb{R}^{n \times n}$ is symmetric.

Theorem 5.6.4. If $\mu \in \mathbb{R}$ is large enough, then the combinatorial quadratic problem

$$\min(\langle c, x \rangle + \tfrac{1}{2}\langle x, Cx \rangle), \quad \text{s.t.} \quad Ax \geq b, \; x \in \mathbb{B}^n \tag{5.6.10}$$

is equivalent to the linear complementarity problem

$$\min\langle \tilde{c}, y \rangle, \quad \text{s.t.} \quad \tilde{A}y + t = \tilde{b}, \; y \geq O, \; t \geq O, \; \langle y, t \rangle = 0, \tag{5.6.11}$$

where $y, t \in \mathbb{R}^{2n+m}$ and

$$\tilde{c} := \frac{1}{2}\begin{pmatrix} c + \mu e \\ b \\ -e \end{pmatrix}, \quad \tilde{b} := \begin{pmatrix} c + \mu e \\ -b \\ e \end{pmatrix}, \quad \tilde{A} := \begin{pmatrix} -C + 2\mu I_n & A^T & -I_n \\ -A & O & O \\ I_n & O & O \end{pmatrix}.$$

Proof. Because of Theorem 5.6.3, whose hypotheses are trivially satisfied, (5.7.10) is equivalent to the quadratic problem:

$$\min[\langle c + \mu e, x \rangle + \tfrac{1}{2}\langle x, (C - 2\mu I_n)x \rangle], \quad \text{s.t.} \quad Ax \geq b, \; O \leq x \leq e, \tag{5.6.12}$$

if μ is large enough. According to Theorem 5.3.1 (where we consider (1.1.1), and set $p = 0$, $\mathfrak{J}^o = \varnothing$, $\mathfrak{J} = \mathfrak{J}^+$, $D = \mathbb{R}^{m+2n}$, so that (5.3.15a)''''-(5.3.15b,c) is to be adopted), a necessary condition for x to be a m.p. of (5.6.12) is that there exist multipliers $\lambda \in \mathbb{R}^m_+$, $\nu \in \mathbb{R}^n_+$, $\tau \in \mathbb{R}^n_+$ (associated, respectively, with the constraints $Ax \geq b$, $x \leq e$, $x \geq 0$), s.t.

$$c + \mu e + (C - 2\mu I_n)x - A^T\lambda + \nu - \tau = O, \tag{5.6.13a}$$

$$Ax - r = b, \qquad x + s = e, \tag{5.6.13b}$$

$$x, s, \nu, \tau \in \mathbb{R}^n_+, \qquad r, \lambda \in \mathbb{R}^m_+, \tag{5.6.13c}$$

$$\langle x, \tau \rangle = \langle \lambda, r \rangle = \langle \nu, s \rangle = 0, \tag{5.6.13d}$$

where r and s are slack variables. Solving (5.6.12) is equivalent to finding, among the solutions of the complementarity system (5.6.13) — which are stationary points (see Sect.3.3) — , those which minimize the function in square brackets of (5.6.12). Such a function, evaluated at the stationary points, *"becomes linear"*:

$$\tfrac{1}{2}[\langle c + \mu e, x\rangle + \langle b, \lambda\rangle - \langle e, \nu\rangle].$$

In fact, (5.6.13a) implies:

$$Cx - 2\mu x = -(c + \mu e) - A^T\lambda - \nu + \tau;$$

from (5.6.13d) we have:

$$
\begin{aligned}
\langle x, \tau\rangle &= 0, \\
0 &= \langle\lambda, r\rangle = \langle\lambda, b\rangle - \langle\lambda, Ax\rangle, \\
0 &= \langle\nu, s\rangle = \langle\nu, e\rangle - \langle\nu, x\rangle,
\end{aligned}
$$

and therefore:

$$
\begin{aligned}
\langle\lambda, b\rangle &= \langle\lambda, Ax\rangle, \\
\langle\nu, e\rangle &= \langle\nu, x\rangle.
\end{aligned}
$$

Now, to achieve the thesis, it is sufficient to set:

$$
y = \begin{pmatrix} x \\ \lambda \\ \nu \end{pmatrix}, \qquad t = \begin{pmatrix} \tau \\ r \\ s \end{pmatrix}. \qquad\qquad \square
$$

Note that no assumption has been made on the matrix C, so that the convex case, as well as the nonconvex one, have been considered. See also [73] for a reduction of the so-called mixed integer feasibility problem to a linear complementarity problem.

The previous results can be extended to a generalized system of type (1.3.16). The great development of the theory of constrained extrema (Sect. 1.1) and, more recently, that of Variational Inequalities and Complementarity Systems (Sect. 1.3) has led to search for mathematical models which embody both theories. A possible answer is offered by (1.3.16), which here is written in the following form:

$$A(x; \xi) \in \mathcal{H}, \qquad x \in G \cap Z, \tag{5.6.14}$$

where A is the mapping of (1.3.16) (and not the matrix of Theorem 5.6.4), $\mathcal{H}$ is the cone of (1.3.16), G, Z and X are as in (5.6.1)-(5.6.2); X differs formally from that of Sect. 1.3, since here we aim to perform a relaxation like for (5.6.1).

Let us call $\mathcal{P}$ the problem, which consists in finding $\xi \in G \cap Z$, such that the system (in the unknown x) (5.6.14) be impossible.

Consider the vector-valued function $\Phi : \mathbb{R}^n \times \Xi \to \mathbb{R}^\nu$, and the family $\{\mathcal{P}(\mu)\}_{\mu\in\mathbb{R}}$ of problems, where $\mathcal{P}(\mu)$ consists in finding $\xi \in G \cap X$ such that the system (in the unknown x):

$$A(x; \xi) + \mu\Phi(x; \xi) \in \mathcal{H}, \quad x \in G \cap X, \tag{5.6.15}$$

be impossible (Φ is different from that of (3.2.28)).

It is easy to see that (3.2.1) is a special case of (5.6.14), so that this system can be considered as an extension of (5.6.1), in the sense that the optimality of (5.6.1) can be reduced to the problem $\mathcal{P}$, by following the approach of Sect. 3.2. System (3.2.26) escapes from the format (5.6.14); however, it can be reduced to (5.6.14) by means of the selection function (3.2.28); see also Sect.4.7.

The replacement of $G \cap Z$ with $G \cap X$ represents a relaxation of the domain of $\mathcal{P}$; of course, this may change the solutions of $\mathcal{P}$; the purpose is to counterbalance this drawback with the penalization given by Φ; namely, we want to state conditions under which $\mathcal{P}$ and $\mathcal{P}(\mu)$ are equivalent, in the sense that they have the same set of solutions (if any, or none of them has solutions). In the sequel, $\forall x \in X$, $z(x)$ will denote a vector belonging to the set $\mathrm{proj}_Z x$, where $\mathrm{proj}_Z : X \rightrightarrows Z$ is a multivalued function, which projects x on Z.

Theorem 5.6.5. Let $G \subset \mathbb{R}^n$ be a closed set, $Z \subseteq X \subset \mathbb{R}^n, Z$ and X be compact, and let the following hypotheses hold:

(H_1) $A : X \times X \to \mathbb{R}^\nu$ is bounded on $X \times X$, and there exist an open set $\Omega \supset Z$ and

real numbers $\alpha, L > 0$, such that:

$$\|A(z(x); x)\| \le L\|x - z(x)\|^\alpha, \quad \forall x \in \Omega \cap X, \ \forall z(x) \in \mathrm{proj}_Z x.$$

(H_2) It is possible to find $\Phi : X \times X \to \mathbb{R}^\nu$, such that:

(i) Φ is continuous on $X \times X$;

(ii) $\forall x, y \in Z$, $\Phi(x; y) = O$;

(3i) there exists a closed cone $\mathcal{H}^+$ with $\varnothing \ne (\mathcal{H}^+ \backslash \{O\}) \subseteq \mathrm{int} \ \mathcal{H}$, such that:

$$\Phi(x; y) \in (\mathcal{H}^+ \backslash \{O\}), \quad \forall x \in Z, \quad \forall y \in X \backslash Z;$$

(4i) $\forall z \in Z$, there exist a neighbourhood $N(z)$ of z and a real $\varepsilon(z) > 0$, such

that:

$$\|\Phi(z(x); x)\| \ge \varepsilon(z)\|x - z(x)\|^\alpha, \forall x \in N(z) \cap (X \backslash Z),$$

$$\forall z(x) \in \mathrm{proj}_Z x.$$

Then, a real μ_0 exists, such that, $\forall \mu > \mu_0$, a solution of $\mathcal{P}(\mu)$ is a solution of $\mathcal{P}$.

Proof. To prove the thesis, it is sufficient to show that $\exists \mu_0 \in \mathbb{R}$, s.t. $\forall \mu > \mu_0$ a solution of $\mathcal{P}(\mu)$ is achieved necessarily on $G \cap Z$; because of (H_2)(ii), this claim assures that a solution of $\mathcal{P}(\mu)$ is a solution of $\mathcal{P}$ too. Let us introduce the sets $X_G := G \cap X$, $Z_G := G \cap Z$, $\overline{N}(z) := \Omega \cap N(z)$, where $N(z)$ is precisely that of (H_2)(4i). The family $\{\overline{N}(z)\}_{z \in Z_G}$ is obviously a cover of Z_G; since Z is compact and Z_G is a closed subset of Z, there is a finite subfamily, say $\{\overline{N}(z^i)\}_1^k$, which is still a cover of Z_G. Set:

$$S := \bigcup_{i=1}^k \overline{N}(z^i), \qquad \rho := \max\left\{ \frac{L}{\varepsilon(z^i)}, i = 1, ..., k \right\}.$$

Because of (H_1) and (H_2) (4i), we have:

$$\left\| \frac{1}{\|\Phi(z(x);x)\|} A(z(x);x) \right\| \leq \rho, \quad \forall x \in S \cap (X_G \backslash Z_G).$$

Consider the set $S_= := \{ x \in \mathbb{R}^\nu : \|x\| = 1 \}$; because of $(H_2)(3i)$, we have:

$$\frac{1}{\|\Phi(z(x);x)\|} \Phi(z(x);x) \in \mathcal{H}^+ \cap S_=, \quad \forall x \in X_G \backslash Z_G. \tag{5.6.16}$$

We can apply Theorem 2.1.5: $n, K, \tilde{K}, V_1$ and V_2 are identified with $\nu, \mathcal{H}, \mathcal{H}^+$,

$$\frac{1}{\|\Phi(z(x);x\|)} A(z(x);x) \quad \text{and} \quad \frac{1}{\|\Phi(z(x);x)\|} \Phi(z(x);x),$$

respectively. The assumptions of Theorem 2.1.5 being fulfilled, we achieve the existence of a real η_0, s.t. (2.1.21) holds, namely, $\forall \eta > \eta_0$ and $\forall x \in S \cap (X_G \backslash Z_G)$, we have:

$$\frac{1}{\|\Phi(z(x);x)\|} A(z(x);x) + \eta \frac{1}{\|\Phi(z(x);x)\|} \Phi(z(x);x) \in \mathcal{H}. \tag{5.6.17}$$

It follows that, $\forall \mu > \eta_0$, $\mathcal{P}(\mu)$ cannot have solutions in $S \cap (X_G \backslash Z_G)$. Now, let us introduce the compact set $X_0 := X_G \backslash S$, and fix $\hat{z} \in Z_G$. Because of $(H_2)(i,3i)$, Φ is continuous and different from the null vector on the compact set $\{\hat{z}\} \times X_0$. Then we find:

$$M_\Phi := \min_{x \in X_0} \|\Phi(\hat{z};x)\| > 0.$$

Set

$$M_A := \sup_{(x,y) \in X \times X} \|A(x;y)\|.$$

We can apply Theorem 2.1.5: we choose $\rho = M_A / M_\Phi$; $n, K, \tilde{K}, V_1$ and V_2 are identified with $\nu, \mathcal{H}, \mathcal{H}^+$,

$$\frac{1}{\|\Phi(\hat{z};x)\|} A(\hat{z};x) \quad \text{and} \quad \frac{1}{\|\Phi(\hat{z};x)\|} \Phi(\hat{z};x),$$

respectively. Then, the hypotheses of Theorem 2.1.5 being satisfied, we achieve the existence of η_1, s.t. $\forall \eta > \eta_1$ and $\forall x \in X_0$, we have:

$$\frac{1}{\|\Phi(\hat{z};x)\|} A(\hat{z};x) + \eta \frac{1}{\|\Phi(\hat{z};x)\|} \Phi(\hat{z};x) \in \mathcal{H}. \tag{5.6.18}$$

Hence, $\forall \mu > \eta_1$, $\mathcal{P}(\mu)$ cannot have solutions in X_0. If $\mu > \mu_0 := \max\{\eta_0, \eta_1\}$, account taken of (5.6.17) and (5.6.18), $\mathcal{P}(\mu)$ cannnot have solutions in $X_G \backslash Z_G$. $\quad \square$

Let the impossibility of (5.6.14) express (like in (3.2.1)) the optimality of (1.1.8) in the particular case where $R = G \cap Z$ with $Z = \mathbb{B}^n$. In (5.6.14) set $\nu = \ell$, $\xi = \overline{x}$, $\mathcal{H} = C_0$. In this case:

$$A(x;\xi) = A(x;\overline{x}) = f(\overline{x}) - f(x). \tag{5.6.19}$$

Then, the impossibility of (5.6.14) is equivalent to (1.1.9) (and to (3.2.1), if $\nu = \ell = 1$). Let f be bounded; it is trivial to check that, if f fulfils the Hölder Condition on Ω. i.e.

$\exists \alpha, L > 0$, s.t.

$$\|f(x) - f(y)\| \leq L \|x - y\|^{\alpha}, \qquad \forall x, y \in \Omega \cap X, \tag{5.6.20}$$

then (5.6.19) satisfies (H_1) of Theorem 5.6.5; the converse is not true as shown by Example 5.6.2. An analogous remark holds for (1.1.10).

Example 5.6.2. In (5.6.14), set $n = \nu = 1, G = \mathbb{R}, Z = [0,1]$ and $X = [-1, 1]$, and in (5.6.19) set:

$$f(x) = \begin{cases} x \sin \frac{1}{x}, & \text{if } x \neq 0, \\ 0, & \text{if } x = 0. \end{cases}$$

(H_1) of Theorem 5.6.5 is satisfied for $\alpha = 1, L = 1$ and $\Omega =]-2, 2[$. In fact, if $x \in Z$ — so that $z(x) = x$ —, then $A(z(x); x) = 0$; if $x \in (\Omega \cap X) \backslash Z = X \backslash Z = [-1, 0[$, then $|A(0; x)| = |f(x)| = |x \sin \frac{1}{x}| \leq |x|$. Thus (H_1) holds. Of course, f does not fulfil (5.6.20). $\square$

With the notation of Sect. 1.3, let us now consider two special cases for the map A (with $\nu = \ell$) :

$$A(x; \overline{x}) = F(\overline{x})(\overline{x} - x), \tag{5.6.21}$$

$$A(x; \overline{x}) = F(x)(\overline{x} - x), \tag{5.6.22}$$

which correspond to (1.3.8) and (1.3.9), respectively. If F is bounded on X, then functions (5.6.21)-(5.6.22) fulfil (H_1) of Theorem 5.6.5 (for $\alpha = 1$ and $L = \|F\|_\infty$), as it is easy to check. Example 5.6.3 shows that the converse is not true.

In (5.6.14), set (5.6.21) (or (5.6.22)), $\xi = \overline{x}$ and $\mathcal{H} = C_0$, so that $\overline{x}$ is a solution of $\mathcal{P}$, iff it is a solution of (1.3.8) (or (1.3.9)), where $\mathbb{K} = G \cap Z$. The application of Theorem 5.6.5 allows us to relax the domain of (1.3.8) (or (1.3.9)) without modifying the set of solutions, on condition that the left-hand side of (1.3.8) (or (1.3.9)) receives a suitable change; this will be exploited in Vol. 2. The same can be done for (1.3.10) and (1.3.11).

Example 5.6.3. In (5.6.14), set $n = \nu = 1, \mathcal{H} = \mathbb{R}_+ \backslash \{0\}, G = \mathbb{R}, Z = [0, 1]$ and $X = [0, 2]$; in (5.6.21), set $\ell = 1$ and

$$F(x) = \begin{cases} \frac{1}{\sqrt{|x-1|}}, & \text{if } x \neq 1, \\ 0, & \text{if } x = 1. \end{cases}$$

In order to agree with the notation of Theorem 5.6.5, $\overline{x}$ and x are replaced by x and $z(x)$, respectively. In case (5.6.21), we find:

$$|A(z(x); x)| \leq \sqrt{|x - z(x)|}, \qquad \forall x \in [1, 2].$$

In fact, if $x \in Z$ so that $z(x) = x$, then $A(z(x); x) = 0$; if $x \in]1, 2]$, then $z(x) = 1$ so that $|A(z(x); x)| = (x-1)^{1/2}$. Therefore (H_1) of Theorem 5.6.5 is satisfied (with $\Omega = \mathbb{R}$, for instance), but F is not bounded on X. In case (5.6.22), we find $A(z(x); x) \equiv 0$ on X. Again (H_1) is fulfilled, while F is not bounded. $\square$

Example 5.6.4. Let $A : \mathbb{R} \times \mathbb{R} \to \mathbb{R}$ with $A(x;y) = \sqrt{|x-y|}(x-1)$, $Z = [0,1]$, $X = [0,2]$. Such a function does not fulfil the following condition: there exist a constant L and an open set $\Omega \supset Z$, s.t.

$$|A(x;y)| \le |x-y|, \quad \forall x \in \Omega \cap X, \quad \forall y \in Z.$$

It is immediate to see that such a function fulfils hypothesis (H_1) of Theorem 5.6.5 for $\Omega = \mathbb{R}, L = \sqrt{2}, \alpha = 3/2$. $\qquad\square$

Example 5.6.5. Let us set $n = \nu = \ell = 1, G = \mathbb{R}, \mathcal{H} = \mathbb{R}_+ \backslash \{0\}, Z = [0,1], X = [0,2]$, $A : X \times X \to \mathbb{R}$ with $A(x;\xi) = (x-\xi)^2(1-\xi)(x-1)$. Such a function A fulfils (H_1) of Theorem 5.6.5 for $\Omega =]-1,2[, L = 1, \alpha = 2$. In fact, A is bounded; moreover, $x \in]1,2[$ implies $z(x) = 1$ and $A(1;x) = 0$, $x \in Z$ implies $z(x) = x$ and $A(x;x) = 0$. Each $\xi \in [0,1]$ is a solution of the following particular case of $\mathcal{P}$: find $\xi \in Z$, s.t.

$$A(x;\xi) \in \mathcal{H}, \quad x \in [0,1]$$

be impossible. The penalty function Φ of (5.6.15) is now of type $\Phi : X \times X \to \mathbb{R}$; let it be given by

$$\Phi(x;\xi) = \begin{cases} -(1-x)^2, & \text{if } (x;\xi) \in]1,2] \times [0,1], \\ 0, & \text{if } (x;\xi) \in [0,1] \times [0,1], \\ (1-\xi)^2, & \text{if } (x;\xi) \in [0,1] \times [1,2], \\ -(x-\xi)^2, & \text{if } (x;\xi) \in]1,2] \times]1,2], \ \xi \le x, \\ (x-\xi)^2, & \text{if } (x;\xi) \in]1,2] \times]1,2], \ \xi > x. \end{cases}$$

Such a Φ fulfils (H_2) of Theorem 5.6.5: it is enough to choose $\mathcal{H}^+ = \mathbb{R}_+, \varepsilon(z) = 1$ $\forall z \in Z, \alpha = 2, L = 1$. We show that, $\forall \mu \in \mathbb{R}_+, \xi \in [0,1]$ is not a solution of $\mathcal{P}(\mu)$. In fact, $\forall x \in [0,,2]$,

$$A(x;\xi) + \mu\Phi(x;\xi) = (x-1)[(x-\xi)^2(1-\xi) - \mu(x-1)].$$

Observe that $\lim\limits_{x\downarrow 1}[(x-\xi)^2(1-\xi) - \mu(x-1)] = (1-\xi)^3 > 0$. We conclude that $\xi \in [0,1]$ is not a solution of $\mathcal{P}(\mu)$. Observe that A does not fulfil (H_1) of the following Theorem 5.6.6 for $\alpha = 2, L = 1$. In fact, for $x \ge 1$, $z(x) = 1$; then, the inequality

$$|A(z(x);\xi) - A(x;\xi)| \le (x-1)^2, \quad \forall x \in X \cap \Omega, \quad \forall z(x) \in \text{proj}_Z x$$

holds iff

$$(x-\xi)^2(1-\xi) \le x-1, \quad \forall x > 1, \quad \forall \xi \in [0,1].$$

This is impossible for $x = 5/4, \xi = 1/4$. Moreover, note that $\mathcal{P}(\mu)$ has not solutions: $\forall \xi \in]1,2]$ and $\forall x \in [0,1]$, we have:

$$A(x;\xi) + \mu\Phi(x;\xi) \in \mathcal{H}, \quad \forall \mu > 0. \qquad\square$$

If Z is finite, then the inequality in $(H_2)(4i)$ of Theorem 5.6.5 can be equivalently replaced (in the sense that the thesis of Theorem 5.6.5 is still achieved and the class of the penalty functions Φ which fulfil it is non-empty) with the following condition:

(H_2)(4i)′ $\forall z \in Z$, there exist a neighbourhood $N(z)$ of z and a real $\varepsilon(z) > 0$, s.t.

$$\|\Phi(z; x)\| \geq \varepsilon(z)\|x - z\|, \quad \forall x \in N(z) \cap (X \backslash Z).$$

In fact, by choosing a suitable neighbourhood $N(z)$ of z, we have $z(x) = z$.

If Z is not finite, then the above condition might be in contrast with assumptions (H_2)(i,ii) of theorem 5.6.5.

Now, let us consider the following condition:

(H_2)″ It possible to find a vector-valued function $\phi : X \to \mathbb{R}^\nu$, such that:

    (i) ϕ is continuous on X;

    (ii) $\forall z \in Z$, $\phi(z) = O$;

    (3i) there exists a closed cone $\mathcal{H}^+$ with $\varnothing \neq (\mathcal{H}^+ \backslash \{O\}) \subseteq \operatorname{int} \mathcal{H}$, such that:

$$\phi(x) \in (\mathcal{H}^+ \backslash \{O\}), \quad \forall x \in X \backslash Z;$$

    (4i) $\forall z \in Z$, there exist a neighbourhood $N(z)$ and a real $\varepsilon(z) > 0$, such that:

$$\|\phi(x)\| \geq \varepsilon(z)\|x - z(x)\|^\alpha,$$

$$\forall x \in N(z) \cap (X \backslash Z), \quad \forall z(x) \in \operatorname{proj}_Z x.$$

Note that, if we set $\Phi(x; \xi) = \phi(\xi) - \phi(x)$ $\forall x, \xi \in X$, then (H_2) of Theorem 5.6.5 is fullfilled, if (H_2)″ holds.

The following theorem deals with the special case of problem $\mathcal{P}$, where $\mathcal{H} = \mathbb{R}^\nu_+$; it gives a condition which assures that a solution of $\mathcal{P}$ is a solution also of a suitable problem $\mathcal{P}(\mu)$ for μ large enough.

Theorem 5.6.6. Let $G \subset \mathbb{R}^n$ be a closed set, $Z \subseteq X \subset \mathbb{R}^n$, Z and X be compact, and let the following hypotheses hold:

(H_1) $A : X \times X \to \mathbb{R}^\nu$ is bounded on $X \times X$, and there exist an open set $\Omega \supset Z$ and

    real numbers $\alpha, L > 0$, such that:

$$\|A(x; \xi) - A(z(x); \xi)\| \leq L\|x - z(x)\|^\alpha, \quad \forall x, \xi \in X, \quad \forall z(x) \in \operatorname{proj}_Z x.$$

(H_2) It is possible to find $\Phi : X \times X \to \mathbb{R}^\nu$, such that:

    (i) Φ is continuous on $X \times X$;

    (ii) $\forall x, \xi \in Z, \Phi(x; \xi) = O$; $\forall x \in X$, $\Phi(x; \bullet)$ is constant on Z;

    (3i) there exists a closed cone $\mathcal{H}^-$, with $\varnothing \neq (\mathcal{H}^- \backslash \{O\}) \subseteq \operatorname{int}(-\mathcal{H})$, such that:

$$\Phi(x; \xi) \in (\mathcal{H}^- \backslash \{O\}), \quad \forall x \in X \backslash Z, \quad \forall \xi \in Z;$$

(4i) $\forall z \in Z$, there exist a neighbourhood $N(z)$ of z and a real $\varepsilon(z)$, such that:

$$\|\Phi(x; z(x))\| \geq \varepsilon(z)\|x - z(x)\|^{\alpha}, \forall x \in N(z) \cap (X\backslash Z),$$

$$\forall z(x) \in \mathrm{proj}_Z x.$$

Then, a real μ_0 exists, such that, $\forall \mu > \mu_0$, a solution of $\mathcal{P}$ is a solution of $\mathcal{P}(\mu)$.

Proof. Let us introduce the sets $X_G := G \cap X, Z_G := G \cap Z, \overline{N}(z) := \Omega \cap N(z),$ $S := \overset{k}{\underset{i=1}{\cup}} \overline{N}(z)$, where $N(z)$ is the neighbourhood in $(H_2)(4i)$ and S is a cover of Z_G. Because of $(H_2)(4i)$, $\forall x \in S \cap (X_G\backslash Z_G)$, we have:

$$\frac{1}{\|\Phi(x; z(x))\|}\Phi(x; z(x)) \in \mathcal{H}^- \cap U,$$

where $U := \{x \in \mathbb{R}^\nu : \|x\| = 1\}$. We can apply Theorem 2.1.5. To this end, let us set:

$$\rho := \max\left\{\frac{L}{\varepsilon(z^i)}, i = 1, ..., k\right\};$$

$n, K, \tilde{K}, V_1$ and V_2 are identified with $\nu, -\mathcal{H}, \mathcal{H}^-$,

$$\frac{1}{\|\Phi(x; z(x))\|}[A(x; \xi) - A(z(x); \xi)] \quad \text{and} \quad \frac{1}{\|\Phi(x; z(x))\|}\Phi(x; z(x)),$$

respectively. Hence, the assumptions of Theorem 2.1.5 being satisfied, we achieve the existence of $\eta_0 \in \mathbb{R}$, s.t. $\forall \eta > \eta_0$, we have:

$$A(x; \xi) - A(z(x); \xi) + \eta\Phi(x; z(x)) \in \mathrm{int}\,(-\mathcal{H}),$$

$$\forall x \in S \cap (X_G\backslash Z_G), \quad \forall \xi \in X, \quad \forall z(x) \in \mathrm{proj}_Z x. \tag{5.6.23}$$

Then, because of $(H_2)(ii)$, $\forall \eta > \eta_0$, we have:

$$A(x; \xi) - A(z(x); \xi) + \eta\Phi(x; \xi) \in (-\mathcal{H}),$$

$$\forall x \in S \cap (X_G\backslash Z_G), \quad \forall \xi \in Z_G, \quad \forall z(x) \in \mathrm{proj}_Z x. \tag{5.6.24}$$

Now we will prove that $\exists \eta_1 \in \mathbb{R}$, s.t., $\forall \eta > \eta_1$, we have:

$$A(x; \xi) + \eta\Phi(x; \xi) \in (-\mathcal{H}), \quad \forall x \in X_0 := X_G\backslash S, \quad \forall \xi \in Z_G. \tag{5.6.25}$$

In fact, Φ is continuous and different from the null vector on $(X\backslash Z) \times Z$; this fact and $(H_2)(3i)$ imply:

$$\frac{1}{\|\Phi(x; \xi)\|}\Phi(x; \xi) \in \mathcal{H}^- \cap U.$$

We can apply Theorem 2.1.5 with $\rho = \frac{1}{M}\|A\|$, where

$$M := \min_{(x,\xi)\in X_0 \times Z}\|\Phi(x; \xi)\|;$$

$n, K, \tilde{K}, V_1$ and V_2 are identified with $\nu, -\mathcal{H}, \mathcal{H}^-$,

$$\frac{1}{\|\Phi(x;\xi)\|}A(x;\xi) \quad \text{and} \quad \frac{1}{\|\Phi(x;\xi)\|}\Phi(x;\xi),$$

respectively. Hence, the assumptions of Theorem 2.1.5 being satisfied, we achieve the existence of η_1, s.t. $\forall \eta > \eta_1$, (5.6.25) holds. Now, let $\overline{\xi}$ be a solution of $\mathcal{P}$. Then, $\forall x \in X$ and $\forall z(x) \in \text{proj}_Z x$, account taken of $\mathcal{H} = \mathbb{R}_+^\nu$, there exists an index i, s.t.

$$A(z(x);\overline{\xi})_i \;<\; 0. \tag{5.6.26}$$

We conclude that, $\forall \mu > \mu_0 := \max\{\eta_0, \eta_1\}$, $\forall x \in X_G$, we have:

$$A(x;\overline{\xi}) + \mu\Phi(x;\overline{\xi}) \notin \mathcal{H}. \tag{5.6.27}$$

In fact, since $\overline{\xi}$ is a solution of $\mathcal{P}$ and Φ is null on $Z \times Z$, (5.6.27) holds for $x \in Z_G$; (5.6.25) implies (5.6.27) for $x \in X_0$; conditions (5.6.24) and (5.6.26) imply (5.6.27) for $x \in S \cap (X_G \backslash Z_G)$. $\qquad\qquad\qquad\qquad\qquad\qquad\qquad\qquad\qquad\qquad\qquad \square$

Note that hypothesis (H_1) of Theorem 5.6.6 can be weakened by replacing the condition "$x, \xi \in \Omega \cap X$" with "$x \in \Omega \cap X, \xi \in Z$."

Now let us consider the case $Z = \mathbb{B}^n$, which extends (5.6.6). To this end, set:

$$\mathcal{H} = \mathbb{R}_+^\nu \backslash \{O\}, Z = \mathbb{B}^n, \mathcal{H}^+ = \{h \in \mathbb{R}_+^\nu : h_1 = h_2 = ... = h_\nu\}, \mathcal{H}^- = -\mathcal{H}^+. \tag{5.6.28}$$

Like for (5.6.6), the relaxation of Z we consider is X_Q in (5.6.5). As penalty term Φ, we choose:

$$\Phi(x;\xi) = \begin{pmatrix} \varphi(\xi) & - & \varphi(x) \\ & \vdots & \\ \varphi(\xi) & - & \varphi(x) \end{pmatrix}, \tag{5.6.29}$$

where φ is that in (5.6.5).

Under the assumptions (5.6.28), the function Φ defined in (5.6.29) fulfils, for $\alpha = 1$, the conditions (H_2) of Theorem 5.6.5 and (H_2) of Theorem 5.6.6. In fact, (H_2)(i,ii,3i) of Theorem 5.6.5 and (H_2)(i,ii,3i) of Theorem 5.6.6 are obvious; as concerns (H_2)(4i) of both theorems, let us note that, if $N(z)$ is small enough — so that $z(x) = x$ — , then φ satisfies the following condition; "$\forall z \in Z$, there exist a neighbourhood $N(z)$ of z and a real $\overline{\varepsilon}(z)$, s.t.

$$\varphi(x) \geq \overline{\varepsilon}(z)\|x - z\|, \qquad \forall x \in N(z) \cap (X \backslash Z)",$$

as it has been proved in Theorem 5.6.3; therefore, $\forall z \in Z$ and $\forall x \in N(z) \cap (X \backslash Z)$, we find:

$$\|\Phi(z(x);x)\| = \|\Phi(x;z(x))\| = \left\| \begin{pmatrix} \varphi(x) \\ \vdots \\ \varphi(x) \end{pmatrix} \right\| = \sqrt{\nu}|\varphi(x)| \geq \sqrt{\nu}\,\overline{\varepsilon}(z)\|x - z\|,$$

which proves (H_2)(4i) of both Theorems 5.6.5 and 5.6.6, by setting $\varepsilon(z) = \sqrt{\nu}\,\overline{\varepsilon}(z)$. Hence, we have proved the following:

Theorem 5.6.7. In the case (5.6.28)-(5.6.29), let the function A verify, for $\alpha = 1$, the hypothesis (H_1) of Theorem 5.6.6, and, furthermore, be such that $A(x; x) = O$, $\forall x \in \Omega$. Then, there exists $\mu_0 \in \mathbb{R}$, such that, $\forall \mu > \mu_0$, $\mathcal{P}$ and $\mathcal{P}(\mu)$ have the same solutions.

When A is of kinds (5.6.19), (5.6.21), (5.6.22), then it fulfils the condition $A(x; x) = O$, $\forall x \in \Omega$.

In Theorem 5.6.7, we have considered the hypothesis (H_1) for $\alpha = 1$, since this is enough for the special Φ we have chosen. Concerning such a choice, note that the above theorem is still valid, if we select any strictly concave functions $\varphi_1, ..., \varphi_\nu$, s.t., $\forall i, \varphi_i : \mathbb{R}^n \to \mathbb{R}, \varphi_i(x) = 0 \ \forall x \in Z$, and $\varphi_i(x) > 0 \ \forall x \in X \backslash Z$; moreover, $\forall \xi \in Z$, there exist a neighbourhood $N^i(\xi)$ and a real $\varepsilon_i(\xi) > 0$, s.t.

$$|\varphi_i(x)| \geq \varepsilon_i(\xi)\|x - \xi\|^\alpha, \quad \forall x \in N^i(\xi) \cap (X \backslash Z).$$

Note that, $\forall i$, the above condition is a slight generalization of the condition on the function φ of Theorem 5.6.1, and it is equivalent to (H_2) of Theorem 5.6.5 for $\nu = 1$. Then, we can put $\phi = (\varphi_1, ..., \varphi_\nu)$ and, $\forall x, \xi \in X$,

$$\Phi(x; \xi) = \phi(\xi) - \phi(x).$$

Condition $(H_2)(4i)$ for Φ follows by choosing, $\forall z \in Z$, $N(z) = \overset{\nu}{\underset{i=1}{\cap}} N^i(z)$ and $\varepsilon(z) = \sqrt{\nu} \min\{\varepsilon_i(\xi), i = 1, ..., \nu\}$.

The following theorem gives a condition which assures that, $\forall \xi \in X$, the function $A(\bullet; \xi) + \mu\Phi(\bullet; \xi)$ is component-wise strictly convex. This is a straightforward extension of Theorem 5.6.2.

Theorem 5.6.8. In the case (5.6.28)-(5.6.29), let the function A fulfil the hypotheses of Theorem 5.6.7 for $\alpha = 1$. If $A \in [C^2(X \times X)]^\nu$, then $\exists \mu_0 \in \mathbb{R}$, such that, $\forall \mu > \mu_0$, $\mathcal{P}$ and $\mathcal{P}(\mu)$ have the same solutions, and, $\forall \xi \in X$, the function $A(\bullet; \xi) + \mu\Phi(\bullet; \xi)$ is component-wise strictly convex.

Proof. $\forall \xi \in X$, let $H_i(x; \xi)$ and $\hat{H}_i(x; \xi)$ be the Hessian matrices at x of the ith component of $A(\bullet; \xi)$ and of $A(\bullet; \xi) + \mu\Phi(\bullet; \xi)$, respectively. Hence, $\forall i = 1, ..., \nu$, we have $\hat{H}_i(x; \xi) = H_i(x; \xi) + 2\mu I_n$. Now, $\forall i \in \{1, ..., \nu\}$ and $\forall r \in \{1, ..., n\}$, let $\lambda_{ir} : X \times X \to \mathbb{R}$ be the function where $\lambda_{ir}(x; \xi), r = 1, ..., n$ are the eigenvalues of the Hessian $H_i(x; \xi)$. Because of the continuity of A, λ_{ir} is continuous. Hence,

$$\eta_0 := \max_{i,r} \max_{X \times X} |\lambda_{ir}(x; \xi)| < +\infty.$$

Moreover, let us observe that, $\forall x, \xi, \nu_i(x; \xi)$ is an eigenvalue of $\hat{H}(x; \xi)$ iff $\nu_i(x; \xi) - 2\mu$ is an eigenvalue of $H_i(x; \xi)$. Then, denoting by μ'_0 and μ''_0 the reals μ_0 which appear, respectively, in Theorems 5.6.5 and 5.6.6, we have that, $\forall \mu > \mu_0 := \frac{1}{2} \max\{\eta_0, \mu'_0, \mu''_0\}$, $\forall (x, \xi) \in X \times X$,

$$\nu_{ir}(x; \xi) = \lambda_{ir}(x; \xi) + 2\mu$$

is eigenvalue of $\hat{H}(x; \xi)$ and it results $\nu_{ir}(x; \xi) > 0$. $\qquad \square$

Let us now continue the analysis of case (5.6.19), and consider the VOP:

$$\min_{C_0}[f(x) + \mu\phi(x)], \quad \text{s.t.} \quad x \in G \cap X_Q, \qquad (5.6.30)$$

where $\phi : \mathbb{R}^n \to \mathbb{R}^\nu$ with $\phi = (\varphi, ..., \varphi)$, φ and X_Q being given by (5.6.5), and $\mu \in \mathbb{R}$.

Corollary 5.6.1. Let the following hypothesis be satisfied.

(H) $f : \mathbb{R}^n \to \mathbb{R}^\ell$ is bounded on X_Q, and there exist an open set $\Omega \supset \mathbb{B}^n$ and a real

$L > 0$, which make true the inequality:

$$|f_i(x) - f_i(\overline{x})| \;\leq\; L||x - \overline{x}||, \quad \forall x, \overline{x} \in \Omega \cap X_Q, i = 1, ..., \ell,$$

where f_i denotes the ith component of f.

Then , a real μ_0 exists, such that, $\forall \mu > \mu_0$, (1.1.8) and (5.6.30) have the same solutions. If, in addition, $f \in [C^2(X)]^\ell$, then $\exists \mu_1 \in \mathbb{R}$, such that, $\forall \mu > \hat{\mu} := \max\{\mu_0, \mu_1\}, f + \mu\phi$ is component-wise strictly concave.

Proof. Set $\Phi : X \times X \to \mathbb{R}^\ell$ with $\Phi(x; \overline{x}) = \phi(\overline{x}) - \phi(x)$. According to what stated just after (5.6.19), hypothesis (H) implies that $A(x; \overline{x}) = f(\overline{x}) - f(x)$ satisfies (H_1) of Theorem 5.6.5 and (H_1) of Theorem 5.6.6. Moreover, hypothesis (H_2) of Theorem 5.6.5 and (H_2) of Theorem 5.6.6 are fulfilled by the present Φ given by (5.6.29), because of what has been shown after (5.6.29). As concerns the 2nd part of the thesis, it is enough to note that (1.1.8) and (5.6.30) are equivalent to $\mathcal{P}$ and $\mathcal{P}(\mu)$, respectively. Hence, Theorem 5.6.8 can be applied. $\qquad\qquad\qquad\qquad\qquad\qquad\qquad\qquad\qquad\qquad\qquad\qquad\qquad$ $\square$

In the special $-$ but important $-$ case where G is a polyhedron, note that Corollary 5.6.1 shows a class of VOP with strictly concave function, i.e. (5.6.30) having a v.m.p. necessarily at a vertex of the feasible region; in fact, because of the equivalence between (1.1.8) and (5.6.30), the solutions of the latter are obviously vertices. In general, this is not true, as the following Theorem 5.6.9 and Examples 5.6.7 and 5.6.8 show.

Theorem 5.6.9. Let $f : \mathbb{R}^n \to \mathbb{R}^\ell$ be component-wise concave, and $G \subset \mathbb{R}^n$ be a non-empty polytope. Then, at least a vector minimum point of (1.1.8) happens at a vertex of G.

Proof. Consider the sets:

$$S_0 := G, \quad S_i := \operatorname*{argmin}_{x \in S_{i-1}} f_i(x), \quad i = 1, ..., \ell$$

and the related problems:

$$f_i^\downarrow := \min_{x \in S_{i-1}} f_i(x), \quad i = 1, ..., \ell.$$

We obviously have:

$$S_i \subseteq S_{i-1}, \quad i = 1, ..., \ell,$$

and, because of the concavity of f, $S_1, ..., S_\ell$ are unions of faces of G. We will show that each element of S_ℓ is a v.m.p. of (1.1.8), so that the thesis will follow. Consider any $x^0 \in S_\ell$. Ab absurdo, suppose that x^0 be not a solution of (1.1.8). Then, $\exists y(x^0) \in G$, s.t.:

$$f(y(x^0)) \;\leq_{C_0}\; f(x^0),$$

so that $\exists i(x^0) \in \{1, ..., \ell\}$, s.t.

$$f_{i(x^0)}(y(x^0)) \ < \ f_{i(x^0)}(x^0) = f_{i(x^0)}^{\downarrow}; \tag{5.6.31a}$$

$$f_i(y(x^0)) \ \leq \ f_i(x^0), \quad \forall i = 1, ..., \ell, \ i \neq i(x^0). \tag{5.6.31b}$$

$y(x^0)$ must belong to S_ℓ. In fact, $\forall i = 1, ..., \ell - 1$,

$$y(x^0) \in S_i \backslash S_{i+1} \ \Rightarrow \ f_{i+1}(y(x^0)) \ > \ f_{i+1}(x^0) = f_{i+1}^{\downarrow},$$

which contradicts (5.6.31b). Since

$$f(x) = (f_1^{\downarrow}, ..., f_\ell^{\downarrow}), \quad \forall x \in S_\ell,$$

we have $f(y(x^0)) = (f_1^{\downarrow}, ..., f_\ell^{\downarrow})$ which contradicts (5.6.31a). Finally, observe that the sets $S_1, ... S_\ell$ are unions of faces of G. $\qquad\square$

Unlike the case $\ell = 1$, when $\ell > 1$ a v.m.p. of (1.1.8) is not necessarily a vertex of G (in spite of the strict concavity of f), as Example 5.6.6 shows; this conclusion does not change, if we make the further assumption that the (global) maximum points of the several f_i's fall in int G, as Example 5.6.7 shows.

Example 5.6.6. Set $n = 1, G = [0, 1], f_1(x) = 1 - x^2, f_2(x) = x(2 - x)$. It is easy to chek that every element of G is a v.m.p. of (1.1.8). $\qquad\square$

Example 5.6.7. Set $n = 1, G = [-3, 3], f_1(x) = (x + 3)(7 - x), f_2(x) = (3 - x)(x + 7)$. It is easy to chek that the v.m.p. of (1.1.8) are now $x = \pm 3$ and all the elements of $]-1, 1[$. $\qquad\square$

The case were a v.m.p. of (1.1.8) is necessarily a vertex of G is a very special one. For instance, it happens if the function f is component-wise strictly concave and vert $G \subseteq \text{lev}_{=\beta} f_i, \ i = 1, ..., \ell$.

5.7. Comments

1.It has been already noted that any superset of the set of minimum (or maximum) points represents a necessary optimality condition. In Sect. 5.3, we have met the most classic and important way of establishing a necessary condition: the Lagrangian one. After the fundamental Works of Euler and Lagrange, several other necessary conditions have been stated; sometimes they are called minimum or maximum principles, since some of them contains − as part of the necessary condition − a minimization or maximization. In as much as the IS Analysis has shown to be the source for deriving the Lagrangian Principle (Theorem 5.3.1), it is conceivable and extremely interesting to try to draw the known maximum (or minimum) principle starting from the IS. Here we shortly outline some hints for deriving the celebrated so-called Pontryagin Maximum

Principle. To this end, consider ta following special case of (1.1.5) namely (1.1.12) in the autonomous case:

$$\min \left[f(x) := \int_T f^0(x(t); \xi(t)) dt \right] \tag{5.7.1a}$$

$$\text{s.t.} \qquad \frac{dx_i}{dt} = f^i(x; \xi), \qquad i \in I := \{1, ..., n\}, \tag{5.7.1b}$$

$$(x, \xi) \in X \times \Xi \tag{5.7.1c}$$

The correspondence between (5.7.1) and (1.1.5) is obvious: the above pair (x, ξ) corresponds to x of (1.1.5); f^0 corresponds to ψ_0 of (1.1.5), which is here independent of t and of the derivatives; now we have only bilateral constraints, and g_i of (1.1.5) is here $\frac{dx_i}{dt} - f^i(x; \xi)$; X of (1.1.5) is now replaced by $X \times \Xi$, which must be meant as in (1.2.12). Our proposal is to try to apply to (5.7.1) the approach developed in Sect. 3.2 for (1.1.5). What follows is merely a set of temptative steps towards Pontryagin Maximum Principle; they need to be deepened and transformed into propositions, which have to be proved. Apart from the intrinsic interest in achieving such a principle by a different way, the success in the enterprise should lead to extend the approach to other optimal control problems. Let us note that system (3.2.26) becomes here:

$$\begin{cases} u := \int_T [f^0(\overline{x}; \overline{\xi}) - f^0(x; \xi)] dt > 0, \\ \\ v_i := \frac{dx_i}{dt} - f^i(x; \xi) = 0, \quad i \in I, \ x \in X, \ \xi \in \Xi. \end{cases} \tag{5.7.2}$$

Of course, the admissible pair $(\overline{x}, \overline{\xi})$ is a (global) m.p. of (5.7.1), iff (5.7.2) is impossible. The *image set* of (5.7.1), denoted now by $\mathcal{K}_{\overline{x},\overline{\xi}}$, is the set of $(u, v_i, i \in I)$ given by (5.7.2), and is the image of a point-to-set map, here denoted by $A_{\overline{x},\overline{\xi}}(x; \xi)$. By following the approach Sects. 3.2 and 4.7, we have now to introduce a function $\Phi : X \times \Xi \times \Omega \to \mathbb{R}^{1+n}$, defined, $\forall x \in X$ and $\forall \xi \in \Xi$, by:

$$\Phi(A_{\overline{x},\overline{\xi}}(x; \xi), \omega) := \left(f^0(\overline{x}; \overline{\xi}) - f^0(x; \xi), \int_T \omega_i(t) \left[\frac{dx_i}{dt} - f^i(x; \xi) \right] dt, i \in I \right). \tag{5.7.3}$$

The above (5.7.3) is the present form of (3.2.28). The first of the previously mentioned steps consists in proving that (5.7.3) is a GSF, or that (3.2.29) holds for the present maps; in other words, we have to prove that Theorem 3.2.4 and Proposition 3.2.11 hold here too. The set

$$\mathcal{K}_{\overline{x},\overline{\xi}}(\Omega) := \Phi(A_{\overline{x},\overline{\xi}}(X; \Xi), \Omega)$$

is the *selected image* of (5.7.1) and ω is the *selection multiplier*. As said in Sect.3.2, following this approach, the IS is finite dimensional; indeed, the set $\mathcal{H}$ is now:

$$\mathcal{H} = \{(u, v) \in \mathbb{R} \times \mathbb{R}^n : u > 0, \ v_i = 0, \ i \in I\},$$

where $v = (v_i, ..., v_n)$. However, the infinite dimensionality of the images of the functions

g_i (which now are differential operators) has not disappeared; it has been transformed into the selection, through ω, of an element of $A_{\bar{x},\bar{\xi}}(x;\xi)$, and then postponed to the writing of a necessary condition. In as much as we have proved that (5.7.3) is a GSF, then the image problem (3.2.3) becomes now:

$$\max(u), \quad \text{s.t.} \quad (u,v) \in \mathcal{K}_{\bar{x},\bar{\xi}}(\omega), \quad v = O, \quad \omega \in \Omega. \tag{5.7.9}$$

A further step consists now in establishing equivalence between (5.7.9) and the problem:

$$\min_{\omega \in \Omega} \max_{\substack{(u,v)\in\mathcal{K}_{\bar{x},\bar{\xi}} \\ v=O}} (u). \tag{5.7.10}$$

Of course, here and in the sequel, the existence of maxima and minima or suprema and infima must be discussed. (5.7.10) shows that, for ω fixed, we have a problem of the same type of (3.2.3); for such a problem it is conceivable to repeat the development which has led from (3.2.3) to Theorem 5.3.1; in this sense, we work in a finite dimensional IS. The infinite dimensionality of the images of the $g_i's$ has been postponed to the "external" operator min. A subsequent step consists now in applying, to the selected image, the separation scheme, which has been developed previously for problems (1.1.1) and (1.1.4). This should lead to replace (5.7.10) with the problem:

$$\min_{\substack{\omega \in \Omega \\ \lambda \in \mathbb{R}^m}} \max_{(u,v)\in\mathcal{K}_{\bar{x},\bar{\xi}}} \left[u - \sum_{i\in I} \lambda_i \omega_i(t) v_i \right]. \tag{5.7.11}$$

$\lambda := (\lambda_1, ..., \lambda_n)$ is the gradient of the hyperplane $u - \langle \lambda, v \rangle = 0$ by means of which we try to separate $\mathcal{H}$ and the selected image of (5.7.1); apart from the difference of notation and dimensions, we have to do, on the selected image, what has been done on $\mathcal{K}_{\bar{x}}$ associated to (1.1.1) and (1.1.4). A further step consists now in coming back from the IS to the given space. If we replace u and v with their expressions in terms of the given data of (5.7.1), and if we neglect the constant term, then (5.7.11) becomes:

$$\max_{\substack{\omega \in \Omega \\ \lambda \in \mathbb{R}^m}} \min_{(x,\xi)\in R} \int_T \left\{ f^0(x;\xi) - \sum_{i\in I} \lambda_i \omega_i(t) \left[\frac{dx_i}{dt} - f^i(x;\xi) \right] \right\} dt, \tag{5.7.12}$$

where R denotes here the set of $(x,\xi) \in X \times \Xi$. By setting :

$$\lambda_0(t) :\equiv -1; \quad \lambda_i(t) := \lambda_i \omega_i(t), \, i \in I; \quad H(x;\xi;\lambda) := \sum_{i=0}^{n} \lambda_i(t) f^i(x,\xi),$$

where $\lambda = \lambda(t) := (\lambda_0(t), ..., \lambda_n(t))$, problem (5.7.12) becomes:

$$-\min_{\lambda \in \Lambda} \max_{(x,\xi)\in R} \int_T \left[H(x;\xi;\lambda) - \sum_{i\in I} \lambda_i \frac{dx_i}{dt} \right] dt, \tag{5.7.13}$$

where

$$\Lambda := \{ \lambda \in \{-1\} \times C^0(T)^n : \lambda_i(t) = \lambda_i \omega_i(t), \, i \in I \}.$$

H is called *Hamiltonian function* (it is different from H of (2.3.2a)$'$, even if similar), while that in square brackets of (5.7.13) is called *Lagrangian auxiliary function*; note that the former is an addend of the latter. The vector $\lambda(t)$ is callef *vector of auxiliary variables* in the Theory of Optimal Control; in the present approach, unlike the existing literature, each of its component has been factorized into a constant term, namely λ_i, and a variable term, namely $\omega_i(t)$; λ_i is precisely a Lagrange multiplier in the sense of Sect. 5.3, while $\omega_i(t)$ has the sense of a weight. Such a factorization should lead to useful information in the applications of (5.7.1). Now a further step consists in showing that the maximum in (5.7.13) is not achieved if, for each x and λ, H does not achieve, in (5.7.13), its maximum with respect to ξ, i.e.

$$\mathcal{M}(x;\lambda) := \max_{\xi \in \Xi} H(x;\xi;\lambda). \tag{5.7.14}$$

A step consists now in showing that the gradient, with respect to λ, of the integrand of (5.7.13) must vanish, or

$$\frac{\partial H}{\partial \lambda_i} - \frac{dx_i}{dt} = 0, \quad i \in I. \tag{5.7.15}$$

By integrating by parts (5.7.13), we find:

$$-\min_{\lambda \in \Lambda} \max_{(x,\xi) \in R} \int_T \left[H(x;\xi;\lambda) + \sum_{i \in I} x_i \frac{d\lambda_i}{dt} \right] dt - \sum_{i \in I} [\lambda_i x_i]_{t_0}^{t_1}. \tag{5.7.16}$$

Now, we have to show that the gradient, with respect to x, of the integrand of (5.7.16) must vanish, or

$$\frac{\partial H}{\partial x_i} + \frac{d\lambda_i}{dt} = 0, \quad i \in I. \tag{5.7.17}$$

(5.7.14), (5.7.15) and (5.7.17) form the Pontryagin necessary condition. Once the above approach has been proved, the extension to kinds of control problems other than (5.7.1) are conceivable. In this sense, the development of Sect. 3.2 (for instance, the conic extension) might be useful.

2. In Example 5.4.1, the image sets of the two reciprocal particular problems have been constructed. As said in Sect. 3.2, the fact that they have been obtained beginning from the optimal point $\bar{x}$ does not affects their properties. We might have started from any other point; $\bar{x}$ has been chosen to simplify the exposition. A systematic investigation of the relationships between the image sets of two reciprocal problems, and consequently the perturbation functions, would be useful.

3. In Examples 1.2.2 and 5.4.1, elementary isoperimetric problems have been considered in order to discuss some aspects of the IS Analysis. In passing, we have found classic *isoperimetric inequalities*. Their systematic investigation, besides being interesting by itself, might help in deepening the relationships between reciprocal problems. Indeed, if we have enough information on the perturbation function, then Theorem 5.5.1 gives such a relationship. Otherwise, we might be unable to write (5.5.9) or the

other similar equalities. The extrema of reciprocal problems might not be equal, and then it would be useful to introduce and analyse their difference, call it *reciprocity gap*.

4. Theorem 5.6.1 deals with strict maximum and strict minimum of the perturbation function $f^\dagger(\xi)$ on the interior of Ξ. The approach can be extended to the cases of non-strict extrema of $f^\dagger$ with ξ not necessarily interior point, or to other types of critical values of $f^\dagger$.

5. Theorems 5.5.1-5.5.4 are preliminary general results on reciprocal problems. It is necessary to address the investigation in several directions: to weaken the assumptions of strict monotonicity; to specialize the results to particular classes of constrained extremum problems; to analyse the connections among (5.5.1)-(5.5.4), (5.5.19), (5.5.20) and their dual problems; to investigate the properties of their image sets and of the corresponding conic extensions (see Sect.3.2).

6. Theorem 5.5.4 is a preliminary result. It shows that, by means of a reciprocity result, it is conceivable to replace a problem with a less difficult one. Full investigation in this area would be, therefore, extremely important.

7. Since the results of Sect. 5.5 do not require differentiability properties and do not exploit any structure of the space where the problems are located, it would be interesting to investigate reciprocity properties for problem (1.1.7).

8. As Sect. 3.2 has shown, the image set of a constrained extremum problem, like (1.1.1) or (1.1.4), is associated to a system, like (3.2.1). Therefore, every type of problem $-$ whether it be an extremum problem or not $-$, which leads to (3.2.1), receives the same IS Analysis. This must be taken into account, when the reciprocity is investigated through the IS. One of the consequences of this is that the reciprocity properties of (1.1.1) and (1.1.4) can be extended to vector problems (1.1.8), (1.1.10) and to the complex problems (1.1.21)-(1.1.23), and, more generally, to system (1.3.16).

9. By means of the selection approach introduced in Sects. 3.2 and 4.8, it has been possible to keep finite the dimensionality of the IS associated with (1.1.5) (restricting the infinite dimensionality to the selection). Therefore, it is conceivable and interesting to extend the results of Sect. 5.5 and the related questions to problems of type (1.1.5).

10. With reference to the subject of Sect. 5.6, a suitable terminology for identifying the two fields might be "discrete structures of mathematics" and "continuous structures of mathematics" In every syllabus at every level, since the elementary schools, we should have both fields in order to avoid a unilateral mathematical education.

11. Theorem 5.6.1 has been exploited in the special case where $Z = \mathbb{B}^n$ and its relaxation is $X = X_Q = \operatorname{conv} \mathbb{B}^n$; the purpose has been to replace a combinatorial minimization problem with an equivalent minimization on a real space. Of course, Theorem 5.6.2 might be used in the opposite sense: given a problem in a real space, to obtain an equivalent combinatorial problem or one in a discrete space. However, we must note that Theorem 5.6.1 is a first result in this field, and it requires to be generalized in several directions. One obvious direction consists in trying to weaken the assumptions of Theorem 5.6.1. Another aspect is offered by the function φ, which

appears in (5.6.2): the fact that its argument be exactly x is a limitation, as next example shows.

Example 5.7.1. Let (1.3.15) be a complementarity problem and consider the particular case $\mathbb{K} = \mathbb{R}^n_+$, so that the constraining system is (1.3.7) (for $\mathbb{K} = \mathbb{R}^n_+$; $\overline{x}$ is replaced by x to stress the fact that the argument of (1.3.7) is now a variable); therefore (1.3.15) becomes:

$$\min f(x), \quad \text{s.t.} \ \ x \geq O, \ \ F(x) \geq O, \ \ \langle F(x), x \rangle = 0. \tag{5.7.18}$$

With the positions

$$G = \{x \in \mathbb{R}^n_+ : F(x) \geq O\}, \quad Z = \{x \in \mathbb{R}^n : \langle F(x), x \rangle = 0\},$$

(5.7.18) becomes a special case of (5.6.1). Let $s = (s_1, ..., s_n), F(x) = (F_1(x), ..., F_n(x))$, and consider the function $\varphi : \mathbb{R}^n \times \mathbb{B}^n \to \mathbb{R}$, given by:

$$\varphi(x, s) = \sum_{i=1}^{n} \left[(1 - s_i)x_i + s_i F_i(x)\right]. \tag{5.7.19}$$

With $X = \mathbb{R}^n$, we consider the problem:

$$\min[f(x) + \mu\varphi(x, s)], \quad \text{s.t.} \ \ x \in G \cap X, \ s \in \mathbb{B}^n, \tag{5.7.20}$$

which must be considered in place of (5.6.2). If μ is large enough, at $s_i = 0$ (or at $s_i = 1$) the above minimization pushes x_i (or $F_i(x)$) to be zero and hence to satisfy the orthogonality constraint. $\qquad\qquad\qquad\qquad\qquad\qquad\qquad\qquad\qquad\qquad\qquad$ $\square$

This shows the importance to enlarge (5.6.2) to embed at least (5.7.20), and hence to extend Theorem 5.6.1 to achieve equivalence for at least (5.7.18) and (5.7.20). The proposal to reduce a problem in a real space, like (5.7.18), to a partially combinatorial one, like (5.7.20), might seem against the stream. Indeed, it may happen that a concrete situation, like that of Sect. 1.4, leads to a mathematical formulation in a real space with an orthogonality constraint, like (1.4.7e); this hides a combinatorial nature; to maintain it hidden may lead to (at least) computational drawbacks; to pass to problem (5.7.20) means to lay bare the combinatorial nature. This happens also when (1.1.1) has many local (but not global) minima and we search for them through the necessary condition of Theorem 5.3.1: we are faced with (5.3.15c) which, again, hides a combinatorial nature.

For a development of Theorem 5.6.1 in Combinatorial Optimization, see [73] and C. Larsen and J. Tind, "Lagrangian duality for facial programs with applications to integer and complementarity problems" in "Operations Research Letters" Vol. 11, 1992, pp. 293-302.

12. Besides the previously discussed extensions of Theorem 5.6.1, it would be interesting to consider problems of type (1.1.4) and, by means of the selection approach of Sect. 3.2, also problems of type (1.1.5). A first result in this direction is contained in V. F. Dem'yanov, F. Giannessi and V. V. Karelin, "Optimal control problems via exact penalty functions" Jou. of Global Optimiz., vol.12, 1998, pp. 215-223; see also Vol. 2.

13. For minimizing a (strictly) concave function on a convex set of $\mathbb{R}^n$, a very interesting method has been proposed by H. Tuy [102]; see also [I38]. Through Theorem 5.6.2, such a method can be transferred to solve combinatorial problems of type (1.1.7). A general method for solving this way (1.1.7), as well as its specializations for particular classes of (1.1.7), might be extremely interesting.

14. The proof of Theorem 5.6.9 shows that the set of solutions to (1.1.8) contains a union of faces of the polytope G. If, $\forall k \in \{1, ..., \ell\}, S_k$ is a singleton, then obviously its (unique) element is a v.m.p. of (1.1.8), and the subsequent $S_i's$ are equal to S_k. Such a proof suggest a method for finding a solution of (1.1.8); indeed, this method does not require the concavity of f. However, it does not necessarily find all v.m.p.; for instance, if f is component-wise strictly concave, then the method does not find the v.m.p. (if any) which fall in $\mathrm{int}\, G$, whatever the ordering of the components of f may be. Moreover, note that the thesis of Theorem 5.6.9 can be achieved by means of the same proof under the assumption that only one component of f (which in the proof must be considered as f_1) be strictly concave. This comments hold for (1.1.10) too.

15. Corollary 5.6.1 suggests a method for solving (1.1.8), which is based on the theory introduced in [102] (see also [56]), and will be shortly outlined. To this end, we will consider the special, but important, case where G is a polytope. Because of Corollary 5.6.1, (1.1.8) can be replaced by (5.6.30). If μ is large enough, then, because of Theorem 5.6.9, a v.m.p. of (5.6.30) is a vertex of $G \cap X_Q$. Therefore, a method can start by finding a vertex, say x^0, of $G \cap X_Q$. It is not restrictive to assume that x^0 be a local v.m.p. of (5.6.30); otherwise, this can be achieved by jumping from one vertex to an adjacent one until it has been obtained. Now, consider the family of strictly concave (scalar) problems:

$$\min f_i(x; \mu), \quad \text{s.t.} \quad x \in G \cap X_Q; \quad i = 1, ..., \ell, \tag{5.7.21}$$

where $f_i(x; \mu) := f_i(x) + \mu\phi(x)$. If i is s.t. x^0 be a local (scalar) m.p. of (5.7.21), then the above mentioned theory gives us a "cutting halfspace" say H_i^+, s.t.

$$x^0 \in H_i^+; \quad \left\{ \begin{array}{c} x \in G \cap X_Q \\ f_i(x; \mu) < f_i(x^0; \mu) \end{array} \right\} \;\Rightarrow\; x \in (G \cap H_i^+) \cap X_Q. \tag{5.7.22}$$

This condition means that, if there exists an x at which f_i takes a value less than $f_i(x^0; \mu)$, then x must belong to H_i. For all other indexes i, H_i^+ collapses to a supporting halfspace of $G \cap X_Q$; it will be denoted again by H_i^+. The former (latter) set of indexes will be denoted by I^+ (respectively I^-). If $I^+ \neq \varnothing$, then from (5.6.22) we easily draw that:

$$\left\{ \begin{array}{c} x \in G \cap X_Q \\ f(x; \mu) <_{C_0} f(x^0; \mu) \end{array} \right\} \;\Rightarrow\; x \in G \cap \left(\bigcap_{i \in I^+} H_i^+ \right) \cap X_Q. \tag{5.7.23}$$

This condition means that, if there exists an x at which f takes a value less (in vector sense; with respect to C_0) than $f(x^0; \mu)$, then x must belong to $\bigcap_{i \in I^+} H_i^+$; hence, such an intersection plays a role for VOP as the theory by Tuy [112, 56] has played for the scalar ones. The case $I^+ = \varnothing$ is a degenerate one for all the Tuy's cuts, and requires a special analysis: the present vertex can be replaced by any of the adjacent vertices, since they are alternative local v.m.p. From (5.7.23) we have that the condition:

$$I^+ = \varnothing, \qquad G_1 \cap X_Q = \varnothing, \tag{5.7.24}$$

where

$$G_1 := G \cap \left(\bigcap_{i \in I^+} H_i^+ \right),$$

is a sufficient condition for x^0 to be a v.m.p. of (1.1.8) in the case of G polytope. If (5.7.24) is not satisfied, then we can replace, in (5.7.21), G with G_1 and repeat the above reasoning. Note that the set

$$G \cap \left(\bigcap_{i \in I^+} \sim H_i^+ \right) \cap X_Q$$

does not contain any alternative v.m.p. of (5.6.30); while such points might happen in the sets

$$G \cap (\sim H_r^+) \cap \left(\bigcap_{i \in I^+ \setminus \{r\}} H_i^+ \right) \cap X_Q, \quad r \in I^+.$$

According to the previous comment, an alternative method for finding a v.m.p. of (1.1.8) may consist in solving ℓ scalar problems, having a strictly concave objective function and a union of vertices of G as feasible region. The suggestions contained in this comment extend, of course, to (1.1.10) too.

16. The particular cases (5.6.21)-(5.6.22) deserve special attention. As an instance, consider again (1.3.8) in the particular case where $\mathbb{K} = G \cap Z$ with $Z = \mathbb{B}^n$, namely the following VVI:

$$F(\overline{x})(x - \overline{x}) \not\leq_{C_0} O, \quad \forall x \in G \cap \mathbb{B}^n. \tag{5.7.25}$$

With the positions $Z = \mathbb{B}^n, \xi = \overline{x}$, (5.6.21) and $\mathcal{H} = C_0$, (5.7.25) is equivalent to the impossibility of (5.6.14). Now, consider the problem, which consists in finding $\overline{x} \in G \cap X_Q$, s.t.

$$F(\overline{x})(x - \overline{x}) - \mu \Phi(x; \overline{x}) \not\leq_{C_0} O, \quad \forall x \in G \cap X_Q, \tag{5.7.26}$$

where Φ is the function in (5.6.29) and $\mu \in \mathbb{R}$.

Proposition 5.7.1. Let $F : \mathbb{R}^n \to \mathbb{R}^{\ell \times n}$ be bounded on X_Q. Then, there exists a real $\mu_0 \in \mathbb{R}$, such that, $\forall \mu > \mu_0$, (5.7.25) and (5.7.26) have the same solutions.

Proof. Let A be defined by (5.6.21). According to the remark (5.6.19)-(5.6.20), such a function A fulfils (H_1) of Theorem 5.6.5. (H_1) of Theorem 5.6.6 holds, since

$$\|A(x^1; \overline{x}) - A(x^2; \overline{x})\| = \|F(\overline{x})(\overline{x} - x^1) - F(\overline{x})(\overline{x} - x^2)\|$$

$$\leq \|F(\overline{x})\| \, \|x^2 - x^1\|, \quad \forall x^1, x^2, \overline{x} \in X_Q.$$

The present Φ fulfils (H_2) of Theorems 5.6.5 and 5.6.6. as shown by Theorem 5.6.7. Hence, Theorems 5.6.5 and 5.6.6 give the existence of a real μ_0, s.t. (5.7.25) and (5.7.26) have the same solutions. $\qquad\qquad\qquad\qquad\qquad\qquad\qquad\qquad\qquad\qquad\qquad\qquad\qquad\square$

In the special — but important — case where G is a polyhedron, note that the above proposition shows a class of VVI with bounded operator, i.e. (5.7.26), having — because of the equivalence with (5.7.25) — a solution necessarily at a vertex of the domain. In general, this is not true, as simple examples show.

17. Let us consider a special case of (1.1.8): f is component-wise convex, and $R = \bigcup_{k=1}^{r} Z_k$, with each Z_k convex and compact. Of course, $\bar{x}$ is a v.m.p., iff the system (in the unknown x):

$$f(\bar{x}) - f(x) \in C_0, \qquad x \in \bigcup_{k=1}^{r} Z_k \qquad\qquad (5.7.27)$$

is impossible. With the positions $G = \mathbb{R}^n, Z = \bigcup_{k=1}^{r} Z_k, \mathcal{H} = C_0, \xi = \bar{x}$ and $A(x;\xi) = A(x;\bar{x}) = f(\bar{x}) - f(x)$, (5.7.27) becomes a particular case of (5.6.14). Assume we are given the compact sets $X_k \supseteq Z_k$ and the functions $\phi_k : X \to \mathbb{R}, k = 1, ..., r$, where $X = \bigcup_{k=1}^{r} X_k$, s.t. $\exists \alpha \in \mathbb{R}$ for which each ϕ_k fulfils $(H_2)''$ of Sect. 5.6 at $\ell = 1, \mathcal{H}^+ = \mathbb{R}_+$. It is easily seen that $\varphi : X \to \mathbb{R}$, with

$$\varphi(x) := \prod_{k=1}^{r} \phi_k(x)^{\frac{\alpha}{r}} \qquad\qquad (5.7.28)$$

fulfils $(H_2)''$ of Sect. 5.6. In fact, it is trivial to verify (i), (ii) and (3i). In the following formulas, k as index will denote that we are referred to Z_k and X_k instead of Z and X. φ fulfils also (4i):

$$\varphi(x) = \prod_{k=1}^{r} \phi_k(x)^{\frac{\alpha}{r}} \geq \prod_{k=1}^{r} (\varepsilon_k(z)\|x - \text{proj}_{Z_k} x\|)^{\frac{\alpha}{r}} =$$

$$= \prod_{k=1}^{r} \varepsilon_k(z)^{\frac{\alpha}{r}} \cdot \prod_{k=1}^{r} \|x - \text{proj}_{Z_k} x\|^{\frac{\alpha}{r}} \geq \prod_{k=1}^{r} \varepsilon_k(z)^{\frac{\alpha}{r}} \cdot \|x - \text{proj}_Z x\|^{\alpha} =$$

$$= \tilde{\varepsilon}(z)\|x - \text{proj}_Z x\|^{\alpha}, \qquad \tilde{\varepsilon}(z) := \prod_{k=1}^{r} \varepsilon_k(z)^{\frac{\alpha}{r}},$$

where the 1st inequality comes from $(H_2)''$ of Sect. 5.6 at $\phi = \phi_k, k = 1, ..., r$, and the 2nd inequality is due to the fact that $\|x - \text{proj}_Z x\| = \min\{\|x - \text{proj}_{Z_k} x\|, k = 1, ..., r\}$. The function (5.6.29) for $\xi = \bar{x}$ can be chosen as "penalty term". The decomposition (5.7.28) may help in setting up φ, and in conceiving methods for solving (1.1.8). For instance, if Z_k is a polytope defined by:

$$Z_k := \{x \in \mathbb{R}^n : M_k x \geq b^k\}, \quad k = 1, ..., r,$$

where $M_k \in \mathbb{R}^{m_k \times n}, b^k \in \mathbb{R}^{m_k}$, then, with obvious notation, we can set

$$\phi_k(x) := \max \left\{ 0, \ \exp\left(-\alpha_i^k \sum_{j=1}^{n} m_{ij}^k x_j - b_i^k\right) - 1, i = 1, ..., m_k \right\},$$

where $M_k = (m_{ij}^k, i = 1, ..., m_k, j = 1, ..., n)$, $b^k = (b_1^k, ..., b_{m_k}^k)^T$, α_i^k being a positive parameter. A particularly interesting case is that where Z is not convex, while the sets Z_k, X_k and $\bigcup_{k=1}^{r} X_k$ are all convex. For instance, for $n = r = 2$, it happens to the sets:

$$Z = ([0,2] \times [0,1]) \cup ([0,1] \times [0,2], \quad Z_1 = [0,2] \times [0,1],$$
$$Z_2 = [0,1] \times [0,2], \quad X_1 = Z_1,$$
$$X_2 = \{(x_1, x_2) \in \mathbb{R}_+^2 : x_2 \leq 2, \ x_1 - x_2 \leq 1, \ x_1 + x_2 \leq 3\}.$$

It would be interesting to define a decomposition of the method, proposed with Proposition 5.7.1, which exploits the decomposition (5.7.28). To this end, it might be useful to investigate about the properties of φ given by (5.7.28) and those of the ϕ_k's; in particular, as concerns the (strict) concavity and the fulfilment of hypotheses (H$_2$) of Theorems 5.6.5 and 5.6.6. An interesting application of the above decomposition should be to the case where (1.1.8) in case (5.7.27) is replaced by (1.3.8) or (1.3.9).

18. The possible extensions discussed in the previous 8 comments extend, obviously, to all the other formats of Sects. 1.1 and 1.3. To this end, we stress the importance of extending the theorems of Sect. 5.6.

19. When a VI, like (1.3.1), escapes from known classes (the operator is not isotone or the domain is not convex), then a result like Theorem 5.6.8 would lead to a strictly antitone operator and a decomposition like that proposed in a previous comment would split the domain into convex ones. The new Variational Inequality or Inequalities might be less worse than the given one.

20. The IS Analysis (see Sect. 3.2) should produce useful results also for the connection between different classes of extremum problems or generalized systems. The approach of Sect. 5.6 might, for instance, allow us to work on a "continuous image set" for a combinatorial optimization problem: given (5.6.6), by means of Theorem 5.6.3, it is replaced by (5.6.7), whose image set is, in general, not a discrete set (if, of course, g is not discrete-valued).

21. Consider the family $\mathcal{F}$ of all the sets K satisfying the following properties with reference to (5.6.1)-(5.6.2):

(i) K is convex and compact;

(ii) $K \supset X$;

(iii) $(\mathrm{frt}\, K) \cap X_G = Z_G$;

where $X_G := G \cap X, Z_G := G \cap Z$ as in the proof of Theorem 5.6.1. Note that the above conditions are fulfilled by $Z = \mathbb{B}^n, X = X_Q$ and $\varphi(x)$ of (5.6.5), and $K = \mathrm{lev}_{\geq 0}\, \varphi$. Now, with respect to any K satisfying the above condition, consider the function ϕ defined by (2.1.8)* which, to within a translation, is a gauge function, whatever y may be (see also (2.3.19)). Theorem 5.6.1 $-$ as well as the subsequent ones $-$ should be proved with ϕ of (2.1.8)*: for $\alpha = 1$, by exploiting y, the function ϕ in (2.1.8)* can be chosen in order to fulfil the hypotheses of Theorem 5.6.1 with μ_0 as small as possible (the question of keeping μ_0 under control is crucial for the numerical applications; it would be extremely important to find conditions under which an upper bound for μ_0 can be established and found even if for particular classes of problems). Such extensions may take advantage from the wideness of $\mathcal{F}$. For instance, besides the above $K = \text{lev}_\geq \varphi$, another interesting set $K \in \mathcal{F}$ is the following. Again with $Z = \mathbb{B}^n, X = X_Q$ and φ as above, setting $K_\varphi := \text{lev}_{\geq 0}\, \varphi = \{x \in \mathbb{R}^n : ||x - \frac{1}{2}e||_2 \leq \sqrt{n}/2\}$, we consider the set:

$$\tilde{K}_\varphi := \left\{ y \in \mathbb{R}^n : \left\langle x - \frac{1}{2}e,\, y - \frac{1}{2}e \right\rangle \leq \frac{n}{4},\ \ \forall x \in G \cap Z \cap K_\varphi \right\}.$$

It is not restrictive to assume that $\tilde{K}_\varphi$ be compact; if not, it can be intersected with the simplex $\{x \in \mathbb{R}^n_+ : \langle a, x \rangle \leq b\}$, with a, b suitable. It will be shown that $\tilde{K}_\varphi$ satisfies the above conditions. **(i)** Let $y^1, y^2 \in \tilde{K}_\varphi$. Because of the obvious identity $-\frac{1}{2}e = -\frac{1-\alpha}{2} - \frac{\alpha}{2}e$, we have:

$$\left\langle x - \frac{1}{2}e,\, (1 - \alpha)y^1 + \alpha y^2 - \frac{1}{2}e \right\rangle = \left\langle x - \frac{1}{2}e,\, (1 - \alpha)y^1 - \frac{1 - \alpha}{2}e + \alpha y^2 - \frac{\alpha}{2}e \right\rangle =$$

$$= (1 - \alpha)\left\langle x - \frac{1}{2}e,\, y^1 - \frac{1}{2}e \right\rangle + \alpha \left\langle x - \frac{1}{2}e,\, y^2 - \frac{1}{2}e \right\rangle \leq$$

$$\leq (1 - \alpha)\frac{n}{4} + \alpha\frac{n}{4} = \frac{n}{4},\ \ \ \ \ \forall \alpha \in [0, 1],$$

which proves the convexity of $\tilde{K}_\varphi$. **(ii)** Since $\tilde{K}_\varphi$ is a linear transformation of the polar (see (2.2.8)) of K_φ, we can apply Theorem 2.2.6 (ii). **(iii)** $z \in Z_G$ implies $z \in K_\varphi$, $||z - \frac{1}{2}e||_2 = \sqrt{n}/2$, and

$$\text{frt}\,\tilde{K}_\varphi = \left\{ y \in \tilde{K}_\varphi : \left\langle x - \frac{1}{2}e,\, y - \frac{1}{2}e \right\rangle = \frac{n}{4}\ \text{ for at least an }\ x \in K_\varphi \right\},$$

because the above scalar product equals $||z = \frac{1}{2}e||_2^2 = \frac{n}{4}$. Now,

$$z \in \text{frt}\,\tilde{K}_\varphi\ \Rightarrow\ ||z - \frac{1}{2}e||_2 \geq \frac{\sqrt{n}}{2}\ \ (\text{since } \tilde{K}_\varphi \supset K_\varphi),$$

$$z \in X_G\ \Rightarrow\ ||z - \frac{1}{2}e||_2 \leq \frac{\sqrt{n}}{2}.$$

Therefore, $||z - \frac{1}{2}e||_2 = \sqrt{n}/2$, and

$$Z_G = \left\{ x \in X_G : ||x - \frac{1}{2}e||_2 = \frac{\sqrt{n}}{2} \right\},$$

which completes the proof.

22. The possible developments outlined in most of the previous comments extend, obviously, to all the other formats of Sects. 1.1 and 1.3. To this end, we stress the

importance of generalizing the theorems of Sect. 5.6, both weakening the hypotheses and considering non-Euclidean spaces.

References

[1] Abadie J., "On the Kuhn-Tucker Theorem". In [I1], pp.19-36.

[2] Antoni C. and Giannessi F., "On the equivalence, via relaxation-penalization between vector generalized systems" Acta Mathematica Vietnamica, Vol.22, No.2, 1997, pp.567-588.

[3] Auslender A., "Optimization. Méthodes numeriques". Masson, Paris, 1976.

[4] Auslender A., "Noncoercive optimization problems". Mathem. of Oper.Research, Vol.21, No.4, 1966, pp.769-782.

[5] Balas E., "Nonconvex quadratic programming via generalized polars". SIAM Jou. on Appl.Mathem., Vol.28, No.2, 1975, pp.335-349.

[6] Bank B., Guddat J., Klatte D., Kummer B. and Tammer K., "Non-linear Parametric Optimization". Akademie-Verlag, Berlin, 1982.

[7] Ben-Tal A. and Zowe J., "Necessary and sufficient optimality conditions for a class of nonsmooth minimization problems". Mathematical Programming, Vol.24, 1982, pp.70-91.

[8] Bigi G. and Pappalardo M., "Regularity conditions for the linear separation of sets". Jou.Optimiz. Th.Appls., Vol.99, No.2, 1998, pp.533-540.

[9] Bigi G. and Pappalardo M., "Generalized Lagrange multipliers: regularity and boundedness". In [I19], pp.1-14.

[10] Bonnesen T., "Les problémes des isopérimétres et des isépiphanes". Gauthier-Villars, Paris, 1929, pp.11-13.

[11] Borwein J., "Multivalued convexity and Optimization: a unified approach to inequality and equality constraints". Mathematical Programming, Vol.13, 1977, pp.183-199.

[12] Bussotti P., "On the Genesis of the Lagrange Multipliers". Jou. of Optimiz. Theory and Appls., Vol.117, No.3, 2003, pp.453-459.

[13] Castellani M., Mastroeni G. and Pappalardo M., "On regularity for generalized systems and applications". In [I18], pp.13-26.

[14] Castellani M., Mastroeni G. and Pappalardo M., "Separation of Sets, Lagrange Multipliers and Totally Regular Extremum Problems". Jou.Optiz.Th.Appls., Vol.92, No.2, pp.249-261.

[15] Castellani M. and Pappalardo M., "First order cone approximations and necessary optimality conditions". Optimization, Vol.35, 1995, pp.113-126.

[16] Castellani M. and Pappalardo M., "Local second-order approximations and applications in optimization". Optimization, Vol.37, 1996, pp.305-321.

[17] Cesari L., "Optimization Theory and Applications". Springer-Verlag, New York, 1983.

[18] Castellani M. and Pappalardo M., "Unifying approach for higher-order necessary optimality conditions". Communications Appl.Analysis, Vol.3, No.1, 1999, pp.15-28.

[19] Chaney R.W., "On sufficient conditions in nonsmooth optimization". Mathem.of Oper.Research, Vol.7, pp.463-475.

[20] Clarke F.H., "A new approach to Lagrange multipliers". Mathem.of Oper.Research, Vol.1, 1979, pp.165-174.

[21] Cohen G., "Auxiliary Problem Principle and Decomposition of Optimization Problems". Jou.Optimiz.Th.Appls., Vol.32, No.3, 1980, pp.277-305.

[22] Cottle R.W., "Theorem of Fritz John in mathematical Programming". Report RM-3858-PR of The Rand Corporation (Santa Monica, California), 1963, pp. I-V+1-10.

[23] Craven B.D., "Avoiding a constraint qualification". Optimization, Vol.41, 1997, pp.291-302.

[24] Dem'yanov V.F. and Vasiliev L.V., "Nondifferentiable Optimization". Nauka, Moscow, 1981.

[25] Ekeland I., "On the Variational Principle". Jou. of Mathem. Analysis and Appls., Vol.47, 1974, pp.324-353.

[26] Elster K.H. and Sutti C. (Eds.), "Mathematical Optimization. Theory Methods and Applications". Proc. Workshop Days (Verona, Dec.9,1992), Published by Univ.of Verona, Via dell'Artigliere, 19-Verona, Italy, 1993.

[27] Elster K.H. and Thierfelder J., "Abstrac cone approximations and generalized directional derivatives". Optimization, Vol.19, 1998, pp.315-341.

[28] Everett H., "Generalized Lagrange multiplier method for solving problems of optimum allocation of resources". Operations Research, Vol.11, 1963, pp.399-417.[29] Evtushenko Yu.G., Rubinov A.M. and Zhadan V.G., "General Lagrange-type functions in constrained global optimization. Part I: auxiliary functions and optimality conditions". Optimization Methods ad Software, Vol.16, 2001, pp.193-230.

[30] Evtushenko Yu.G., Rubinov A.M. and Zadhan V.G., "General Lagrange-type functions in constrained global optimization. Part. II: exact auxiliary functions". OPtimization Methods and Software, Vol.16, pp.231-256.

[31] Ferrero O., "On nonlinear Programming in Complex Spaces". Jou. Mathem.Analysis Appls., Vol.164, No.2, pp.399-416.

[32] Foudas C.A. and Pardalos P.M. (Eds.), "Encyclopedia of Optimization". Kluver Academie Publishers, Dordrecht, 2001, Vols.I-V.

[33] Gao Y., "Dem'yanov Difference of Two Sets and Optimality Conditions of Lagrange Multiplier Type for Constrained Quasidifferentiable Optimization". Jou.Optimiz.Th. Appls., Vol.104, No.2, 2000, pp.337-394.

[34] Giannessi F. "Sulla legge di reciprocità nei problemi di massimo e minimo condizionati" ("On the reciprocity principle for maximum and minimum constrained problems"; in Italian). Proceedings of the Institute of Mathematics of Univ.of Venice, 1970, published by CEDAM, Padova, Italy, pp.71-95.

[35] Giannessi F., "Functional aspects od Dynamic Programming". Control and Cybernetics, Vol.2, No.3/4, 1977, pp.31-42.

[36] Giannessi F., "On Lagrangian Non-Linear Multipliers Theory for Constrained Optimization and related topics", Tech.Report No.123, Dept. of Mathem., Univ.of Pisa, Sect. of Optimization, 1984, pp.1-79. Published as "General Optimality Conditions via a Separation Scheme". In "Algorithms for Continuous Optimization", E. Spedicato (Ed.), Kluver Acad.Publishers, Dordrecht, Boston, 1994, pp.1-23.

[37] Giannessi F., "A common understanding or a common misunderstanding?". Numer.Funct.Analysis and Optimiz., Vol.16, No.9-10, 1995, pp.1359-1363.

[38] Giannessi F., "Some Remarks on Minimum Priciples". In [47], pp.75-103.

[39] Giannessi F., "On the existence of Lagrange multipliers". Proceedings of "Seminario Matematico" of Univ.of Messina, Series II, No.7, pp.1-20.

[40] Giannessi F., Mastroeni G. and Uderzo A., "A multifunction approach to extremum problems having infinite dimensional image. Necessary coonditions for unilateral constraints". Kibernetika (National Academy of Sciences of Ucraina), Vol.3, May 2002, pp.39-51.

[41] Giannessi F. and Niccolucci F., "Connections between nonlinear and integer programming problems". In "Symposia Mathematica", Vol.XIX, Academic Press, London, 1976, pp.161-176.

[42] Giannessi F., Pappalardo M. and Pellegrini L., "Necessary Optimality Conditions via Image Problem". In [I9], pp.185-217.

[43] Giannessi F. and Tardella F., "Connections between Nonlinear Programming and Discrete Optimization". In "Handbook of Combinatorial Optimization" edited by Du D.-Z. and Pardalos P., Vol.I, Kluver Acad.Publ., 1998, pp.149-188.

[44] Giannessi F. and Tomasin E., "Nonconvex quadratic programs, linear complementarity problems, and integer linear programs". In" Mathematical Programming in Theory and Practice", P.L.Hammer and G.Zoutendijk Eds., North-Holland, 1974, pp.161-199.

[45] Giorgi G., "On Sufficient Optimality Conditions for a Quasiconvex Programming Problem". Jou.Optimiz.Th.Appls., Vol.81, No.2, 1994, pp.401-405.

[46] Goh C.J. and Yang X.Q., "Nonlinear Lagrangian Theory for Nonconvex Optimization". Jou.Optimiz.Th.Appls., Vol.109, No.1, 2001, pp.99-121.

[47] Hadijsavvas and Pardalos P.M. (Eds.), "Advances in Convex Analysis and Global Optimization". Series Nonconvex Optimiz. and its Appls., Vol.54, Kluver, Dordrecht, 2001.

[48] Hanson M.A., "On sufficiency of the Kuhn-Tucker conditions". Jou.Mathem. Analysis Appls., Vol.80, 1981, pp.545-550.

[49] Hanson M.A., "A generalization of the Kuhn-Tucker sufficiency conditions". Jou. Mathem.Analysis Appls., Vol.184, 1994, pp.146-155.

[50] Hanson M.A. and Mond B., "Necessary and sufficient conditions in constrained optimization". Mathematical Programming, Vol.37, 1987, pp.51-58.

[51] Hestenes M.R., "Multiplier and gradient methods". Jou. of Optimiz. Theory and Appls., Vol.4, No.5, 1969, pp.303-320.

[52] Hiriart-Urruty J.-B., "On optimality conditions in nondifferentiable programming". Mathematical Programming, Vol.14, 1978, pp.73-86.

[53] Hiriart-Urruty J.-B., "From convex optimization to nonconvex optimization. Necessary and sufficient conditions for global optimality". In[I9], pp.219-239.

[54] Hiriart-Urruty J.-B., "Testing necessary and sufficient conditions for global optimality in the problem of maximizing a convex quadratic function over a convex polyhedron". Report of Univ.Paul Sabatier, Seminar of Numerical Analysis, 1990, pp.1-34.

[55] Hiriart-Urruty J.-B., "Boris Nicolaevich Pshenichnyi: two examples of his scientific works". Cybernetics and Systems Analysis, No.3, 2002, pp.68-73.

[56] Horst R. and Tuy H., "Global Optimization". Springer-Verlag, Berlin, 1990.

[57] Joffe A.D., "Necessary conditions in nonsmooth optimization". Mathem.of Oper. Research, Vol.9, No.2, 1984, pp.159-189.

[58] Joffe A.D., "Nonconvex subdifferentials". In [I9], pp.241-253.

[59] Jeyakumar V., "On optimality conditions in nonsmooth inequality constrained minimization". Numer.Funct.Analysis and Optimiz., Vol.9, 1987, pp.535-546.

[60] Komlósi S. and Pappalardo M., "A general scheme for first order approximations in optimization". Optimiz.Meth. and Soft., Vol.3, 1994 pp.143-152.

[61] Leitmann G., "The Calculus of Variations and Optimal Control". Plenum Publ. Co., New York, 1981.

[62] Mangasarian O.L. and Fromovitz S., "A Maximum Principle in Mathematical Programming". In "Mathematical Theory of Control", A.V. Balakrishnan and L.W. Neustadt Eds., Academic Press, 1967, pp85-95.

[63] Martein L., "Regularity Conditions for Constrained Extremum Problems". Jou. Optimiz.Th.Appls., Vol.47, 1985, pp.217-233.

[64] Maugeri A., "Convex programming, variational inequalities and applications to traffic equilibrium problems". Appl.Mathem.Optimiz., Vol.16, 1987, pp.169-185.

[65] Maugeri A., "Optimization problems with side constraints and generalized equilibrium principles". Le Matematiche, Vol.XLIX, Fasc.II, published by Dept.of Mathematics, Univ.of Catania, Italy, 1994, pp.305-312.

[66] Miele A., "Theory of Optimum Aerodynamic Shapes". Academic Press, New York, 1965.

[67] Mordukhovich B.S. and Outrata J., "On second-order subdifferentials and their applications". SIAM Jou. on Optimiz., Vol.12, No.1, 2001, pp.139-169.

[68] Ngai H.V. and Théra M., "On Necessary Conditions for NonLipschitz Optimization Problems". SIAM Jou.on Optimization, Vol.12, No.3, 2002, pp.565-668.

[69] Pappalardo M., "A necessary optimality condition for nondifferentiable constrained extremum problems". Optimization, Vol.22, No.6, 1991, pp.869-883.

[70] Pappalardo M., "Stability sudies in parametric optimization via the image space approach". In "Mathematical Research, Parametric Optimization and Related Topics. II", J.Guddat, H.Th.Jongen, B.Kummer and F.Nožička Eds., Vol.62, Akademie Verlag, 1991, pp.137-145.

[71] Pappalardo M., "Error bounds for generalized Lagrange multipliers in locally Lipschitz programming". Jou.Optimiz.Th.Appls., Vol.73, 1992, pp.205-210.

[72] Pappalardo M., "Sufficient optimality conditions in nondifferentiable otimization". Optimization, Vol.50, 2001, pp.413-426.

[73] Pardalos P.M., "Continuous approaches to discrete optimization problems". In. [I18], pp.313-328.

[74] Pellegrini L., "A sufficient condition for semistationarity in constrained optimization". In [I26], pp..

[75] Penot J.-P., "Optimality conditions for midly nonsmooth constrained optimization". Optimization, Vol.43, 1998, pp.323-337.

[76] Penot J.-P., "Characterization of Solution Set of Quasiconvex Program". Jou. Optimiz.Th.Appls., Vol.117, No.3, 2003, pp.627-636.

[77] Penot J.-P., "Lagrangian Approach to Quasiconvex Programming". Jou.Optimiz. Th.Appls., Vol.117, No.3, 2003, pp.637-647.

[78] Pesch H.J. and Bulirsch R., "The Maximum Principle, Bellman's Equation, and Carathéodory's Work". Jou.Optimiz.Th.Appls., Vol.80, No.2, 1994, pp.199-225.

[79] Poljack B.T., "A general method for solving extremum problems". Dokl.Akad.Nauk SSSR, Tom.174, No.1, 1967, pp.593-597.

[80] Pourciau B.H., "Multipliers rules". Amer.Mathem.Monthly, Vol.87, 1980, pp.443-452.

[81] Pourciau B.H., "Multipliers rules and separation of convex sets". Jou.Optimiz.Th. Appls., Vol.40,1983, pp.321-331.

[82] Qi L.Q., "On an Extended Lagrange Claim". Jou.Optimiz.Th.Appls., Vol.108, No.3, 2001, pp.685-688.

[83] Quang P.H., "Lagrangian multiplier rules via image space analysis". In [I 25], pp.354-378.

[84] Rademacher H., "Uber partielle und totale Differenzierbarkeit". Mathematische Annalen, Vol.79, No.1, 1919, pp.340-359.

[85] Rapcsák T., "Geodesic convexity in nonlinear optimization". Jou.Optimiz.Th.Appls., Vol.69, 1991, pp.169-183.

[86] Rapcsák T., "On nonlinear coordinate representation of nonsmooth optimization problems". Jou.Optimiz.Th.Appls., Vol.86, No2, 1995, pp.459-489.

[87] Rapcsák T., "Global Lagrange multiplier rule and nonsmooth exact penalty functions for equally constraints". In [I 19], pp.351-358.

[88] Rapcsák T. and Thang T.T., "Nonlinear Coordinate Representations of Smooth Optimization Problems". Jou.Optimiz.Th.Appls., Vol.86, No2, 1995, pp.459-489.

[89] Robinson S.M., "First order conditions for general nonlinear optimization". SIAM Jou.Appl.Math., Vol.30, No4, 1976, pp.597-607.

[90] Robinson S.M., "Local structure of feasible sets in nonlinear programming, Part II: stability and sensitivity". Mathematical Programming Study, No.30, North-Holland, 1987, pp.45-66.

[91] Rockafellar R.T., "The Theory of Subgradients and its Applications to Problems of Optimixation". Convex and Nonconvex Functions. Heldermann Verlag, Berlin, 1981.

[92] Rockafellar R.T., "Perturbation of generalized Kuhn-Tucker points in finite-dimensional optimization". In [I19], pp.393-402.

[93] Rockafellar R.T., "Extended nonlinear programming". In [I19], pp.381-399.

[94] Rockafellar R.T., "First and second order epi-differentiability in nonlinear programming". Trans.Amer.Mathem.Soc., Vol.325, 1991, pp.39-72.

[95] Rubinov A.M., "Differences of convex compact sets and their applications in nonsmooth analysis". In [I25], pp.336-378.

[96] Rubinov A.M. and Uderzo A., "On Global Optimality Conditions via Separation Functions". Jou.Optimiz.Th.Appls., Vol.109, No.2, 2001, pp.345-370.

[97] Rubinov. A.M. and Yang X.Q., "Lagrange-type Functions in Constrained Nonconvex Optimization". Kluver, Dordrecht, to appear.

[98] Scorza Dragoni G., "Sui minimi e massimi parziali per le funzioni di più variabili.". Rendiconti Accademia Naz.Lincei, Roma, Vol.VI, serie 12, 1927, pp.579-583.

[99] Scorza Dragoni G., "Un problema sui minimi e massimi parziali di una funzione". Rendiconti Accademia Naz.Lincei, Roma, Vol.XI, serie 6, 1930, pp.865-872.

[100] Smith R.H. and Vandenlinde V.D., "A saddle-point optimality criterion for nonconvex programming in normed spaces". SIAM Jou. on Appl.Mathem., Vol.23, no.2, 1972, pp.203-213.

[101] Sutti C., "On a monotone generalized derivative". In [I26], pp...

[102] Tuy H., "Concave programming under linear constraints". Soviet Mathematics, Vol.5, 1964, pp.1437-1440.

[103] Tuy H. and Oettli W., "On necessary and sufficient conditions for global optimality". Revista de Matemáticas Aplicadas, Vol.15, Universidad de Chile, 1994, pp.39-41.

[104] Uderzo A., "Quasi-multiplier rules for quasidifferentiable extremum problems". Optimization, Vol.51, No.6, 2002, pp.761-795.

[105] Warga J., "Controllability and Necessary Conditions in Unilateral Problems without differentiabilty assumptions". SIAM Jou. on Control and Optimiz., Vol.14, 1976, pp.546-573.

[106] Yang X.Q. and Huang X.X., "A nonlinear Lagrangian approach to constrained optimization problems". SIAM Jou. on Optimiz., Vol.11, No.4, 2001, pp.1119-1144.

[107] Yang X.M., Yang X.Q. and Teo K.L., "Characterizations and Applications of Prequasi-Invex Functions". Jou.Optimiz.Th.Appls., Vol.110, No.3, 2001, pp.645-668.

[108] Zălinescu C., "Stability for a class of nonlinear optimization problems". In [19], pp.437-458.

[109] Zălinescu C., "On a new stability condition in mathematical programming". In [125], pp.429-438.

GLOSSARY OF NOTATION

General Notation

$a := b$	a equals b by definition
$a \equiv b$	a equals b identically
$a \not\equiv b$	a does not equal b identically
$a \cong b$	a is approximately equal to
$a \Rightarrow b$	a implies b
$a \Leftrightarrow b$	a implies b and is implicated by b
$\exists$	there exist(s)
$\not\exists$	there is (are) no
$\exists!$	there exists and is unique
$\forall$	for each
$\{x : P\}$	set of all x with the property P
$\varnothing$	empty set
$a \in A$	a is an element of set A
$\operatorname{card} A$	cardinality of set A
$\sim A$	complement of set A
$\operatorname{int} A$ or $\overset{o}{A}$	topological interior of set A
$\operatorname{r}\partial A$	relative boundary of set A, i.e. boundary with respect to aff A (∂ is used also for denoting subgradient)
$\operatorname{ri} A$	relative topological interior of set A
∂A or $\operatorname{frt} A$	boundary (or frontier) of set A (∂ is used also for denoting subgradient)
$\operatorname{cl} A$	closure of set A
$\operatorname{ext} A$	exterior of set A; $\operatorname{ext} A = \operatorname{int}(\sim A) = \sim(\operatorname{cl} A)$
$\operatorname{vert} A$	set of extreme points of set A; set of vertices of polyhedron A
$A \subseteq B$	the set A is contained in the set B (A is subset of B)
$A \supseteq B$	the set A contains the set B (A is superset of B)
$A \subset B$	the set A is contained in the set B, but $A \neq B$ (A is proper subset of B)
$A \supset B$	the set A contains the set B, but $A \neq B$ (A is proper subset of B)
$\cup, \cap, \setminus$	denote union, intersection, difference of sets, respectively
$\dim A$	dimension of set A
$\operatorname{aff} A$	affine hull of set A; $\operatorname{aff} \varnothing := \{O\}$
$A \times B$	Cartesian product of sets A and B
$A^n := \underset{i=1}{\overset{n}{\times}} A_i = A_1 \times ... \times A_n$	Cartesian product of sets $A_1, ..., A_n$
$A - B$	denotes vector difference between sets A and B
2^A	denotes the power set of A
$\operatorname{conv} A$	convex hull of set A

cl conv A	convex closure of set A
$\sum, \prod$	denote continued summand and continued product, respectively
$N(x)$	denotes neighbourhood of x
$N_\varepsilon(x)$	denotes open hypersphere with centre at x and radius ε
sgn (x)	signum of x
$f : X \to Y$	denotes function with domain within X and image within Y
gr f	graph of function f; i.e. $\{(x,y) : y = f(x)\}$
epi f	epigraph of function f; i.e. $\{(x,y) : y \geq f(x)\}$
hypo f	hypograph of function f; i.e. $\{(x,y) : y \leq f(x)\}$
dom f	effective domain of f; i.e. $\{x : \exists y$ such that $(x,y) \in$ epi $f\} = \{x : f(x) < +\infty\}$
Im f	image of function f
$\mathrm{lev}_{\substack{\leq \\ \geq}\alpha} f$	denotes the various level sets of function f, defined, respectively, by $\{x : f(x) \substack{\leq \\ \geq} \alpha\}$
$\mathrm{lev}_Z f$	denotes the generalized level set of function f, defined by $\{x : f(x) \in Z\}$
$\liminf\limits_{x\to\bar{x}} f(x)$	denotes the lower limit of function f, defined by $$\sup_{\varepsilon>0}\ \inf_{x\in N_\varepsilon(\bar{x})\setminus\{O\}} f(x)$$
$\limsup\limits_{x\to\bar{x}} f(x)$	denotes the upper limit of function f, defined by $$\inf_{\varepsilon>0}\ \sup_{x\in N_\varepsilon(\bar{x})\setminus\{O\}} f(x)$$
cl f	denotes closure of the function f: $$\mathrm{cl}\, f(x) = \limsup_{y\to x} f(y)$$
∂f	denotes subgradient of function f (∂ is used also to denote boundary)
$f \circ g = f(g(x))$	composition of functions
$f'(x)$ or $\nabla f(x), f''(x)$ or $\nabla^2 f(x)$	denote gradient, Hessian matrix of function f at x, respectively; (the symbol ∇ is called also nabla)
$f^{(r)}(x)$	denotes rth derivative at x
$f(\bullet; y)$	denotes the restriction of f to the 1st argument at the fixed value y for the 2nd argument
$f_{/A}$	denotes the restriction of function f to the set A
$C^k(T)$	set of all continuous functions $x : T \to \mathbb{R}$, having the first k derivatives continuous on T; $C^0(T)$ is the set of all continuous function on T
$C^k(T)^k := \overset{n}{\underset{i=1}{X}} C^k(T)$	Cartesian product (n times)
$f : X \rightrightarrows Y$	denotes the point-to-set map (multifunction) with domain within X and image within 2^Y
$\{x^i\}_1^\infty$	denotes the sequence $x^1, x^2, ..., x, ...; x^i = (x_1^i, ..., x_n^i)$ is the ith vector; a subfix denotes scalar; a superfix relates to vector
$\{A(\xi)\}_{\xi\in\Xi}$	denotes a family of sets
dist (x, A)	distance of the point x from the set A

dist (A, B)	distance between the sets A and B
$A^\perp$	orthogonal complement of A; in particular: $$\{O\}^\perp = \mathbb{R}^n, \ (\mathbb{R}^n)^\perp = \{O\}.$$
A^*	polar (or dual) of cone (or set) A (a star as apex denotes polarity, iff it is applied to a set); we stipulate that $A^{**} := (A^*)^*$
D_C^*	denotes vector polar of cone D with respect to cone C
cone $(\overline{x}; S)$, cone S	denote the cones generated by the set S from $\overline{x}$ or from the origin, respectively
$\mathrm{proj}_B A$	denotes projection of set A upon set B
M^T	denotes the transpose of matrix M
det M	denotes the determinant of the (square) matrix M
rank M	denotes the rank of the (square) matrix M
diag M	denotes the vector whose entries are, respectively, those in the main diagonal of the (square) matrix M
$\mathbb{R}^{m \times n}$	is the set of matrices of dimension $m \times n$, and with real entries
e_n	is the n-tuple whose entries are all equal to 1
$\mathbb{N}, \mathbb{Z}, \mathbb{Q}, \mathbb{R}, \mathbb{C}, \mathbb{B}$	sets of (positive) natural, integer, rational, real, complex, zero-one numbers, respectively
$\overline{\mathbb{R}} := \mathbb{R} \cup \{\pm\infty\} = [-\infty, +\infty]$	set of extended reals
$\mathbb{Z}_+, \mathbb{R}_+ = [0, +\infty[$	sets of non-negative integer, real numbers, respectively
$\overline{\mathbb{R}}_+ := \mathbb{R}_+ \cup \{+\infty\} = [0, +\infty]$	set of extended non-negative reals
$\mathbb{Z}^n, \mathbb{R}^n, \mathbb{B}^n$	sets of n-tuples with integer, real, zero-one entries, respectively; $\mathbb{Z}^1 = \mathbb{Z}$, $\mathbb{R}^1 = \mathbb{R}$, $\mathbb{B}^1 = \mathbb{B}$
$\overline{\mathbb{R}}^n := \mathbb{R}^n \cup \{\pm\infty\}$	set of extended n-tuples with real entries
$\mathbb{Z}_+^n, \mathbb{R}_+^n$	sets of n-tuples with non-negative integer, real entries, respectively; $\mathbb{Z}_+ = \mathbb{N} \cup \{0\}$
$\overset{o}{\mathbb{R}_+^n} = \mathrm{int}\, \mathbb{R}_+^n$	interior of $\mathbb{R}_+^n$
$\overline{\mathbb{R}}_+^n := \mathbb{R}_+^n \cup \{+\infty\}$	set of extended n-tuples of non-negative reals
$\mathbb{Z}_-^n, \mathbb{R}_-^n$	sets of n-tuples with nonpositive integer, real entries, respectively; $\mathbb{Z}_- = -\mathbb{N} \cup \{0\}$
$\overset{o}{\mathbb{R}_-^n} := \mathrm{int}\, \mathbb{R}_-^n$	interior of $\mathbb{R}_-^n$
$B, \mathcal{B}$	Banach spaces
$\mathcal{B}_+$	closed and convex cone of $\mathcal{B}$ with apex at the origin
O_n	n-tuple, whose entries are zero; when there is no fear of confusion, the subfix is omitted; for $n = 1$, without no fear of confusion, the 1-tuple O_1 is identified with its elements, namely we set $O_1 = 0$
$\lfloor x \rfloor$	denotes the vector of the lower integer parts of the elements of $x \in \mathbb{R}^n$, namely the vector, whose elements are $\max\{z_i \in \mathbb{Z} : z_i \leq x_i\}$, $i = 1, ..., n$; for $n = 1, \lfloor x \rfloor$ denotes the lower integer part of the real x
$\lceil x \rceil$	denotes the vector of the upper integer parts of the elements of $x \in \mathbb{R}^n$, namely the vector, whose ele-

	ments are $\min \{z_i \in \mathbb{Z} : z_i \geq x_i\}$, $i = 1, ..., n$; for $n = 1, \lceil x \rceil$ denotes the upper integer part of the real x
$\lvert x \rvert$	denotes absolute value of $x \in \mathbb{R}$
$\lVert x \rVert, \langle \bullet, \bullet \rangle$	denote generically norm and scalar product
$\lVert x \rVert_p := \left(\sum_{i=1}^{n} \lvert x_i \rvert^p \right)^{\frac{1}{p}}, \, p \geq 1,$	denotes the p-norm of $x = (x_1, ..., x_n) \in \mathbb{R}^n$;
$\lVert x \rVert_2 = \langle x, x \rangle^{\frac{1}{2}}$	is the Euclidean norm;
$\lVert x \rVert_\infty = \max\{\lvert x_1 \rvert, ..., \lvert x_n \rvert\}$	is the Tchebycheff norm
$\lVert x \rVert_A := \langle x, Ax \rangle^{\frac{1}{2}}$	with A positive definite matrix, denotes the elliptic norm of A
norm $\lVert A \rVert_p := \sup_{\lVert x \rVert_p} \frac{\lVert Ax \rVert_p}{\lVert x \rVert_p}$	denotes the norm of the square matrix A

$[a,b] := \{x \in \mathbb{R}^n : a \leq x \leq b\}$, with $a, b \in \overline{\mathbb{R}}^n$

$]a,b[:= \{x \in \mathbb{R}^n : a < x < b\} = \mathrm{ri}\,[a,b]$, with $a, b \in \overline{\mathbb{R}}^n$

$]a,b] := \{x \in \mathbb{R}^n : a < x \leq b\}$, with $a, b \in \overline{\mathbb{R}}^n$

$[a,b[:= \{x \in \mathbb{R}^n : a \leq x < b\}$, with $a, b \in \overline{\mathbb{R}}^n$

$(a_i \quad i \in \mathfrak{I}) := (a_1, ..., a_m)$ with $\mathfrak{I} = \{1, ..., m\}$

a^r denotes either rth vector (of a sequence of vectors) or rth power of the real a; the context should resolve the alternative

Special Symbols

$\mathfrak{I}$	set of indices of constraining functions
$\mathfrak{I}^0$	set of indices of constraining functions of bilateral constraints
$\mathfrak{I}^+$	set of indices of constraining functions of unilateral constraints
$\mathcal{H}$	set of Image Space which identifies the kind of constraints
$\mathcal{K}_{\overline{x}}$	image set
$\mathcal{K}_{\overline{x}}^h$	homogenization of the image set
$\mathcal{E}(\mathcal{K}_{\overline{x}})$	conic extension of $\mathcal{K}_{\overline{x}}$
$\mathcal{E}(\mathcal{K}_{\overline{x}}^h)$	conic extension of $\mathcal{K}_{\overline{x}}^h$
z	element of the Image Space; $\overline{z}$ image of $\overline{x}$
$\mathcal{C}$	set of sublinear functions
$\mathcal{D}_{\mathcal{C}}f$	denotes $\mathcal{C}$-derivative of f
$\mathcal{N}(z)$	neighbourhood of z (in the Image Space)
$\mathcal{G}$	denotes a set of positively homogeneous functions
$\mathcal{L}$	denotes either a set of linear functions or the generalized Lagrangian function
L	denotes the Lagrangian function, unless differently said
L_p	denotes the Lebesgue integration space (Banach space), $1 \leq p < \infty$
argmin $\bullet$	set of minimum points of $\bullet$

argmax $\bullet$	set of maximum points of $\bullet$
C	denotes usually a cone; $C_0 := C\backslash\{O\}$; $\overset{o}{C} := \operatorname{int} C$
$A_{\overline{x}}(x)$	is the map which sends the variable x of a problem into its image variable z
cone $(\overline{x}; X)$	is the cone generated by X from $\overline{x}$; cone $(X) := := \operatorname{cone}(O; X)$
$TC(\overline{x}; X)$	is the tangent cone to X with apex at $\overline{x}$; $TC(X) := TC(O; X)$
$RC(\overline{x}; X)$	is the reachable cone to X with apex at $\overline{x}$; $RC(X) := RC(O; X)$
$AC(\overline{x}; X)$	is the admissible cone to X with apex at $\overline{x}$; $AC(X) := AC(O; X)$
$IC(\overline{x}; X)$	is the interior cone to X with apex at $\overline{x}$; $IC(X) := IC(O; X)$
$NC(\overline{x}; X)$	is the normal cone to X with apex at $\overline{x}$; $NC(X) := NC(O; X)$
$HC(\overline{x}; X)$	is the hypertangent cone to X with apex at $\overline{x}$; $HC(X) := HC(O; X)$
$x \leq_C y$	means $y - x \in C$
$x \geq_C y$	means $x - y \in C$
$x \not\leq_C y$	means $y - x \notin C$
$x \not\geq_C y$	means $x - y \notin C$
$\min_C$	denotes vector minimum with respect to cone C

Acronyms

s.t.	subject to or such that
w.r.t	with respect to
iff	if and only if
a.a.	almost averywhere
$\square$	marks the end of a proof or an example
l.s.c.	lower semicontinuous
u.s.c	upper semicontinuous
min, max	minimum, maximum
inf, sup	infimum, supremum
m.p.	minimum point(s)
v.m.p.	vector minimum point(s)
IS	Image Space
ST	Separation Theorem(s)
TA	Theorem(s) of Alternative
VOP	Vector Optimization Problems
VI	Variational Inequality/Inequalities
VII	Vector VI
CS	Complementarity Systems
SVVI	Stampacchia VVI

MVVI	Minty VVI
LMM	Lagrange Method of Multipliers
GSF	Generalized Selected Function(s)
SM	Selection Multiplier(s)

Comments

References will be quoted, either with an Arabic numeral within square brackets if they belong to the same chapter as the text, or with a Roman numeral — which identifies the chapter — and an Arabic numeral if they belong to a chapter different from that of the text. For instance, in Chapter 2, [5] means reference No. 5 of the same chapter, while [III 8] means reference No. 8 of Chapter 3.

Even if the subjunctive is an obsolete mood, being convinced that its survival "be" an important cultural fact, and not a mere formal aspect, it is used as as much as possible. Therefore, to meet in this book, for instance, "be" instead of "is" or "have" instead of "has" does not mean to have found a mistake.

SUBJECT INDEX